C# 13 and .NET 9 – Cross-Platform Development Fundamentals

Ninth Edition

Start building websites and services with ASP.NET Core 9, Blazor, and EF Core 9

Mark J. Price

C# 13 and .NET 9 – Modern Cross-Platform Development Fundamentals

Ninth Edition

Senior Publishing Product Manager: Suman Sen
Acquisition Editor – Peer Reviews: Jane D'Souza
Project Editor: Janice Gonsalves
Content Development Editor: Shazeen Iqbal
Copy Editor: Safis Editing
Technical Editor: Simanta Rajbangshi
Proofreader: Safis Editing
Indexer: Rekha Nair
Presentation Designer: Pranit Padwal
Developer Relations Marketing Executive: Priyadarshini Sharma

First published: March 2016
Second edition: March 2017
Third edition: November 2017
Fourth edition: October 2019
Fifth edition: November 2020
Sixth edition: November 2021
Seventh edition: November 2022
Eighth edition: November 2023
Ninth edition: November 2024

Production reference: 1071124

Published by Packt Publishing Ltd.
Grosvenor House
11 St Paul's Square
Birmingham
B3 1RB, UK.

ISBN 978-1-83588-122-4

`www.packt.com`

Contributors

About the author

Mark J. Price is a Microsoft Specialist: Programming in C# and Architecting Microsoft Azure Solutions, with over 20 years of experience. Since 1993, he has passed more than 80 Microsoft programming exams and specializes in preparing others to pass them. Between 2001 and 2003, Mark was employed to write official courseware for Microsoft in Redmond, USA. His team wrote the first training courses for C# while it was still an early alpha version. While with Microsoft, he taught "train-the-trainer" classes to get Microsoft Certified Trainers up-to-speed on C# and .NET. Mark has spent most of his career training a wide variety of students from 16-year-old apprentices to 70-year-old retirees, with the majority being professional developers. Mark holds a BSc in Computer Science.

Thank you to all my readers. Your support means I get to write these books and celebrate your successes.

Special thanks to the readers who give me actionable feedback via my GitHub repository and email and interact with me and the book communities on Discord. You help make my books even better with every edition.

Extra special thanks to Alex, who brings joy and comfort to my life.

About the reviewer

Kieran Foot is a self-taught C# developer with a deep passion for learning new technologies and techniques. He is constantly exploring the latest developments in C#, with a particular focus on web technologies. As the lead software developer at ConnX Business Solutions, a small software company based in the UK, he has the opportunity to apply his knowledge in practice and assist others in acquiring new skills. He enjoys helping others and is an active member of the Packt community.

Learn more on Discord

To join the Discord community for this book – where you can share feedback, ask questions to the author, and learn about new releases – follow the QR code below:

`https://packt.link/csharp13dotnet9`

Table of Contents

Chapter 7: Packaging and Distributing .NET Types 369

Chapter 8: Working with Common .NET Types 417

Chapter 9: Working with Files, Streams, and Serialization 477

Chapter 10: Working with Data Using Entity Framework Core 527

Chapter 15: Building and Consuming Web Services 733

Epilogue 779

Index 785

Preface

There are programming books that are thousands of pages long that aim to be comprehensive references to the C# language, the .NET libraries, and app models like websites, services, and desktop and mobile apps.

This book is different. It is concise and aims to be a brisk, fun read that is packed with practical hands-on walk-throughs of each subject. The breadth of the overarching narrative comes at the cost of some depth, but you will find many signposts to explore further if you wish.

This book is simultaneously a step-by-step guide to learning modern C# and proven practices using cross-platform .NET, and a brief introduction to the fundamentals of modern web development, along with the creation of websites and services that can be built with these technologies. This book is most suitable for beginners to C# and .NET, as well as programmers who have worked with C# in the past but may feel left behind by the changes in the past few years.

If you already have experience with older versions of the C# language, then in the first topic of *Chapter 2, Speaking C#*, you can review the tables of new language features in an online section.

If you already have experience with older versions of the .NET libraries, then, in the first section of *Chapter 7, Packaging and Distributing .NET Types*, you can review the tables of the new library features in an online section.

I will point out the cool corners and gotchas of C# and .NET so that you can impress colleagues and get productive fast. Rather than slowing down and boring some readers by explaining every little thing, I will assume that you are smart enough to Google an explanation for topics that are related but not necessary to include in a beginner-to-intermediate guide that has limited space in a printed book.

Some chapters have links to additional related online-only content for those readers who would like more details. For example, *Chapter 1, Hello, C#! Welcome, .NET!*, has an online section about the history and background of .NET.

Where to find the code solutions

You can download solutions for the step-by-step guided tasks and exercises from the GitHub repository at the following link:

`https://github.com/markjprice/cs13net9`

If you don't know how to download or clone a GitHub repository, then I provide instructions at the end of *Chapter 1, Hello, C#! Welcome, .NET!*.

What this book covers

Chapter 1, Hello, C#! Welcome, .NET!, is about setting up your development environment to use either Visual Studio or VS Code with C# Dev Kit. Then you will learn how to use them to create the simplest application possible with C# and .NET. For simplified console apps, you will see the use of the top-level program feature introduced in C# 9, which is then used by default in the project templates for C# 10 onwards. You will also learn about some good places to look for help, including AI tools like ChatGPT and GitHub Copilot, and ways to contact me to get help with an issue or give me feedback to improve the book today through its GitHub repository and in future print editions.

Chapter 2, Speaking C#, introduces the versions of C# and has tables showing which version introduced new features in an online section. I will explain the grammar and vocabulary that you will use every day to write the source code for your applications. In particular, you will learn how to declare and work with variables of different types.

Chapter 3, Controlling Flow, Converting Types, and Handling Exceptions, covers using operators to perform simple actions on variables, including comparisons, writing code that makes decisions, pattern matching, repeating a block of statements, and converting between types. This chapter also covers writing code defensively to handle exceptions when they inevitably occur, including using guard clauses like `ThrowIfLessThan` on the `ArgumentOutOfRangeException` class introduced with .NET 8.

Chapter 4, Writing, Debugging, and Testing Functions, is about following the **Don't Repeat Yourself** (**DRY**) principle by writing reusable functions using both imperative and functional implementation styles. You will also learn how to use debugging tools to track down and remove bugs, use Hot Reload to make changes while your app is running, monitor your code while it executes to diagnose problems, and rigorously test your code to remove bugs, ensuring stability and reliability before it gets deployed into production.

Chapter 5, Building Your Own Types with Object-Oriented Programming, discusses all the different categories of members that a type like a class can have, including fields to store data and methods to perform actions. You will use **Object-Oriented Programming** (**OOP**) concepts, such as aggregation and encapsulation, and how to manage namespaces for types, including the ability to alias any type introduced with C# 12. You will learn language features such as tuple syntax support and `out` variables, local functions, and default literals and inferred tuple names. You will also learn how to define and work with immutable types using the `record` keyword, `init`-only properties, and `with` expressions, introduced in C# 9. Finally, we look at how C# 11 introduced the `required` keyword to help avoid the overuse of constructors to control initialization, how C# 12 introduced primary constructors for non-record types, and how C# 13 expands the supported types for a `params` parameter and how you can now define partial properties as well as partial methods.

Chapter 6, Implementing Interfaces and Inheriting Classes, explains deriving new types from existing ones using OOP. You will learn how to define operators, delegates, and events, how to implement interfaces about base and derived classes, how to override a member of a type, how to use polymorphism, how to create extension methods, how to cast between classes in an inheritance hierarchy, and about the big changes in C# 8 with the introduction of nullable reference types, along with the switch to make this the default in C# 10 and later. In an optional online-only section, you can learn how analyzers can help you write better code.

Chapter 7, Packaging and Distributing .NET Types, introduces the versions of .NET and includes tables showing which version introduced new library features in an online section. I will then present the .NET types that are compliant with .NET Standard and explain how they relate to C#. Throughout this chapter, you will learn how to write and compile code on any of the supported operating systems, including the Windows, macOS, and Linux variants. You will learn how to package, deploy, and distribute your own apps and libraries. In three optional online-only sections, you can learn how to use legacy .NET Framework libraries in .NET libraries, about the possibility of porting legacy .NET Framework code bases to modern .NET, and about source generators and how to create them.

Chapter 8, Working with Common .NET Types, discusses the types that allow your code to perform common practical tasks, such as manipulating numbers and text, storing items in collections, and, in an optional online-only section, working with a network using low-level types. You will also learn about regular expressions and the improvements that make writing them easier, as well as how to use source generators to improve their performance. Introduced with .NET 9 are new classes for Base64-encoded URLs, performing multiplication on big integers, ordered dictionaries, read-only sets, improvements to searching in strings, and being able to generate regular expressions on a partial property.

Chapter 9, Working with Files, Streams, and Serialization, covers interacting with a filesystem, reading and writing to files and streams, text encoding, and serialization formats like JSON and XML, including the improved functionality and performance of the `System.Text.Json` classes. If you use Linux, then you will be interested in how to programmatically work with `tar` archives, which you can learn about in an online-only section. .NET 9 (finally!) removes the dangerous legacy `BinaryFormatter` serialization class and adds a class to export JSON schemas.

Chapter 10, Working with Data Using Entity Framework Core, explains reading and writing to relational databases, such as SQL Server and SQLite, using the **object-relational mapping** (**ORM**) technology named **Entity Framework Core** (**EF Core**). You will learn how to define entity models that map to existing tables in a database using **Database First** models. In three optional online-only sections, you can also learn how to define **Code First** models that can create tables and databases at runtime, how to insert, update, and delete data, and how to group multiple changes together using transactions.

Chapter 11, Querying and Manipulating Data Using LINQ, teaches you about **Language INtegrated Queries (LINQ)**—language extensions that add the ability to work with sequences of items and filter, sort, and project them into different outputs. This chapter includes LINQ methods introduced in .NET 6, like `TryGetNonEnumeratedCount` and `DistinctBy`, in .NET 7, like `Order` and `OrderDescending`, and in .NET 9, like `CountBy`, `AggregateBy`, and `Index`. Optional online-only sections cover using multiple threads with parallel LINQ, working with LINQ to XML, and creating your own LINQ extension methods.

Chapter 12, Introducing Modern Web Development Using .NET, introduces you to the types of web projects that can be built using C# and .NET. You will also build an EF Core model to represent the database for a fictional organization named *Northwind* that will be used throughout the rest of the chapters in the book. You will learn how to efficiently manage NuGet package version using Central Package Management. Finally, you will be introduced to common web technologies like HTML and CSS.

Chapter 13, Building Websites Using ASP.NET Core, is about learning the basics of building websites with a modern HTTP architecture on the server side, using ASP.NET Core. You will learn how to implement the ASP.NET Core feature known as Blazor static **Server-Side Rendering (SSR)**, which replaces the legacy Razor Pages technology for creating data-driven web pages. You will learn how to optimize static assets like JavaScript and stylesheets using a new method introduced with .NET 9, `MapStaticAssets`. In two optional online-only sections, you'll see how to customize the HTTP request and response pipeline, and you'll gain insight into enabling HTTP/3 in your website project.

Chapter 14, Building Interactive Web Components Using Blazor, introduces how to build web user interface components using Blazor that can be executed either on the server side or on the client side inside the web browser. You will see how to build components that are easy to switch between the client and the server, with the new hosting model introduced with .NET 8.

Chapter 15, Building and Consuming Web Services, explains building backend REST architecture web services using the ASP.NET Core Minimal APIs. We will cover how to document them using the new Microsoft-implemented OpenAPI documentation generator. Introduced with .NET 9 is a hybrid cache that you will learn how to use to get the best of in-memory and distributed caching. Then we will see how to properly consume them using factory-instantiated HTTP clients. In two optional online-only sections, you will be introduced to advanced features, like health checks and adding security HTTP headers, and how Minimal APIs projects can use native **ahead-of-time (AOT)** compilation during the publishing process to improve startup time and memory footprint.

Epilogue describes your options for further study about C# and .NET.

Appendix, Answers to the Test Your Knowledge Questions, has the answers to the test questions at the end of each chapter. You can read the appendix at the following link: `https://packt.link/LsQtz`.

What you need for this book

You can develop and deploy C# and .NET apps using VS Code and the command-line tools on most operating systems, including Windows, macOS, and many varieties of Linux. An operating system that supports VS Code and an internet connection is all you need to follow along with this book.

If you prefer alternatives, then the choice is yours whether to use Visual Studio, or a third-party tool like JetBrains Rider.

Downloading the color images of this book

We also provide you with a PDF file that has color images of the screenshots and diagrams used in this book. The color images will help you better understand the changes in the output.

You can download this file from `https://packt.link/gbp/9781835881224`.

Conventions

In this book, you will find a number of text styles that distinguish between different kinds of information. Here are some examples of these styles and an explanation of their meaning.

`CodeInText`: Indicates code words in text, database table names, folder names, filenames, file extensions, pathnames, dummy URLs, user input, and Twitter handles. For example: "The `Controllers`, `Models`, and `Views` folders contain ASP.NET Core classes and the `.cshtml` files for execution on the server."

A block of code is set as follows:

```
// storing items at index positions
names[0] = "Kate";
names[1] = "Jack";
names[2] = "Rebecca";
names[3] = "Tom";
```

When we wish to draw your attention to a particular part of a code block, the relevant lines or items are highlighted:

```
// storing items at index positions
names[0] = "Kate";
names[1] = "Jack";
names[2] = "Rebecca";
names[3] = "Tom";
```

Any command-line input or output is written as follows:

```
dotnet new console
```

Bold: Indicates a new **term**, an important **word**, or words that you see on the screen, for example, in menus or dialog boxes. For example: "Clicking on the **Next** button moves you to the next screen."

Important notes and links to external sources for further reading appear in a box like this.

Good Practice: Recommendations for how to program like an expert appear like this.

Get in touch

Feedback from our readers is always welcome.

General feedback: If you have questions about any aspect of this book, mention the book title in the subject of your message and email us at `customercare@packtpub.com`.

Errata: Although we have taken every care to ensure the accuracy of our content, mistakes do happen. If you have found a mistake in this book, we would be grateful if you would report this to us. Please visit `www.packtpub.com/support/errata`, select your book, click on the **Errata Submission Form** link, and enter the details.

Piracy: If you come across any illegal copies of our works in any form on the internet, we would be grateful if you would provide us with the location address or website name.

Please contact us at `copyright@packt.com` with a link to the material.

If you are interested in becoming an author: If there is a topic that you have expertise in and you are interested in either writing or contributing to a book, please visit `authors.packtpub.com`.

Leave a Review!

Thank you for purchasing this book from Packt Publishing—we hope you enjoy it! Your feedback is invaluable and helps us improve and grow. Once you've completed reading it, please take a moment to leave an Amazon review; it will only take a minute, but it makes a big difference for readers like you.

https://packt.link/r/1835881238

Scan the QR code below to receive a free ebook of your choice.

https://packt.link/NzOWQ

Download a free PDF copy of this book

Thanks for purchasing this book!

Do you like to read on the go but are unable to carry your print books everywhere?

Is your eBook purchase not compatible with the device of your choice?

Don't worry, now with every Packt book you get a DRM-free PDF version of that book at no cost.

Read anywhere, any place, on any device. Search, copy, and paste code from your favorite technical books directly into your application.

The perks don't stop there, you can get exclusive access to discounts, newsletters, and great free content in your inbox daily.

Follow these simple steps to get the benefits:

1. Scan the QR code or visit the link below:

https://packt.link/free-ebook/9781835881224

2. Submit your proof of purchase.
3. That's it! We'll send your free PDF and other benefits to your email directly.

1

Hello, C#! Welcome, .NET!

In this first chapter, the first goal is setting up your development environment to use Visual Studio 2022, Visual Studio Code, or JetBrains Rider.

Throughout this book, I will use the names **Visual Studio**, **VS Code**, and **Rider** to refer to these three code editors, respectively.

The other goals are understanding the similarities and differences between modern .NET, .NET Core, .NET Framework, Mono, Xamarin, and .NET Standard; creating the simplest application possible with C# 13 and .NET 9, using various code editors; and then discovering good places to look for help.

My style of writing ensures that you can get started from my book and then switch to online documentation and other online resources, like videos, blogs, and forums, which are the main ways that developers like to learn, as shown at the following link: `https://survey.stackoverflow.co/2024/developer-profile#learning-to-code`.

This chapter covers the following topics:

- Introducing this book and its contents
- Setting up your development environment
- Understanding .NET
- Building console apps using Visual Studio
- Building console apps using VS Code
- Making good use of the GitHub repository for this book
- Looking for help

Introducing this book and its contents

Let's get started by introducing you to the code solutions, structure, and content of this book.

Getting code solutions for this book

The GitHub repository for this book has solutions that use full application projects for all code tasks and exercises, found at the following link:

`https://github.com/markjprice/cs13net9`

After navigating to the GitHub repository in your web browser, press the `.` (dot) key on your keyboard, or manually change `.com` to `.dev` in the link to convert the repository into a live code editor based on VS Code, called GitHub Codespaces, as shown in *Figure 1.1*:

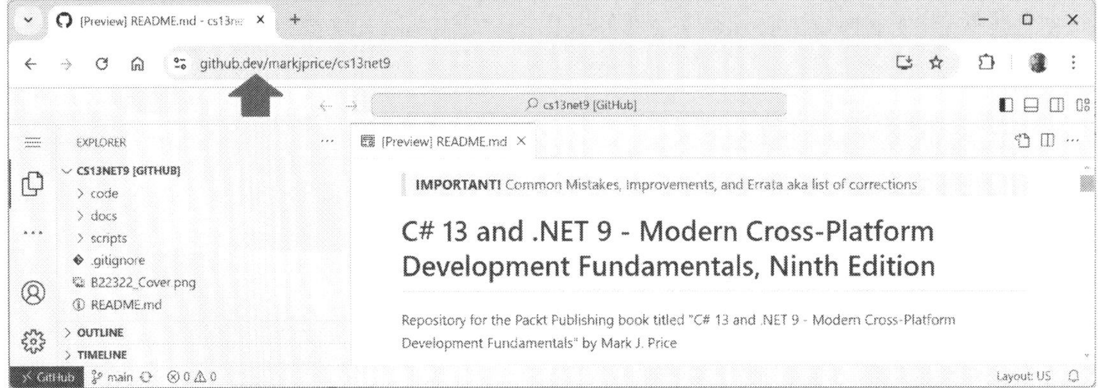

Figure 1.1: GitHub Codespaces live-editing the book's GitHub repository

Warning! When you access a repository using GitHub Codespaces, its GitHub Repositories extension will attempt to download the repository's index. By default, the maximum size can be up to 50MB but the repository exceeds this limit, so you will see the following warning message: `Repository exceeds download size limit! Continuing with partial text services.` You can ignore this message because you are not using the full features.

We provide you with a PDF file that has color images of the screenshots and diagrams used in this book. You can download this file from `https://packt.link/gbp/9781837635870`.

VS Code in a web browser is great to run alongside your chosen local code editor as you work through the book's coding tasks. You can compare your code to the solution code and easily copy and paste parts if needed.

You do not need to use or know anything about Git to get the solution code of this book. You can download a ZIP file containing all the code solutions by using the following direct link and then extract the ZIP file into your local filesystem: `https://github.com/markjprice/cs13net9/archive/refs/heads/main.zip`.

.NET terms used in this book

Throughout this book, I use the term **modern .NET** to refer to .NET 9 and its predecessors, like .NET 6, that derive from .NET Core. I use the term **legacy .NET** to refer to .NET Framework, Mono, Xamarin, and .NET Standard.

Modern .NET is a unification of those legacy platforms and standards.

The structure and style of this book

After this first chapter, the book will be divided into three parts: language, libraries, and web development.

First, we will cover the grammar and vocabulary of the C# language; second, we will cover the types available in the .NET libraries for building app features; and third, we will cover the fundamentals of cross-platform websites, services, and browser apps that you can build using C# and .NET.

Most people learn complex topics best by imitation and repetition, rather than reading a detailed explanation of the theory; therefore, I will not overload you with detailed explanations of every step throughout this book. The idea is to get you to write some code and see it run.

You don't need to know all the nitty-gritty details immediately. That will be something that comes with time as you build your own apps and go beyond what any book can teach you.

In the words of Samuel Johnson, author of the English dictionary in 1755, I have committed *"a few wild blunders, and risible absurdities, from which no work of such multiplicity is free."* I take sole responsibility for these and hope you appreciate the challenge of my attempt to lash the wind by writing this book about rapidly evolving technologies, like C# and .NET, and the apps that you can build with them.

If you have a complaint about this book, then please contact me before writing a negative review on Amazon. Authors cannot respond to Amazon reviews, so I cannot contact you to resolve the problem and help you or listen to your feedback and try to do better in the next edition. Please ask a question on the Discord channel for this book at `https://packt.link/csharp13dotnet9`, email me at `markjprice@gmail.com`, or raise an issue in the GitHub repository for the book at the following link: `https://github.com/markjprice/cs13net9/issues`.

Topics covered by this book

The following topics are covered in this book:

- **Language fundamentals**: Fundamental features of the C# language, from declaring variables to writing functions and object-oriented programming.
- **Library fundamentals**: Fundamental features of the .NET base class library, as well as some important optional packages for common tasks like database access.

- **Modern web development fundamentals:** Modern features of the ASP.NET Core framework for server-side and client-side website and web service development. This includes Blazor and Minimal APIs and excludes controller-based features like MVC and Web API, or legacy features like Razor Pages.

This book, *C# 13 and .NET 9 – Modern Cross-Platform Development Fundamentals*, is best read linearly, chapter by chapter, because it builds up fundamental skills and knowledge.

Topics covered by Apps and Services with .NET 8

The following topics are available in a companion book, *Apps and Services with .NET 8*:

- **Data:** SQL Server and Azure Cosmos DB.
- **Specialized libraries:** Dates, times, time zones, and internationalization; common third-party libraries for image handling, logging, mapping, and generating PDFs; multitasking and concurrency; and many more.
- **Services:** Caching, queuing, background services, gRPC, GraphQL, Azure Functions, SignalR, and Minimal APIs.
- **User interfaces:** ASP.NET Core, Blazor, and .NET MAUI.

This book can be read more like a cookbook, so if you are especially interested in building gRPC services, then you can read that chapter without the preceding chapters about minimal API services.

Topics covered by Tools and Skills for .NET 8

The following topics are available in a companion book, *Tools and Skills for .NET 8*:

- Debugging and memory analysis.
- All the important types of testing, from unit and integration to performance and web UI testing.
- Docker and .NET Aspire for local distributed development and testing.
- Design patterns and solution architecture.
- Preparing for an interview to get the .NET developer career that you want.

My books for .NET 8 are equally applicable to .NET 9. You can make your projects target .NET 9 and all the code examples will work. You should not need to wait for packages to upgrade to also target .NET 9, due to backward compatibility.

Finding all my books

To see a list of all the books I have published with Packt, you can use the following link:

`https://subscription.packtpub.com/search?query=mark+j.+price`

A similar list is available on Amazon:

`https://www.amazon.com/Mark-J-Price/e/B071DW3QGN/`

You can search other book-selling sites for my books too.

Setting up your development environment

Before you start programming, you'll need a code editor for C#, either from Microsoft or a third party.

Microsoft has a family of code editors and **Integrated Development Environments** (**IDEs**), which include:

- Visual Studio for Windows
- VS Code for Windows, Mac, or Linux
- VS Code for the Web or GitHub Codespaces

Third parties have created their own C# code editors; for example, JetBrains has the cross-platform Rider, which is available for Windows, Mac, or Linux and since October 2024 is free for non-commercial use. Rider is popular with more experienced .NET developers.

Warning! Although JetBrains is a fantastic company with great products, both Rider and the ReSharper extension for Visual Studio are software, and all software has bugs and quirky behavior. For example, they might show errors like *Cannot resolve symbol* in your Razor Pages, Razor views, and Blazor components. Yet you can build and run those files because there is no actual problem. If you have installed the Unity Support plugin, then it will complain about boxing operations (which are a genuine problem for Unity game developers) but in projects that are not Unity; hence, the warning is not applicable.

Most readers use Visual Studio, which is a large and complex tool that can do many things. But Visual Studio likes to provide its own mechanism to do as much as possible, and a .NET developer who uses it could easily think that Visual Studio is the only way to complete a .NET-related task, like modifying project configuration or editing a code file.

Always try to remember that Visual Studio and all the other code editors are just tools that do work for you that you could do manually. They just show you a view above what is really happening in the files you're working on, like the project file and all the C# code files.

You could just use a plain text editor to manually edit the project and code files. Ultimately, you use the `dotnet` command-line interface to compile– aka build– the project files into a runnable assembly packaged as either a `.dll` or `.exe` file, as shown in *Figure 1.2*:

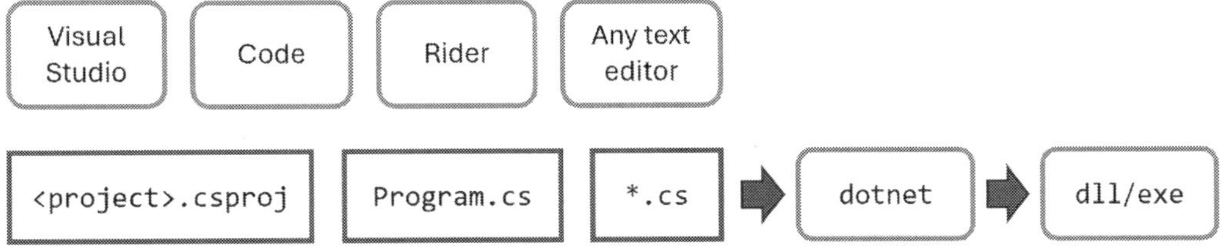

Figure 1.2: All code editors ultimately just change underlying files

Choosing the appropriate tool and application type for learning

What is the best tool and application type for learning C# and .NET?

When learning, the best tool is one that helps you write code and configuration but does not hide what is really happening. IDEs provide graphical user interfaces that are friendly to use, but what are they doing for you underneath? A more basic code editor that is closer to the action while providing help to write your code can be better while you are learning.

Having said that, you could make the argument that the best tool is the one you are already familiar with or that you or your team will use as your daily development tool. For that reason, I want you to be free to choose any C# code editor or IDE to complete the coding tasks in this book, including VS Code, Visual Studio, and even Rider.

In this book, I give detailed step-by-step instructions in *this chapter* on how to create multiple projects in both Visual Studio and VS Code. There are also links to online instructions for other code editors, as shown at the following link: `https://github.com/markjprice/cs13net9/blob/main/docs/code-editors/README.md`.

In subsequent chapters, I will only give the names of projects along with general instructions, so you can use whichever tool you prefer.

The best application type for learning the C# language constructs and many of the .NET libraries is one that does not distract with unnecessary application code. For example, there is no need to create an entire Windows desktop application or a website just to learn how to write a `switch` statement.

For that reason, I believe the best method for learning the C# and .NET topics in *Chapters 1* to *11* is to build console apps. Then, in *Chapters 12* to *15*, which are about web development, you will build websites and services using the modern parts of ASP.NET Core, including Blazor and Minimal APIs.

VS Code for cross-platform development

The most modern and lightweight code editor to choose from, and the only one from Microsoft that is cross-platform, is VS Code. It can run on all common operating systems, including Windows, macOS, and many varieties of Linux, including **Red Hat Enterprise Linux** (**RHEL**) and Ubuntu.

VS Code is a good choice for modern cross-platform development because it has an extensive and growing set of extensions to support many languages beyond C#. The most important extension for C# and .NET developers is the **C# Dev Kit** that was released in preview in June 2023, as it turns VS Code from a general-purpose code editor into a tool optimized for C# and .NET developers.

More Information: You can read about the **C# Dev Kit** extension in the official announcement at the following link: `https://devblogs.microsoft.com/visualstudio/announcing-csharp-dev-kit-for-visual-studio-code/`.

Being cross-platform and lightweight, VS Code and its extensions can be installed on all platforms that your apps will be deployed to for quick bug fixes and so on. Choosing VS Code means a developer can use a cross-platform code editor to develop cross-platform apps. VS Code is supported on ARM processors so that you can develop on Apple Silicon computers and Raspberry Pi computers.

VS Code has strong support for web development, although it currently has weak support for mobile and desktop development.

VS Code is by far the most popular code editor or IDE, with over 73% of professional developers selecting it in a Stack Overflow survey, which you can read at the following link: `https://survey.stackoverflow.co/2024/`.

GitHub Codespaces for development in the cloud

GitHub Codespaces is a fully configured development environment, based on VS Code, that can be spun up in an environment hosted in the cloud and accessed through any web browser. It supports Git repos, extensions, and a built-in command-line interface, so you can edit, run, and test from any device.

But note that for your GitHub Codespaces experience to be fully functional and practically useful, it does have a license cost.

More Information: You can learn more about GitHub Codespaces at the following link: `https://github.com/features/codespaces`.

Visual Studio for general development

Visual Studio can create most types of applications, including console apps, websites, web services, and desktop apps. Although you can use Visual Studio to write a cross-platform mobile app, you still need macOS and Xcode to compile it.

Visual Studio only runs on Windows 10 version 1909 or later, Home, Professional, Education, or Enterprise; or on Windows 11 version 21H2 or later, Home, Pro, Pro Education, Pro for Workstations, Enterprise, or Education. Windows Server 2016 and later are also supported. 32-bit operating systems and Windows S mode are *not* supported.

Warning! Visual Studio for Mac does not officially support .NET 8 or later, and it reached its end of life in August 2024. If you have been using Visual Studio for Mac, then you should switch to VS Code for Mac, Rider for Mac, or use Visual Studio for Windows in a virtual machine on your local computer or in the cloud, using a technology like Microsoft Dev Box. The retirement announcement can be read here: `https://devblogs.microsoft.com/visualstudio/visual-studio-for-mac-retirement-announcement/`.

What I used

To write and test the code for this book, I used the following hardware and software:

- Windows 11 on a Surface Laptop 7 Copilot+ PC with Visual Studio, VS Code, and Rider.
- macOS on an Apple Silicon Mac mini (M1) desktop with VS Code and Rider.

I hope that you have access to a variety of hardware and software too, as seeing the differences in platforms deepens your understanding of development challenges, although any one of the above combinations is enough to learn the fundamentals of C# and .NET and how to build practical apps and websites.

Deploying cross-platform

Your choice of code editor and operating system for development does not limit where your code gets deployed.

.NET 9 supports the following platforms for deployment:

- **Windows:** Windows 10 version 1607 or later, Windows 11 version 22000 or later, Windows Server 2012 R2 SP1 or later, and Nano Server version 2019 or 2022.
- **Mac:** macOS Catalina version 10.15 or later and in the Rosetta 2 x64 emulator.
- **Linux:** Alpine Linux 3.19 or 3.20, CentOS Stream 9, Debian 12, Fedora 40, openSUSE 15.5 or 15.6, RHEL 8 or 9, SUSE Enterprise Linux 15.5 or 15.6, and Ubuntu 20.04, 22.04, or 24.04.
- **Android:** API 21 or later is the minimum SDK target. Versions 12, 12.1, 13, and 14.
- **iOS** and **iPadOS:** 15, 16, or 17. iOS 12.2 is used as the minimum SDK target.
- **Mac Catalyst:** 12, 13, or 14.

Warning! .NET support for Windows 7 and 8.1 ended in January 2023: `https://github.com/dotnet/core/issues/7556`.

Windows Arm64 support in .NET 5 and later means you can develop on, and deploy to, Windows Arm devices like Microsoft's Windows Dev Kit 2023 (formerly known as Project Volterra) and Surface Pro 11 and Surface Laptop 7.

You can review the latest supported operating systems and versions at the following link: `https://github.com/dotnet/core/blob/main/release-notes/9.0/supported-os.md`.

All versions of .NET that are supported can be automatically patched via Microsoft Update on Windows.

Downloading and installing Visual Studio

Many professional .NET developers use Visual Studio in their day-to-day development work. Even if you choose to use VS Code to complete the coding tasks in this book, you might want to familiarize yourself with Visual Studio too. It is not until you have written a decent amount of code with a tool that you can really judge if it fits your needs.

If you do not have a Windows computer, then you can skip this section and continue to the next section, where you will download and install VS Code on macOS or Linux.

Since October 2014, Microsoft has made a professional-quality edition of Visual Studio available to students, open-source contributors, and individuals for free. It is called Community Edition. Any of the editions are suitable for this book. If you have not already installed it, let's do so now:

1. Download the latest version of Visual Studio from the following link: `https://visualstudio.microsoft.com/downloads/`.

Visual Studio vNext: At the time of writing, Visual Studio is version 17.12 and branded as Visual Studio 2022. I expect the next major version of Visual Studio to be version 18.0 and be branded as Visual Studio 2025. It is likely to be released in the first half of 2025, after this book is published. Visual Studio 2025 will have mostly the same features as the 2022 edition, although the user interface might move things around a bit.

2. Run the installer to start the installation.
3. On the **Workloads** tab, select the following:
    - **ASP.NET and web development**.
    - **.NET desktop development** (because this includes console apps).
    - **Desktop development with C++** with all default components (because this enables you to publish console apps and web services that start faster and have smaller memory footprints).
4. Click **Install** and wait for the installer to acquire the selected software, and then install it.
5. When the installation is complete, click **Launch**.
6. The first time that you run Visual Studio, you will be prompted to sign in. If you have a Microsoft account, you can use that account. If you don't, then register for a new one at the following link: `https://signup.live.com/`.
7. The first time that you run Visual Studio, you will be prompted to configure your environment. For **Development Settings**, choose **Visual C#**. For the color theme, I chose **Blue**, but you can choose whatever tickles your fancy.
8. If you want to customize your keyboard shortcuts, navigate to **Tools** | **Options...**, and then select the **Keyboard** section.

Keyboard shortcuts for Visual Studio

In this book, I will avoid showing keyboard shortcuts, since they are often customized. Where they are consistent across code editors and commonly used, I will try to show them.

If you want to identify and customize your keyboard shortcuts, then you can, as shown at the following link: `https://learn.microsoft.com/en-us/visualstudio/ide/identifying-and-customizing-keyboard-shortcuts-in-visual-studio`.

Downloading and installing VS Code

VS Code has rapidly improved over the past couple of years and has pleasantly surprised Microsoft with its popularity. If you are brave and like to live on the bleeding edge, then there is the **Insiders** edition, which is a daily build of the next version.

Even if you plan to only use Visual Studio for development, I recommend that you download and install VS Code and try the coding tasks in this chapter using it, and then decide if you want to stick with just using Visual Studio for the rest of the book.

Let's now download and install VS Code, the .NET SDK, and the **C# Dev Kit** extension:

1. Download and install either the Stable build or the Insiders edition of VS Code from the following link: `https://code.visualstudio.com/`.

More Information: If you need more help installing VS Code, you can read the official setup guide at the following link: `https://code.visualstudio.com/docs/setup/setup-overview`.

2. Download and install the .NET SDK for version 9.0 and version 8.0 from the following link: `https://www.microsoft.com/net/download`.

In real life, you are extremely unlikely to only have one .NET SDK version installed on your computer. To learn how to control which .NET SDK version is used to build a project, we need multiple versions installed. .NET 8 and .NET 9 are the only supported versions at the time of publishing in November 2024. You can safely install multiple SDKs side by side. The most recent SDK will be used to build your projects.

3. To install the **C# Dev Kit** extension with a user interface, you must first launch the VS Code application.
4. In VS Code, click the **Extensions** icon or navigate to **View | Extensions**.
5. **C# Dev Kit** is one of the most popular extensions available, so you should see it at the top of the list, or you can enter `C#` in the search box.

C# Dev Kit has a dependency on the **C#** extension version 2.0 or later, so you do not have to install the **C#** extension separately. Note that **C#** extension version 2.0 or later no longer uses OmniSharp, since it has a new **Language Server Protocol (LSP)** host. **C# Dev Kit** also has dependencies on the **.NET Install Tool for Extension Authors** and **IntelliCode for C# Dev Kit** extensions, so they will be installed too.

6. Click **Install** and wait for the supporting packages to download and install.

Good Practice: Be sure to read the license agreement for **C# Dev Kit.** It has a more restrictive license than the **C#** extension: `https://aka.ms/vs/csdevkit/license`.

Installing other extensions

In later chapters of this book, you will use more VS Code extensions. If you want to install them now, all the extensions that we will use are shown in *Table 1.1*:

Extension name and identifier	Description
C# Dev Kit `ms-dotnettools.csdevkit`	Official C# extension from Microsoft. Helps you manage your code with a solution explorer and test your code with integrated unit test discovery and execution, elevating your C# development experience wherever you like to develop (Windows, macOS, Linux, and even in a codespace).
C# `ms-dotnettools.csharp`	Provides rich language support for C# and is shipped along with C# Dev Kit. Powered by a **Language Server Protocol** (**LSP**) server, this extension integrates with open source components like Roslyn and Razor to provide rich type information and a faster, more reliable C# experience.
IntelliCode for C# Dev Kit `ms-dotnettools.vscodeintellicode-csharp`	Provides AI-assisted development features for Python, TypeScript/JavaScript, C#, and Java developers.
MSBuild project tools `tintoy.msbuild-project-tools`	Provides IntelliSense for MSBuild project files, including autocomplete for `<PackageReference>` elements.
Markdown All in One `yzhang.markdown-all-in-one`	All you need for Markdown (keyboard shortcuts, table of contents, auto preview, and more).

Polyglot Notebooks `ms-dotnettools.dotnet-interactive-vscode`	This extension adds support for using .NET and other languages in a notebook. It has a dependency on the **Jupyter** extension (`ms-toolsai.jupyter`), which itself has dependencies.
ilspy-vscode `icsharpcode.ilspy-vscode`	Decompile MSIL assemblies – support for modern .NET, .NET Framework, .NET Core, and .NET Standard.
REST Client `humao.rest-client`	Send an HTTP request and view the response directly in VS Code.

Table 1.1: VS Code extensions for .NET development

You can install a VS Code extension at the command prompt or terminal, as shown in *Table 1.2*:

Command	**Description**
`code --list-extensions`	List installed extensions.
`code --install-extension <extension-id>`	Install the specified extension.
`code --uninstall-extension <extension-id>`	Uninstall the specified extension.

Table 1.2: Managing VS Code extensions at the command prompt

For example, to install the **C# Dev Kit** extension, enter the following at the command prompt:

```
code --install-extension ms-dotnettools.csdevkit
```

I have created PowerShell scripts to install and uninstall the VS Code extensions in the preceding table. You can find them at the following link: `https://github.com/markjprice/cs13net9/tree/main/scripts/extension-scripts/`. PowerShell scripts are cross-platform, as you can read about at the following link: `https://learn.microsoft.com/en-us/powershell/scripting/overview`.

Understanding VS Code versions

Microsoft releases a new feature version of VS Code (almost) every month and bug-fix versions more frequently. For example:

- Version 1.93.0, August 2024 feature release
- Version 1.93.1, August 2024 bug fix release

The version used in this book is 1.93.0, the August 2024 feature release, but the version of VS Code is less important than the version of the **C# Dev Kit** or **C#** extension that you install. I recommend **C# Dev Kit** v1.10.18 or later with **C#** extension v2.45.20 or later.

While the **C#** extension is not required, it provides IntelliSense as you type, code navigation, and debugging features, so it's something that's very handy to install and keep updated to support the latest C# language features.

Keyboard shortcuts for VS Code

If you want to customize your keyboard shortcuts for VS Code, then you can, as shown at the following link: `https://code.visualstudio.com/docs/getstarted/keybindings`.

I recommend that you download a PDF of VS Code keyboard shortcuts for your operating system from the following list:

- Windows: `https://code.visualstudio.com/shortcuts/keyboard-shortcuts-windows.pdf`
- macOS: `https://code.visualstudio.com/shortcuts/keyboard-shortcuts-macos.pdf`
- Linux: `https://code.visualstudio.com/shortcuts/keyboard-shortcuts-linux.pdf`

Understanding .NET

> *"Those who cannot remember the past are condemned to repeat it."*
>
> *– George Santayana*

.NET, .NET Core, .NET Framework, .NET Standard, and Xamarin are related and overlapping platforms for developers used to build applications and services.

If you are not familiar with the history of .NET, then I will introduce you to each of these .NET concepts at the following link:

`https://github.com/markjprice/cs13net9/blob/main/docs/ch01-dotnet-history.md`

As time moves on, more and more readers already know the history of .NET, so it would be a waste of space in the book to continue to include it. But if you are new to it, then make sure you read all the extras that I provide online, like the preceding one.

Understanding .NET support

.NET versions are either **Long-Term Support** (**LTS**), **Standard-Term Support** (**STS**) (formerly known as **Current**), or **Preview**, as described in the following list:

- **LTS** releases are a good choice for applications that you do not intend to update frequently, although you must update the .NET runtime for your production code monthly. LTS releases are supported by Microsoft for 3 years after **General Availability** (**GA**), or 1 year after the next LTS release ships, whichever is longer.
- **STS** releases include features that may change based on feedback. These are a good choice for applications that you are actively developing because they provide access to the latest improvements. STS releases are supported by Microsoft for 18 months after GA, or 6 months after the next STS or LTS release ships, whichever is longer.
- **Preview** releases are for public testing. These are a good choice for adventurous programmers who want to live on the bleeding edge, or programming book authors who need to have early access to new language features, libraries, and app and service platforms. Preview releases are not usually supported by Microsoft, but some preview or **Release Candidate** (**RC**) releases may be declared **Go Live**, meaning they are supported by Microsoft in production.

STS and LTS releases receive critical patches throughout their lifetime for security and reliability.

Good Practice: You must stay up to date with the latest patches to get support. For example, if a system is running on .NET runtime version 9.0.0 and then version 9.0.1 is released, you must install version 9.0.1 to get support. These updates are released on the second Tuesday of each month, aka **Patch Tuesday**.

To better understand your choices of STS and LTS releases, it is helpful to see them visualized, with 3-year-long black bars for LTS releases, and 1½-year-long gray bars for STS releases, as shown in *Figure 1.3*:

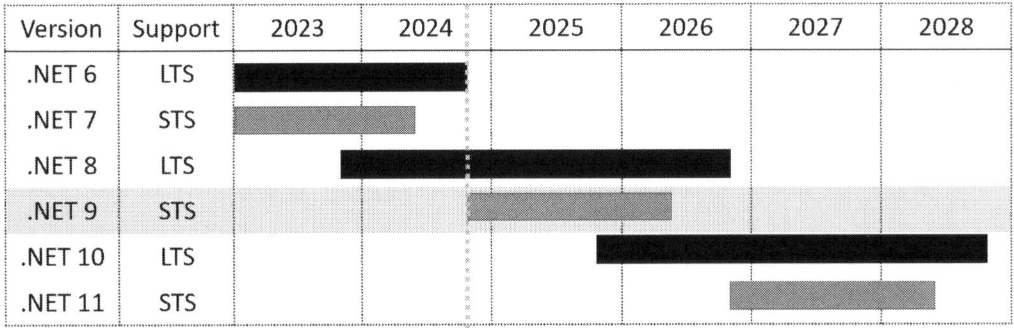

Figure 1.3: Support durations for recent and planned STS and LTS releases

During the lifetime of .NET 9, .NET 8 will still be supported and .NET 10 will be released. I have tried to be cognizant that you might choose to use .NET 8 or .NET 10 with this book; however, obviously, the book cannot cover new features of .NET 10, since I don't know what they will be!

If you need LTS from Microsoft, then set your .NET projects to target .NET 8 today and then migrate to .NET 10 after it is released in November 2025. This is because .NET 9 is an STS release, and therefore, it will lose support in May 2026, before .NET 8 does in November 2026. As soon as .NET 10 is released, start upgrading your .NET 8 projects to it. You will have a year to do so before .NET 8 reaches its end of life.

Good Practice: Remember that with all releases, you must upgrade to bug-fix releases like .NET runtime 9.0.1 and .NET SDK 9.0.101, which are expected to release in December 2024, as updates are released every month.

At the time of publishing in November 2024, all versions of modern .NET have reached their EOL, except those shown in the following list, which are ordered by their EOL dates:

- .NET 9 will reach EOL in May 2026.
- .NET 8 will reach EOL in November 2026.
- .NET 10 will be available from November 2025 and it will reach EOL in November 2028.

You can check which .NET versions are currently supported and when they will reach EOL at the following link: `https://github.com/dotnet/core/blob/main/releases.md`.

Understanding end of life (EOL)

End of support or **end of life** (**EOL**) means the date after which bug fixes, security updates, or technical assistance are no longer available from Microsoft.

For example, now that .NET 6 has reached end of support on November 12, 2024, you can expect the following:

- Projects that use .NET 6 will continue to run.
- No new security updates will be issued for .NET 6 and therefore continuing to use an unsupported version will increasingly expose you to security vulnerabilities.
- You might not be able to access technical support for any .NET 6 applications that you continue to use.
- You will get `NETSDK1138` build warnings when targeting .NET 6 from a later SDK like the .NET 9 SDK.
- You will get warnings in Visual Studio when targeting .NET 6.

Understanding .NET support phases

The lifetime of a version of .NET passes through several phases, during which they have varying levels of support, as described in the following list:

- **Preview**: These are not supported at all. .NET 9 Preview 1 to Preview 7 were in this support phase from February 2024 to August 2024.
- **Go Live**: These are supported until GA, and then they become immediately unsupported. You must upgrade to the final release version as soon as it is available. .NET 9 Release Candidate 1 and Release Candidate 2 were in this support phase in September and October 2024, respectively.
- **Active**: .NET 9 will be in this support phase from November 2024 to November 2025.
- **Maintenance**: Supported only with security fixes for the last 6 months of its lifetime. .NET 9 will be in this support phase from November 2025 to May 2026.
- **EOL**: Not supported. .NET 9 will reach its EOL in May 2026.

Understanding .NET runtime and .NET SDK versions

If you have not built a standalone app, then the .NET runtime is the minimum you need to install so that an operating system can run a .NET application. The .NET SDK includes the .NET runtime, as well as the compilers and other tools needed to build .NET code and apps.

.NET runtime versioning follows semantic versioning – that is, a major increment indicates breaking changes, minor increments indicate new features, and patch increments indicate bug fixes.

.NET SDK versioning does not follow semantic versioning. The major and minor version numbers are tied to the runtime version they are matched with. The third number follows a convention that indicates the minor and patch versions of the SDK. The third number starts at 100 for the initial version (equivalent to 0.0 for the minor and patch numbers). The first digit increments with minor increments, and the other two digits increment with patch increments.

You can see an example of this in *Table 1.3*:

Change	Runtime	SDK
Initial release	9.0.0	9.0.100
SDK bug fix	9.0.0	9.0.101
Runtime and SDK bug fix	9.0.1	9.0.102
SDK new feature	9.0.1	9.0.200

Table 1.3: Examples of changes and versions for a .NET runtime and SDK

Listing and removing versions of .NET

.NET runtime updates are compatible with a major version such as 9.x, and updated releases of the .NET SDK maintain the ability to build applications that target previous versions of the runtime, which enables the safe removal of older versions.

You can see which SDKs and runtimes are currently installed using the following commands:

```
dotnet --list-sdks
dotnet --list-runtimes
dotnet --info
```

Good Practice: To make it easier to enter commands at the command prompt or terminal, the following link lists all commands throughout the book that can be easily copied and pasted: `https://github.com/markjprice/cs13net9/blob/main/docs/command-lines.md`.

On Windows, use the **Apps & features** section to remove .NET SDKs.

On Linux, there is no single mechanism, but you can learn more at the following link:

`https://learn.microsoft.com/en-us/dotnet/core/install/remove-runtime-sdk-versions?pivots=os-linux`

You could use a third-party tool like Dots, the friendly .NET SDK manager, found at the following link: `https://johnnys.news/2023/01/Dots-a-dotnet-SDK-manager`. At the time of writing, you must build the app from source on its GitHub repository, so I only recommend that for advanced developers.

Understanding intermediate language

The C# compiler (named **Roslyn**) used by the `dotnet` CLI tool converts your C# source code into **intermediate language** (IL) code and stores the IL in an **assembly** (a DLL or EXE file). IL code statements are like assembly language instructions, which are executed by .NET's virtual machine, known as CoreCLR, the newer name for the **Common Language Runtime (CLR)** in modern .NET. The legacy .NET Framework has a CLR that is Windows-only, and modern .NET has one for each OS, like Windows, macOS, and Linux. These days, they are all commonly referred to as CLRs.

At runtime, CoreCLR loads the IL code from the assembly, the **just-in-time** (JIT) compiler compiles it into native CPU instructions, and then it is executed by the CPU on your machine.

The benefit of this two-step compilation process is that Microsoft can create **CLRs** for Linux and macOS, as well as for Windows. The same IL code runs everywhere because of the second compilation step, which generates code for the native operating system and CPU instruction set.

Regardless of which language the source code is written in (for example, C#, Visual Basic, or F#), all .NET applications use IL code for their instructions stored in an assembly. Microsoft and others provide disassembler tools that can open an assembly and reveal this IL code, such as the ILSpy .NET Decompiler extension. You will learn more about this in *Chapter 7, Packaging and Distributing .NET Types*, in an online section found at the following link: `https://github.com/markjprice/cs13net9/blob/main/docs/ch07-decompiling.md`.

So, the compilation process typically involves translating source code into IL, which is then compiled into machine code at runtime by the CLR using JIT compilation. **Ahead-of-Time (AOT)** compilation is an alternative to this approach, and you will learn about it in *Chapter 7, Packaging and Distributing .NET Types*.

Comparing .NET technologies

We can summarize and compare the current .NET technologies, as shown in *Table 1.4*:

Technology	Description	Host operating systems
Modern .NET	A modern feature set, with full C# 8 to C# 13 language support. It can be used to port existing apps or create new desktop, mobile, and web apps and services.	Windows, macOS, Linux, Android, iOS, tvOS, and Tizen
.NET Framework	A legacy feature set with limited C# 8 support and no C# 9 or later support. It should be used to maintain existing applications only.	Windows only
Xamarin	Mobile and desktop apps only.	Android, iOS, and macOS

Table 1.4: Comparison of .NET technologies

Visual Studio, Rider, and even VS Code (with the **C# Dev Kit** extension installed) all have a concept called a **solution** that allows you to open and manage multiple projects simultaneously. We will use a solution to manage the two projects that you will create in this chapter.

Building console apps using Visual Studio

The goal of this section is to showcase how to build a console app using Visual Studio.

If you do not have a Windows computer or want to use VS Code, then you can skip this section, since the code will be the same; just the tooling experience is different. However, I recommend that you review this section because it does explain some of the code and how top-level programs work, and that information applies to all code editors.

This section is also available in the GitHub repository (so it can be updated after publishing if needed) at the following link:

`https://github.com/markjprice/cs13net9/blob/main/docs/code-editors/vs.md`

If you want to see similar instructions for using Rider, they are available in the GitHub repository at the following link:

`https://github.com/markjprice/cs13net9/blob/main/docs/code-editors/rider.md`

Writing code using Visual Studio

Let's get started writing code:

1. Start Visual Studio.
2. In the **Create a new project** dialog, select the C# language to filter the project templates, enter `console` in the **Search for templates** box, and then select **Console App**.

Make sure that you have chosen the cross-platform project template, not the one for .NET Framework, which is Windows-only, and the **C#** project template rather than another language, such as Visual Basic or TypeScript.

3. Click **Next**.
4. In the **Configure your new project** dialog, enter `HelloCS` for the project name, `C:\cs13net9` for the location, and `Chapter01` for the solution name.

Screenshots of Visual Studio when creating new projects can be found in the GitHub repository at the following link: `https://github.com/markjprice/cs13net9/blob/main/docs/ch01-project-options.md`.

5. Click **Next**.

6. In the **Additional information** dialog, in the **Framework** drop-down list, note that your .NET SDK choices indicate if that version is **Standard Term Support**, **Long Term Support**, **Preview**, or **Out of support**, and then select **.NET 9.0 (Standard Term Support)**.

You can install as many .NET SDK versions as you like. If you are missing a .NET SDK version, then you can install it from the following link: `https://dotnet.microsoft.com/en-us/download/dotnet`.

7. Leave the checkbox labeled **Do not use top-level statements** clear. (Later in this chapter, you will create a console app that selects this option, so you will see the difference.)
8. Leave the checkbox labeled **Enable native AOT publish** clear. You will learn what this option does in *Chapter 7, Packaging and Distributing .NET Types*.
9. Click **Create**.
10. If you cannot see **Solution Explorer**, then navigate to **View** | **Solution Explorer**.
11. If code is not shown, then in **Solution Explorer**, double-click the file named `Program.cs` to open it, and note that **Solution Explorer** shows the **HelloCS** project, as shown in *Figure 1.4*:

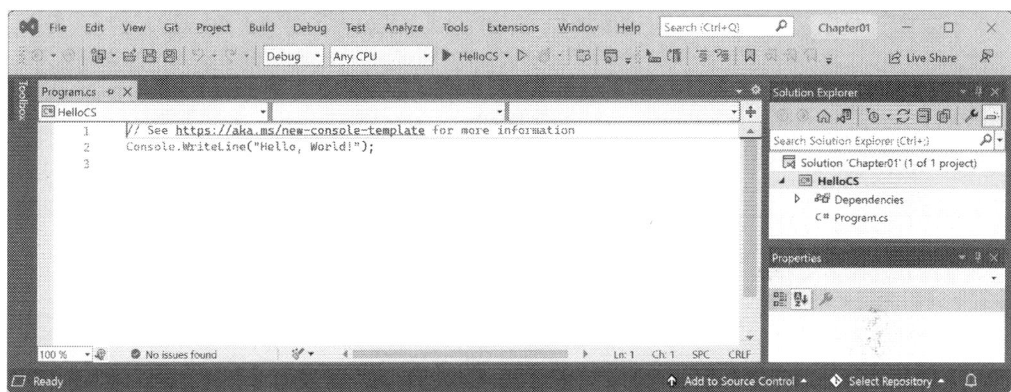

Figure 1.4: Editing Program.cs in Visual Studio

12. In `Program.cs`, note that the code consists of only a comment and a single statement, as shown in the following code:

```
// See https://aka.ms/new-console-template for more information
Console.WriteLine("Hello, World!");
```

This template uses the top-level program feature introduced in C# 9, which I will explain later in this chapter. As the comment in the code says, you can read more about this template at the following link: `https://aka.ms/new-console-template`.

13. In `Program.cs`, modify line 2 so that the text that is being written to the console says `Hello, C#!`.

All code examples and commands that you must review or type are shown in plain text, so you will never have to read code or commands from a screenshot, like in *Figure 1.4*, which might be too small or too faint in print.

Compiling and running code using Visual Studio

The next task is to compile and run the code:

1. In Visual Studio, navigate to **Debug | Start Without Debugging**.

Good Practice: When you start a project in Visual Studio, you can choose whether to attach a debugger or not. If you do not need to debug, then it is better not to attach one because attaching a debugger requires more resources and slows everything down. Attaching a debugger also limits you to only starting one project. If you want to run more than one project, each with a debugger attached, then you must start multiple instances of Visual Studio. In the toolbar, click the green outline triangle button (to the right of **HelloCS** in the top bar shown in *Figure 1.5*) to start without debugging, instead of the green solid triangle button (to the left of **HelloCS** in the top bar shown in *Figure 1.5*), unless you need to debug.

2. The output in the console window will show the result of running your application, as shown in *Figure 1.5*:

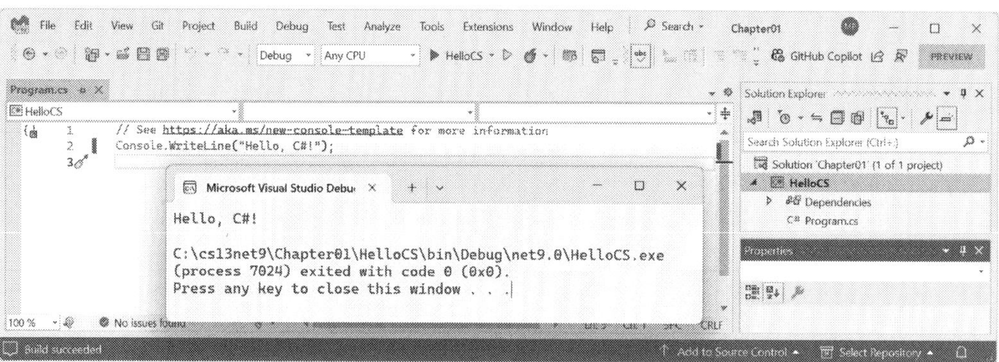

Figure 1.5: Running the console app on Windows

3. Press any key to close the console app window and return to Visual Studio.
4. Optionally, close the **Properties** pane to make more vertical space for **Solution Explorer**.
5. Double-click the **HelloCS** project, and note that the `HelloCS.csproj` project file shows that this project has its target framework set to `net9.0`, as shown in *Figure 1.6*.
6. In the **Solution Explorer** toolbar, toggle on the **Show All Files** button, and note that the compiler-generated `bin` and `obj` folders are visible, as shown in *Figure 1.6*:

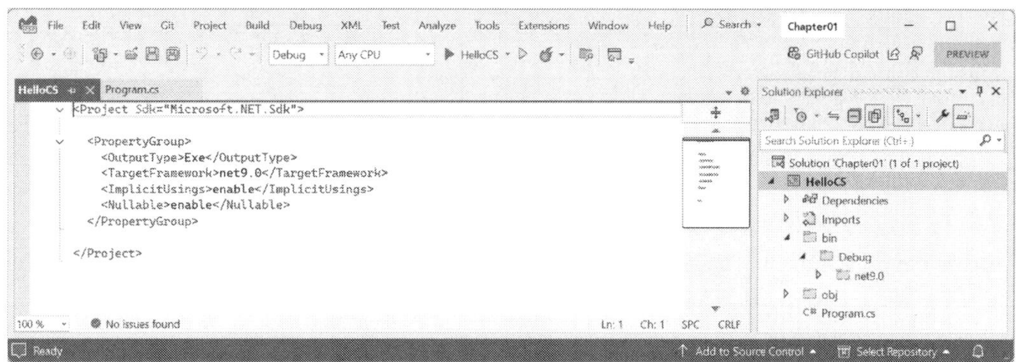

Figure 1.6: Showing the compiler-generated folders and files

Understanding the compiler-generated folders and files

Two compiler-generated folders were created, named `obj` and `bin`, as described in the following list:

- The `obj` folder contains one compiled *object* file for each source code file. These objects haven't been linked together into a final executable yet.
- The `bin` folder contains the *binary* executable for the application or class library. We will look at this in more detail in *Chapter 7, Packaging and Distributing .NET Types.*

You do not need to look inside these folders or understand their files yet (but feel free to browse around if you are curious).

Just be aware that the compiler needs to create temporary folders and files to do its work. You could delete these folders and their files, and they will be automatically recreated the next time you "build" or run the project. Developers often delete these temporary folders and files to "clean" a project. Visual Studio even has a command on the **Build** menu named **Clean Solution** that deletes some of these temporary files for you. The equivalent command with the CLI is `dotnet clean`.

Understanding top-level programs

If you have seen older .NET projects before, then you might have expected more code, even just to output a simple message. This project has minimal statements because some of the required code is written for you by the compiler when you target .NET 6 or later.

If you had created the project with .NET SDK 5 or earlier, or if you had selected the checkbox labeled **Do not use top-level statements**, then the `Program.cs` file would have more statements, as shown in the following code:

```
using System;

namespace HelloCS
{
  class Program
  {
```

```
      static void Main(string[] args)
      {
        Console.WriteLine("Hello, World!");
      }
    }
  }
```

During compilation with .NET SDK 6 or later, all the boilerplate code to define the `Program` class and its `Main` method is generated and wrapped around the statements you write.

This uses a feature introduced in .NET 5 called **top-level programs**, but it was not until .NET 6 that Microsoft updated the project template for console apps to use top-level statements by default. Then, in .NET 7 and later, Microsoft added options to use the older style if you prefer:

- If you are using Visual Studio, select the checkbox labeled **Do not use top-level statements**.
- If you are using the `dotnet` CLI at the command prompt, add a switch:

  ```
  dotnet new console --use-program-main
  ```

Warning! One functional difference is that the auto-generated code does not define a namespace, so the `Program` class is implicitly defined in an empty namespace with no name, instead of a namespace that matches the name of the project.

Requirements for top-level programs

Key points to remember about top-level programs include the following:

- There can be only one file like the file you use for top-level program code in a project.
- Any `using` statements must be at the top of the file.
- If you declare any classes or other types, they must be at the bottom of the file.
- Although you should name the entry-point method `Main` if you explicitly define it, the method is named `<Main>$` when created by the compiler.

Implicitly imported namespaces

The `using System;` statement at the top of the file imports the `System` namespace. This enables the `Console.WriteLine` statement to work. But why do we not have to import it in our project?

The trick is that we still need to import the `System` namespace, but it is now done for us using a combination of features introduced in C# 10 and .NET 6. Let's see how:

1. In **Solution Explorer**, expand the `obj`, `Debug`, and `net9.0` folders, and open the file named `HelloCS.GlobalUsings.g.cs`.
2. Note that this file is automatically created by the compiler for projects that target .NET 6 or later and uses a feature introduced in C# 10, called **global namespace imports**, which imports some commonly used namespaces like `System` for use in all code files, as shown in the following code:

```
// <autogenerated />
global using global::System;
global using global::System.Collections.Generic;
global using global::System.IO;
global using global::System.Linq;
global using global::System.Net.Http;
global using global::System.Threading;
global using global::System.Threading.Tasks;
```

3. In **Solution Explorer**, click the **Show All Files** button to hide the `bin` and `obj` folders.

I will explain more about the implicit imports feature in the next chapter. For now, just note that a significant change that happened between .NET 5 and .NET 6 is that many of the project templates, like the one for console apps, use new SDK and language features to hide what is really happening.

Revealing the hidden code by throwing an exception

Now let's discover how the hidden code has been written:

1. In `Program.cs`, after the statement that outputs the message, add a statement to throw a new exception, as shown in the following code:

```
throw new Exception();
```

2. In Visual Studio, navigate to **Debug | Start Without Debugging.** (Do not start the project with debugging, or the exception will be caught by the debugger!)
3. The output in the console window will show the result of running your application, including that a hidden `Program` class was defined by the compiler, with a method named `<Main>$` that has a parameter named `args` to pass in arguments, as shown in *Figure 1.7* and the following output:

```
Hello, C#!
Unhandled exception. System.Exception: Exception of type 'System.
Exception' was thrown.
   at Program.<Main>$(String[] args) in C:\cs13net9\Chapter01\HelloCS\
Program.cs:line 3
```

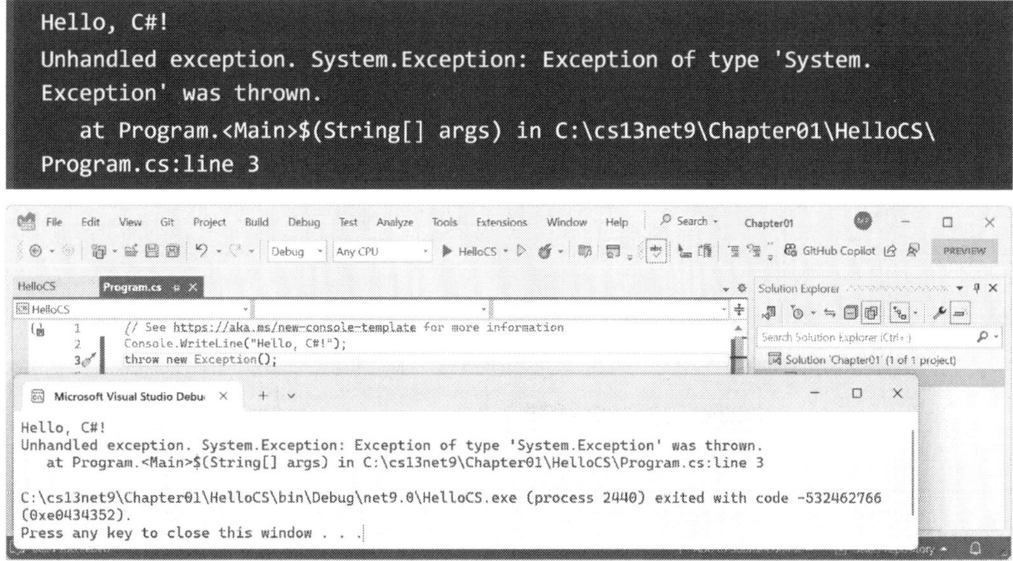

Figure 1.7: Throwing an exception to reveal the hidden Program.<Main>$ method

4. Press any key to close the console app window and return to Visual Studio.

Revealing the namespace for the Program class

Now, let's discover what namespace the `Program` class has been defined within:

1. In `Program.cs`, before the statement that throws an exception, add statements to get the name of the namespace of the `Program` class, and then write it to the console, as shown in the following code:

```
string name = typeof(Program).Namespace ?? "<null>";
Console.WriteLine($"Namespace: {name}");
```

`??` is the **null-coalescing operator.** The first statement means, "If the namespace of `Program` is `null`, then return `<null>`; otherwise, return the name." You will see more explanations of these keywords and operators throughout the book. For now, just enter the code and run it to see what it does.

Good Practice: Code editors have a feature named **code snippets.** These allow you to insert pieces of code that you commonly use, by typing a shortcut and pressing *Tab* twice. For example, in Visual Studio, to enter `Console.WriteLine()` and leave the cursor in the middle of the parentheses ready for you to type what you want to output, type `cw`, and then press *Tab*, *Tab*. Read the documentation for your code editor to learn how to insert code snippets using shortcuts.

2. In Visual Studio, navigate to **Debug** | **Start Without Debugging**.
3. The output in the console window will show the result of running your application, including that the hidden `Program` class was defined without a namespace, as shown in the following output:

```
Namespace: <null>
```

4. Press any key to close the console app window and return to Visual Studio.

Adding a second project using Visual Studio

Let's add a second project to our solution to explore how to work with multiple projects:

1. In Visual Studio, navigate to **File** | **Add** | **New Project...**.

Warning! The above step adds a new project to the existing solution. Do NOT navigate to **File** | **New** | **Project...**, which instead is meant to be used to create a new project and solution (although the dialog box has a dropdown to choose to add to an existing solution too).

2. In the **Add a new project** dialog, in **Recent project templates**, select **Console App [C#]**, and then click **Next**.
3. In the **Configure your new project** dialog, for **Project name**, enter `AboutMyEnvironment`, leave the location as `C:\cs13net9\Chapter01`, and then click **Next**.
4. In the **Additional information** dialog, select **.NET 9.0 (Standard Term Support)** and select the **Do not use top-level statements** checkbox.

Warning! Make sure you have selected the **Do not use top-level statements** checkbox so that we get to see the older style of `Program.cs`.

5. Click **Create**.
6. In the `AboutMyEnvironment` project, in `Program.cs`, note the statements to define a namespace that matches the project name, an internal class named `Program`, and a static method named `Main` with a parameter named `args` that returns nothing (`void`), as shown in the following code:

```
namespace AboutMyEnvironment
{
  internal class Program
  {
    static void Main(string[] args)
    {
      Console.WriteLine("Hello, World!");
    }
  }
}
```

7. In `Program.cs`, in the `Main` method, replace the existing `Console.WriteLine` statement with statements to output the current directory, the version of the operating system, and the namespace of the `Program` class, as shown in the following code:

```
Console.WriteLine(Environment.CurrentDirectory);
Console.WriteLine(Environment.OSVersion.VersionString);
Console.WriteLine("Namespace: {0}",
  typeof(Program).Namespace ?? "<null>");
```

8. In **Solution Explorer**, right-click the **Chapter01** solution, and then select **Configure Startup Projects...**.
9. In the **Solution 'Chapter01' Property Pages** dialog box, set **Startup Project** to **Current selection**, and then click **OK**.
10. In **Solution Explorer**, click the `AboutMyEnvironment` project (or any file or folder within it), and note that Visual Studio indicates that `AboutMyEnvironment` is now the startup project by making the project name bold.

Good Practice: I recommend this way of setting the startup project because it then makes it very easy to switch startup projects by simply clicking a project (or any file in a project) to make it the startup project. Although you can right-click a project and set it as a startup project, if you then want to run a different project, you must manually change it again. Simply clicking anywhere in the project is easier. In most chapters, you will only need to run one project at a time. In *Chapter 15, Building and Consuming Web Services*, I will show you how to configure multiple startup projects.

11. Navigate to **Debug | Start Without Debugging** to run the `AboutMyEnvironment` project, and note the result, as shown in the following output and *Figure 1.8*:

```
C:\cs13net9\Chapter01\AboutMyEnvironment\bin\Debug\net9.0
Microsoft Windows NT 10.0.26100.0
Namespace: AboutMyEnvironment
```

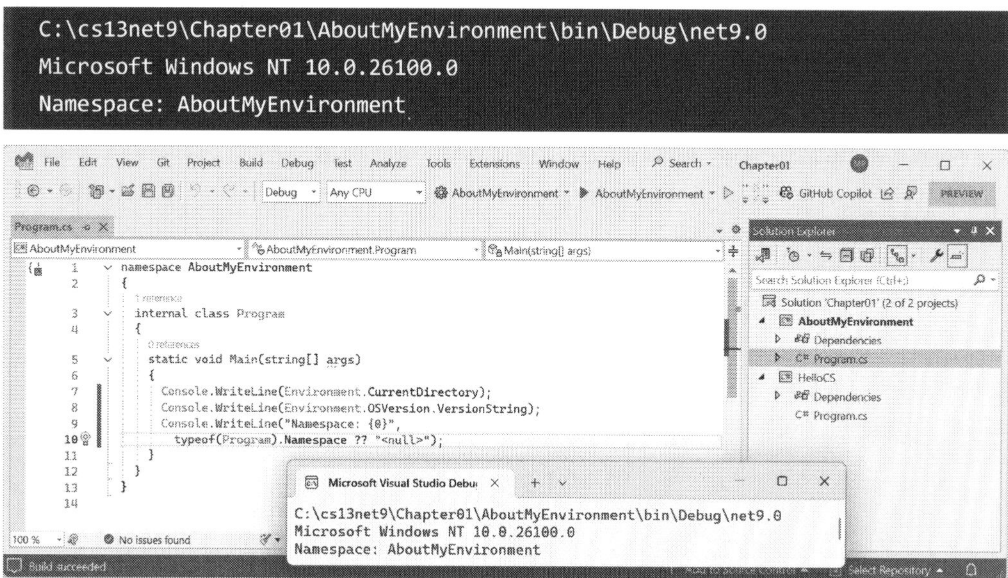

Figure 1.8: Running a console app in a Visual Studio solution with two projects

Windows 11 is just branding. Its official name is Windows NT, and its major version number is still 10! But its patch version is 22000 or higher.

12. Press any key to close the console app window and return to Visual Studio.

When Visual Studio runs a console app, it executes it from the `<projectname>\bin\Debug\net9.0` folder. It will be important to remember this when we work with the filesystem in later chapters. When using VS Code, or more accurately, the `dotnet` CLI, it has different behavior, as you are about to see.

Building console apps using VS Code

The goal of this section is to showcase how to build a console app using VS Code and the `dotnet` CLI.

If you never want to try VS Code or the `dotnet` command-line tool, then please feel free to skip this section, and then continue with the *Making good use of the GitHub repository for this book* section.

Both the instructions and screenshots in this section are for Windows, but the same actions will work with VS Code on the macOS and Linux variants.

The main differences will be native command-line actions such as deleting a file; both the command and the path are likely to be different on Windows, macOS, and Linux. Luckily, the `dotnet` CLI tool itself and its commands are identical on all platforms.

Writing code using VS Code

Let's get started writing code!

1. Start your favorite tool for working with the filesystem, for example, **File Explorer** on Windows or **Finder** on Mac.
2. Navigate to your `C:` drive on Windows, your user folder on macOS or Linux (mine are named `markjprice` and `home/markjprice`), or any directory or drive in which you want to save your projects.
3. Create a new folder named `cs13net9`. (If you completed the section for Visual Studio, then this folder will already exist.)
4. In the `cs13net9` folder, create a new folder named `Chapter01-vscode`.

If you did not complete the section for Visual Studio, then you could name this folder `Chapter01`, but I will assume you will want to complete both sections and, therefore, need to use a non-conflicting name.

5. In the `Chapter01-vscode` folder, open the command prompt or terminal. For example, on Windows, right-click on the folder and then select **Open in Terminal**.
6. At the command prompt or terminal, use the `dotnet` CLI to create a new solution named `Chapter01`, as shown in the following command:

```
dotnet new sln --name Chapter01
```

You can use either `-n` or `--name` as the switch to specify a name. If you do not explicitly specify a solution name with one of these switches, then the default would match the name of the folder, for example, `Chapter01-vscode`.

7. Note the result, as shown in the following output:

```
The template "Solution File" was created successfully.
```

8. At the command prompt or terminal, use the `dotnet` CLI to create a new subfolder and project for a console app named `HelloCS`, as shown in the following command:

```
dotnet new console --output HelloCS
```

You can use either `-o` or `--output` as the switch to specify the folder and project name. The **`dotnet new console`** command targets your latest .NET SDK version by default. To target a different version, use the `-f` or `--framework` switch to specify a target framework. For example, to target .NET 8, use the following command: `dotnet new console -f net8.0`.

9. At the command prompt or terminal, use the `dotnet` CLI to add the project to the solution, as shown in the following command:

```
dotnet sln add HelloCS
```

10. Note the results, as shown in the following output:

```
Project `HelloCS\HelloCS.csproj` added to the solution.
```

11. At the command prompt or terminal, start VS Code and open the current folder, indicated with a `.` (dot), as shown in the following command:

```
code .
```

12. If you are prompted with **Do you trust the authors of the files in this folder?**, select the **Trust the authors of all files in the parent folder 'cs13net9'** checkbox, and then click **Yes, I trust the authors**.
13. In VS Code, in **EXPLORER**, in the **CHAPTER01-VSCODE** folder view, expand the `HelloCS` folder, and you will see that the `dotnet` command-line tool created two files, `HelloCS.csproj` and `Program.cs`, and the `bin` and `obj` folders, as shown in *Figure 1.9*:

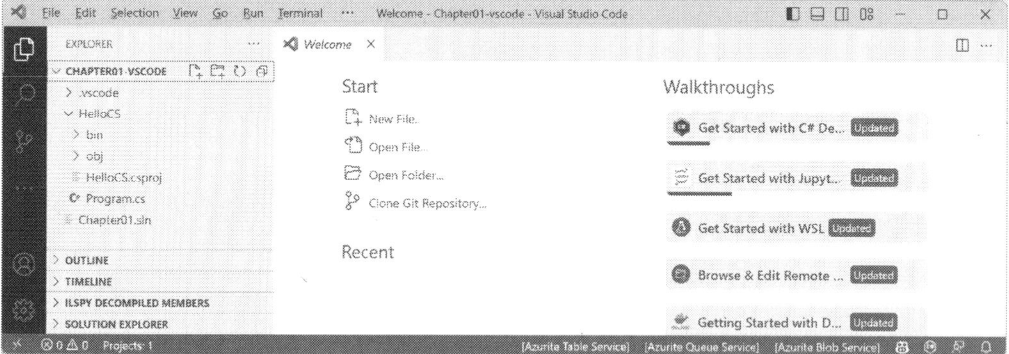

Figure 1.9: EXPLORER shows that two files and temporary folders have been created

14. Navigate to **View** | **Output**.

15. In the **OUTPUT** pane, select **C# Dev Kit**, and note that the tool has recognized and processed the solution.
16. At the bottom of **EXPLORER**, note **SOLUTION EXPLORER**.
17. Drag **SOLUTION EXPLORER** to the top of the **EXPLORER** pane and expand it.
18. In **SOLUTION EXPLORER**, expand the **HelloCS** project, and then click the file named `Program.cs` to open it in the editor window.
19. In `Program.cs`, modify line 2 so that the text that is being written to the console says `Hello, C#!`.

Good Practice: Navigate to **File** | **Auto Save**. This toggle will avoid the annoyance of remembering to save before rebuilding your application each time.

In the preceding steps, I showed you how to use the `dotnet` CLI to create solutions and projects. Finally, with the August 2024 or later releases of the C# Dev Kit, VS Code has an improved project creation experience that provides you access to the same options you can use when creating a new project through the `dotnet` CLI.

To enable this ability, you must change a setting, as shown in the following configuration:

```
"csharp.experimental.dotnetNewIntegration": true
```

In VS Code, navigate to **File** | **Preferences** | **Settings**, search for `dotnet new`, and then select the **Csharp > Experimental: Dotnet New Integration** checkbox.

You can learn more at the following link:

https://devblogs.microsoft.com/dotnet/whats-new-in-csharp-dev-kit-august-2024/#create-new-project-configuration-options

Compiling and running code using the dotnet CLI

The next task is to compile and run the code:

1. In **SOLUTION EXPLORER**, right-click on any file in the `HelloCS` project and choose **Open In Integrated Terminal**.
2. In **TERMINAL**, enter the following command: `dotnet run`.
3. The output in the **TERMINAL** window will show the result of running your application.
4. In `Program.cs`, after the statement that outputs the message, add statements to get the name of the namespace of the `Program` class, write it to the console, and then throw a new exception, as shown in the following code:

```
string name = typeof(Program).Namespace ?? "<null>";
Console.WriteLine($"Namespace: {name}");
throw new Exception();
```

5. In **TERMINAL**, enter the following command: `dotnet run`.

In **TERMINAL**, you can press the up and down arrows to loop through previous commands, and then press the left and right arrows to edit the commands before pressing *Enter* to run them.

6. The output in the **TERMINAL** window will show the result of running your application, including that a hidden `Program` class was defined by the compiler, with a method named `<Main>$` that has a parameter named `args` to pass in arguments, and that it does not have a namespace, as shown in the following output:

```
Hello, C#!
Namespace: <null>
Unhandled exception. System.Exception: Exception of type 'System.
Exception' was thrown.
   at Program.<Main>$(String[] args) in C:\cs13net9\Chapter01-vscode\
HelloCS\Program.cs:line 7
```

Adding a second project using VS Code

Let's add a second project to explore how to work with multiple projects:

1. In **TERMINAL**, change to the `Chapter01-vscode` directory, as shown in the following command:

```
cd ..
```

2. In **TERMINAL**, create a new console app project named `AboutMyEnvironment`, using the older non-top-level program style, as shown in the following command:

```
dotnet new console -o AboutMyEnvironment --use-program-main
```

Good Practice: Be careful when entering commands in **TERMINAL**. Ensure that you are in the correct folder before entering potentially destructive commands!

3. In **TERMINAL**, use the `dotnet` CLI to add the new project folder to the solution, as shown in the following command:

```
dotnet sln add AboutMyEnvironment
```

4. Note the results, as shown in the following output:

```
Project `AboutMyEnvironment\AboutMyEnvironment.csproj` added to the
solution.
```

5. In **SOLUTION EXPLORER**, in the `AboutMyEnvironment` project, open `Program.cs`, and then in the `Main` method, change the existing statement to output the current directory, the operating system version string, and the namespace of the `Program` class, as shown in the following code:

```
Console.WriteLine(Environment.CurrentDirectory);
Console.WriteLine(Environment.OSVersion.VersionString);
Console.WriteLine("Namespace: {0}",
  typeof(Program).Namespace ?? "<null>");
```

6. In **SOLUTION EXPLORER**, right-click on any file in the `AboutMyEnvironment` project and choose **Open In Integrated Terminal**.
7. In **TERMINAL**, enter the command to run the project, as shown in the following command: `dotnet run`.
8. Note the output in the **TERMINAL** window, as shown in the following output:

```
C:\cs13net9\Chapter01-vscode\AboutMyEnvironment
Microsoft Windows NT 10.0.26100.0
Namespace: AboutMyEnvironment
```

Once you open multiple terminal windows, you can toggle between them by clicking their names in the panel on the right-hand side of **TERMINAL**. By default, the name will be one of the common shells like **pwsh**, **powershell**, **zsh**, or **bash**. Right-click and choose **Rename** to set something else.

When VS Code, or more accurately, the `dotnet` CLI, runs a console app, it executes it from the `<projectname>` folder. Visual Studio executes the app from the `<projectname>\bin\Debug\net9.0` folder. It will be important to remember this when we work with the filesystem in later chapters.

If you were to run the program on macOS Ventura, the environment operating system would be different, as shown in the following output:

```
Unix 13.5.2
```

Good Practice: Although the source code, like the `.csproj` and `.cs` files, is identical, the `bin` and `obj` folders that are automatically generated by the compiler could have mismatches that give you errors. If you want to open the same project in both Visual Studio and VS Code, delete the temporary `bin` and `obj` folders before opening the project in the other code editor. This potential problem is why I asked you to create a different folder for the VS Code projects in this chapter.

Summary of steps for VS Code

Follow these steps to create a solution and projects using VS Code, as shown in *Table 1.5*:

Step Description	Command
1. Create a folder for the solution.	`mkdir <solution_folder_name>`
2. Change to the folder.	`cd <solution_folder_name>`
3. Create a solution file in the folder.	`dotnet new sln`
4. Create a folder and project using a template.	`dotnet new console -o <project_folder_name>`
5. Add the folder and its project to the solution.	`dotnet sln add <project_folder_name>`
6. Repeat steps 4 and 5 to create and add any other projects.	
7. Open the current folder path (`.`) containing the solution using VS Code.	`code .`

Table 1.5: Summary of steps to create a solution and projects using VS Code

Summary of other project types used in this book

A **Console App** / `console` project is just one type of project template. In this book, you will also create projects using the following project templates, as shown in *Table 1.6*:

Visual Studio	dotnet new	Rider – Type
Console App	`console`	Console Application
Class Library	`classlib`	Class Library
xUnit Test Project	`xunit`	Unit Test Project – xUnit
ASP.NET Core Empty	`web`	ASP.NET Core Web Application – Empty
Blazor Web App	`blazor`	ASP.NET Core Web Application – Blazor Web App
ASP.NET Core Web API	`webapi`	ASP.NET Core Web Application – Web API
ASP.NET Core Web API (native AOT)	`webapiaot`	ASP.NET Core Web Application – Web API (native AOT)

Table 1.6: Project template names for various code editors

The steps for adding any type of new project to a solution are the same. Only the type name of the project template differs and, sometimes, some command-line switches to control options. I will always specify what those switches and options should be if they differ from the defaults.

A summary of project template defaults, options, and switches can be found here: `https://github.com/markjprice/cs13net9/blob/main/docs/ch01-project-options.md`.

Making good use of the GitHub repository for this book

Git is a commonly used source code management system. **GitHub** is a company, website, and desktop application that makes it easier to manage Git. Microsoft purchased GitHub in 2018, so it will continue to get closer integration with Microsoft tools.

I created a GitHub repository for this book, and I use it for the following:

- To store the solution code for the book that can be maintained after the print publication date.
- To provide extra materials that extend the book, like errata fixes, small improvements, lists of useful links, and optional sections about topics that cannot fit in the printed book.
- To provide a place for readers to get in touch with me if they have issues with the book.

Good Practice: I strongly recommend that you all review the errata, improvements, post-publication changes, and common errors pages before attempting any coding task in this book. You can find them at the following link: `https://github.com/markjprice/cs13net9/blob/main/docs/errata/README.md`.

Understanding the solution code on GitHub

The solution code in the GitHub repository for this book includes folders for each chapter that can be opened with any of the following code editors:

- Visual Studio or Rider: Open the `.sln` solution file.
- VS Code: Open the folder that contains the solution file.

Chapters 1 to *11* each have their own solution file named `ChapterXX.sln`, where `XX` is the chapter number *01* to *11*. *Chapters 12* to *15* share a single solution file named `ModernWeb.sln`.

The `.sln` solution file format is a Microsoft proprietary file format that is verbose, hard to read, and uses **Globally Unique Identifiers (GUIDs)** to reference projects and other components of a solution. A new format based on XML, designed to be simpler and easier to read, is coming soon and will use the `.slnx` file extension. You can learn more about this at the following link: `https://github.com/dotnet/sdk/issues/40913`.

All the code solutions can be found at the following link:

`https://github.com/markjprice/cs13net9/tree/main/code`

Good Practice: If you need to, return to this chapter to remind yourself how to create and manage multiple projects in the code editor of your choice. The GitHub repository has step-by-step instructions for three code editors (Visual Studio, VS Code, and Rider), along with additional screenshots: `https://github.com/markjprice/cs13net9/tree/main/docs/code-editors/`.

Raising issues with the book

If you get stuck following any of the instructions in this book, or if you spot a mistake in the text or the code in the solutions, please raise an issue in the GitHub repository:

1. Use your favorite browser to navigate to the following link: `https://github.com/markjprice/cs13net9/issues`.
2. Click **New Issue**.
3. Enter as much detail as possible that will help me to diagnose the issue. For example:
    - The specific section title, page number, and step number.
    - As much of your code and configuration that you feel is relevant and necessary.
    - A description of the expected behavior and the behavior experienced.
    - Screenshots (you can drag and drop image files into the **Issue** box).

The following is less relevant but might be useful:

- Your code editor, for example, Visual Studio, VS Code, Rider, or something else, including the version number.
- Your operating system, for example, Windows 11 64-bit or macOS Ventura version 13.5.2
- Your hardware, for example, Intel, Apple Silicon, or ARM CPU

I cannot always respond immediately to issues. But I want all my readers to find success with my book, so if I can help you (and others) without too much trouble, then I will gladly do so.

Giving me feedback

If you'd like to give me more general feedback about the book, then either email me at `markjprice@gmail.com` or ask me a question on Discord in the book channel. You can provide the feedback anonymously, or if you would like a response from me, then you can supply an email address. I will only use this email address to answer your feedback.

Please join me and your fellow readers on Discord using this invite: `https://packt.link/csharp13dotnet9`.

I love to hear from my readers about what they like about my book, as well as suggestions for improvements and how they are working with C# and .NET, so don't be shy. Please get in touch!

Thank you in advance for your thoughtful and constructive feedback.

Avoiding common mistakes

After working through the step-by-step tasks in this book, readers often then strike out on their own and attempt to write similar code, but sometimes, they hit problems and either raise an issue in the GitHub repository or post a question to the Discord channel for the book.

From these, I have noted some common mistakes, so I maintain a page in the repository to highlight and explain these potential traps and how to fix them:

`https://github.com/markjprice/cs13net9/blob/main/docs/errata/common-mistakes.md`

Downloading solution code from the GitHub repository

If you just want to download all the solution files without using Git, click the green <> **Code** button and then select **Download ZIP**, as shown in *Figure 1.10*:

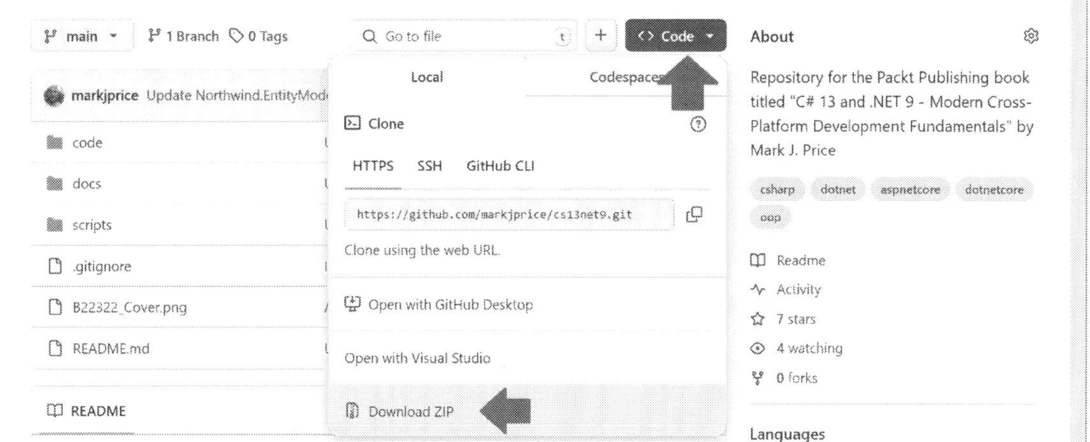

Figure 1.10: Downloading the repository as a ZIP file

Good Practice: It is best to clone or download the code solutions to a short folder path, like `C:\cs13net9\` or `C:\book\`, to avoid build-generated files that exceed the maximum path length. You should also avoid special characters like #. For example, do not use a folder name like `C:\C# projects\`. That folder name might work for a simple console app project, but once you start adding features that automatically generate code, you are likely to have strange issues. Keep your folder names short and simple.

Using Git with VS Code and the command prompt

VS Code has integrations with Git, but it will use your operating system's Git installation, so you must install Git 2 or later first before you get these features.

You can install Git from the following link: `https://git-scm.com/download`.

If you like to use a GUI, you can download GitHub Desktop from the following link: `https://desktop.github.com`.

Cloning the book solution code repository

Let's clone the book solution code repository. In the steps that follow, you will use the VS Code terminal, but you can enter the commands at any command prompt or terminal window:

1. Create a folder named `Repos-vscode` in your `user` or `Documents` folder, or wherever you want to store your Git repositories.

2. Open the `Repos-vscode` folder at the command prompt or terminal, and then enter the following command:

```
git clone https://github.com/markjprice/cs13net9.git
```

Note that cloning all the solutions for all the chapters will take a minute or so, so please be patient.

Looking for help

This section is all about how to find quality information about programming on the web.

Microsoft Learn documentation and Ask Learn

The definitive resource for getting help with Microsoft developer tools and platforms is in the technical documentation on Microsoft Learn, and you can find it at the following link: `https://learn.microsoft.com/en-us/docs`.

> *"One of the most ambitious and impactful projects our engineers have built recently is Ask Learn, an API that provides generative AI capabilities to Microsoft Q&A." - Bob Tabor, Microsoft's Skilling organization*

You can read about it at the following link:

`https://devblogs.microsoft.com/engineering-at-microsoft/how-we-built-ask-learn-the-rag-based-knowledge-service/`

Documentation links in this book

The official Microsoft documentation for .NET needs to cover all versions. The default version shown in the documentation is always the most recent GA version.

For example, between November 2024 and November 2025, the default version of .NET shown in the documentation pages will be for .NET 9. Between November 2025 and November 2026, the default version of .NET will be for .NET 10. The following link will automatically direct to the current version, depending on the current date:

`https://learn.microsoft.com/en-us/dotnet/api/system.diagnostics.codeanalysis.stringsyntaxattribute`

After November 2025, to view the documentation page specifically for .NET 9, append `?view=net-9.0` to the end of a link. For example, use the following link:

`https://learn.microsoft.com/en-us/dotnet/api/system.diagnostics.codeanalysis.stringsyntaxattribute?view=net-9.0`

All documentation links in this book do not specify a version. If you want to force the documentation to show the version for .NET 8 because you need to target an LTS version (three years) instead of .NET 9, which is an STS version (18 months), then append `?view=net-8.0` to the end of a link.

You can check what versions a .NET feature supports by appending `#applies-to` to the end of a link, for example:

`https://learn.microsoft.com/en-us/dotnet/api/system.diagnostics.codeanalysis.stringsyntaxattribute#applies-to`

We can, therefore, see that the `StringSyntax` attribute is only available in .NET 7 or later.

Getting help from the dotnet tool

At the command prompt, you can ask the `dotnet` tool for help with its commands. The syntax is:

```
dotnet help <command>
```

This will cause your web browser to open a page in the documentation about the specified command. Common `dotnet` commands include `new`, `build`, `run`, and many more.

Warning! The `dotnet help new` command worked with .NET Core 3.1 to .NET 6, but it returns an error with .NET 7 or later: `Specified command 'new' is not a valid SDK command. Specify a valid SDK command. For more information, run dotnet help.` Hopefully, .NET will fix that bug soon!

Another type of help is command-line documentation. It follows this syntax:

```
dotnet <command> -?|-h|--help
```

For example, `dotnet new -?` or `dotnet new -h` or `dotnet new --help` outputs documentation about the `new` command at the command prompt.

As you should now expect, `dotnet help help` opens a web browser for the `help` command, and `dotnet help -h` outputs documentation for the `help` command at the command prompt!

Let's try some examples:

1. To open the official documentation in a web browser window for the `dotnet build` command, enter the following at the command prompt or in the VS Code terminal, and note the page opened in your web browser:

   ```
   dotnet help build
   ```

2. To get help output at the command prompt, use the `-?` or `-h` or `--help` flag, as shown in the following command:

```
dotnet build -?
```

3. You will see the following partial output:

```
Description:
  .NET Builder

Usage:
  dotnet build [<PROJECT | SOLUTION>...] [options]

Arguments:
  <PROJECT | SOLUTION>  The project or solution file to operate on. If a
file is not specified, the command will search the current directory for
one.

Options:
  --ucr, --use-current-runtime          Use current runtime as the target
runtime.
  -f, --framework <FRAMEWORK>           The target framework to build for.
The target framework must also be specified in the project file.
...
  -?, -h, --help                        Show command line help.
```

4. Repeat both types of help requests for the following commands: `add`, `help`, `list`, `new`, and `run`, remembering that `new` might not show its web page, due to a bug introduced in .NET 7.

Getting definitions of types and their members

One of the most useful features of a code editor is **Go To Definition** (*F12*). It is available in VS Code, Visual Studio, and Rider. It will show what the public definition of the type or member looks like by reading the metadata in the compiled assembly.

Some tools, such as ILSpy .NET Decompiler, will even reverse-engineer the metadata and IL code back into C# or another language for you.

A similar and related feature is named **Go To Implementation** (*Ctrl + F12*). Instead of reading the metadata or decompiling, this will show the actual source code if that is embedded, using the optional source link feature.

Warning! Go To Definition should go to the decompiled metadata for a member or type. But if you have previously viewed the source link, then it goes to that. **Go To Implementation** should go to the source link implementation for a member or type. But if you have disabled the source link, then it goes to the decompiled metadata.

Let's see how to use the **Go To Definition** feature:

1. In your preferred code editor, open the solution/folder named `Chapter01`.

   If you are using Visual Studio:

   - Navigate to **Tools | Options.**
   - In the search box, enter `navigation to source`.
   - Select **Text Editor | C# | Advanced.**
   - Clear the **Enable navigation to Source Link and Embedded sources** checkbox, and then click **OK**, as shown in *Figure 1.11*:

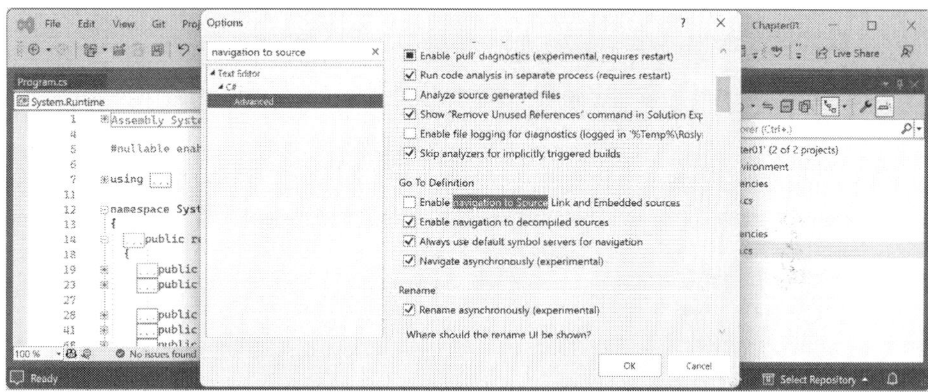

Figure 1.11: Disabling Source Link for the Go To Definition feature

Definitions can be either reverse-engineered from metadata or loaded from the original source code if that is enabled. Personally, I find the code from metadata more useful, as you are about to see. At the end of this section, try switching the Source Link option back on to see the difference.

2. In the `HelloCS` project, at the bottom of `Program.cs`, enter the following statement to declare an integer variable named `z`:

```
int z;
```

3. Click on `int`, right-click on `int`, and then choose **Go To Definition** in Visual Studio or VS Code. In Rider, choose **Go to | Go to Declaration or Usages.**
4. In the code window that appears, you can see how the `int` data type is defined, as shown in the following code:

```
#region Assembly System.Runtime, Version=9.0.0.0, Culture=neutral,
PublicKeyToken=b03f5f7f11d50a3a
// C:\Program Files\dotnet\packs\Microsoft.NETCore.App.Ref\9.0.0\ref\
net9.0\System.Runtime.dll
#endregion
```

```
#nullable enable

using System.Diagnostics.CodeAnalysis;
using System.Globalization;
using System.Numerics;
using System.Runtime.CompilerServices;

namespace System
{
  public readonly struct Int32 : IComparable,
    IComparable<Int32>, IConvertible, ...
  {
    public const Int32 MaxValue = 2147483647;
    public const Int32 MinValue = -2147483648;
    ...
  }
}
```

You can see that `int`:

- Is defined using the `struct` keyword.
- Is in the `System.Runtime` assembly.
- Is in the `System` namespace.
- Is named `Int32`.
- Is, therefore, an alias for the `System.Int32` type.
- Implements interfaces such as `IComparable`.
- Has constant values for its maximum (`2147483647`) and minimum (`-2147483648`) values.
- Has methods such as `Parse` (not visible in the preceding code).

Currently, the **Go To Definition** feature is not that useful to you because you do not yet know what all of this information means. By the end of the first part of this book, which consists of *Chapters 2* to *6* and teaches you about the C# language, you will know enough for this feature to become very handy.

5. In the code editor window, scroll down to find the `Parse` method with a single `string` parameter, as shown in the following code:

```
public static Int32 Parse(string s)
```

6. Expand the code and review the comments that document this method, as shown in *Figure 1.12*:

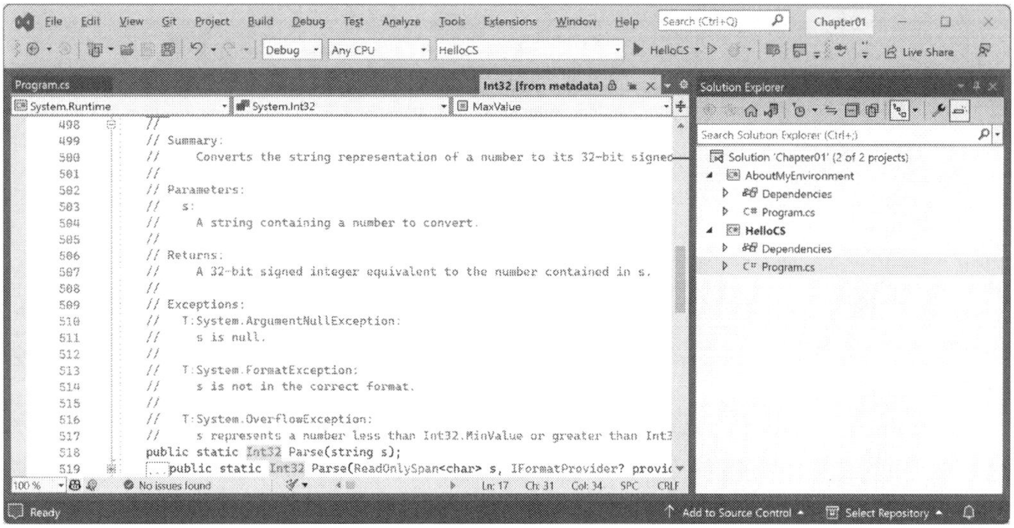

Figure 1.12: The comments for the Parse method with a single string parameter

In the comments, you will see that Microsoft has documented the following:

- A summary that describes the method.
- Parameters like the `string` value that can be passed to the method.
- The return value of the method, including its data type.
- Three exceptions that might occur if you call this method, including `ArgumentNullException`, `FormatException`, and `OverflowException`. Now, we know that we could choose to wrap a call to this method in a `try` statement and which exceptions to catch.

Configuring inline aka inlay hints

Throughout the code in this book, when calling a method, I often explicitly specify named parameters to help you learn what is going on. For example, I have specified the names of the parameters `format` and `arg0` in the following code:

```
Console.WriteLine(format: "Value is {0}.", arg0: 19.8);
```

Inline hints, aka **inlay hints**, show the names of parameters without you having to type them, as shown in *Figure 1.13*:

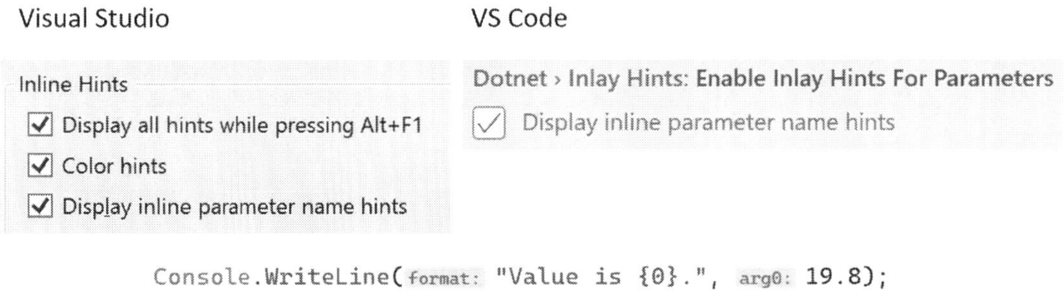

Figure 1.13: Configuring inline hints, aka inlay hints

Most code editors have this feature, which you can enable permanently or only when a key combination like *Alt* + *F1* or *Ctrl* is held down:

- In Visual Studio, navigate to **Tools** | **Options**, then **Text Editor** | **C#** | **Advanced**, scroll down to the **Inline Hints** section, select the **Display inline parameter name hints** checkbox, and then click **OK**.
- In VS Code, navigate to **File** | **Preferences** | **Settings**, search for `inlay`, select the **C#** filter, and then select the **Display inline parameter name hints** checkbox.
- In Rider, in **Settings**, navigate to **Editor** | **Inlay Hints** | **C#** | **Parameter Name Hints**.

Hopefully, you are getting impatient to learn what all this means!

Be patient for a little longer. You are almost at the end of this chapter, and in the next chapter, you will dive into the details of the C# language. But first, let's see where else you can look for help.

Looking for answers on Stack Overflow

Stack Overflow is the most popular third-party website for getting answers to difficult programming questions. Let's see an example:

1. Start your favorite web browser.
2. Navigate to `stackoverflow.com`; in the search box, enter `securestring` and note the search results.

Searching for answers using Google

You can search Google with advanced search options to increase the likelihood of finding what you need:

1. Navigate to Google at the following link: `https://www.google.com/`.
2. Search for information about `garbage collection` using a simple Google query, and note that you will probably see a lot of ads for garbage collection services in your local area before you see the Wikipedia definition of garbage collection in computer science!

3. Improve the search by restricting it to a useful site such as Stack Overflow, by removing languages that we might not care about, such as C++, Rust, and Python, or by adding C# and .NET explicitly, as shown in the following search query:

```
garbage collection site:stackoverflow.com +C# -Java
```

Getting help on Discord and other chat forums

Asking questions in programming forums and Discord channels is an art as much as it is a science. To maximize your chances of receiving a helpful answer, there's a blend of clarity, specificity, and community awareness that you should aim for.

Here are some tips for asking questions:

- **Ask in a public channel, not in private. Please do not direct message an author with a question or a friend request.** Remember, every question asked and answered builds the collective knowledge and resourcefulness of the whole community. Asking in public also allows other readers to help you, not just the author. The community that Packt and I have built around my books is friendly and smart. Let us *all* help you.
- **Research before asking:** It's important to look for answers yourself before turning to the community. Use search engines, official documentation, and the search function within the forum or Discord server. This not only respects the community's time but also helps you learn more effectively. Another place to look first is the errata and improvements section of the book, found at the following link: `https://github.com/markjprice/cs13net9/blob/main/docs/errata/README.md`.
- **Be specific and concise:** Clearly state what you're trying to achieve, what you've tried so far, and where you're stuck. A concise question is more likely to get a quick response.
- **Specify the book location:** If you are stuck on a particular part of the book, specify the page number and section title so that others can look up the context of your question.
- **Show your work:** Demonstrating that you've made an effort to solve the problem yourself not only provides context but also helps others understand your thought process and where you might have gone down a wrong path.
- **Prepare your question:** Avoid too broad or vague questions. Screenshots of errors or code snippets (with proper formatting) can be very helpful.

Oddly, I've been seeing more and more examples of readers taking photos of their screens and posting those. These are harder to read and limited in what they can show. It's better to copy and paste the text of your code or the error message so that others can copy and paste it themselves. Alternatively, at least take a high-resolution screenshot instead of a photo with your phone camera at a jaunty angle!

- **Format your code properly**: Most forums and Discord servers support code formatting using Markdown syntax. Use formatting to make your code more readable. For example, surround code keywords in single backticks, like `` `public void` ``, and surround code blocks with three backticks with optional language code, as shown in the following code:

````
```cs
using static System.Console;
WriteLine("This is C# formatted code.");
```
````

Good Practice: After the three backticks that start a code block in Markdown, specify a language short name like `cs`, `csharp`, `js`, `javascript`, `json`, `html`, `css`, `cpp`, `xml`, `mermaid`, `python`, `java`, `ruby`, `go`, `sql`, `bash`, or `shell`.

More Information: To learn how to format text in Discord channel messages, see the following link: `https://support.discord.com/hc/en-us/articles/210298617-Markdown-Text-101-Chat-Formatting-Bold-Italic-Underline`.

- **Be polite and patient:** Remember, you're asking for help from people who are giving their time voluntarily. A polite tone and patience while waiting for a response go a long way. Channel participants are often in a different time zone, so you may not see your question answered until the next day.
- **Be ready to actively participate:** After asking your question, stay engaged. You might receive follow-up questions for clarification. Responding promptly and clearly can significantly increase your chances of getting a helpful answer. When I ask a question, I set an alarm for three hours later to go back and see if anyone has responded. If there hasn't been a response yet, then I set another alarm for 24 hours later.

Incorporating these approaches when asking questions not only increases your likelihood of getting a useful response but also contributes positively to the community, by showing respect for others' time and effort.

Good Practice: Never just say "Hello" as a message on any chat system. You can read why at the following link: `https://nohello.net/`. Similarly, don't ask to ask: `https://dontasktoask.com/`.

Searching the .NET source code

Sometimes, you can learn a lot from seeing how the Microsoft teams have implemented .NET. The source for the entire code base for .NET is available in public GitHub repositories. For example, you might know that there is a built-in attribute to validate an email address.

Let's search the repositories for the word "email" and see if we can find out how it works:

1. Use your preferred web browser to navigate to `https://github.com/search`.
2. Click **advanced search**.
3. In the search box, type `email`.
4. In the **In these respositories** box, type `dotnet/runtime`. (Other repositories that you might want to search include `dotnet/core`, `dotnet/aspnetcore`, `dotnet/wpf`, and `dotnet/winforms`.)
5. In the **Written in this language** box, select **C#**.
6. At the top right of the page, note how the advanced query has been written for you. Click **Search**, then the **Code** filter, and note that the results include `EmailAddressAttribute`, as shown in *Figure 1.14*:

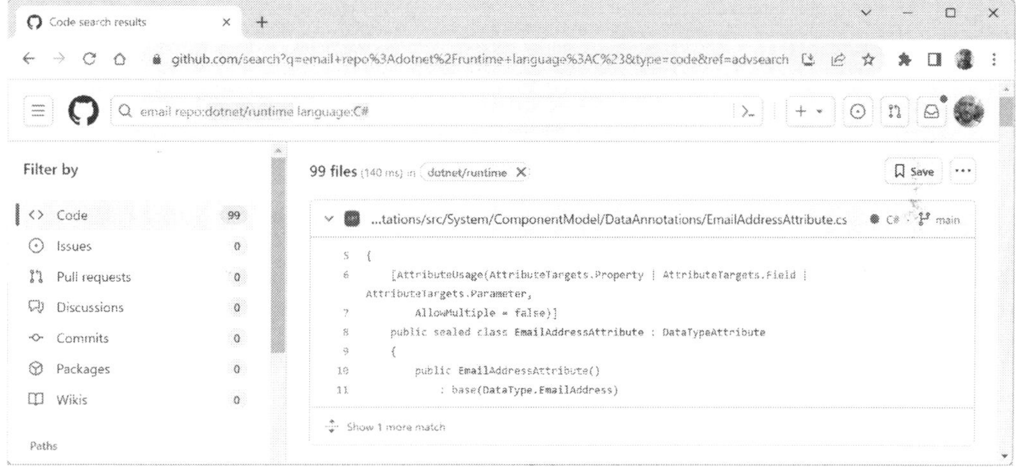

Figure 1.14: Advanced search for email in the dotnet/runtime repository

7. Click the source file, and note that it implements email validation by checking that the `string` value contains an @ symbol but not as the first or last character, as shown in the following code:

```
// only return true if there is only 1 '@' character
// and it is neither the first nor the last character
int index = valueAsString.IndexOf('@');

return
    index > 0 &&
    index != valueAsString.Length - 1 &&
    index == valueAsString.LastIndexOf('@');
```

8. Close the browser.

For your convenience, you can do a quick search for other terms by replacing the search term `email` in the following link: `https://github.com/search?q=%22email%22+repo%3Adotnet%2Fruntime+language%3AC%23&type=code&ref=advsearch`.

Source code in documentation

When you read API reference documentation, you often want to review the actual source code. For .NET APIs that have Source Link enabled, have an accessible PDB, and are hosted in a public GitHub repository, links to source code are included in the definition metadata. For example, the `String` class documentation page now has this new **Source** link, and its `IndexOf` method has a **Source** link to another of its source files, as shown in *Figure 1.15*:

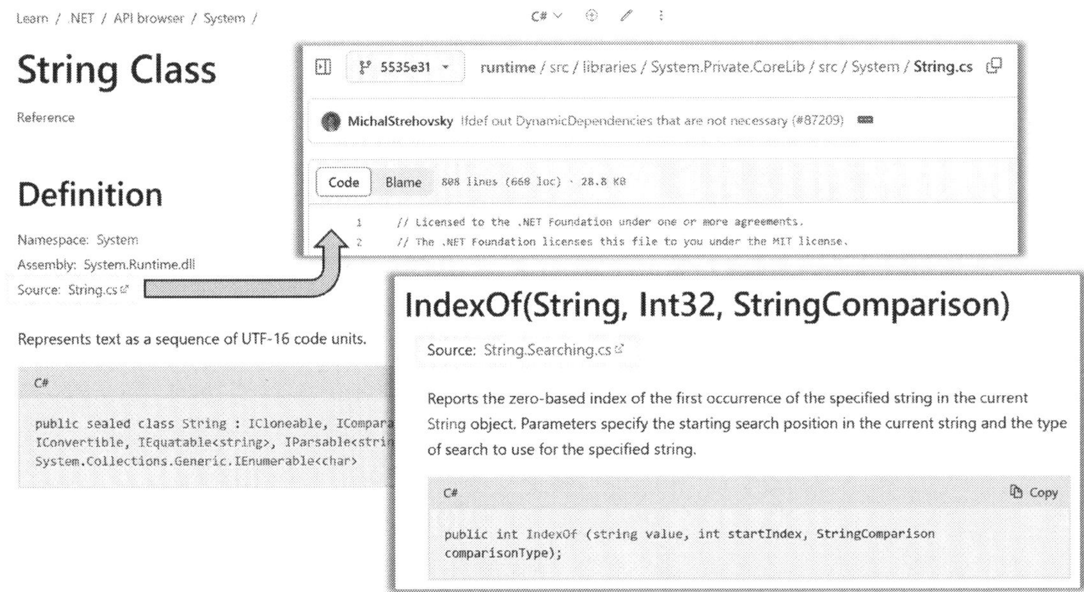

Figure 1.15: Documentation with links to source files

You can read more about how the Microsoft team achieved this in the article **Introducing links to source code for .NET API Docs**, found at the following link: `https://devblogs.microsoft.com/dotnet/dotnet-docs-link-to-source-code/`.

Official .NET blog, standups, and news

To keep up to date with .NET, an excellent blog to subscribe to is the official .NET blog, written by the .NET engineering teams, and you can find it at the following link: `https://devblogs.microsoft.com/dotnet/`.

To watch .NET team members walk through previews of new features, you can watch monthly "stand-ups," available at the following link:

https://dotnet.microsoft.com/en-us/live/community-standup

You can subscribe to the latest .NET news at the following link:

https://github.com/dotnet/core/discussions/categories/news

Watching Scott Hanselman's videos

Scott Hanselman from Microsoft has an excellent YouTube channel about computer stuff that they didn't teach you at school: `http://computerstufftheydidntteachyou.com/`.

I recommend it to everyone working with computers.

AI tools like ChatGPT and GitHub Copilot

One of the biggest changes in coding and development in the past few years is the emergence of generative **artificial intelligence** (**AI**) tools that can help with coding tasks, like completing a code statement, implementing an entire function, writing unit tests, and suggesting debugging fixes for existing code.

You can read what developers say about AI tools in the 2023 Stack Overflow Developer Survey: "44% of them use AI tools in their development process now, and 26% plan to soon" (`https://stackoverflow.blog/2023/06/14/hype-or-not-developers-have-something-to-say-about-ai/`):

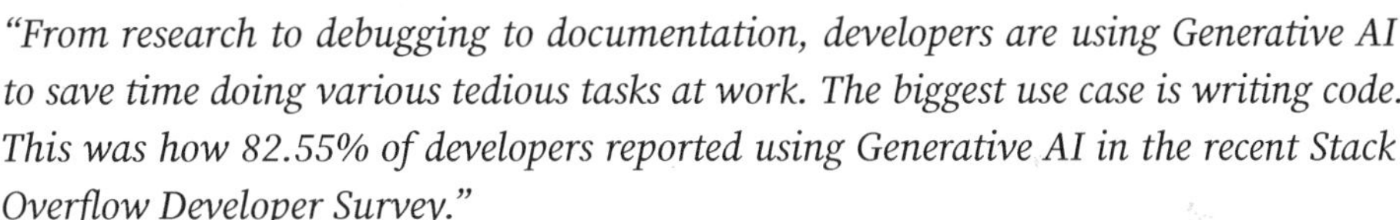

> *"From research to debugging to documentation, developers are using Generative AI to save time doing various tedious tasks at work. The biggest use case is writing code. This was how 82.55% of developers reported using Generative AI in the recent Stack Overflow Developer Survey."*

ChatGPT currently has several models for individuals: 4o mini (free), 4o (free but limited), and 4o with 5x more requests and other benefits, like early access to new features and DALL·E image generation ($20 per month). It also has pricing for enterprises. You can check their pricing at the following link: `https://openai.com/chatgpt/pricing/`.

ChatGPT example

Let's say you need to write a C# function to validate an email address. You might go to ChatGPT and enter the following prompt:

```
write a c# function to validate an email address
```

It responds with a complete class with methods, as shown in *Figure 1.16*:

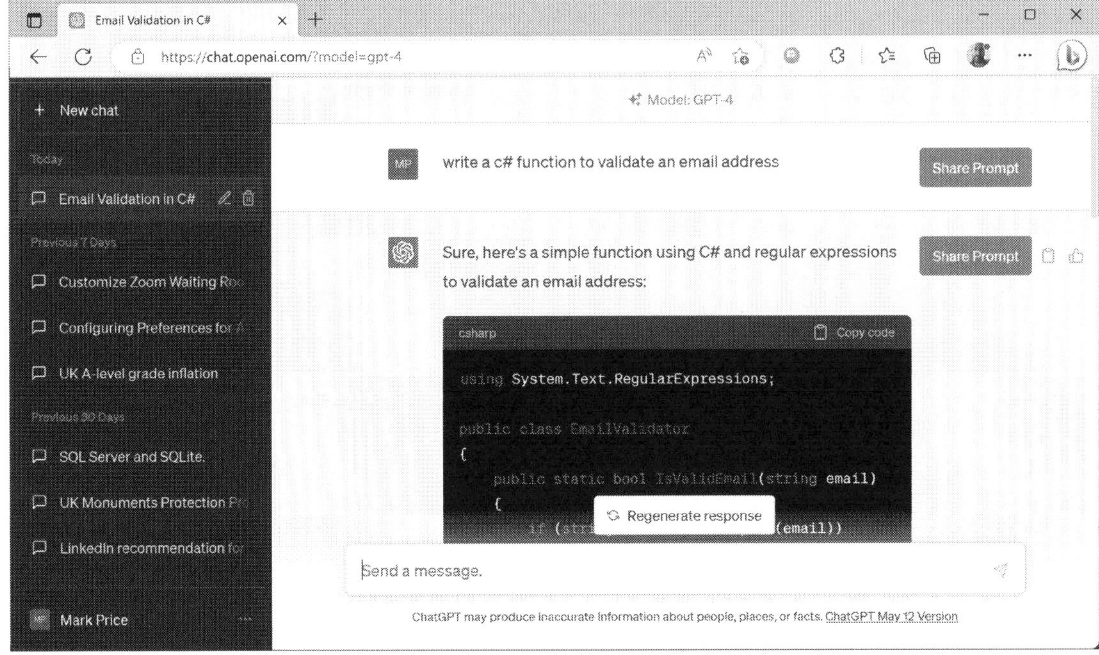

Figure 1.16: ChatGPT writes a function to validate an email address

It then provides an explanation of the code and examples of how to call the function, as shown in the following code:

```
bool isValid = EmailValidator.IsValidEmail("test@example.com");
Console.WriteLine(isValid ? "Valid" : "Invalid");
```

But is a general-purpose generative AI like ChatGPT the best partner for a C# programmer?

GitHub Copilot for programmers

Microsoft has a service specifically for programmers, named GitHub Copilot, that can help autocomplete code directly in your code editor. It has plugins for code editors, including Visual Studio, VS Code, and JetBrains IntelliJ-based IDEs.

Personally, I really like the Copilot branding. It makes it clear that you are the pilot. You are ultimately responsible for "flying the plane." But for the easy or boring bits, you can hand it over to your co-pilot for a bit, while being actively ready to take back control if needed.

GitHub Copilot is free for students, teachers, and some open-source project maintainers. For everyone else, it has a 30-day free trial, and then it costs $10 per month or $100 per year for individuals. Once you have an account, you can then sign up for waiting lists to get the more advanced experimental GitHub Copilot features.

You should check online which Copilot features are available for various code editors. As you can imagine, this is a fast-changing world, and some of what I might write in the book today will be out of date by the time you read it: `https://github.com/features/copilot`.

JetBrains has its own GitHub Copilot equivalent, named AI Assistant, which you can read about at the following link: `https://blog.jetbrains.com/idea/2023/06/ai-assistant-in-jetbrains-ides/`.

So what can GitHub Copilot do for you today?

Imagine that you have just added a new class file named `Product.cs`. You click inside the `Product` class, press *Enter* to insert a blank line, and then pause for a second as you think about what you need to type... and then GitHub Copilot generates some sample code in gray, as shown in *Figure 1.17*:

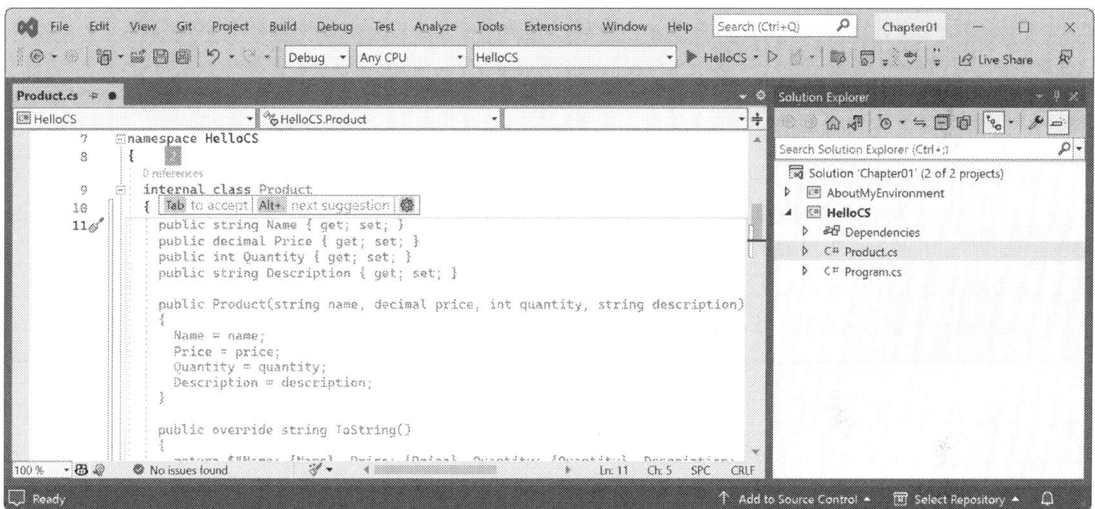

Figure 1.17: GitHub Copilot suggesting how to define a Product class

At this point, you can glance over the code and, if it is close to what you want, just press *Tab* to insert it all, or press *Alt* + . (dot) to toggle between other suggestions.

Sometimes, it is too far from what you need, and you'd be better off ignoring its suggestion completely and just writing it yourself. But usually, there's something there that's usable or reminds you of the syntax you need to use. And sometimes, it feels like magic, writing dozens of lines of exactly what you need.

Microsoft feeds its AI tools with code from public GitHub repositories, including all the repositories I have created since 2016 for all the editions of this book. This means that it can suggest code completions for the readers of this book that are surprisingly accurate predictions, including my frequent use of pop culture references in my code. It's like I, Mark J. Price, am the "ghost in the machine" guiding your coding.

It's easy to imagine a custom ChatGPT that has ingested all the official Microsoft .NET documentation, every public blog article written about .NET, and perhaps even hundreds of books about .NET, and then having a conversation with it to find a bug or suggest how to solve a programming problem.

You can sign up for GitHub Copilot at the following link: `https://github.com/github-copilot/signup/`.

Good Practice: Learn more about how to use Copilot as your coding GPS at the following link: `https://devblogs.microsoft.com/visualstudio/using-github-copilot-as-your-coding-gps/`.

Here are some more links about using AI for coding:

- A Beginner's Guide to Prompt Engineering with GitHub Copilot: `https://dev.to/github/a-beginners-guide-to-prompt-engineering-with-github-copilot-3ibp`
- The Register article about AI: `https://www.theregister.com/2024/01/27/ai_coding_automatic/`
- StackOverflow 2024 survey – Code editor / IDE: `https://survey.stackoverflow.co/2024/technology#1-integrated-development-environment`
- StackOverflow 2024 survey – AI search tools: `https://survey.stackoverflow.co/2024/technology#1-ai-search-and-developer-tools`

Disabling tools when they get in the way

Although these tools can be helpful, they can also get in your way, especially when learning, because they sometimes do work for you without telling you. If you do not do that work for yourself at least a few times, you won't learn fully.

To configure IntelliSense for C# in Visual Studio:

1. Navigate to **Tools** | **Options**.
2. In the **Options** dialog box tree view, navigate to **Text Editor** | **C#** | **IntelliSense**.
3. Click the **?** button in the caption bar to view the documentation.

To configure GitHub Copilot in Visual Studio:

1. Navigate to **Tools** | **Options**.
2. In the **Options** dialog box tree view, navigate to **GitHub** | **Copilot**.
3. Set **Enable Globally** to **True** or **False**, and then click **OK**.

To disable GitHub Copilot in VS Code:

1. In the status bar, on the right, to the left of the notification icon, click the GitHub Copilot icon.
2. In the popup, click **Disable Globally**.
3. To enable it, click the GitHub Copilot icon again and then click **Enable Globally**.

For help with Rider IntelliSense, please see the following link: `https://www.jetbrains.com/help/rider/Auto-Completing_Code.html`.

Practicing and exploring

Let's now test your knowledge and understanding by trying to answer some questions, getting some hands-on practice, and going into the topics covered throughout this chapter in greater detail.

Exercise 1.1 – Online material

Online material can be extra content written by me for this book, or it can be references to content created by Microsoft or third parties.

Current versions of .NET

You can check what the latest version of .NET is for various platforms and what is officially supported at the following link: `https://versionsof.net`.

If you need to know more about support for the legacy .NET Framework, you can read about it at the following link: `https://learn.microsoft.com/en-us/lifecycle/products/microsoft-net-framework`.

Upgrade to a new .NET version

.NET releases new versions annually. Some developers begin upgrading as soon as a new version drops, while others prefer to wait until their current version reaches end-of-support. Upgrading involves several important considerations that you can read about at the following link:

`https://learn.microsoft.com/en-us/dotnet/core/install/upgrade`

freeCodeCamp and C# certification

For many years, Microsoft had an exam for C# 5, *Exam 70-483: Programming in C#*. I taught hundreds of developers the skills needed to get qualified and pass it. Sadly, that exam was retired a few years ago.

In August 2023, Microsoft announced a new foundational certification for C# alongside a free 35-hour online course. You can read more about how to qualify for the certification at the following link:

`https://www.freecodecamp.org/learn/foundational-c-sharp-with-microsoft/`

C# certification signup: `https://aka.ms/learningseries/csharp/signup`

Explore Polyglot Notebooks

Complete the following online-only section to explore how you can use Polyglot Notebooks with its .NET Interactive engine:

`https://github.com/markjprice/cs13net9/blob/main/docs/ch01-polyglot-notebooks.md`

Windows development

This book is about modern cross-platform development. But what about **graphical user interface (GUI)** development especially for Windows?

You can read an article discussing this at the following link:

`https://visualstudiomagazine.com/articles/2024/02/13/desktop-dev.aspx`

C# versus other languages

Even Microsoft does not use C# and .NET for all its own development needs. You can read an interesting discussion about C# and Rust at the following link:

`https://www.reddit.com/r/dotnet/comments/1aezqmg/comment/ko8lnf2/`

The key quote at the preceding Reddit link is this:

> *"There's been a lot of hype around Rust, and for good reason. But it's a system language. It's not like Microsoft is about to go rewrite millions and millions of lines of code and toss out C# (for anyone getting nervous 😅). They're just being pragmatic and using an effective tool for the job."*

Free computer science course

Harvard CS50: Introduction to Computer Science: `https://pll.harvard.edu/course/cs50-introduction-computer-science`

.NET newsletters

There are email newsletters that you can subscribe to get daily or weekly updates, hints, and tips. Some of the better ones include the following:

- Milan Jovanovic's The .NET Weekly: `https://www.milanjovanovic.tech/blog`
- C# Digest: `https://csharpdigest.net/`
- Alvin Ashcraft's The Morning Dew: `https://www.alvinashcraft.com/`
- Mukesh's .NET Newsletter: `https://newsletter.codewithmukesh.com/`
- JetBrains' dotInsights: `https://www.jetbrains.com/lp/dotinsights-monthly/`

Exercise 1.2 – Practice exercises

The following practice exercises will go deeper into the topics for this chapter.

Practice C# anywhere with a browser

You don't need to download and install VS Code or even Visual Studio to write C#. You can start coding online at any of the following links:

- SharpLab: `https://sharplab.io/`
- C# Online Compiler | .NET Fiddle: `https://dotnetfiddle.net/`
- W3Schools C# Online Compiler: `https://www.w3schools.com/cs/cs_compiler.php`

Alpha versions of .NET

You can (but probably shouldn't) download future versions of .NET, including alpha versions from the following link:

`https://github.com/dotnet/sdk/blob/main/documentation/package-table.md`

Warning! Alpha versions are designed to be used only internally by Microsoft employees. Beta versions (official previews) are designed to be used externally and become available from mid-February each year.

For more about using .NET 10 with this book, please see the following link: `https://github.com/markjprice/cs13net9/blob/main/docs/dotnet10.md`.

Exercise 1.3 – Test your knowledge

Try to answer the following questions, remembering that although most answers can be found in this chapter, you should do some online research or code writing to answer others:

1. Is Visual Studio better than VS Code?
2. Are .NET 5 and later versions better than .NET Framework?
3. What is .NET Standard, and why is it still important?
4. Why can a programmer use different languages (for example, C# and F#) to write applications that run on .NET?
5. What is a top-level program, and how do you access any command-line arguments?
6. What is the name of the entry point method of a .NET console app, and how should it be explicitly declared if you are not using the top-level program feature?
7. What namespace is the `Program` class defined in with a top-level program?
8. Where would you look for help for a C# keyword?
9. Where would you look first for solutions to common programming problems?
10. What should you do after getting AI to write code for you?

Appendix, Answers to the Test Your Knowledge Questions, is available to download from a link in the README file in the GitHub repository: `https://github.com/markjprice/cs13net9`.

Exercise 1.4 – Explore topics

A printed book is a curated experience. I have tried to find the right balance of topics to include in this book. Other content that I have written can be found in the GitHub repository for this book.

I believe that this book covers all the fundamental knowledge and skills a C# and .NET developer should have or be aware of. Some longer examples are best included as links to Microsoft documentation or third-party article authors.

Use the links on the following page to learn more details about the topics covered in this chapter:

`https://github.com/markjprice/cs13net9/blob/main/docs/book-links.md#chapter-1---hello-c-welcome-net`

Summary

In this chapter, we:

- Set up your development environment.
- Discussed the similarities and differences between modern .NET, .NET Core, .NET Framework, Xamarin, and .NET Standard in an online article.
- Used Visual Studio and VS Code with the .NET SDK CLI to create a couple of simple console apps, grouped in a solution.
- Learned how to download the solution code for this book from its GitHub repository.
- Learned how to find help. This could be in the traditional way, by using help command switches, documentation, and articles, or the modern way, by having a conversation with coding expert AI or using an AI-based tool to perform "grunt work."

In the next chapter, you will learn how to "speak" C#.

Learn more on Discord

To join the Discord community for this book – where you can share feedback, ask questions to the author, and learn about new releases – follow the QR code below:

`https://packt.link/csharp13dotnet9`

Leave a Review!

Thank you for purchasing this book from Packt Publishing—we hope you enjoy it! Your feedback is invaluable and helps us improve and grow. Once you've completed reading it, please take a moment to leave an Amazon review; it will only take a minute, but it makes a big difference for readers like you.

Scan the QR or visit the link to receive a free ebook of your choice.

`https://packt.link/NzOWQ`

2

Speaking C#

This chapter is all about the basics of the C# programming language. Over the course of this chapter, you'll learn how to write statements using the grammar of C#, as well as be introduced to some of the common vocabulary words that you will use every day. In addition to this, by the end of the chapter, you'll feel confident in knowing how to temporarily store and work with information in your computer's memory.

This chapter covers the following topics:

- Introducing the C# language
- Discovering your C# compiler version
- Understanding C# grammar and vocabulary
- Working with variables
- Exploring more about console apps

Introducing the C# language

This part of the book is about the C# language—the grammar and vocabulary that you will use every day to write the source code for your applications.

Programming languages have many similarities to human languages, except that in programming languages, you can make up your own words, just like Dr. Seuss!

In a book written by Dr. Seuss in 1950, *If I Ran the Zoo*, he states this:

> *"And then, just to show them, I'll sail to Ka-Troo And Bring Back an It-Kutch, a Preep, and a Proo, A Nerkle, a Nerd, and a Seersucker, too!"*

C# language versions and features

This part of the book covers the C# programming language and is written primarily for beginners, so it covers the fundamental topics that all developers need to know, including declaring variables, storing data, and how to define your own custom data types.

This book covers features of the C# language from version 1 up to the latest version, C# 13. You can read a summary of what's new in C# 13 at the following link:

`https://learn.microsoft.com/en-us/dotnet/csharp/whats-new/csharp-13`

If you already have some familiarity with older versions of C# and are excited to find out about the new features in the most recent versions of C#, I have made it easier for you to jump around by listing language versions and their important new features below, along with the chapter number and topic title where you can learn about them.

You can read this information in the GitHub repository at the following link: `https://github.com/markjprice/cs13net9/blob/main/docs/ch02-features.md`.

Understanding C# standards

Over the years, Microsoft has submitted a few versions of C# to ECMA standards bodies. Microsoft made C# open source in 2014. You can read the latest C# standard document at the following link: `https://learn.microsoft.com/en-us/dotnet/csharp/specification/`. More practically useful than the ECMA standards are the public GitHub repositories for making the work on C# and related technologies as open as possible, as shown in *Table 2.1*:

Description	Link
C# language design	`https://github.com/dotnet/csharplang`
Compiler implementation	`https://github.com/dotnet/roslyn`
Standard to describe the language	`https://github.com/dotnet/csharpstandard`

Table 2.1: Public GitHub repositories for C#

Discovering your C# compiler version

The .NET language compiler for C# and Visual Basic, also known as **Roslyn**, along with a separate compiler for F#, is distributed as part of the .NET SDK. To use a specific version of C#, you must have at least that version of the .NET SDK installed, as shown in *Table 2.2*:

.NET SDK	Roslyn compiler	Default C# language
1.0.4	2.0–2.2	7.0
1.1.4	2.3–2.4	7.1
2.1.2	2.6–2.7	7.2
2.1.200	2.8–2.10	7.3

3.0	3.0–3.4	8.0
5.0	3.8	9.0
6.0	4.0	10.0
7.0	4.4	11.0
8.0	4.8	12.0
9.0	4.12	13.0

Table 2.2: .NET SDK versions and their C# compiler versions

When you create class libraries, you can choose to target .NET Standard as well as versions of modern .NET. They have default C# language versions, as shown in *Table 2.3*:

.NET Standard	C#
2.0	7.3
2.1	8.0

Table 2.3: .NET Standard versions and their default C# compiler versions

Although you must have a minimum version of the .NET SDK installed to have access to a specific compiler version, the projects that you create can target older versions of .NET and still use a modern compiler version. For example, if you have the .NET 9 SDK or later installed, then you can use C# 13 language features in a console app that targets .NET 8.

How to output the SDK version

Let's see what .NET SDK and C# language compiler versions you have available:

1. On Windows, start **Windows Terminal** or **Command Prompt**. On macOS, start **Terminal**.
2. To determine which version of the .NET SDK you have available, enter the following command:

```
dotnet --version
```

3. Note that the version at the time of publishing is 9.0.100, indicating that it is the initial version of the SDK without any bug fixes or new features yet, as shown in the following output:

```
9.0.100
```

Enabling a specific language version compiler

Developer tools like Visual Studio and the `dotnet` command-line interface assume that you want to use the latest major version of a C# language compiler by default. Before C# 8 was released, C# 7 was the latest major version and was used by default.

To use the improvements in a C# point release like 7.1, 7.2, or 7.3, you had to add a `<LangVersion>` configuration element to the project file, as shown in the following markup:

```
<LangVersion>7.3</LangVersion>
```

After the release of C# 13 with .NET 9, if Microsoft releases a C# 13.1 compiler and you want to use its new language features, then you will have to add a configuration element to your project file, as shown in the following markup:

```
<LangVersion>13.1</LangVersion>
```

Potential values for the `<LangVersion>` are shown in *Table 2.4*:

<LangVersion>	**Description**
`7`, `7.1`, `7.2`, `7.3`, `8`, `9`, `10`, `11`, `12`, `13`	Entering a specific version number will use that compiler if it has been installed.
`latestmajor`	Uses the highest major number, for example, 7.0 in August 2019, 8 in October 2019, 9 in November 2020, 10 in November 2021, 11 in November 2022, 12 in November 2023, and 13 in November 2024.
`latest`	Uses the highest major and highest minor number, for example, 7.2 in 2017, 7.3 in 2018, 8 in 2019, and perhaps 13.1 in the first half of 2025.
`preview`	Uses the highest available preview version, for example, 14 in mid-July 2025 with .NET 10 Preview 6 installed.

Table 2.4: LangVersion settings for a project file

Using preview C# compiler versions

In February 2025, Microsoft is likely to release the first public preview of .NET 10 with the C# 14 compiler. You will be able to install its SDK from the following link:

`https://dotnet.microsoft.com/en-us/download/dotnet/10.0`

Warning! The link will give a `404 Missing resource` error until February 2025, so do not bother using it until then!

After you've installed a .NET 10 SDK preview, you will be able to use it to create new projects and explore the new language features in C# 14.

After creating a new project, you must edit the `.csproj` file and add the `<LangVersion>` element set to `preview` to use the preview C# 14 compiler, as shown highlighted in the following markup:

```
<Project Sdk="Microsoft.NET.Sdk">
  <PropertyGroup>
    <OutputType>Exe</OutputType>
    <TargetFramework>net10.0</TargetFramework>
    <LangVersion>preview</LangVersion>
  </PropertyGroup>
</Project>
```

Good Practice: You should only set `<LangVersion>` to `preview` for exploration, not production projects, because it is not supported by Microsoft and it is more likely to have bugs. Microsoft makes previews available because they want to hear feedback from developers like you. In this way, you can be a part of C#'s development and improvement.

Switching the C# compiler for .NET 9 to a future version

.NET 9 comes with the C# 13 compiler but that does not mean that you are stuck with the C# 13 compiler. Once the .NET 10 SDK is made generally available in November 2025, you will be able to get the best of both worlds.

You can use the .NET 10 SDK and its C# 14 compiler while your projects continue to target .NET 9. To do so, set the target framework to `net9.0` and add a `<LangVersion>` element set to `14`, as shown highlighted in the following markup:

```
<Project Sdk="Microsoft.NET.Sdk">
  <PropertyGroup>
    <OutputType>Exe</OutputType>
    <TargetFramework>net9.0</TargetFramework>
    <ImplicitUsings>enable</ImplicitUsings>
    <Nullable>enable</Nullable>
    <LangVersion>14</LangVersion>
  </PropertyGroup>
</Project>
```

The preceding project targets `net9.0`, so it is supported until May 2026 when run on a monthly patched version of the .NET 9 runtime. If the preceding project is built using .NET 10 SDK, then it can have the `<LangVersion>` set to `14`, meaning C# 14.

If you target `net10.0`, which new projects will by default if you have installed the .NET 10 SDK, then the default language will be C# 14 so it would not need to be explicitly set.

In February 2026, Microsoft is likely to release the first preview of .NET 11, and in November 2026, it will likely release .NET 11 for general availability in production. You will be able to install its SDK from the following link and explore C# 15 in the same way as described above for C# 14 with .NET 10:

`https://dotnet.microsoft.com/en-us/download/dotnet/11.0`

Again, the preceding link is for future use! It will give a `404 Missing resource` error until February 2026, so do not bother using the preceding link until then.

Warning! Some C# language features depend on changes in the underlying .NET libraries. Even if you use the latest SDK with the latest compiler, you might not be able to use all the new language features while targeting an older version of .NET. For example, C# 11 introduced the `required` keyword, but it cannot be used in a project that targets .NET 6 because that language feature requires new attributes that are only available in .NET 7. Luckily, the compiler will warn you if you try to use a C# feature that is not supported. Just be prepared for that eventuality.

Showing the compiler version

We will start by writing code that shows the compiler version:

1. If you've completed *Chapter 1, Hello, C#! Welcome, .NET!*, then you will already have a `cs13net9` folder. If not, then you'll need to create it.
2. Use your preferred code editor to create a new project, as defined in the following list:
    - Project template: **Console App [C#]** / `console`
    - Project file and folder: `Vocabulary`
    - Solution file and folder: `Chapter02`
    - **Do not use top-level statements:** Cleared
    - **Enable native AOT publish:** Cleared

Good Practice: If you have forgotten how, or did not complete the previous chapter, then step-by-step instructions for creating a solution with multiple projects are given in *Chapter 1, Hello, C#! Welcome, .NET!*.

3. In the `Vocabulary` project, in `Program.cs`, after the comment, add a statement to show the C# version as an error, as shown in the following code:

```
#error version
```

4. Run the console app:
    - If you are using Visual Studio, then navigate to **Debug | Start Without Debugging**. When prompted to continue and run the last successful build, click **No**.
    - If you are using VS Code, then in a terminal for the `Vocabulary` folder, enter the `dotnet run` command. Note that we are expecting a compiler error, so do not panic when you see it!
5. Note that the compiler version and the language version appear as compiler error message number `CS8304`, as shown in *Figure 2.1*:

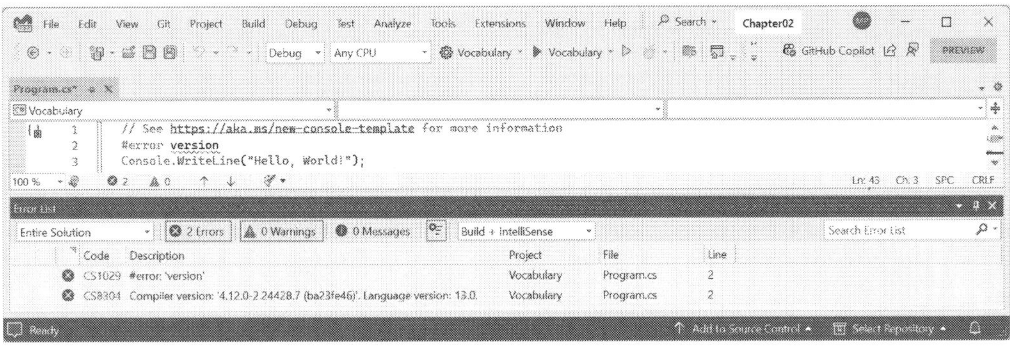

Figure 2.1: A compiler error that shows the C# language version

The error message in the VS Code **PROBLEMS** window or Visual Studio **Error List** window says `Compiler version: '4.12.0...'` with language version `default (13.0)`.

6. Comment out the statement that causes the error, as shown in the following code:

```
// #error version
```

7. Note that the compiler error messages disappear.

Understanding C# grammar and vocabulary

Let's start by looking at the basics of the grammar and vocabulary of C#. Throughout this chapter, you will create multiple console apps, with each one showing related features of the C# language.

Understanding C# grammar

The grammar of C# includes statements and blocks. To document your code, you can use comments.

Good Practice: Comments should not be the only way that you document your code. Choosing sensible names for variables and functions, writing unit tests, and creating actual documents are other ways to document your code.

Statements

In English, we indicate the end of a sentence with a period. A sentence can be composed of multiple words and phrases, with the order of words being part of the grammar. For example, in English, we say "the black cat."

The adjective, *black*, comes before the noun, *cat*. However, French grammar has a different order; the adjective comes after the noun: "le chat noir." What's important to take away from this is that the order matters.

C# indicates the end of a **statement** with a semicolon. A statement can be composed of multiple **types**, **variables**, and **expressions** made up of **tokens**. Each token is separated by white space or some other recognizably different token, like an operator, for example, = or +.

For example, in the following statement, `decimal` is a type, `totalPrice` is a variable, and `subtotal + salesTax` is an expression:

```
decimal totalPrice = subtotal + salesTax;
```

The expression is made up of an operand named `subtotal`, an operator `+`, and another operand named `salesTax`. The order of operands and operators matters because the order affects the meaning and result.

Comments

Comments are the primary method of documenting your code to enhance the understanding of how it works, for other developers to read, or for you to read even when you come back to it months later.

In *Chapter 4, Writing, Debugging, and Testing Functions*, you will learn about XML comments that start with three slashes, `///`, and work with a tool to generate web pages to document your code.

You can add comments to explain your code using a double slash, `//`. The compiler will ignore everything after the `//` until the end of the line, as shown in the following code:

```
// Sales tax must be added to the subtotal.
decimal totalPrice = subtotal + salesTax;
```

To write a multiline comment, use `/*` at the beginning and `*/` at the end of the comment, as shown in the following code:

```
/*
This is a
multi-line comment.
*/
```

Although `/* */` is mostly used for multiline comments, it is also useful for commenting in the middle of a statement, as shown in the following code:

```
decimal totalPrice = subtotal /* for this item */ + salesTax;
```

Good Practice: Well-designed code, including function signatures with well-named parameters and class encapsulation, can be somewhat self-documenting. When you find yourself putting too many comments and explanations in your code, ask yourself: can I rewrite, aka refactor, this code to make it more understandable without long comments?

Your code editor has commands to make it easier to add and remove comment characters, as shown in the following list:

- Visual Studio: Navigate to **Edit** | **Advanced** | **Comment Selection** or **Uncomment Selection.**
- VS Code: Navigate to **Edit** | **Toggle Line Comment** or **Toggle Block Comment.**
- Rider: Navigate to **Code** | **Comment with Line Comment** or **Comment with Block Comment.**

Good Practice: You **comment** code by adding descriptive text above or after code statements. You **comment out** code by adding comment characters before or around statements to make them inactive. **Uncommenting** means removing the comment characters.

Blocks

In English, we indicate a new paragraph by starting a new line. C# indicates a **block** of code with the use of curly brackets, `{ }`.

Blocks start with a declaration to indicate what is being defined. For example, a block can define the start and end of many language constructs, including namespaces, classes, methods, or statements like `foreach`.

You will learn more about namespaces, classes, and methods later in this chapter and subsequent chapters, but to briefly introduce some of those concepts now:

- A **namespace** contains types like classes to group them together.
- A **class** contains the members of an object, including methods.
- A **method** contains statements that implement an action that an object can take.

Code editors like Visual Studio, Rider, and VS Code provide a handy feature to collapse and expand blocks by toggling the arrow symbol pointing down or right when you move your mouse cursor over the left margin of the code, as shown in *Figure 2.2*:

Visual Studio

```
// Loop through all the types in the assembly.
foreach (TypeInfo t in a.DefinedTypes)
{
  // Add up the counts of all the methods.
  methodCount += t.GetMethods().Length;
}
```

```
// Loop through all the types in the assembly.
foreach (TypeInfo t in a.DefinedTypes)[...]
```

VS Code

```
// Loop through all the types in the assembly.
foreach (TypeInfo t in a.DefinedTypes)
{
  // Add up the counts of all the methods.
  methodCount += t.GetMethods().Length;
}
```

```
// Loop through all the types in the assembly.
foreach (TypeInfo t in a.DefinedTypes)...
```

Figure 2.2: Code editors with expanded and collapsed blocks

Regions

You can define your own labeled regions around any statements you want and then most code editors will allow you to collapse and expand them in the same way as blocks, as shown in the following code:

```
#region Three variables that store the number 2 million.

int decimalNotation = 2_000_000;
int binaryNotation = 0b_0001_1110_1000_0100_1000_0000;
int hexadecimalNotation = 0x_001E_8480;

#endregion
```

In this way, regions can be treated as commented blocks that can be collapsed to show a summary of what the block does.

I will use `#region` blocks throughout the solution code in the GitHub repository, especially for the early chapters before we start defining functions that act as natural collapsible regions, but I won't show them in the print book to save space. Use your own judgment to decide if you want to use regions in your own code.

Examples of statements and blocks

In a simple console app that does not use the top-level program feature, I've added some comments to the statements and blocks, as shown in the following code:

```
using System; // A semicolon indicates the end of a statement.

namespace Basics
{ // An open brace indicates the start of a block.
  class Program
  {
    static void Main(string[] args)
    {
      Console.WriteLine("Hello World!"); // A statement.
    }
  }
} // A close brace indicates the end of a block.
```

Note that C# uses a brace style where both the open and close braces are on their own line and are at the same indentation level, as shown in the following code:

```
if (x < 3)
{
  // Do something if x is less than 3.
}
```

Other languages like JavaScript use curly braces but format them differently. They put the open curly brace at the end of the declaration statement, as shown in the following code:

```
if (x < 3) {
  // Do something if x is less than 3.
}
```

You can use whatever style you prefer because the compiler does not care.

Sometimes, to save vertical space in a print book, I use the JavaScript brace style, but mostly I stick with the C# brace style. I use two spaces instead of the more common four spaces for indenting because my code will be printed in a book and therefore has a narrow width available.

More Information: The official coding style conventions can be found at the following link: `https://learn.microsoft.com/en-us/dotnet/csharp/fundamentals/coding-style/coding-conventions`.

Regardless of any official guidelines, I recommend that you conform to whatever standards have been adopted by your development team, unless you are a solo developer, in which case as long as your code compiles, you can use any conventions you like. Be kind to your future self, though, by being consistent one way or the other!

Good Practice: The brace style used in the Microsoft official documentation is the most commonly used for C#. For example, see the `for` statement, as found at the following link: `https://learn.microsoft.com/en-us/dotnet/csharp/language-reference/statements/iteration-statements`.

Formatting code using white space

White space includes the space, tab, and newline characters. You can use white space to format your code however you like because extra white space has no effect on the compiler.

Warning! Unless a step-by-step instruction tells the reader to enter code, code examples are written to be read and understood, not typed into a code editor.

The following four statements are all equivalent. Since all four statements are equivalent, they all have the same variable name, and therefore cannot be all declared in the same code block, so please do not try to type this code unless you also change the three `sum` variable names:

```
int sum = 1 + 2; // Most developers would prefer this format.

int
```

```
sum=1+
2; // One statement over three lines.

int      sum=    1     +2;int sum=1+2; // Two statements on one line.
```

The only white space character required in the preceding statements is one between `int` and `sum` to tell the compiler they are separate tokens. Any single white space character (for example, a space, tab, or newline) would be acceptable.

More Information: You can read the formal definition of C# white space at the following link: `https://learn.microsoft.com/en-us/dotnet/csharp/language-reference/language-specification/lexical-structure#634-white-space`.

Understanding C# vocabulary

The C# vocabulary is made up of **keywords**, **symbol characters**, and **types**.

Some of the predefined, reserved keywords that you will see in this book and use frequently include `using`, `namespace`, `class`, `static`, `int`, `string`, `double`, `bool`, `if`, `switch`, `break`, `while`, `do`, `for`, `foreach`, `this`, and `true`.

Warning! The `this` keyword can be used in multiple ways, including to refer to the current instance of an object, to call a constructor on the current object instance, and to define an indexer. Examples of all three will be covered in *Chapter 5, Building Your Own Types with Object-Oriented Programming*.

Some of the symbol characters that you will see are `"`, `'`, `+`, `-`, `*`, `/`, `%`, `@`, and `$`.

"Bracket" symbols include the following:

- `()` are called **parentheses**. They are used to call a function, define an expression or condition, and cast between types.
- `{}` are called **braces** (aka curly brackets). They are used to define blocks and perform object and collection initialization.
- `[]` are called **brackets** (aka square brackets). They are used to access items in an array or collection, and around attributes decorating elements of code.
- `<>` are called **angle brackets**. They are used for generic types, in XML and HTML files, and individually as *less than* or *greater than* tokens in an expression.

There are other contextual keywords that only have a special meaning in a specific context, like `and`, `or`, `not`, `record`, and `init`.

However, that still means that there are only about 100 actual C# keywords in the language.

Good Practice: C# keywords use all lowercase. Although you can use all lowercase for your own type names, you should not. With C# 11 and later, the compiler will give a warning if you do, as shown in the following output: `Warning CS8981 The type name 'person' only contains lower-cased ascii characters. Such names may become reserved for the language.`

If you want to use a C# keyword to name a variable, then you can do so by prefixing it with the @ symbol, as shown in the following code:

```
// Prefix with @ to use a C# keyword as a variable name.
string @class = "A poorly named variable!";
int @if = 5;
```

Good Practice: Although you can use C# keywords for variable names, you should avoid doing this because it is poor practice. For multiple C# versions, Microsoft has wanted to add a new keyword, `field`, to the language, but they are reluctant to do so because some developers are likely to have variables named `field` and this would cause a breaking change to those projects.

Comparing programming languages to human languages

The English language has more than 250,000 distinct words, so how does C# get away with only having about 100 keywords? Moreover, why is C# so difficult to learn if it has only 0.0416% of the number of words in the English language?

One of the key differences between a human language and a programming language is that developers need to be able to define the new "words" with new meanings. Apart from the (approximately) 100 keywords in the C# language, this book will teach you about some of the hundreds of thousands of "words" that other developers have defined, but you will also learn how to define your own "words."

Programmers all over the world must learn English because most programming languages use English words such as "if" and "break." There are programming languages that use other human languages, such as Arabic, but they are rare. If you are interested in learning more, this YouTube video shows a demonstration of an Arabic programming language: `https://www.youtube.com/watch?v=EwYIZBL4sAU`.

Changing the color scheme for C# syntax

By default, Visual Studio and VS Code show C# keywords in blue to make them easier to differentiate from other code, which defaults to black. Both tools allow you to customize the color scheme.

In Visual Studio:

1. Navigate to **Tools** | **Options**.
2. In the **Options** dialog box, in the **Environment** section, select **Fonts and Colors**, and then select the display items that you would like to customize. You can also search for the section instead of browsing for it.

In VS Code:

1. Navigate to **File** | **Preferences** | **Theme** | **Color Theme**. It is in the **Code** menu on macOS.
2. Select a color theme. For reference, I'll use the **Light+ (default light)** color theme so that the screenshots look better in a printed book.

In Rider, navigate to **File** | **Settings** | **Editor** | **Color Scheme**.

Help for writing correct code

Plain text editors such as Notepad don't help you write correct English. Likewise, Notepad won't help you write the correct C# either.

Microsoft Word can help you write English by highlighting spelling mistakes with red squiggles, with Word saying that "icecream" should be ice-cream or ice cream, and grammatical errors with blue squiggles, such as a sentence should have an uppercase first letter.

Similarly, Visual Studio and VS Code's C# extension helps you write C# code by highlighting spelling mistakes, such as the method name needing to be `WriteLine` with an uppercase `L`, and grammatical errors, such as statements that must end with a semicolon.

The C# extension constantly watches what you type and gives you feedback by highlighting problems with colored squiggly lines, like that of Microsoft Word.

Let's see it in action:

1. In `Program.cs`, change the `L` in the `WriteLine` method to lowercase.
2. Delete the semicolon at the end of the statement.
3. In VS Code, navigate to **View** | **Problems**; in Visual Studio, navigate to **View** | **Error List**; or in Rider, navigate to **View** | **Tool Windows** | **Problems**, and note that a red squiggle appears under the code mistakes and details are shown, as you can see in *Figure 2.3*:

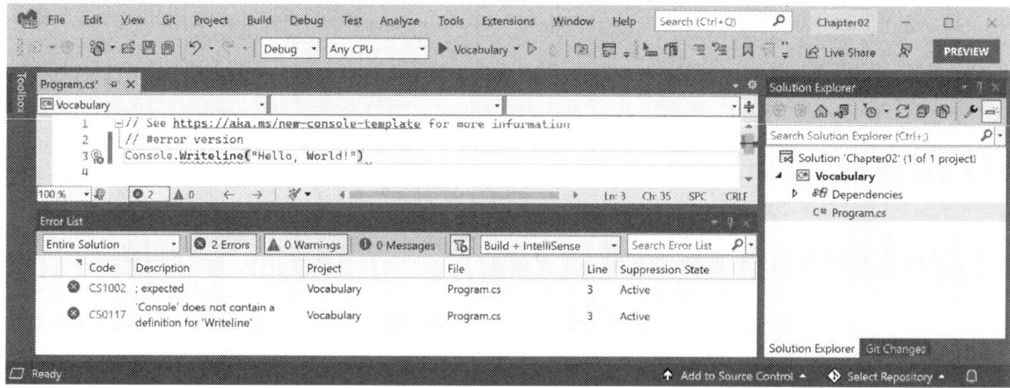

Figure 2.3: The Error List window showing two compile errors

4. Fix the two coding errors.

Importing namespaces

`System` is a namespace, which is like an address for a type. To refer to someone's location exactly, you might use `Oxford.HighStreet.BobSmith`, which tells us to look for a person named Bob Smith on the High Street in the city of Oxford.

`System.Console.WriteLine` tells the compiler to look for a method named `WriteLine` in a type named `Console` in a namespace named `System`.

To simplify our code, the **Console App** project template for every version of .NET before 6.0 added a statement at the top of the code file to tell the compiler to always look in the `System` namespace for types that haven't been prefixed with their namespace, as shown in the following code:

```
using System; // Import the System namespace.
```

We call this *importing the namespace*. The effect of importing a namespace is that all available types in that namespace will be available to your program without needing to enter the namespace prefix. All available types in that namespace will be seen in IntelliSense while you write code.

Implicitly and globally importing namespaces

Traditionally, every `.cs` file that needs to import namespaces would have to start with `using` statements to import those namespaces. Namespaces like `System` and `System.Linq` are needed in almost all `.cs` files, so the first few lines of every `.cs` file often had at least a few `using` statements, as shown in the following code:

```
using System;
using System.Linq;
using System.Collections.Generic;
```

When creating websites and services using ASP.NET Core, there are often dozens of namespaces that each file would have to import.

C# 10 introduced a new keyword combination and .NET SDK 6 introduced a new project setting that work together to simplify importing common namespaces.

The `global using` keyword combination means you only need to import a namespace in one `.cs` file and it will be available throughout all `.cs` files instead of having to import the namespace at the top of every file that needs it. You could put `global using` statements in the `Program.cs` file, but I recommend creating a separate file for those statements named something like `GlobalUsings.cs` with the contents being all your `global using` statements, as shown in the following code:

```
global using System;
global using System.Linq;
global using System.Collections.Generic;
```

Good Practice: As developers get used to this new C# feature, I expect one naming convention for this file to become the de facto standard. As you are about to see, the related .NET SDK feature uses a similar naming convention.

Any projects that target .NET 6 or later, and therefore use the C# 10 or later compiler, generate a `<ProjectName>.GlobalUsings.g.cs` file in the `obj\Debug\net9.0` folder to implicitly globally import some common namespaces like `System`. The specific list of implicitly imported namespaces depends on which SDK you target, as shown in *Table 2.5*:

SDK	Implicitly imported namespaces

`Microsoft.NET.Sdk`	`System`
	`System.Collections.Generic`
	`System.IO`
	`System.Linq`
	`System.Net.Http`
	`System.Threading`
	`System.Threading.Tasks`

`Microsoft.NET.Sdk.Web`	Same as `Microsoft.NET.Sdk`, plus:
	`System.Net.Http.Json`
	`Microsoft.AspNetCore.Builder`
	`Microsoft.AspNetCore.Hosting`
	`Microsoft.AspNetCore.Http`
	`Microsoft.AspNetCore.Routing`
	`Microsoft.Extensions.Configuration`
	`Microsoft.Extensions.DependencyInjection`
	`Microsoft.Extensions.Hosting`
	`Microsoft.Extensions.Logging`

`Microsoft.NET.Sdk.Worker`	Same as `Microsoft.NET.Sdk`, plus:
	`Microsoft.Extensions.Configuration`
	`Microsoft.Extensions.DependencyInjection`
	`Microsoft.Extensions.Hosting`
	`Microsoft.Extensions.Logging`

Table 2.5: .NET SDKs and their implicitly imported namespaces

Let's see the current autogenerated implicit imports file:

1. In **Solution Explorer**, toggle on the **Show All Files** button, and note the compiler-generated `bin` and `obj` folders are now visible.
2. In the `Vocabulary` project, expand the `obj` folder, expand the `Debug` folder, expand the `net9.0` folder, and then open the file named `Vocabulary.GlobalUsings.g.cs`.

The naming convention for this file is `<ProjectName>.GlobalUsings.g.cs`. Note the g for **generated** to differentiate it from developer-written code files.

3. Remember that this file is automatically created by the compiler for projects that target .NET 6 and later and that it imports some commonly used namespaces, including `System.Threading`, as shown in the following code:

```
// <autogenerated />
global using global::System;
global using global::System.Collections.Generic;
global using global::System.IO;
global using global::System.Linq;
global using global::System.Net.Http;
global using global::System.Threading;
global using global::System.Threading.Tasks;
```

4. Close the `Vocabulary.GlobalUsings.g.cs` file.
5. In **Solution Explorer**, open the `Vocabulary.csproj` project file, and then add additional entries to the project file to control which namespaces are implicitly imported, as shown highlighted in the following markup:

```
<Project Sdk="Microsoft.NET.Sdk">

  <PropertyGroup>
    <OutputType>Exe</OutputType>
    <TargetFramework>net9.0</TargetFramework>
    <Nullable>enable</Nullable>
    <ImplicitUsings>enable</ImplicitUsings>
  </PropertyGroup>

  <ItemGroup>
    <Using Remove="System.Threading" />
    <Using Include="System.Numerics" />
    <Using Include="System.Console" Static="true" />
    <Using Include="System.Environment" Alias="Env" />
  </ItemGroup>

</Project>
```

Note that `<ItemGroup>` is different from `<ImportGroup>`. Be sure to use the correct one! Also, note that the order of elements in a project group or item group does not matter. For example, `<Nullable>` can be before or after `<ImplicitUsings>`.

6. Save the changes to the project file.
7. Expand the `obj` folder, expand the `Debug` folder, expand the `net9.0` folder, and open the file named `Vocabulary.GlobalUsings.g.cs`.
8. Note that this file now imports `System.Numerics` instead of `System.Threading`, the `Environment` class has been imported and aliased to `Env`, and we have statically imported the `Console` class, as shown highlighted in the following code:

```
// <autogenerated />
global using global::System;
global using global::System.Collections.Generic;
global using global::System.IO;
global using global::System.Linq;
global using global::System.Net.Http;
global using global::System.Numerics;
global using global::System.Threading.Tasks;
global using Env = global::System.Environment;
global using static global::System.Console;
```

9. In `Program.cs`, add a statement to output a message from the computer and note that because we statically imported the `Console` class, we can call its methods like `WriteLine` without prefixing them with `Console`, and we can reference the `Environment` class using its alias, `Env`, as shown in the following code:

```
WriteLine($"Computer named {Env.MachineName} says \"No.\"");
```

10. Run the project and note the message, as shown in the following output:

```
Computer named DAVROS says "No."
```

Your computer name will be different unless you name your computers after characters from Doctor Who like I do.

You can disable the implicitly imported namespaces feature for all SDKs by removing the `<ImplicitUsings>` element completely from the project file, or changing its value to `disable`, as shown in the following markup:

```
<ImplicitUsings>disable</ImplicitUsings>
```

Good Practice: You might choose to do this if you want to manually create a single file with all the `global using` statements instead of potentially having one generated automatically and others created manually. But my recommendation is to leave the feature enabled and modify the project file to change what is included in the auto-generated class file in the `obj` folder hierarchy.

Verbs are methods

In English, verbs are doing or action words, like "run" and "jump." In C#, doing or action words are called **methods**. There are hundreds of thousands of methods available to C#. In English, verbs change how they are written based on when in time the action happens. For example, Amir *was jumping* in the past, Beth *jumps* in the present, they *jumped* in the past, and Charlie *will jump* in the future.

In C#, methods such as `WriteLine` change how they are called or executed based on the specifics of the action. This is called overloading, which we'll cover in more detail in *Chapter 5, Building Your Own Types with Object-Oriented Programming*. But for now, consider the following example:

```
// Outputs the current line terminator.
// By default, this is a carriage-return and line feed.
Console.WriteLine();

// Outputs the greeting and the current line terminator.
Console.WriteLine("Hello Ahmed");

// Outputs a formatted number and date and the current line terminator.
Console.WriteLine(
  "Temperature on {0:D} is {1}°C.", DateTime.Today, 23.4);
```

When I show code snippets without numbered step-by-step instructions, I do not expect you to enter them as code, so they won't execute out of context.

A different and not quite exact analogy is that some verbs are spelled the same but have different meanings depending on the context, for example, you can lose a game, lose your place in a book, or lose your keys.

Nouns are types, variables, fields, and properties

In English, nouns are names that refer to things. For example, Fido is the name of a dog. The word "dog" tells us the type of thing that Fido is, and so to order Fido to fetch a ball, we would use his name.

In C#, their equivalents are **types**, **variables**, **fields**, and **properties**. For example:

- `Animal` and `Car` are types; they are nouns for categorizing things.
- `Head` and `Engine` might be fields or properties; they are nouns that belong to `Animal` and `Car`.
- `Fido` and `Bob` are variables; they are nouns for referring to a specific object.

There are tens of thousands of types available to C#, though have you noticed how I didn't say, "There are tens of thousands of types *in* C#"? The difference is subtle but important. The language of C# only has a few keywords for types, such as `string` and `int`, and strictly speaking, C# doesn't define any types. Keywords such as `string` that look like types are **aliases**, which represent types provided by the platform on which C# runs.

It's important to know that C# cannot exist alone; after all, it's a language that runs on variants of .NET. In theory, someone could write a compiler for C# that uses a different platform, with different underlying types. In practice, the platform for C# is .NET, which provides tens of thousands of types to C#, including `System.Int32`, which is the C# keyword alias that `int` maps to, as well as many more complex types, such as `System.Xml.Linq.XDocument`.

It's worth taking note that the term **type** is often confused with **class**. Have you ever played the parlor game *Twenty Questions*, also known as *Animal, Vegetable, or Mineral*? In the game, everything can be categorized as an animal, vegetable, or mineral. In C#, every **type** can be categorized as a `class`, `struct`, `enum`, `interface`, or `delegate`. You will learn what these mean in *Chapter 6, Implementing Interfaces and Inheriting Classes*. As an example, the C# keyword `string` is a `class`, but `int` is a `struct`. So, it is best to use the term **type** to refer to both.

Revealing the extent of the C# vocabulary

We know that there are more than 100 keywords in C#, but how many types are there? Let's write some code to find out how many types (and their methods) are available to C# in our simple console app.

Don't worry about exactly how this code works for now, but know that it uses a technique called **reflection**:

1. Comment out all the existing statements in `Program.cs`.
2. We'll start by importing the `System.Reflection` namespace at the top of the `Program.cs` file so that we can use some of the types in that namespace like `Assembly` and `TypeName`, as shown in the following code:

```
using System.Reflection; // To use Assembly, TypeName, and so on.
```

Good Practice: We could use the implicit imports and `global using` features to import this namespace for all `.cs` files in this project, but since there is only one file, it is better to import the namespace in the one file in which it is needed.

3. Write statements to get the compiled console app and loop through all the types that it has access to, outputting the names and number of methods each has, as shown in the following code:

```
// Get the assembly that is the entry point for this app.
Assembly? myApp = Assembly.GetEntryAssembly();

// If the previous line returned nothing then end the app.
if (myApp is null) return;

// Loop through the assemblies that my app references.
foreach (AssemblyName name in myApp.GetReferencedAssemblies())
{
  // Load the assembly so we can read its details.
  Assembly a = Assembly.Load(name);

  // Declare a variable to count the number of methods.
  int methodCount = 0;

  // Loop through all the types in the assembly.
  foreach (TypeInfo t in a.DefinedTypes)
  {
    // Add up the counts of all the methods.
    methodCount += t.GetMethods().Length;
  }

  // Output the count of types and their methods.
  WriteLine("{0:N0} types with {1:N0} methods in {2} assembly.",
    arg0: a.DefinedTypes.Count(),
    arg1: methodCount,
    arg2: name.Name);
}
```

`N0` is uppercase `N` followed by the digit zero. It is not uppercase `N` followed by uppercase `O`. It means "format a number (`N`) with zero (`0`) decimal places."

4. Run the project. You will see the actual number of types and methods that are available to you in the simplest application when running on your **operating system** (**OS**). The number of types and methods displayed will be different depending on the OS that you are using, as shown in the following output on Windows:

```
0 types with 0 methods in System.Runtime assembly.
41 types with 639 methods in System.Console assembly.
112 types with 1,190 methods in System.Linq assembly.
```

Why does the `System.Runtime` assembly contain zero types? This assembly is special because it contains only **type-forwarders** rather than actual types. A type-forwarder represents a type that has been implemented outside of .NET or for some other advanced reason.

5. Add statements to the top of the file (after importing the `System.Reflection` namespace) to declare some variables, as shown in the following code:

```
// Declare some unused variables using types in
// additional assemblies to make them load too.
System.Data.DataSet ds = new();
HttpClient client = new();
```

6. Run the console app again and view the results, which should look like the following output:

```
0 types with 0 methods in System.Runtime assembly.
406 types with 7,139 methods in System.Data.Common assembly.
425 types with 4,681 methods in System.Net.Http assembly.
41 types with 639 methods in System.Console assembly.
112 types with 1,190 methods in System.Linq assembly.
```

Now, you have a better sense of why learning C# is a challenge – because there are so many types and methods to learn. Methods are only one category of a member that a type can have, and you and other programmers are constantly defining new types and members!

Example of asking ChatGPT to explain code

The code we wrote in the `Vocabulary` project is a good example of when you might want to ask ChatGPT to explain something. If you enter a prompt of "Explain the following code:" followed by the code, here's what it responded with:

"This C# code performs an analysis of the assemblies (essentially libraries or modules) referenced by the current application. It tallies up the total number of types and methods in each referenced assembly.

Let's break down the code in more detail:

1. `System.Reflection` namespace is used to work with assemblies, modules, types, methods, and so on, at runtime.
2. The `System.Data.DataSet` and `HttpClient` instances are declared but not used. The comment implies that these are declared to ensure their respective assemblies are loaded for the subsequent analysis. Creating instances of classes from certain assemblies can cause those assemblies to be loaded into memory.
3. `Assembly.GetEntryAssembly()` gets the assembly that is the entry point of the application.
4. If `myApp` is null (meaning there's no entry assembly, which should be impossible in a normally running .NET app), the application exits.

5. The `foreach` loop iterates over each assembly name in the referenced assemblies of the entry point assembly.
6. `Assembly.Load(name)` loads an assembly given the assembly name. This allows you to interact with the assembly programmatically.
7. A counter variable, `methodCount`, is declared and initialized to `0`. This is used to count the total number of methods in the assembly.
8. Another nested `foreach` loop then iterates over each type in the assembly.
9. For each type, it gets the number of methods `(t.GetMethods().Length)` and adds it to the `methodCount`.
10. Finally, the number of types and their methods in the assembly are output to the console using `WriteLine()`. The `{0:N0}` format specifier in the string is a placeholder for a number (`0` refers to the argument index, `N` is for number format, and 0 after N specifies the number of decimal places).

So, in essence, this script counts how many types and methods exist in each assembly referenced by the application. This might be useful to understand the complexity or size of the dependencies the application has."

And now back to the flesh and blood, Mark. That's a pretty good explanation with no mistakes or hallucinations!

Good Practice: Tools like GitHub Copilot and ChatGPT are especially useful for programmers when they are learning new things. And a good programmer is always learning new things! These tools are not just for writing code for you. They can explain existing code written by others and even suggest improvements.

Working with variables

All applications process data. Data comes in, data is processed, and then data goes out.

Data usually comes into our program from files, databases, or user input, and it can be put temporarily into variables, which will be stored in the memory of the running program. When the program ends, the data in memory is lost. Data is usually output to files and databases, or to the screen or a printer. When using variables, you should think about, firstly, how much space the variable takes up in the memory, and, secondly, how fast it can be processed.

We control this by picking an appropriate type. You can think of simple common types such as `int` and `double` as being different-sized storage boxes, where a smaller box would take less memory but may not be as fast at being processed; for example, adding 16-bit numbers might not be processed as quickly as adding 64-bit numbers on a 64-bit operating system. Some of these boxes may be stacked close by, and some may be thrown into a big heap further away.

Naming things and assigning values

There are naming conventions for things, and it is a good practice to follow them, as shown in *Table 2.6*:

Naming convention	Examples	Used for
Camel case	`cost`, `orderDetail`, and `dateOfBirth`	Local variables and private fields
Title case, aka Pascal case	`String`, `Int32`, `Cost`, `DateOfBirth`, and `Run`	Types, non-private fields, and other members like methods

Table 2.6: Naming conventions and what they should be used for

Some C# programmers like to prefix the names of private fields with an underscore, for example, `_dateOfBirth` instead of `dateOfBirth`. The naming of private members of all kinds is not formally defined because they will not be visible outside the class, so writing them either with or without an underscore prefix is valid.

Good Practice: Following a consistent set of naming conventions will enable your code to be easily understood by other developers (and yourself in the future!).

The following code block shows an example of declaring a named local variable and assigning a value to it with the = symbol. You should note that you can output the name of a variable using a keyword introduced in C# 6, `nameof`:

```
// Let the heightInMetres variable become equal to the value 1.88.
double heightInMetres = 1.88;
Console.WriteLine($"The variable {nameof(heightInMetres)} has the value
{heightInMetres}.");
```

Warning! The message in double quotes in the preceding code wraps onto a second line because the width of a printed page is too narrow. When entering a statement like this in your code editor, type it all in a single line.

In C# 12 and later, `nameof` can now access instance data from a static context. You will learn the difference between instance and static data in *Chapter 5, Building Your Own Types with Object-Oriented Programming*.

Literal values

When you assign to a variable, you often, but not always, assign a **literal** value. But what is a literal value? A literal is a notation that represents a fixed value. Data types have different notations for their literal values, and over the next few sections, you will see examples of using literal notation to assign values to variables.

More Information: You can read the formal definition of literals in the C# language specification: `https://learn.microsoft.com/en-us/dotnet/csharp/language-reference/language-specification/lexical-structure#645-literals`.

Storing text

For text, a single letter, such as an `A`, is stored as a `char` type.

Good Practice: Actually, it can be more complicated than that. Egyptian Hieroglyph A002 (U+13001) needs two `System.Char` values (known as surrogate pairs) to represent it: `\uD80C` and `\uDC01`. Do not always assume one `char` equals one letter or you could introduce hard-to-notice bugs into your code.

A `char` is assigned using single quotes around the literal value, or assigning the return value of a function call, as shown in the following code:

```
char letter = 'A'; // Assigning literal characters.
char digit = '1';
char symbol = '$';
char userChoice = GetChar(); // Assigning from a fictitious function.
```

For text, multiple letters, such as `Bob`, are stored as a `string` type and are assigned using double quotes around the literal value, or by assigning the return value of a function call or constructor, as shown in the following code:

```
string firstName = "Bob"; // Assigning literal strings.
string lastName = "Smith";
string phoneNumber = "(215) 555-4256";

// Assigning a string returned from the string class constructor.
string horizontalLine = new('-', count: 74); // 74 hyphens.

// Assigning a string returned from a fictitious function.
string address = GetAddressFromDatabase(id: 563);

// Assigning an emoji by converting from Unicode.
string grinningEmoji = char.ConvertFromUtf32(0x1F600);
```

Outputting emojis

To output emojis at a command prompt on Windows, you must use Windows Terminal because Command Prompt does not support emojis, and set the output encoding of the console to use UTF-8, as shown in the following code:

```
Console.OutputEncoding = System.Text.Encoding.UTF8;
string grinningEmoji = char.ConvertFromUtf32(0x1F600);
Console.WriteLine(grinningEmoji);
```

Verbatim strings

An escape character is a special character in programming and text processing that is used to introduce an **escape sequence**. Escape sequences allow the representation of characters that are difficult or impossible to include directly in a string. These sequences often start with a backslash, \, followed by one or more characters.

When storing text in a `string` variable, you can include escape sequences, which represent special characters like tabs and newlines using a backslash, as shown in the following code:

```
string fullNameWithTabSeparator = "Bob\tSmith";
```

But what if you are storing the path to a file on Windows, and one of the folder names starts with a `T`, as shown in the following code?

```
string filePath = "C:\televisions\sony\bravia.txt";
```

The compiler will convert the `\t` into a tab character and you will get errors!

You must prefix it with the `@` symbol to use a verbatim literal `string`, as shown in the following code:

```
string filePath = @"C:\televisions\sony\bravia.txt";
```

With C# 13 and later, you can represent the ESC character (Unicode U+001B) using the character literal escape sequence `\e`, as shown in the following code:

```
// C# 13 or later.
char esc = '\e';
// C# 12 or earlier.
char esc = '\u001b';
```

In the past, you might have used `\u001b` or `\x1b`. However, using `\x1b` was not recommended because any valid hexadecimal digits following `1b` would be incorrectly interpreted as part of the escape sequence.

Raw string literals

Introduced in C# 11, raw string literals are convenient for entering any arbitrary text without needing to escape the contents. They make it easy to define literals containing other languages like XML, HTML, or JSON.

Raw string literals start and end with three or more double-quote characters, as shown in the following code:

```
string xml = """
             <person age="50">
               <first_name>Mark</first_name>
             </person>
             """;
```

Why three *or more* double-quote characters? This is for scenarios where the content itself needs to have three double-quote characters; you can then use four double-quote characters to indicate the beginning and end of the content. Where the content needs to have four double-quote characters, you can then use five double-quote characters to indicate the beginning and end of the content. And so on.

In the previous code, the XML is indented by 13 spaces. The compiler looks at the indentation of the last three or more double-quote characters, and then automatically removes that level of indentation from all the content inside the raw string literal. The results of the previous code would therefore not be indented as in the defining code but, instead, will be aligned with the left margin, as shown in the following markup:

```
<person age="50">
  <first_name>Mark</first_name>
</person>
```

If the end three double-quote characters are aligned with the left margin, as shown in the following code:

```
string xml = """
             <person age="50">
               <first_name>Mark</first_name>
             </person>
""";
```

then the 13-space indentation would not be removed, as shown in the following markup:

```
             <person age="50">
               <first_name>Mark</first_name>
             </person>
```

Raw interpolated string literals

You can mix interpolated strings that use curly braces { } with raw string literals. You specify the number of braces that indicate a replaced expression by adding that number of dollar signs to the start of the literal. Any fewer braces than that are treated as raw content.

For example, if we want to define some JSON, single braces will be treated as normal braces, but the two dollar symbols tell the compiler that any two curly braces indicate a replaced expression value, as shown in the following code:

```
var person = new { FirstName = "Alice", Age = 56 };

string json = $$"""
              {
                "first_name": "{{person.FirstName}}",
                "age": {{person.Age}},
                "calculation": "{{{ 1 + 2 }}}"
              }
              """;

Console.WriteLine(json);
```

The previous code would generate the following JSON document:

```
{
  "first_name": "Alice",
  "age": 56,
  "calculation": "{3}"
}
```

The number of dollar symbols tells the compiler how many curly braces are needed to become recognized as an interpolated expression.

Note the braces {} around the 3. This is intentional. In this example, the JSON document must generate a calculation that contains braces. To show this, the code uses three braces: the first open brace will output as a literal character. The next two braces will be interpreted as the beginning of an expression. The first two close braces will be interpreted as the end of an expression. The last close brace will be a literal character.

If the code only used two braces, then those are treated as a delimiter for the expression `1 + 2` and do not appear in the output:

```
var person = new { FirstName = "Alice", Age = 56 };

string json = $$"""
              {
                "first_name": "{{person.FirstName}}",
                "age": {{person.Age}},
                "calculation": "{{ 1 + 2 }}"
              }
```

```
                """;

Console.WriteLine(json);
```

Now it produces the following output:

```
{
  "first_name": "Alice",
  "age": 56,
  "calculation": "3"
}
```

Summarizing options for storing text

To summarize:

- **Literal string:** Characters enclosed in double-quote characters. They can use escape characters like `\t` for tab. To represent a backslash, use two: `\\`.
- **Raw string literal:** Characters enclosed in three or more double-quote characters.
- **Verbatim string:** A literal string prefixed with `@` to disable escape characters so that a backslash is a backslash. It also allows the `string` value to span multiple lines because the whitespace characters are treated as themselves instead of instructions to the compiler.
- **Interpolated string:** A literal string prefixed with `$` to enable embedded formatted variables. You will learn more about this later in this chapter.

Storing numbers

Numbers are data that we want to perform an arithmetic calculation on, for example, multiplying. A telephone number is not a quantifiable number. To decide whether a variable should be stored as a number or not, ask yourself whether you need to perform arithmetic operations on the number or whether the number includes non-digit characters such as parentheses or hyphens to format the number, such as (414) 555-1234. In this case, the "number" is a sequence of characters, so it should be stored as a `string`.

Numbers can be natural numbers, such as 42, used for counting (also called whole numbers); they can also be negative numbers, such as -42 (called integers); or they can be real numbers, such as 3.9 (with a fractional part), which are called single- or double-precision floating-point numbers in computing.

Let's explore numbers:

1. Use your preferred code editor to add a new **Console App** / `console` project named `Numbers` to the `Chapter02` solution:
    - For Visual Studio, configure the startup project to the current selection.

2. In `Program.cs`, delete the existing code, and then type statements to declare some number variables using various data types, as shown in the following code:

```
// An unsigned integer is a positive whole number or 0.
uint naturalNumber = 23;
// An integer is a negative or positive whole number or 0.
int integerNumber = -23;

// A float is a single-precision floating-point number.
// The F or f suffix makes the value a float literal.
// The suffix is required to compile.
float realNumber = 2.3f;

// A double is a double-precision floating-point number.
// double is the default for a number value with a decimal point.
double anotherRealNumber = 2.3; // A double literal value.
```

Storing whole numbers

You might know that computers store everything as bits. The value of a bit is either `0` or `1`. This is called a **binary number system**. Humans use a **decimal number system**.

The decimal number system, also known as Base 10, has 10 as its **base**, meaning there are 10 digits, from 0 to 9. Although it is the number base most used by human civilizations, other number base systems are popular in science, engineering, and computing. The binary number system, also known as Base 2, has 2 as its base, meaning there are two digits, 0 and 1.

The following image shows how computers store the decimal number 10. Take note of the bits with the value 1 in the 8 and 2 columns: 8 + 2 = 10.

128	64	32	16	8	4	2	1
0	0	0	0	1	0	1	0

Figure 2.4: How computers store the decimal number 10

So, `10` in decimal is `00001010` in a binary byte (8 bits).

Improving legibility by using digit separators

Two of the improvements seen in C# 7 and later are the use of the underscore character _ as a digit separator and support for binary literals.

You can insert underscores anywhere into the digits of a number literal, including decimal, binary, or hexadecimal notation, to improve legibility.

For example, you could write the value for 1 million in decimal notation, that is, Base 10, as `1_000_000`.

You can even use the 2/3 grouping common in India: `10_00_000`.

Using binary or hexadecimal notation

To use binary notation, that is, Base 2, using only 1s and 0s, start the number literal with `0b`. To use hexadecimal notation, that is, Base 16, using 0 to 9 and A to F, start the number literal with `0x`.

Exploring whole numbers

Let's enter some code to see some examples:

1. In the `Numbers` project, in `Program.cs`, type statements to declare some number variables using underscore separators, as shown in the following code:

```
int decimalNotation = 2_000_000;
int binaryNotation = 0b_0001_1110_1000_0100_1000_0000;
int hexadecimalNotation = 0x_001E_8480;

// Check the three variables have the same value.
Console.WriteLine($"{decimalNotation == binaryNotation}");
Console.WriteLine(
  $"{decimalNotation == hexadecimalNotation}");

// Output the variable values in decimal.
Console.WriteLine($"{decimalNotation:N0}");
Console.WriteLine($"{binaryNotation:N0}");
Console.WriteLine($"{hexadecimalNotation:N0}");

// Output the variable values in hexadecimal.
Console.WriteLine($"{decimalNotation:X}");
Console.WriteLine($"{binaryNotation:X}");
Console.WriteLine($"{hexadecimalNotation:X}");
```

2. Run the project and note the result is that all three numbers are the same, as shown in the following output:

```
True
True
2,000,000
2,000,000
2,000,000
1E8480
1E8480
1E8480
```

Computers can always exactly represent integers using the `int` type or one of its sibling types, such as `long` and `short`.

Storing real numbers

Computers cannot always represent real, aka decimal or non-integer, numbers precisely. The `float` and `double` types store real numbers using single- and double-precision floating points.

Most programming languages implement the **Institute of Electrical and Electronics Engineers (IEEE)** Standard for Floating-Point Arithmetic. IEEE 754 is a technical standard for floating-point arithmetic established in 1985 by the IEEE.

The following image shows a simplification of how a computer represents the number `12.75` in binary notation. Note the bits with the value `1` in the 8, 4, ½, and ¼ columns:

8 + 4 + ½ + ¼ = 12¾ = 12.75

128	64	32	16	8	4	2	1	.	1/2	1/4	1/8	1/16
0	0	0	0	1	1	0	0		1	1	0	0

Figure 2.5: Computer representing the number 12.75 in binary notation

So, `12.75` in decimal notation is `00001100.1100` in binary notation. As you can see, the number `12.75` can be exactly represented using bits. However, most numbers can't, which is something that we'll be exploring shortly.

Writing code to explore number sizes

C# has an operator named `sizeof()` that returns the number of bytes that a type uses in memory. Some types have members named `MinValue` and `MaxValue`, which return the minimum and maximum values that can be stored in a variable of that type. We are now going to use these features to create a console app to explore number types:

1. In `Program.cs`, at the bottom of the file, type statements to show the size of three number data types, as shown in the following code:

```
Console.WriteLine($"int uses {sizeof(int)} bytes and can store numbers in
the range {int.MinValue:N0} to {int.MaxValue:N0}.");
Console.WriteLine($"double uses {sizeof(double)} bytes and can store
numbers in the range {double.MinValue:N0} to {double.MaxValue:N0}.");
Console.WriteLine($"decimal uses {sizeof(decimal)} bytes and can store
numbers in the range {decimal.MinValue:N0} to {decimal.MaxValue:N0}.");
```

Warning! The width of the printed pages in this book makes the `string` values (in double quotes) wrap over multiple lines. You must type them on a single line, or you will get compile errors.

2. Run the code and view the output, as shown in *Figure 2.6*:

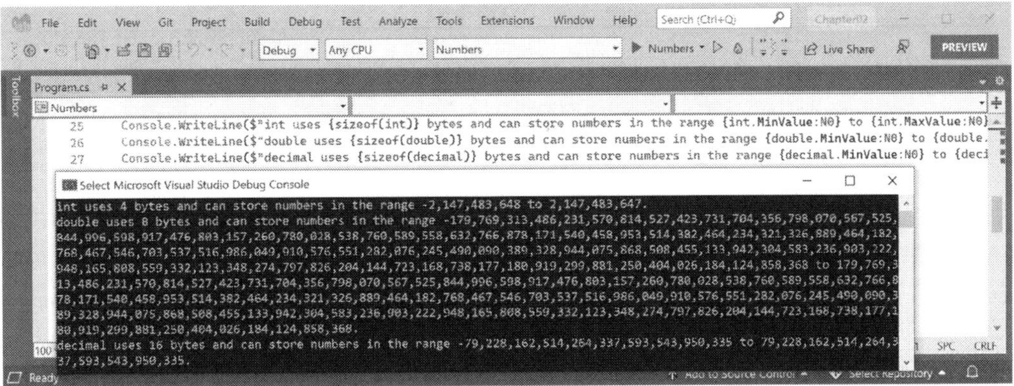

Figure 2.6: Size and range information for common number data types

An `int` variable uses 4 bytes of memory and can store positive or negative numbers up to about 2 billion. A `double` variable uses 8 bytes of memory and can store much bigger values! A `decimal` variable uses 16 bytes of memory and can store big numbers, but not as big as a `double` type.

But you may be asking yourself, why might a `double` variable be able to store bigger numbers than a `decimal` variable, yet it's only using half the space in memory? Well, let's now find out!

Comparing double and decimal types

You will now write some code to compare `double` and `decimal` values. Although it isn't hard to follow, don't worry about understanding the syntax right now:

1. Type statements to declare two `double` variables, add them together, and compare them to the expected result. Then, write the result to the console, as shown in the following code:

```
Console.WriteLine("Using doubles:");
double a = 0.1;
double b = 0.2;
if (a + b == 0.3)
{
  Console.WriteLine($"{a} + {b} equals {0.3}");
}
else
{
  Console.WriteLine($"{a} + {b} does NOT equal {0.3}");
}
```

2. Run the code and view the result, as shown in the following output:

```
Using doubles:
0.1 + 0.2 does NOT equal 0.3
```

In cultures that use a comma for the decimal separator, the result will look slightly different, as shown in the following output: `0,1 + 0,2 does NOT equal 0,3.`

The `double` type is not guaranteed to be accurate because most numbers like `0.1`, `0.2`, and `0.3` literally cannot be exactly represented as floating-point values.

If you were to try different values, like `0.1 + 0.3 == 0.4`, it would happen to return `true` because, with `double` values, some imprecise values happen to be exactly equal in their current representation even though they might not actually be equal mathematically. So, some numbers can be directly compared but some cannot. I deliberately picked `0.1` and `0.2` to compare to `0.3` because they cannot be compared, as proven by the result.

You could compare real numbers stored in the `float` type, which is less accurate than the `double` type, but the comparison would actually appear to be `true` because of that lower accuracy!

```
float a = 0.1F;
float b = 0.2F;
if (a + b == 0.3F) // True because float is less "accurate" than double.
...
```

As a rule of thumb, you should only use `double` when accuracy, especially when comparing the equality of two numbers, is not important. An example of this might be when you're measuring a person's height; you will only compare values using greater than or less than, but never equals.

The problem with the preceding code is illustrated by how the computer stores the number `0.1`, or multiples of it. To represent `0.1` in binary notation, the computer stores 1 in the 1/16 column, 1 in the 1/32 column, 1 in the 1/256 column, 1 in the 1/512 column, and so on.

The number `0.1` in decimal notation is `0.00011001100110011`... in binary notation, repeating forever, as shown in *Figure 2.7*:

4	2	1	.	1/2	1/4	1/8	1/16	1/32	1/64	1/128	1/256	1/512	1/1024	1/2048
0	0	0		0	0	0	1	1	0	0	1	1	0	0

Figure 2.7: Number 0.1 in decimal notation repeating forever in binary notation

Good Practice: Never compare `double` values using `==`. During the Gulf War of 1990-1991, an American Patriot missile battery used `double` values in its calculations. The inaccuracy caused it to fail to track and intercept an incoming Iraqi Scud missile, and 28 soldiers were killed; you can read about this at `https://www.ima.umn.edu/~arnold/disasters/patriot.html`. The Patriot missile system has improved since then. *"Forty years after it was brought into service, the Patriot air-defense system is finally doing what it was designed for." "No one was 100% sure that the Patriot was capable of destroying a Kh-47 hypersonic missile," said Col. Serhiy Yaremenko, commander of the 96th Anti-Aircraft Missile Brigade, which defends Kyiv. "Ukrainians proved it."*: `https://archive.ph/2023.06.11-132200/https://www.wsj.com/amp/articles/u-s-patriot-missile-is-an-unsung-hero-of-ukraine-war-db6053a0`.

Now let's see the same code using the `decimal` number type:

1. Copy and paste the statements that you wrote before (which used the `double` variables).
2. Modify the statements to use `decimal` and rename the variables to `c` and `d`, as shown in the following code:

```
Console.WriteLine("Using decimals:");
decimal c = 0.1M; // M suffix means a decimal literal value
decimal d = 0.2M;

if (c + d == 0.3M)
{
  Console.WriteLine($"{c} + {d} equals {0.3M}");
}
else
{
  Console.WriteLine($"{c} + {d} does NOT equal {0.3M}");
}
```

3. Run the code and view the result, as shown in the following output:

```
Using decimals:
0.1 + 0.2 equals 0.3
```

The `decimal` type is accurate because it stores the number as a large integer and shifts the decimal point. For example, `0.1` is stored as `1`, with a note to shift the decimal point one place to the left. `12.75` is stored as `1275`, with a note to shift the decimal point two places to the left.

Good Practice: Use `int` for whole numbers. Use `double` for real numbers that will not be compared for equality to other values; it is okay to compare `double` values being less than or greater than, and so on. Use `decimal` for money, CAD drawings, general engineering, and wherever the accuracy of a real number is important.

Special real number values

The float and double types have some useful special values: NaN represents not-a-number (for example, the result of dividing zero by zero), Epsilon represents the smallest positive number that can be stored in a float or double, and PositiveInfinity and NegativeInfinity represent infinitely large positive and negative values. They also have methods for checking for these special values like IsInfinity and IsNaN.

Let's write some code statements to see what some of these special values look like:

1. Add statements to output some special double values, as shown in the following code:

```
#region Special float and double values

Console.WriteLine($"double.Epsilon: {double.Epsilon}");
Console.WriteLine($"double.Epsilon to 324 decimal places: {double.
Epsilon:N324}");
Console.WriteLine($"double.Epsilon to 330 decimal places: {double.
Epsilon:N330}");

const int col1 = 37; // First column width.
const int col2 = 6; // Second column width.
string line = new string('-', col1 + col2 + 3);

Console.WriteLine(line);
Console.WriteLine($"{"Expression",-col1} | {"Value",col2}");
Console.WriteLine(line);
Console.WriteLine($"{"double.NaN",-col1} | {double.NaN,col2}");
Console.WriteLine($"{"double.PositiveInfinity",-col1} | {double.
PositiveInfinity,col2}");
Console.WriteLine($"{"double.NegativeInfinity",-col1} | {double.
NegativeInfinity,col2}");
Console.WriteLine(line);
Console.WriteLine($"{"0.0 / 0.0",-col1} | {0.0 / 0.0,col2}");
Console.WriteLine($"{"3.0 / 0.0",-col1} | {3.0 / 0.0,col2}");
Console.WriteLine($"{"-3.0 / 0.0",-col1} | {-3.0 / 0.0,col2}");
Console.WriteLine($"{"3.0 / 0.0 == double.PositiveInfinity",-col1} | {3.0
/ 0.0 == double.PositiveInfinity,col2}");
Console.WriteLine($"{"-3.0 / 0.0 == double.NegativeInfinity",-col1} |
{-3.0 / 0.0 == double.NegativeInfinity,col2}");
Console.WriteLine($"{"0.0 / 3.0",-col1} | {0.0 / 3.0,col2}");
Console.WriteLine($"{"0.0 / -3.0",-col1} | {0.0 / -3.0,col2}");
Console.WriteLine(line);

#endregion
```

2. Run the code and view the result, as shown in the following output:

```
double.Epsilon: 5E-324
double.Epsilon to 324 decimal places: 0.000000000000000000000000000000000
0000000000000000000000000000000000000000000000000000000000000000000000000
0000000000000000000000000000000000000000000000000000000000000000000000000
0000000000000000000000000000000000000000000000000000000000000000000000000
000000000000000000000000000000000000000000000000000000000000000000000005
double.Epsilon to 330 decimal places: 0.000000000000000000000000000000000
0000000000000000000000000000000000000000000000000000000000000000000000000
000000000000000000000000000000000000000000000000000000000000000000000000
000000000000000000000000000000000000000000000000000000000000000000000000
000000000000000000000000000000000000000000000000000000000000000000000000
04940656
----------------------------------------------
Expression                            |  Value
----------------------------------------------
double.NaN                            |    NaN
double.PositiveInfinity               |      8
double.NegativeInfinity               |     -8
----------------------------------------------
0.0 / 0.0                             |    NaN
3.0 / 0.0                             |      8
-3.0 / 0.0                            |     -8
3.0 / 0.0 == double.PositiveInfinity  |   True
-3.0 / 0.0 == double.NegativeInfinity |   True
0.0 / 3.0                             |      0
0.0 / -3.0                            |     -0
----------------------------------------------
```

Note the following:

- `NaN` outputs as `NaN`. (`NaN` means "not a number".) It can be generated from an expression of zero divided by zero.
- The `PositiveInfinity` value outputs as an `8`, which looks like an infinity symbol on its side. It can be generated from an expression of any positive real number divided by zero.
- The `NegativeInfinity` value outputs as `-8`, which looks like an infinity symbol on its side with a negative sign before it. It can be generated from an expression of any negative real number divided by zero.
- Zero divided by any positive real number is zero.
- Zero divided by any negative real number is negative zero.
- `Epsilon` is slightly less than `5E-324` represented using scientific notation: `https://en.wikipedia.org/wiki/Scientific_notation`.

New number types and unsafe code

The `System.Half` type was introduced in .NET 5. Like `float` and `double`, it can store real numbers. It normally uses two bytes of memory. The `System.Int128` and `System.UInt128` types were introduced in .NET 7. Like `int` and `uint`, they can store signed (positive and negative) and unsigned (only zero and positive) integer values. They normally use 16 bytes of memory.

For these new number types, the `sizeof` operator only works in an unsafe code block, and you must compile the project using an option to enable unsafe code. Let's explore how this works:

1. In `Program.cs`, at the bottom of the file, type statements to show the size of the `Half` and `Int128` number data types, as shown in the following code:

```
unsafe
{
  Console.WriteLine($"Half uses {sizeof(Half)} bytes and can store
numbers in the range {Half.MinValue:N0} to {Half.MaxValue:N0}.");
  Console.WriteLine($"Int128 uses {sizeof(Int128)} bytes and can store
numbers in the range {Int128.MinValue:N0} to {Int128.MaxValue:N0}.");
}
```

2. In `Numbers.csproj`, add an element to enable unsafe code, as shown highlighted in the following markup:

```
<PropertyGroup>
  <OutputType>Exe</OutputType>
  <TargetFramework>net9.0</TargetFramework>
  <ImplicitUsings>enable</ImplicitUsings>
  <Nullable>enable</Nullable>
  <AllowUnsafeBlocks>True</AllowUnsafeBlocks>
</PropertyGroup>
```

3. Run the `Numbers` project and note the sizes of the two new number types, as shown in the following output:

```
Half uses 2 bytes and can store numbers in the range -65,504 to 65,504.
Int128 uses 16 bytes and can store numbers in the range -170,141,183,460,
469,231,731,687,303,715,884,105,728 to 170,141,183,460,469,231,731,687,
303,715,884,105,727.
```

More Information: The `sizeof` operator requires an unsafe code block, except for the commonly used types like `int` and `byte`. You can learn more about `sizeof` at the following link: https://learn.microsoft.com/en-us/dotnet/csharp/language-reference/operators/sizeof. Unsafe code cannot have its safety verified. You can learn more about unsafe code blocks at the following link: https://learn.microsoft.com/en-us/dotnet/csharp/language-reference/unsafe-code.

Storing Booleans

Booleans can only contain one of the two literal values `true` or `false`, as shown in the following code:

```
bool happy = true;
bool sad = false;
```

They are most used to branch and loop. You don't need to fully understand them yet, as they are covered more in *Chapter 3, Controlling Flow, Converting Types, and Handling Exceptions*.

Storing any type of object

There is a special type named `object` that can store any type of data, but its flexibility comes at the cost of messier code and possibly poor performance. For those two reasons, you should avoid it whenever possible. The following steps show you how to use object types if you need to use them because you must use a Microsoft or third-party library that uses them:

1. Use your preferred code editor to add a new **Console App** / `console` project named `Variables` to the `Chapter02` solution.
2. In `Program.cs`, delete the existing statements and then type statements to declare and use some variables using the `object` type, as shown in the following code:

```
object height = 1.88; // Storing a double in an object.
object name = "Amir"; // Storing a string in an object.
Console.WriteLine($"{name} is {height} metres tall.");

int length1 = name.Length; // This gives a compile error!
int length2 = ((string)name).Length; // Cast name to a string.
Console.WriteLine($"{name} has {length2} characters.");
```

3. Run the code and note that the fourth statement cannot compile because the data type of the `name` variable is not known by the compiler, as shown in *Figure 2.8*:

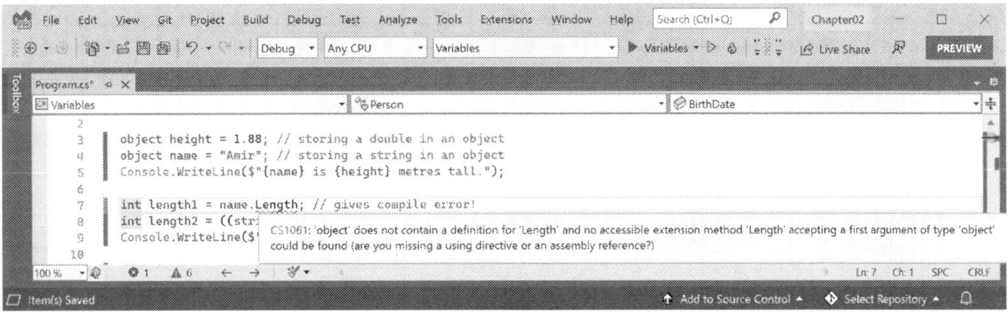

Figure 2.8: The object type does not have a Length property

4. Add double slashes to the beginning of the statement that cannot compile to comment out the statement, making it inactive.

5. Run the code again and note that the compiler can access the `length` of a `string` if the programmer explicitly tells the compiler that the `object` variable contains a `string` by prefixing the `name` variable with a cast expression like `(string)name`. The results can then successfully be written to the console, as shown in the following output:

```
Amir is 1.88 meters tall.
Amir has 4 characters.
```

You will learn about cast expressions in *Chapter 3, Controlling Flow, Converting Types, and Handling Exceptions*.

The `object` type has been available since the first version of C#, but C# 2 and later have a better alternative called **generics**, which we will cover in *Chapter 6, Implementing Interfaces and Inheriting Classes*. This will provide us with the flexibility we want but without the performance overhead.

Good Practice: Avoid declaring variables as a `System.Object`, aka `object` type. There are usually better alternatives that you can use instead, like generics.

Storing dynamic types

There is another special type named `dynamic` that can also store any type of data, but even more than `object`, its flexibility comes at the cost of performance. The `dynamic` keyword was introduced in C# 4. However, unlike `object`, the value stored in the variable can have its members invoked without an explicit cast. Let's make use of a `dynamic` type:

1. Add statements to declare a `dynamic` variable. Assign a `string` literal value, and then an integer value, and then an array of integer values. Finally, add a statement to output the length of the `dynamic` variable, as shown in the following code:

```
dynamic something;

// Storing an array of int values in a dynamic object.
// An array of any type has a Length property.
something = new[] { 3, 5, 7 };
// Storing an int in a dynamic object.
// int does not have a Length property.
something = 12;

// Storing a string in a dynamic object.
// string has a Length property.
something = "Ahmed";

// This compiles but might throw an exception at run-time.
```

```
Console.WriteLine($"The length of something is {something.Length}");

// Output the type of the something variable.
Console.WriteLine($"something is a {something.GetType()}");
```

You will learn about arrays in *Chapter 3, Controlling Flow, Converting Types, and Handling Exceptions.*

2. Run the code and note that it works because the last value assigned to something was a `string` value that does have a `Length` property, as shown in the following output:

```
The length of something is 5
something is a System.String
```

3. Comment out the statement that assigns a `string` value to the `something` variable by prefixing the statement with two slashes, `//`.
4. Run the code and note the runtime error because the last value assigned to `something` is an `int` that does not have a `Length` property, as shown in the following output:

```
Unhandled exception. Microsoft.CSharp.RuntimeBinder.
RuntimeBinderException: 'int' does not contain a definition for 'Length'
```

5. Comment out the statement that assigns an `int` to the `something` variable.
6. Run the code and note the output because an array of three `int` values does have a `Length` property, as shown in the following output:

```
The length of something is 3
something is a System.Int32[]
```

One limitation of `dynamic` is that code editors cannot show IntelliSense to help you write the code. This is because the compiler cannot check what the type is during build time. Instead, the **CLR (Common Language Runtime)** checks for the member at runtime and throws an exception if it is missing.

Exceptions are a way to indicate that something has gone wrong at runtime. You will learn more about them and how to handle them in *Chapter 3, Controlling Flow, Converting Types, and Handling Exceptions.*

Dynamic types are most useful when interoperating with non-.NET systems. For example, you might need to work with a class library written in F#, Python, or some JavaScript. You might also need to interop with technologies like the **Component Object Model** (**COM**), for example, when automating Excel or Word.

Declaring local variables

Local variables are declared inside methods, and they only exist during the execution of that method. Once the method returns, the memory allocated to any local variables is released.

Strictly speaking, value types are released while reference types must wait for garbage collection. You will learn about the difference between value types and reference types and how to make sure that only one garbage collection is needed rather than two when releasing unmanaged resources in *Chapter 6, Implementing Interfaces and Inheriting Classes*.

Specifying the type of a local variable

Let's explore local variables declared with specific types and using type inference.

Type statements to declare and assign values to some local variables using specific types, as shown in the following code:

```
int population = 67_000_000; // 67 million in UK.
double weight = 1.88; // in kilograms.
decimal price = 4.99M; // in pounds sterling.
string fruit = "Apples"; // string values use double-quotes.
char letter = 'Z'; // char values use single-quotes.
bool happy = true; // Booleans can only be true or false.
```

Depending on your code editor and color scheme, it will show green squiggles under each of the variable names and lighten their text color to warn you that the variable is assigned but its value is never used.

Inferring the type of a local variable

You can use the `var` keyword to declare local variables with C# 3 and later. The compiler will infer the type from the value that you assign after the assignment operator, `=`. This happens at compile time so using `var` has no effect on runtime performance.

A literal number without a decimal point is inferred as an `int` variable, that is, unless you add a suffix, as described in the following list:

- `L`: Compiler infers `long`
- `UL`: Compiler infers `ulong`
- `M`: Compiler infers `decimal`
- `D`: Compiler infers `double`
- `F`: Compiler infers `float`

A literal number with a decimal point is inferred as `double` unless you add the `M` suffix (in which case the compiler infers a `decimal` variable), or the `F` suffix, in which case it infers a `float` variable.

Double quotes indicate a `string` variable, single quotes indicate a `char` variable, and the `true` and `false` values infer a `bool` type:

1. Modify the previous statements to use `var`, as shown in the following code:

```
var population = 67_000_000; // 67 million in UK.
var weight = 1.88; // in kilograms.
```

```
var price = 4.99M; // in pounds sterling.
var fruit = "Apples"; // string values use double-quotes.
var letter = 'Z'; // char values use single-quotes.
var happy = true; // Booleans can only be true or false.
```

2. Hover your mouse over each of the `var` keywords and note that your code editor shows a tooltip with information about the type that has been inferred.
3. At the top of `Program.cs`, import the namespace for working with XML to enable us to declare some variables using types in that namespace, as shown in the following code:

```
using System.Xml; // To use XmlDocument.
```

4. At the bottom of `Program.cs`, add statements to create some new objects, as shown in the following code:

```
// Good use of var because it avoids the repeated type
// as shown in the more verbose second statement.
var xml1 = new XmlDocument(); // Works with C# 3 and later.
XmlDocument xml2 = new XmlDocument(); // Works with all C# versions.

// Bad use of var because we cannot tell the type, so we
// should use a specific type declaration as shown in
// the second statement.
var file1 = File.CreateText("something1.txt");
StreamWriter file2 = File.CreateText("something2.txt");
```

Good Practice: Although using `var` is convenient, some developers avoid using it to make it easier for a code reader to understand the types in use. Personally, I use it only when the type is obvious. For example, in the preceding code statements, the first statement is just as clear as the second in stating what the types of the `xml` variables are, but it is shorter. However, the third statement isn't clear in showing the type of the `file` variable, so the fourth is better because it shows that the type is `StreamWriter`. If in doubt, spell it out!

Note that `var` is converted to the actual type by the compiler when you build a project. A variable declared using `var` has a specific, known, fixed data type. This is different from `dynamic`, which the compiler does not change. It remains a `System.Dynamic` type that can reference any object of any data type. The actual type is only checked at runtime, so that's when an exception might be thrown if you access a member on it that does not exist.

You can use a Visual Studio refactoring feature to automatically convert uses of `var` into the specific type, as documented at the following link: `https://learn.microsoft.com/en-us/visualstudio/ide/reference/convert-var-to-explicit-type`.

What does new do?

There have been a few examples of using the C# `new` keyword before this section, but so far, I haven't explicitly explained what it does. The C# `new` keyword is used to allocate and/or initialize memory. To understand when you need to use `new`, you need to know a bit more about types.

Value and reference types and their relationship to memory are explained in more detail in *Chapter 6, Implementing Interfaces and Inheriting Classes*, so I will only introduce the minimum explanation for now.

There are two categories of types: **value types** and **reference types**.

Value types are simple and do not need to use the `new` keyword to explicitly allocate memory. But value types can use the `new` keyword to initialize their value. This is useful when there is no way to use a literal to set the value.

Reference types are more complex and need to use the `new` keyword to explicitly allocate memory. At the same time, they can use the `new` keyword to initialize their state.

For example, when you declare variables, space is only allocated in memory for value types like `int` and `DateTime` but not for reference types like `Person`.

Consider the following code that declares some local variables:

```
/* Value types have memory allocated on the stack automatically. */

short age; // Allocates 2 bytes of memory on the stack to store a System.Int16 value.
long population; // Allocates 8 bytes of memory on the stack to store a System.Int64 value.
DateTime birthdate; // Allocates 8 bytes of memory on the stack to store a System.DateTime value.
Point location; // Allocate 8 bytes of memory on the stack to store a System.Drawing.Point value.

/* Reference types will only have memory allocated on the heap when new is used (but they automatically have some memory allocated on the stack to store information about themselves including the memory address of where they are on the heap). */

Person bob; // Allocates memory in the stack that can point to a Person object in the heap. Initially, bob will have the value null.
```

Note the following about the preceding code:

- age has a value of 0 and 2 bytes of memory have been allocated in stack memory.
- population has a value of 0 and 8 bytes of memory have been allocated in stack memory.
- birthdate has a value of 0001-01-01 and 8 bytes of memory have been allocated in stack memory.
- location has a value of 0, 0 and 8 bytes of memory have been allocated in stack memory.
- bob has a value of null and 4 bytes of memory have been allocated in stack memory. No heap memory has been allocated for the object.

Now let's see when we might choose to use new:

```
age = 45; // Initialize this variable to 45 using a literal value.
population = 68_000_000; // Initialize this variable to 68 million using a
literal value.
birthdate = new(1995, 2, 23); // Initialize this variable to February 23, 1995.
C# does not support literal values for date/time values so we must use new.
location = new(10, 20); // Initialize the X and Y coordinates of this value
type.
bob = new(); // Allocate memory on the heap to store a Person. Any state will
have default values. bob is no longer null.
bob = new("Bob", "Smith", 45); // Allocate memory on the heap to store a Person
and initialize state. bob is no longer null.

// Older syntax with explicit types
birthdate = new DateTime(1995, 2, 23); // Initialize this variable to February
23, 1995.
location = new Point(10, 20); // Initialize the X and Y coordinates of this
value type.
bob = new Person(); // Allocate memory on the heap to store a Person. bob is no
longer null.
bob = new Person("Bob", "Smith", 45); // Allocate memory on the heap to store a
Person and initialize state. bob is no longer null.
```

Note the following about the preceding code:

- age, population, birthdate, and location have already had memory allocated for them on the stack. We only need to use new to initialize their values if we want them to be different from their defaults.
- bob must use new to allocate heap memory for the object. The = assignment stores the memory address of that allocated memory on the stack. Reference types like Person often have multiple constructors that are called by new. A default constructor assigns default values to any state in the object. A constructor with arguments can assign other values to any state in the object.

Constructors are covered in more detail in *Chapter 5, Building Your Own Types with Object-Oriented Programming*, so I have only introduced the minimum explanation for now.

Using target-typed new to instantiate objects

With C# 9, Microsoft introduced another syntax for instantiating objects, known as **target-typed new**. When instantiating an object, you can specify the type first and then use `new` without repeating the type, as shown in the following code:

```
XmlDocument xml3 = new(); // Target-typed new in C# 9 or later.
```

If you have a type with a field or property that needs to be set, then the type can be inferred, as shown in the following code:

```
// In Program.cs.
Person kim = new();
kim.BirthDate = new(1967, 12, 26); // i.e. new DateTime(1967, 12, 26)

// In a separate Person.cs file or at the bottom of Program.cs.
class Person
{
  public DateTime BirthDate;
}
```

This way of instantiating objects is especially useful with arrays and collections because they have multiple objects, often of the same type, as shown in the following code:

```
List<Person> people = new() // Instead of: new List<Person>()
{
  new() { FirstName = "Alice" }, // Instead of: new Person() { ... }
  new() { FirstName = "Bob" },
  new() { FirstName = "Charlie" }
};
```

You will learn about arrays in *Chapter 3, Controlling Flow, Converting Types, and Handling Exceptions*, and collections in *Chapter 8, Working with Common .NET Types*.

Good Practice: Use target-typed new to instantiate objects because it requires fewer characters; when reading a statement from left to right, as in English, you immediately know the type of the variable, and it is not limited to local variables like `var` is, in my opinion, the only reason not to use target-typed new is if you must use a pre-version 9 C# compiler. I do acknowledge that my opinion is not accepted by the whole C# community. I have used target-typed new throughout the remainder of this book. Please let me know if you spot any cases that I missed! There is a Visual Studio refactoring feature to use target-typed new, as documented at the following link: `https://learn.microsoft.com/en-us/visualstudio/ide/reference/use-new`.

Getting and setting the default values for types

Most of the primitive types except `string` are **value types**, which means that they must have a value. You can determine the default value of a type by using the `default()` operator and passing the type as a parameter. You can assign the default value of a type by using the `default` keyword.

The `string` type is a **reference type**. This means that `string` variables contain the memory address of a value, not the value itself. A reference type variable can have a `null` value, which is a literal that indicates that the variable does not reference anything (yet). `null` is the default for all reference types.

You'll learn more about value types and reference types in *Chapter 6, Implementing Interfaces and Inheriting Classes.*

Let's explore default values:

1. Add statements to show the default values of an `int`, a `bool`, a `DateTime`, and a `string`, as shown in the following code:

```
Console.WriteLine($"default(int) = {default(int)}");
Console.WriteLine($"default(bool) = {default(bool)}");
Console.WriteLine($"default(DateTime) = {
  default(DateTime)}");
Console.WriteLine($"default(string) = {
  default(string) ?? "<NULL>"}");
```

The `??` operator means, if `null`, then return the following instead. So if `default(string)` is `null`, then the text, `<NULL>` will be returned.

2. Run the code and view the result. Note that your output for the date and time might be formatted differently if you are not running it in the UK because date and time values are formatted using the current culture of your computer, as shown in the following output:

```
default(int) = 0
default(bool) = False
default(DateTime) = 01/01/0001 00:00:00
default(string) = <NULL>
```

3. Add statements to declare a number, assign a value, and then reset it to its default value, as shown in the following code:

```
int number = 13;
Console.WriteLine($"number set to: {number}");
number = default;
Console.WriteLine($"number reset to its default: {number}");
```

4. Run the code and view the result, as shown in the following output:

```
number set to: 13
number reset to its default: 0
```

Exploring more about console apps

We have already created and used basic console apps, but we're now at a stage where we should delve into them more deeply.

Console apps are text-based and are run at the command prompt. They typically perform simple tasks that need to be scripted, such as compiling a file or encrypting a section of a configuration file.

Equally, they can also have arguments passed to them to control their behavior.

An example of this would be to create a new console app using the F# language with a specified name instead of using the name of the current folder, as shown in the following command:

```
dotnet new console -lang "F#" --name "ExploringConsole"
```

Displaying output to the user

The two most common tasks that a console app performs are writing and reading data. We have already used the `WriteLine` method to output, but if we didn't want a carriage return at the end of a line, for example, if we later wanted to continue to write more text at the end of that line, we could have used the `Write` method.

If you want to write three letters to the console without carriage returns after them, then call the `Write` method, as shown in the following code:

```
Console.Write("A");
Console.Write("B");
Console.Write("C");
```

This will write the three characters on a single line and leave the cursor at the end of the line, as shown in the following output:

```
ABC
```

If you want to write three letters to the console with carriage returns after them, then call the `WriteLine` method, as shown in the following code:

```
Console.WriteLine("A");
Console.WriteLine("B");
Console.WriteLine("C");
```

This will write three lines and leave the cursor on the fourth line:

```
A
B
```

```
C
```

Formatting using numbered positional arguments

One way of generating formatted strings is to use numbered positional arguments.

This feature is supported by methods like `Write` and `WriteLine`. For methods that do not support the feature, the `string` parameter can be formatted using the `Format` method of `string`.

Let's begin formatting:

1. Use your preferred code editor to add a new **Console App** / `console` project named `Formatting` to the `Chapter02` solution.
2. In `Program.cs`, delete the existing statements and then type statements to declare some number variables and write them to the console, as shown in the following code:

```
using System.Globalization; // To use CultureInfo.

// Set current culture to US English so that all readers
// see the same output as shown in the book.
CultureInfo.CurrentCulture = CultureInfo.GetCultureInfo("en-US");

int numberOfApples = 12;
decimal pricePerApple = 0.35M;

Console.WriteLine(
  format: "{0} apples cost {1:C}",
  arg0: numberOfApples,
  arg1: pricePerApple * numberOfApples);

string formatted = string.Format(
  format: "{0} apples cost {1:C}",
  arg0: numberOfApples,
  arg1: pricePerApple * numberOfApples);

//WriteToFile(formatted); // Writes the string into a file.
```

Warning! We used the `CultureInfo` class to force the culture to US English so that the currency will be US dollars for all readers. In *Chapter 4, Writing, Debugging, and Testing Functions*, you will learn how to write a function to control the culture so that they can see US English by default, or override that to show your local computer culture or a specified culture.

The `Write`, `WriteLine`, and `Format` methods can have up to three numbered arguments, named `arg0`, `arg1`, and `arg2`. If you need to pass more than three values, then you cannot name them. The `WriteToFile` method is a nonexistent method used to illustrate the idea.

3. In `Program.cs`, type statements to write three and then five arguments to the console, as shown in the following code:

```
// Three parameter values can use named arguments.
Console.WriteLine("{0} {1} lived in {2}.",
  arg0: "Roger", arg1: "Cevung", arg2: "Stockholm");

// Four or more parameter values cannot use named arguments.
Console.WriteLine(
  "{0} {1} lived in {2} and worked in the {3} team at {4}.",
  "Roger", "Cevung", "Stockholm", "Education", "Optimizely");
```

Good Practice: Once you become more comfortable with formatting strings, you should stop naming the parameters, for example, stop using `format:`, `arg0:`, and `arg1:`. The preceding code uses a non-canonical style to show where the `0` and `1` came from while you are learning.

Rider and its warnings about boxing

If you use Rider and you have installed the Unity Support plugin, then it will complain a lot about boxing. A common scenario when boxing happens is when value types like `int` and `DateTime` are passed as positional arguments to `string` formats. This is a problem for Unity projects because they use a different memory garbage collector than the normal .NET runtime. For non-Unity projects, like all the projects in this book, you can ignore these boxing warnings because they are not relevant. You can read more about this Unity-specific issue at the following link: `https://docs.unity3d.com/Manual/performance-garbage-collection-best-practices.html#boxing`.

Formatting using interpolated strings

C# 6 and later have a handy feature named **interpolated strings**. A `string` prefixed with `$` can use curly braces around the name of a variable or expression to output the current value of that variable or expression at that position in the `string`, as the following shows:

1. Enter a statement at the bottom of the `Program.cs` file, as shown in the following code:

```
// The following statement must be all on one line when using C# 10
// or earlier. If using C# 11 or later, we can include a line break
// in the middle of an expression but not in the string text.
```

```
Console.WriteLine($"{numberOfApples} apples cost {pricePerApple
  * numberOfApples:C}");
```

2. Run the code and view the result, as shown in the following partial output:

```
12 apples cost $4.20
```

For short, formatted `string` values, an interpolated `string` can be easier for people to read. But for code examples in a book, where statements need to wrap over multiple lines, this can be tricky. For many of the code examples in this book, I will use numbered positional arguments. Another reason to avoid interpolated strings is that they can't be read from resource files to be localized.

The next code example is not meant to be entered in your project.

Before C# 10, `string` constants could only be combined by using concatenation with the + operator, as shown in the following code:

```
private const string firstname = "Omar";
private const string lastname = "Rudberg";
private const string fullname = firstname + " " + lastname;
```

With C# 10, interpolated strings (prefixed with `$`) can now be used, as shown in the following code:

```
private const string fullname = $"{firstname} {lastname}";
```

This only works for combining `string` constant values. It cannot work with other types like numbers, which would require runtime data type conversions. You cannot enter `private const` declarations in a top-level program like `Program.cs`. You will see how to use them in *Chapter 5, Building Your Own Types with Object-Oriented Programming*.

Good Practice: If you are writing code that will be part of a Unity project, then interpolated string formats are an easy way to avoid boxing.

Understanding format strings

A variable or expression can be formatted using a format string after a comma or colon.

An `N0` format string means a number with thousands separators and no decimal places, while a `C` format string means currency. The currency format will be determined by the current thread.

For instance, if you run code that uses the number or currency format on a PC in the UK, you'll get pounds sterling with commas as the thousands separators, but if you run it on a PC in Germany, you will get euros with dots as the thousands separators.

The full syntax of a format item is:

```
{ index [, alignment ] [ : formatString ] }
```

Each format item can have an alignment, which is useful when outputting tables of values, some of which might need to be left- or right-aligned within a width of characters. Alignment values are integers. Positive integers mean right-aligned and negative integers mean left-aligned.

For example, to output a table of fruit and how many of each there are, we might want to left-align the names within a column of 10 characters and right-align the counts formatted as numbers with zero decimal places within a column of six characters:

1. At the bottom of `Program.cs`, enter the following statements:

```
string applesText = "Apples";
int applesCount = 1234;
string bananasText = "Bananas";
int bananasCount = 56789;

Console.WriteLine();

Console.WriteLine(format: "{0,-10} {1,6}",
  arg0: "Name", arg1: "Count");

Console.WriteLine(format: "{0,-10} {1,6:N0}",
  arg0: applesText, arg1: applesCount);

Console.WriteLine(format: "{0,-10} {1,6:N0}",
  arg0: bananasText, arg1: bananasCount);
```

2. Run the code and note the effect of the alignment and number format, as shown in the following output:

```
Name        Count
Apples      1,234
Bananas    56,789
```

Custom number formatting

You can take complete control of number formatting using custom format codes, as shown in *Table 2.7*:

Format code	Description
`0`	Zero placeholder. Replaces the zero with the corresponding digit if present; otherwise, it uses zero. For example, `0000.00` formatting the value `123.4` would give `0123.40`.
`#`	Digit placeholder. Replaces the hash with the corresponding digit if present; otherwise, it uses nothing. For example, `####.##` formatting the value `123.4` would give `123.4`.
`.`	Decimal point. Sets the location of the decimal point in the number. Respects culture formatting, so it is a `.` (dot) in US English but a `,` (comma) in French.

`,`	Group separator. Inserts a localized group separator between each group. For example, `0,000` formatting the value `1234567` would give `1,234,567`. Also used to scale a number by dividing by multiples of 1,000 for each comma. For example, `0.00,,` formatting the value `1234567` would give `1.23` because the two commas mean divide by 1,000 twice.
`%`	Percentage placeholder. Multiplies the value by 100 and adds a percentage character.
`\`	Escape character. Makes the next character a literal instead of a format code. For example, `\##,###\#` formatting the value `1234` would give `#1,234#`.
`;`	Section separator. Defines different format strings for positive, negative, and zero numbers. For example, `[0];(0);Zero` formatting: `13` would give `[13]`, `-13` would give `(13)`, and `0` would give `Zero`.
Others	All other characters are shown in the output as is.

Table 2.7: Custom numeric format codes

More Information: A full list of custom number format codes can be found at the following link: `https://learn.microsoft.com/en-us/dotnet/standard/base-types/custom-numeric-format-strings`.

You can apply standard number formatting using simpler format codes, like `C` and `N`. They support a precision number to indicate how many digits of precision you want. The default is two. The most common are shown in *Table 2.8*:

Format code	Description
`C` or `c`	Currency. For example, in US culture, `C` formatting the value `123.4` gives `$123.40`, and `C0` formatting the value `123.4` gives `$123`.
`N` or `n`	Number. Integer digits with an optional negative sign and grouping characters.
`D` or `d`	Decimal. Integer digits with an optional negative sign but no grouping characters.
`B` or `b`	Binary. For example, `B` formatting the value `13` gives `1101`, and `B8` formatting the value `13` gives `00001101`.
`X` or `x`	Hexadecimal. For example, `X` formatting the value `255` gives `FF`, and `X4` formatting the value `255` gives `00FF`.
`E` or `e`	Exponential notation. For example, `E` formatting the value `1234.567` would give `1.234567000E+003`, and `E2` formatting the value `1234.567` would give `1.23E+003`.

Table 2.8: Standard numeric format codes

More Information: A full list of standard number format codes can be found at the following link: `https://learn.microsoft.com/en-us/dotnet/standard/base-types/standard-numeric-format-strings`.

You can take complete control of date and time formatting using custom format codes, as shown in *Table 2.9*:

Format code	Description
`/`	Date part separator. Varies by culture; for example, `en-US` uses `/` but `fr-FR` uses `-` (dash).
`\`	Escape character. Useful if you want to use a special format code as a literal character; for example, `h \h m \m` would format a time of 9:30 AM as `9 h 30 m`.
`:`	Time part separator. Varies by culture; for example, `en-US` uses `:` but `fr-FR` uses `.` (dot).
`d`, `dd`	The day of the month, from `1` to `31`, or with a leading zero from `01` through `31`.
`ddd`, `dddd`	The abbreviated or full name of the day of the week, for example, `Mon` or `Monday`, localized for the current culture.
`f`, `ff`, `fff`	The tenths of a second, hundredths of a second, or milliseconds.
`g`	The period or era, for example, `A.D.`
`h`, `hh`	The hour, using a 12-hour clock from `1` to `12`, or from `01` to `12`.
`H`, `HH`	The hour, using a 24-hour clock from `0` to `23`, or from `01` to `23`.
`K`	Time zone information. `null` for an unspecified time zone, `Z` for UTC, and a value like `-8:00` for local time adjusted from UTC.
`m`, `mm`	The minute, from `0` through `59`, or with a leading zero from `00` through `59`.
`M`, `MM`	The month, from `1` through `12`, or with a leading zero from `01` through `12`.
`MMM`, `MMMM`	The abbreviated or full name of the month, for example, `Jan` or `January`, localized for the current culture.
`s`, `ss`	The second, from `0` through `59`, or with a leading zero from `00` through `59`.
`t`, `tt`	The first or both characters of the AM/PM designator.
`y`, `yy`	The year of the current century, from `0` through `99`, or with a leading zero from `00` through `99`.
`yyy`	The year with a minimum of three digits, and as many as needed. For example, 1 A.D. is `001`. The first sacking of Rome was in `410`. The year the first edition of this book was published in `2016`.
`yyyy`, `yyyyy`	The year as a four- or five-digit number.
`z`, `zz`	Hours offset from UTC, with no leading zeros, or with leading zeros.
`zzz`	Hours and minutes offset from UTC, with a leading zero, for example, `+04:30`.

Table 2.9: Custom format codes for date and time values

More Information: A full list of custom format codes can be found at the following link: `https://learn.microsoft.com/en-us/dotnet/standard/base-types/custom-date-and-time-format-strings`.

You can apply standard date and time formatting using simpler format codes, like the `d` and `D` we used in the code example, as shown in *Table 2.10*:

Format code	Description
`d`	Short date pattern. Varies by culture; for example, `en-US` uses `M/d/yyyy` and `fr-FR` uses `dd/MM/yyyy`.
`D`	Long date pattern. Varies by culture; for example, `en-US` uses `mmmm, MMMM d, yyyy` and `fr-FR` uses `mmmm, dd MMMM yyyy`.
`f`	Full date/time pattern (short time – hours and minutes). Varies by culture.
`F`	Full date/time pattern (long time – hours, minutes, seconds, and AM/PM). Varies by culture.
`m, M`	Month/day pattern. Varies by culture.
`o, O`	A standardized pattern, suitable to serialize date/time values for roundtrips, for example, `2023-05-30T13:45:30.0000000-08:00`.
`r, R`	RFC1123 pattern.
`t`	Short time pattern. Varies by culture; for example, `en-US` uses `h:mm tt` and `fr-FR` uses `HH:mm`.
`T`	Long time pattern. Varies by culture; for example, `en-US` uses `h:mm:ss tt` and `fr-FR` uses `HH:mm:ss`.
`u`	Universal sortable date/time pattern, for example, `2009-06-15 13:45:30Z`.
`U`	Universal full date/time pattern. Varies by culture; for example, `en-US` might be `Monday, June 15, 2009 8:45:30 PM`.

Table 2.10: Standard format codes for date and time values

More Information: A full list of format codes can be found at the following link: `https://learn.microsoft.com/en-us/dotnet/standard/base-types/standard-date-and-time-format-strings`.

Warning! Note the difference between `"m"` (a shorthand for the month/day format, like July 4 in USA or 4 July in UK, which omits the year and presents the month and day in a culture-specific manner) and `"d m"` (a custom format string that displays the day number and then the month number with a space in between, like 4 7 in any country for 4th July).

Let's see how to output a value using multiple formats:

1. Type statements to output a `decimal` value twice, first as currency using the current culture, and then as a percentage, as shown in the following code:

```
decimal value = 0.325M;
WriteLine("Currency: {0:C}, Percentage: {0:0.0%}", value);
```

2. Run the code and note the result, as shown in the following output:

```
Currency: $0.33, Percentage: 32.5%
```

Getting text input from the user

We can get text input from the user using the `ReadLine` method. This method waits for the user to type some text. Then, as soon as the user presses *Enter*, whatever the user has typed is returned as a `string` value.

Let's get input from the user:

1. Type statements to ask the user for their name and age and then output what they entered, as shown in the following code:

```
Console.Write("Type your first name and press ENTER: ");
string firstName = Console.ReadLine();

Console.Write("Type your age and press ENTER: ");
string age = Console.ReadLine();

Console.WriteLine($"Hello {firstName}, you look good for {age}.");
```

By default, with .NET 6 and later, nullability checks are enabled, so the C# compiler gives two warnings because the `ReadLine` method could return a `null` value instead of a `string` value. But there is no practical scenario where this method would actually return `null`, so instead, we will see how to switch off these specific warnings in this scenario.

2. For the `firstName` variable, append a ? after `string`, as shown highlighted in the following code:

```
string? firstName = Console.ReadLine();
```

This tells the compiler that we are expecting a possible `null` value, so it does not need to warn us. If the variable is `null`, then when it is later output with `WriteLine`, it will just be blank, so that works fine in this case. If we were going to access any of the members of the `firstName` variable, then we would need to handle the case where it is `null`.

3. For the age variable, append a ! before the semicolon at the end of the statement, as shown highlighted in the following code:

```
string age = Console.ReadLine()!;
```

This is called the **null-forgiving operator** because it tells the compiler that, in this case, `ReadLine` will not return `null`, so it can stop showing the warning. It is now our responsibility to ensure this is the case. Luckily, the `Console` type's implementation of `ReadLine` always returns a `string` even if it is just an empty `string` value.

4. Run the code, and then enter a name and age, as shown in the following output:

```
Type your name and press ENTER: Gary
Type your age and press ENTER: 34
Hello Gary, you look good for 34.
```

You have now seen two common ways to handle nullability warnings from the compiler. We will cover nullability and how to handle it in more detail in *Chapter 6, Implementing Interfaces and Inheriting Classes.*

When does ReadLine return null?

The `ReadLine` method reads the next line of input from the standard input stream. If the user presses *Enter* without typing anything, `ReadLine` will return an empty string, not `null`.

In typical console apps, `ReadLine` returns `null` only if the **end of stream** (**EOF**) is reached. This is not something that can usually be achieved by user input in a standard console environment since EOF is typically signaled by the console being closed or redirected input being fully consumed.

In the context of the following code:

```
string? name = ReadLine(); // Prompt user to enter their name.
```

Here are the conditions under which `name` might be `null`:

- If the standard input stream is redirected and reaches EOF
- If you are testing in an environment where you can simulate EOF, like some development environments or automated testing setups

However, under normal user input conditions, `null` will "never" be returned by the `ReadLine` method.

Simplifying the usage of the console

In C# 6 and later, the `using` statement can be used not only to import a namespace but also to further simplify our code by importing a static class. Then, we won't need to enter the `Console` type name throughout our code.

Importing a static type for a single file

You can use your code editor's **Find and Replace** feature to remove the times we have previously written `Console`:

1. At the top of the `Program.cs` file, add a statement to **statically import** the `System.Console` class, as shown in the following code:

```
using static System.Console;
```

2. Select the first `Console.` in your code, ensuring that you select the dot after the word `Console` too.
3. In Visual Studio, navigate to **Edit** | **Find and Replace** | **Quick Replace**; in VS Code, navigate to **Edit** | **Replace**; or in Rider, navigate to **Edit** | **Find** | **Replace**, and note that an overlay dialog appears ready for you to enter what you would like to replace `Console.` with, as shown in *Figure 2.9*:

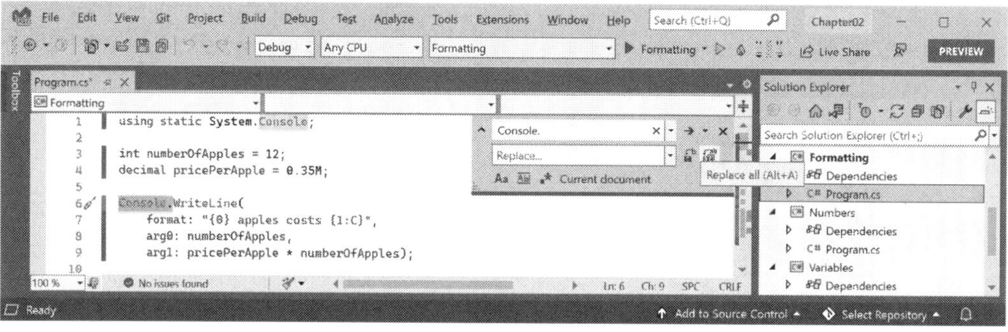

Figure 2.9: Using the Replace feature in Visual Studio to simplify your code

4. Leave the **Replace** box empty, click on the **Replace all** button (the second of the two buttons to the right of the **Replace** box), and then close the **Replace** box by clicking on the cross in its top-right corner.
5. Run the console app and note the behavior is the same as before.

Importing a static type for all code files in a project

Instead of statically importing the `Console` class just for one code file, it would probably be better to import it globally for all code files in the project:

1. Delete the statement to statically import `System.Console`.
2. Open `Formatting.csproj`, and after the `<PropertyGroup>` section, add a new `<ItemGroup>` section to globally and statically import `System.Console` using the implicit `using` .NET SDK feature, as shown in the following markup:

```
<ItemGroup Label="Simplify console app">
  <Using Include="System.Console" Static="true" />
</ItemGroup>
```

Good Practice: The `Label` attribute within an `<ItemGroup>` in .NET projects is optional but useful for several reasons, primarily revolving around organization, customization, and clarity in the build process. When multiple `<ItemGroup>` sections are used, labels provide a clear indication of what each group contains. Using labels, you can conditionally include or exclude certain items based on build configurations or other conditions. In this case, we are just using it for documentation.

3. Run the console app and note that the behavior is the same as before.

Good Practice: In the future, for all console app projects you create for this book, add the section above to simplify the code you need to write in all C# files to work with the `Console` class.

Getting key input from the user

We can get key input from the user using the `ReadKey` method. This method waits for the user to press a key or key combination, which is then returned as a `ConsoleKeyInfo` value.

Let's explore reading key presses:

1. Type statements to ask the user to press any key combination and then output information about it, as shown in the following code:

```
Write("Press any key combination: ");
ConsoleKeyInfo key = ReadKey();
WriteLine();
WriteLine("Key: {0}, Char: {1}, Modifiers: {2}",
  arg0: key.Key, arg1: key.KeyChar, arg2: key.Modifiers);
```

2. Run the code, press the *K* key, and note the result, as shown in the following output:

```
Press any key combination: k
Key: K, Char: k, Modifiers: 0
```

3. Run the code, hold down *Shift* and press the *K* key, and note the result, as shown in the following output:

```
Press any key combination: K
Key: K, Char: K, Modifiers: Shift
```

4. Run the code, press the *F12* key, and note the result, as shown in the following output:

```
Press any key combination:
Key: F12, Char: , Modifiers: 0
```

Warning! When running a console app in a terminal within VS Code, some keyboard combinations will be captured by the code editor before they can be processed by your console app. For example, *Ctrl* + *Shift* + *X* in VS Code activates the **Extensions** view in the sidebar. To fully test this console app, open a command prompt or terminal in the project folder and run the console app from there.

Passing arguments to a console app

When you run a console app, you often want to change its behavior by passing arguments. For example, with the `dotnet` command-line tool, you can pass the name of a new project template, as shown in the following commands:

```
dotnet new console
dotnet new mvc
```

You might have been wondering how to get any arguments that might be passed to a console app.

In every version of .NET prior to version 6, the console app project template made it obvious, as shown in the following code:

```
using System;

namespace Arguments
{
  class Program
  {
    static void Main(string[] args)
    {
      Console.WriteLine("Hello World!");
    }
  }
}
```

The `string[] args` arguments are declared and passed in the `Main` method of the `Program` class. They're an array used to pass arguments into a console app. But in top-level programs, as used by the console app project template in .NET 6 and later, the `Program` class and its `Main` method are hidden, along with the declaration of the `args` array. The trick is that you must know it still exists.

Command-line arguments are separated by spaces. Other characters like hyphens and colons are treated as part of an argument value.

To include spaces in an argument value, enclose the argument value in single or double quotes.

Imagine that we want to be able to enter the names of some colors for the foreground and background and the dimensions of the terminal window at the command line. We would be able to read the colors and numbers by reading them from the `args` array, which is always passed into the `Main` method, aka the entry point of a console app:

1. Use your preferred code editor to add a new **Console App** / `console` project named `Arguments` to the `Chapter02` solution.
2. Open `Arguments.csproj`, and after the `<PropertyGroup>` section, add a new `<ItemGroup>` section to statically import `System.Console` for all C# files using the implicit usings .NET SDK feature, as shown in the following markup:

```
<ItemGroup>
  <Using Include="System.Console" Static="true" />
</ItemGroup>
```

Good Practice: Remember to use the implicit using .NET SDK feature to statically import the `System.Console` type in all future console app projects to simplify your code, as these instructions will not be repeated every time.

3. In `Program.cs`, delete the existing statements and then add a statement to output the number of arguments passed to the application, as shown in the following code:

```
WriteLine($"There are {args.Length} arguments.");
```

4. Run the console app and view the result, as shown in the following output:

```
There are 0 arguments.
```

If you are using Visual Studio:

1. Navigate to **Project | Arguments Properties.**
2. Select the **Debug** tab, click **Open debug launch profiles UI**, and in the **Command line arguments** box, enter the following arguments: `firstarg second-arg third:arg "fourth arg"`, as shown in *Figure 2.10*:

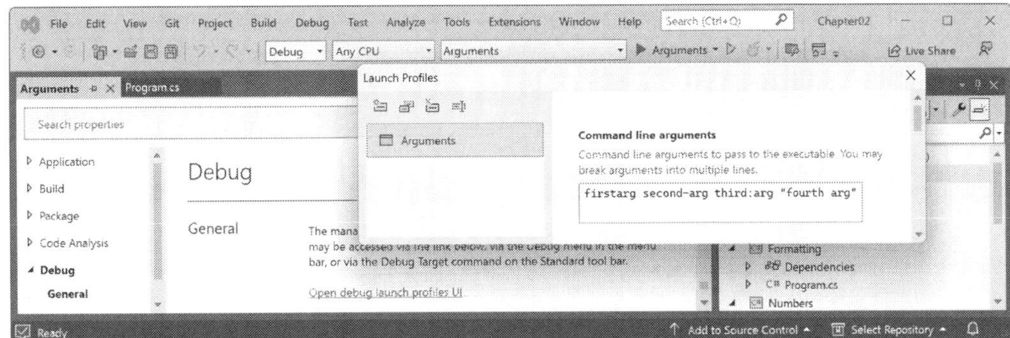

Figure 2.10: Entering command line arguments in the Visual Studio project properties

3. Close the **Launch Profiles** window.

4. In **Solution Explorer**, in the `Properties` folder, open the `launchSettings.json` file and note it defines the command-line arguments when you run the project, as shown highlighted in the following configuration:

```
{
  "profiles": {
    "Arguments": {
      "commandName": "Project",
      "commandLineArgs": "firstarg second-arg third:arg \"fourth arg\""
    }
  }
}
```

The `launchSettings.json` file can also be used by Rider. The equivalent for VS Code is the `.vscode/launch.json` file.

5. Run the console app project.

If you are using VS Code, then in **Terminal**, enter some arguments after the `dotnet run` command, as shown in the following command:

```
dotnet run firstarg second-arg third:arg "fourth arg"
```

For all code editors:

1. Note that the result indicates four arguments, as shown in the following output:

```
There are 4 arguments.
```

2. In `Program.cs`, to enumerate or iterate (that is, loop through) the values of those four arguments, add the following statements after outputting the length of the array:

```
foreach (string arg in args)
{
  WriteLine(arg);
}
```

3. Run the code again and note that the result shows the details of the four arguments, as shown in the following output:

```
There are 4 arguments.
firstarg
second-arg
third:arg
fourth arg
```

Setting options with arguments

We will now use these arguments to allow the user to pick a color for the background, foreground, and cursor size of the output window. The cursor size can be an integer value from 1, meaning a line at the bottom of the cursor cell, up to 100, meaning a percentage of the height of the cursor cell.

We have statically imported the `System.Console` class. It has properties like `ForegroundColor`, `BackgroundColor`, and `CursorSize` that we can now set just by using their names without needing to prefix them with `Console`.

The `System` namespace is already imported so that the compiler knows about the `ConsoleColor` and `Enum` types:

- Add statements to warn the user if they do not enter three arguments, and then parse those arguments and use them to set the color and dimensions of the console window, as shown in the following code:

```
if (args.Length < 3)
{
  WriteLine("You must specify two colors and cursor size, e.g.");
  WriteLine("dotnet run red yellow 50");
  return; // Stop running.
}

ForegroundColor = (ConsoleColor)Enum.Parse(
  enumType: typeof(ConsoleColor),
  value: args[0], ignoreCase: true);

BackgroundColor = (ConsoleColor)Enum.Parse(
  enumType: typeof(ConsoleColor),
  value: args[1], ignoreCase: true);

CursorSize = int.Parse(args[2]);
```

Note the compiler warning that setting the `CursorSize` is only supported on Windows. For now, do not worry about most of this code like `(ConsoleColor)`, `Enum.Parse`, or `typeof`, as it will all be explained in the next few chapters.

- If you are using Visual Studio, change the arguments to `red yellow 50`. Run the console app and note that the cursor is half the size and the colors have changed in the window, as shown in *Figure 2.11*:

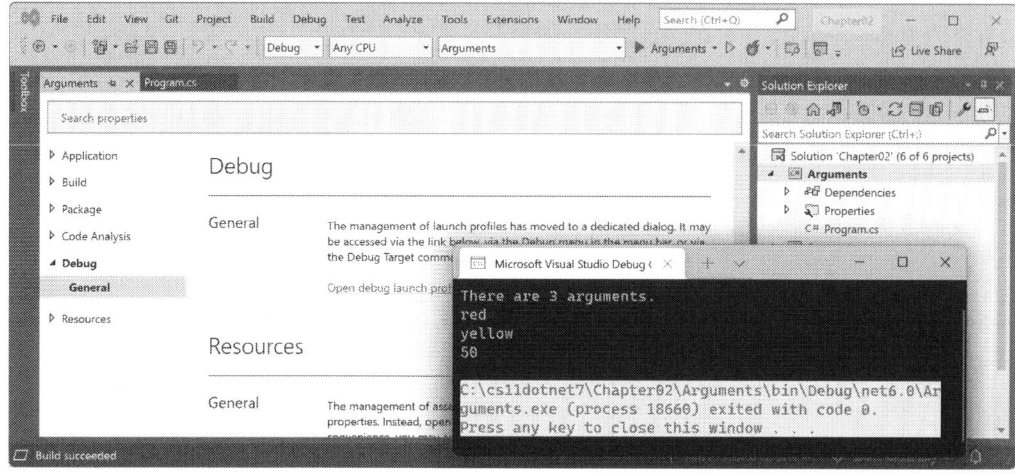

Figure 2.11: Setting colors and cursor size on Windows

- If you are using VS Code, then run the code with arguments to set the foreground color to red, the background color to yellow, and the cursor size to 50%, as shown in the following command:

```
dotnet run red yellow 50
```

On macOS or Linux, you'll see an unhandled exception, as shown in *Figure 2.12*:

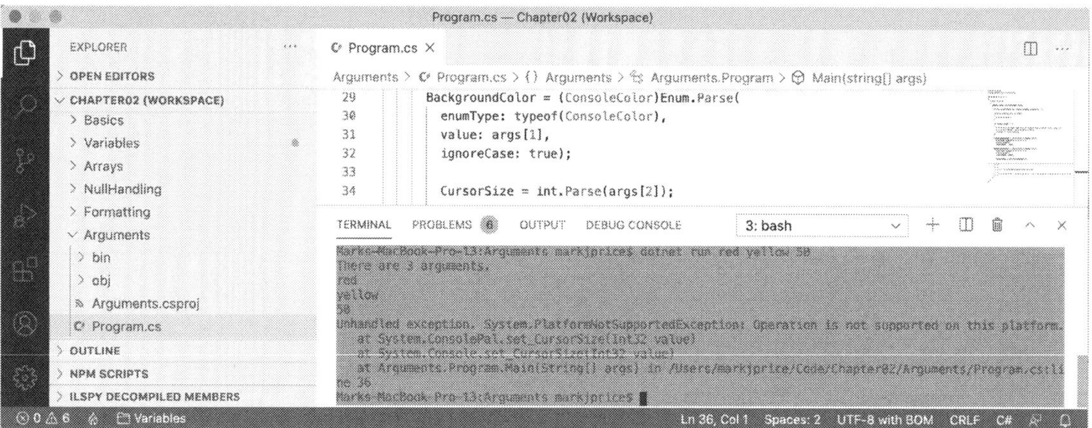

Figure 2.12: An unhandled exception on unsupported macOS

Although the compiler did not give an error or warning, at runtime, some API calls may fail on some platforms. Although a console app running on Windows can change its cursor size, on macOS, it cannot, and it complains if you try.

Handling platforms that do not support an API

So how do we solve this problem? We can solve this by using an exception handler. You will learn more details about the `try-catch` statement in *Chapter 3, Controlling Flow, Converting Types, and Handling Exceptions*, so for now, just enter the code:

1. Modify the code to wrap the lines that change the cursor size in a `try` statement, as shown in the following code:

```
try
{
  CursorSize = int.Parse(args[2]);
}
catch (PlatformNotSupportedException)
{
  WriteLine("The current platform does not support changing the size of
the cursor.");
}
```

2. If you were to run the code on macOS, then you would see the exception is caught, and a friendlier message is shown to the user.

Another way to handle differences in operating systems is to use the `OperatingSystem` class in the `System` namespace, as shown in the following code:

```
if (OperatingSystem.IsWindowsVersionAtLeast(major: 10))
{
  // Execute code that only works on Windows 10 or later.
}
else if (OperatingSystem.IsWindows())
{
  // Execute code that only works earlier versions of Windows.
}
else if (OperatingSystem.IsIOSVersionAtLeast(major: 14, minor: 5))
{
  // Execute code that only works on iOS 14.5 or later.
}
else if (OperatingSystem.IsBrowser())
{
  // Execute code that only works in the browser with Blazor.
}
```

The `OperatingSystem` class has equivalent methods for other common operating systems, like Android, iOS, Linux, macOS, and even the browser, which is useful for Blazor web components.

A third way to handle different platforms is to use conditional compilation statements.

There are four preprocessor directives that control conditional compilation: `#if`, `#elif`, `#else`, and `#endif`.

You define symbols using `#define`, as shown in the following code:

```
#define MYSYMBOL
```

Many symbols are automatically defined for you, as shown in *Table 2.11*:

Target Framework	Symbols
.NET Standard	`NETSTANDARD2_0`, `NETSTANDARD2_1`, and so on
Modern .NET	`NET7_0`, `NET7_0_ANDROID`, `NET7_0_IOS`, `NET7_0_WINDOWS`, and so on

Table 2.11: Predefined compiler symbols

You can then write statements that will compile only for the specified platforms, as shown in the following code:

```
#if NET7_0_ANDROID
// Compile statements that only work on Android.
#elif NET7_0_IOS
// Compile statements that only work on iOS.
#else
// Compile statements that work everywhere else.
#endif
```

Practicing and exploring

Test your knowledge and understanding by answering some questions, getting some hands-on practice, and exploring the topics covered in this chapter with deeper research.

Exercise 2.1 – Online material

Online materials can be extra content written by me for this book, or it can be references to content created by Microsoft or third parties.

Understanding async and await

C# 5 introduced two C# keywords when working with the `Task` type that enable easy multithreading. You can read about them and see a code example at the following link:

`https://github.com/markjprice/cs13net9/blob/main/docs/ch02-async-await.md`

Explore Spectre

No, not the villainous organization from the James Bond films! **Spectre** is a package that enhances console apps. You can read about it at the following link: `https://spectreconsole.net/`.

Exercise 2.2 – Practice exercises

Practice exercises go deeper into the topics for this chapter.

Explore number sizes and ranges

In the `Chapter02` solution, create a console app project named `Exercise_Numbers` that outputs the number of bytes in memory that each of the following number types uses and the minimum and maximum values they can have: `sbyte`, `byte`, `short`, `ushort`, `int`, `uint`, `long`, `ulong`, `Int128`, `UInt128`, `Half`, `float`, `double`, and `decimal`.

The result of running your console app should look something like *Figure 2.13*:

```
Microsoft Visual Studio Debug Console

--------------------------------------------------------------------------------------------------------------
Type     Byte(s) of memory                              Min                                              Max
--------------------------------------------------------------------------------------------------------------
sbyte    1                                             -128                                              127
byte     1                                                0                                              255
short    2                                           -32768                                            32767
ushort   2                                                0                                            65535
int      4                                      -2147483648                                       2147483647
uint     4                                                0                                       4294967295
long     8                             -9223372036854775808                              9223372036854775807
ulong    8                                                0                             18446744073709551615
Int128   16    -170141183460469231731687303715884105728          170141183460469231731687303715884105727
UInt128  16                                               0          340282366920938463463374607431768211455
Half     2                                           -65500                                            65500
float    4                                   -3.4028235E+38                                     3.4028235E+38
double   8                          -1.7976931348623157E+308                           1.7976931348623157E+308
decimal  16                  -79228162514264337593543950335                    79228162514264337593543950335
```

Figure 2.13: The result of outputting number type sizes

Code solutions for all exercises are available to download or clone from the GitHub repository at the following link: `https://github.com/markjprice/cs13net9/tree/main/code`.

Exercise 2.3 – Test your knowledge

To get the best answer to some of these questions, you will need to do your own research. I want you to "think outside the book," so I have deliberately not provided all the answers in the book.

I want to encourage you to get into the good habit of looking for help elsewhere, following the principle of "teach a person to fish."

1. What statement can you type in a C# file to discover the compiler and language version?
2. What are the two types of comments in C#?
3. What is the difference between a verbatim string and an interpolated string?
4. Why should you be careful when using `float` and `double` values?
5. How can you determine how many bytes a type like `double` uses in memory?
6. When should you use the `var` keyword?
7. What is the newest syntax to create an instance of a class like `XmlDocument`?
8. Why should you be careful when using the `dynamic` type?
9. How do you right-align a format string?
10. What character separates arguments for a console app?

Appendix, Answers to the Test Your Knowledge Questions, is available to download from a link in the README in the GitHub repository: `https://github.com/markjprice/cs13net9`.

Test your knowledge of number types

What type would you choose for the following "numbers"?

- A person's telephone number
- A person's height
- A person's age
- A person's salary
- A book's ISBN
- A book's price
- A book's shipping weight
- A country's population
- The number of stars in the universe
- The number of employees in each of the small or medium businesses in the UK (up to about 50,000 employees per business)

Exercise 2.4 – Explore topics

Use the links on the following page to learn more details about the topics covered in this chapter:

`https://github.com/markjprice/cs13net9/blob/main/docs/book-links.md#chapter-2---speaking-c`

Summary

In this chapter, you learned how to:

- Declare variables with a specified or inferred type.
- Use some of the built-in types for numbers, text, and Booleans.
- Choose between number types.
- Control output formatting in console apps.

In the next chapter, you will learn about operators, branching, looping, converting between types, and how to handle exceptions.

3

Controlling Flow, Converting Types, and Handling Exceptions

This chapter is all about writing code that performs simple operations on variables, makes decisions, performs pattern matching, and repeats statements or blocks. You will also learn how to work with arrays to store multiple values, how to convert variable or expression values from one type to another, how to handle exceptions, and how to check for overflows in number variables.

This chapter covers the following topics:

- Operating on variables
- Understanding selection statements
- Understanding iteration statements
- Storing multiple values in an array
- Casting and converting between types
- Handling exceptions
- Checking for overflow

Operating on variables

Operators apply simple operations such as addition and multiplication to **operands** such as variables and literal values. Operators return a new value that is the result of the operation and can be assigned to a variable, and they can also affect the operands.

Understanding binary operators

Most operators are **binary**, meaning that they work on two operands, as shown in the following pseudocode:

```
var resultOfOperation = firstOperand operator secondOperand;
```

Examples of binary operators include adding and multiplying, as shown in the following code:

```
int x = 5;
int y = 3;
int resultOfAdding = x + y;
int resultOfMultiplying = x * y;
```

Understanding unary operators

Some operators are **unary**, meaning they work on a single operand and can be applied before or after the operand, as shown in the following pseudocode:

```
var resultOfOperationAfter = onlyOperand operator;
var resultOfOperationBefore = operator onlyOperand;
```

Examples of unary operators include incrementors and retrieving a type or its size in bytes, as shown in the following code:

```
int x = 5;
int postfixIncrement = x++;
int prefixIncrement = ++x;
Type theTypeOfAnInteger = typeof(int);
string nameOfVariable = nameof(x);
int howManyBytesInAnInteger = sizeof(int);
```

Understanding ternary operators

A **ternary** operator works on three operands, as shown in the following pseudocode:

```
var resultOfOperation = firstOperand firstOperator
  secondOperand secondOperator thirdOperand;
```

An example of a ternary operator is the conditional operator `?:`, which acts like a simplified `if` statement. The first operand is a Boolean expression, the second operand is a value to return if it is `true`, and the third operand is a value to return if it is `false`, as shown in the following code:

```
// Syntax of conditional operator.
var result = boolean_expression ? value_if_true : value_if_false;

// Example of conditional operator.
string result = x > 3 ? "Greater than 3" : "Less than or equal to 3";

// Equivalent using an if statement.
string result;

if (x > 3)
{
```

```
  result = "Greater than 3";
}
else
{
  result = "Less than or equal to 3";
}
```

More experienced C# developers adopt ternary operators as much as possible because they are concise and can result in cleaner code once you are used to reading them.

Exploring unary operators

Two common unary operators are used to increment, ++, and decrement, --, a number. Let us write some example code to show how they work:

1. If you've completed the previous chapters, then you will already have a `cs13net9` folder. If not, then you'll need to create it.
2. Use your preferred code editor to create a new solution and project, as defined in the following list:
   - Project template: **Console App** / `console`
   - Project file and folder: `Operators`
   - Solution file and folder: `Chapter03`
   - **Do not use top-level statements**: Cleared
   - **Enable native AOT publish**: Cleared
3. In `Operators.csproj`, add a new `<ItemGroup>` section to statically import `System.Console` for all C# files using the `implicit usings` .NET SDK feature, as shown in the following markup:

```
<ItemGroup>
  <Using Include="System.Console" Static="true" />
</ItemGroup>
```

4. In `Program.cs`, delete the existing statements and then declare two integer variables named `a` and `b`; set `a` to `3`, increment `a` while assigning the result to `b`, and then output their values, as shown in the following code:

```
#region Exploring unary operators

int a = 3;
int b = a++;
WriteLine($"a is {a}, b is {b}");

#endregion
```

Good Practice: I recommend wrapping the statements for each section in `#region` and `#endregion` (as shown in the preceding code) so that you can easily collapse the sections. But I will not show this in future code tasks to save space.

5. Before running the console app, ask yourself a question: what do you think the value of `b` will be when output? Once you've thought about that, run the code and compare your prediction against the actual result, as shown in the following output:

```
a is 4, b is 3
```

The variable `b` has the value `3` because the `++` operator executes *after* the assignment; this is known as a **postfix operator**. If you need to increment *before* the assignment, then use the **prefix operator**.

6. Copy and paste the statements, and then modify them to rename the variables and use the prefix operator, as shown in the following code:

```
int c = 3;
int d = ++c; // Prefix means increment c before assigning it.
WriteLine($"c is {c}, d is {d}");
```

7. Rerun the code and note the result, as shown in the following output:

```
a is 4, b is 3
c is 4, d is 4
```

Good Practice: Due to the confusion between the prefix and postfix for the increment and decrement operators when combined with an assignment, the Swift programming language designers decided to drop support for this operator in version 3. My recommendation for usage in C# is to never combine the use of the `++` and `--` operators with an assignment operator, `=`. Perform the operations as separate statements.

Exploring binary arithmetic operators

Increment and decrement are unary arithmetic operators. Other arithmetic operators are usually binary and allow you to perform arithmetic operations on two numbers, as the following shows:

1. Add statements to declare and assign values to two integer variables named `e` and `f`, and then apply the five common binary arithmetic operators to the two numbers, as shown in the following code:

```
int e = 11;
int f = 3;
WriteLine($"e is {e}, f is {f}");
WriteLine($"e + f = {e + f}");
WriteLine($"e - f = {e - f}");
WriteLine($"e * f = {e * f}");
WriteLine($"e / f = {e / f}");
WriteLine($"e % f = {e % f}");
```

2. Run the code and note the result, as shown in the following output:

```
e is 11, f is 3
e + f = 14
e - f = 8
e * f = 33
e / f = 3
e % f = 2
```

To understand the divide / and modulo % operators when applied to integers, you need to think back to primary school. Imagine you have eleven sweets and three friends.

How can you divide the sweets between your friends? You can give three sweets to each of your friends, and there will be two left over. Those two sweets are the **modulus**, also known as the **remainder** after dividing. If you had twelve sweets, then each friend would get four of them, and there would be none left over, so the remainder would be zero.

3. Add statements to declare and assign a value to a `double` variable named g to show the difference between whole-number and real-number divisions, as shown in the following code:

```
double g = 11.0;
WriteLine($"g is {g:N1}, f is {f}");
WriteLine($"g / f = {g / f}");
```

4. Run the code and note the result, as shown in the following output:

```
g is 11.0, f is 3
g / f = 3.6666666666666665
```

If the first operand is a floating-point number, such as g with the value `11.0`, then the divide operator returns a floating-point value, such as `3.6666666666665`, rather than a whole number.

Assignment operators

You have already been using the most common assignment operator, =.

To make your code more concise, you can combine the assignment operator with other operators like arithmetic operators, as shown in the following code:

```
int p = 6;
p += 3; // Equivalent to: p = p + 3;
p -= 3; // Equivalent to: p = p - 3;
p *= 3; // Equivalent to: p = p * 3;
p /= 3; // Equivalent to: p = p / 3;
```

Null-coalescing operators

Related operators to the assignment operators are the null-coalescing operators. Sometimes, you want to either assign a variable to a result or, if the variable is `null`, assign an alternative value.

You can do this using the null-coalescing operators, `??` or `??=`, as shown in the following code:

```
string? authorName = GetAuthorName(); // A fictional function.

// The maxLength variable will be the length of authorName if it is
// not null, or 30 if authorName is null.
int maxLength = authorName?.Length ?? 30;

// The authorName variable will be "unknown" if authorName was null.
authorName ??= "unknown";
```

Exploring logical operators

Logical operators operate on Boolean values, so they return either `true` or `false`. Let's explore binary logical operators that operate on two Boolean values, traditionally named `p` and `q` in mathematics:

1. In `Program.cs`, add statements to declare two Boolean variables, `p` and `q`, with values of `true` and `false`, and then output truth tables showing the results of applying AND, OR, and XOR (exclusive OR) logical operators, as shown in the following code:

   ```
   bool p = true;
   bool q = false;
   WriteLine($"AND  | p     | q     ");
   WriteLine($"p    | {p & p,-5} | {p & q,-5} ");
   WriteLine($"q    | {q & p,-5} | {q & q,-5} ");
   WriteLine();
   WriteLine($"OR   | p     | q     ");
   WriteLine($"p    | {p | p,-5} | {p | q,-5} ");
   WriteLine($"q    | {q | p,-5} | {q | q,-5} ");
   WriteLine();
   WriteLine($"XOR  | p     | q     ");
   ```

```
WriteLine($"p    | {p ^ p,-5} | {p ^ q,-5} ");
WriteLine($"q    | {q ^ p,-5} | {q ^ q,-5} ");
```

Remember that `,-5` means left-align within a five-width column.

2. Run the code and note the results, as shown in the following output:

```
AND  | p     | q
p    | True  | False
q    | False | False

OR   | p     | q
p    | True  | True
q    | True  | False

XOR  | p     | q
p    | False | True
q    | True  | False
```

For the AND `&` logical operator, both operands must be `true` for the result to be `true`. For the OR `|` logical operator, either operand can be `true` for the result to be `true`. For the XOR `^` logical operator, either operand can be `true` (but not both!) for the result to be `true`.

Exploring conditional logical operators

Conditional logical operators are like logical operators, but you use two symbols instead of one, for example, `&&` instead of `&`, or `||` instead of `|`.

In *Chapter 4, Writing, Debugging, and Testing Functions*, you will learn about functions in more detail, but I need to introduce functions now to explain conditional logical operators, also known as short-circuiting Boolean operators.

A function executes statements and then returns a value. That value could be a Boolean value like `true` that is used in a Boolean operation. Let's make use of conditional logical operators:

1. At the bottom of `Program.cs`, write statements to declare a function that writes a message to the console and returns `true`, as shown in the following code:

```
static bool DoStuff()
{
  WriteLine("I am doing some stuff.");
  return true;
}
```

Local functions can be anywhere within the statements in `Program.cs` that use the top-level program feature but it is good practice to put them at the bottom of the file.

2. After the previous `WriteLine` statements, perform an AND `&` operation on the `p` and `q` variables, and the result of calling the function, as shown in the following code:

```
WriteLine();
// Note that DoStuff() returns true.
WriteLine($"p & DoStuff() = {p & DoStuff()}");
WriteLine($"q & DoStuff() = {q & DoStuff()}");
```

3. Run the code, view the result, and note that the function was called twice, once for `p` and once for `q`, as shown in the following output:

```
I am doing some stuff.
p & DoStuff() = True
I am doing some stuff.
q & DoStuff() = False
```

4. Copy and paste the three statements and then change the `&` operators into `&&` operators, as shown in the following code:

```
WriteLine();
WriteLine($"p && DoStuff() = {p && DoStuff()}");
WriteLine($"q && DoStuff() = {q && DoStuff()}");
```

5. Run the code, view the result, and note that the function does run when combined with the `p` variable. It does not run when combined with the `q` variable because the `q` variable is `false` so the result will be `false` anyway, so it does not need to execute the function, as shown in the following output:

```
I am doing some stuff.
p && DoStuff() = True
q && DoStuff() = False // DoStuff function was not executed!
```

Good Practice: Now you can see why the conditional logical operators are described as being short-circuiting. They can make your apps more efficient, but they can also introduce subtle bugs in cases where you assume that the function will always be called. It is safest to avoid them when used in combination with functions that cause side effects.

Exploring bitwise and binary shift operators

Bitwise operators compare the bits in the binary representation of a number. Each bit, either the `0` (zero) or `1` (one) value, is compared individually to the bit in the same column.

Binary shift operators can perform some common arithmetic calculations much faster than traditional operators, for example, any multiplication by a factor of 2.

Let's explore bitwise and binary shift operators:

1. In `Program.cs`, add statements to declare two integer variables named `x` and `y` with values `10` and `6`, and then output the results of applying AND, OR, and XOR bitwise operators, as shown in the following code:

```
WriteLine();

int x = 10;
int y = 6;

WriteLine($"Expression | Decimal |   Binary");
WriteLine($"-------------------------------");
WriteLine($"x          | {x,7} | {x:B8}");
WriteLine($"y          | {y,7} | {y:B8}");
WriteLine($"x & y      | {x & y,7} | {x & y:B8}");
WriteLine($"x | y      | {x | y,7} | {x | y:B8}");
WriteLine($"x ^ y      | {x ^ y,7} | {x ^ y:B8}");
```

Remember that `,7` means right-align in a seven-width column and `:B8` means format in binary with eight digits.

2. Run the code and note the results, as shown in the following output:

```
Expression | Decimal |   Binary
-------------------------------
x          |      10 | 00001010
y          |       6 | 00000110
x & y      |       2 | 00000010
x | y      |      14 | 00001110
x ^ y      |      12 | 00001100
```

For `x & y`, only the 2-bit column is set. For `x | y`, the 8-, 4-, and 2-bit columns are set. For `x ^ y`, the 8- and 4-bit columns are set.

3. In `Program.cs`, add statements to output the results of applying the left-shift operator to move the bits of the variable x by three columns, multiplying x by 8, and right-shifting the bits of the variable y by one column, as shown in the following code:

```
// Left-shift x by three bit columns.
WriteLine($"x << 3     | {x << 3,7} | {x << 3:B8}");

// Multiply x by 8.
WriteLine($"x * 8      | {x * 8,7} | {x * 8:B8}");

// Right-shift y by one bit column.
WriteLine($"y >> 1     | {y >> 1,7} | {y >> 1:B8}");
```

4. Run the code and note the results, as shown in the following output:

```
x << 3     |      80 | 01010000
x * 8      |      80 | 01010000
y >> 1     |       3 | 00000011
```

The `80` result is because the bits in it were shifted three columns to the left, so the 1 bits moved into the 64- and 16-bit columns, and 64 + 16 = 80. This is the equivalent of multiplying by 8, but CPUs can perform a bit-shift faster. The `3` result is because the 1 bits in `y` were shifted one column into the 2- and 1-bit columns.

Good Practice: Remember that when operating on integer values, the & and | symbols are bitwise operators, and when operating on Boolean values like `true` and `false`, the & and | symbols are logical operators.

Miscellaneous operators

`nameof` and `sizeof` are convenient operators when working with types:

- `nameof` returns the short name (without the namespace) of a variable, type, or member as a `string` value, which is useful when outputting exception messages.
- `sizeof` returns the size in bytes of simple types, which is useful for determining the efficiency of data storage. Technically, the `sizeof` operator requires an unsafe code block, but the sizes of value types with a C# alias, like `int` and `double`, are hardcoded as constants by the compiler so they do not need an unsafe block.

For example:

```
int age = 50;
WriteLine($"The {nameof(age)} variable uses {sizeof(int)} bytes of memory.");
```

There are many other operators; for example, the dot between a variable and its members is called the **member access operator** and the round brackets at the end of a function or method name are called the **invocation operator**, as shown in the following code:

```
int age = 50;

// How many operators in the following statement?
char firstDigit = age.ToString()[0];

// There are four operators:
// = is the assignment operator
// . is the member access operator
// () is the invocation operator
// [] is the indexer access operator
```

Understanding selection statements

Every application needs to be able to select from choices and branch along different code paths. The two selection statements in C# are `if` and `switch`. You can use `if` for all your code, but `switch` can simplify your code in some common scenarios, such as when there is a single variable that can have multiple values that each require different processing.

Branching with the if statement

The `if` statement determines which branch to follow by evaluating a Boolean expression. If the expression is `true`, then the block executes. The `else` block is optional, and it executes if the `if` expression is `false`. The `if` statement can be nested.

The `if` statement can be combined with other `if` statements as `else if` branches, as shown in the following code:

```
if (expression1)
{
  // Executes if expression1 is true.
}
else if (expression2)
{
  // Executes if expression1 is false and expression2 is true.
}
else if (expression3)
{
  // Executes if expression1 and expression2 are false
  // and expression3 is true.
}
else
```

```
{
  // Executes if all expressions are false.
}
```

Each `if` statement's Boolean expression is independent of the others and, unlike `switch` statements, does not need to reference a single value.

Let's write some code to explore selection statements like `if`:

1. Use your preferred code editor to add a new **Console App** / `console` project named `SelectionStatements` to the `Chapter03` solution.

Remember to statically import `System.Console` in your project file. If you are using Visual Studio, then configure the startup project to be the current selection.

2. In `Program.cs`, delete the existing statements and then add statements to check if a password is at least eight characters long, as shown in the following code:

```
string password = "ninja";

if (password.Length < 8)
{
  WriteLine("Your password is too short. Use at least 8 chars.");
}
else
{
  WriteLine("Your password is strong.");
}
```

3. Run the code and note the result, as shown in the following output:

```
Your password is too short. Use at least 8 chars.
```

Why you should always use braces with if statements

As there is only a single statement inside each block, the preceding code could be written without the curly braces, as shown in the following code:

```
if (password.Length < 8)
  WriteLine("Your password is too short. Use at least 8 chars.");
else
  WriteLine("Your password is strong.");
```

This style of `if` statement should be avoided because it can introduce serious bugs. An infamous example is the #gotofail bug in Apple's iPhone iOS operating system. For 18 months after Apple's iOS 6 was released, in September 2012, it had a bug due to an `if` statement without braces in its **Secure Sockets Layer** (SSL) encryption code. This meant that any user running Safari, the device's web browser, who tried to connect to secure websites, such as their bank, was not properly secure because an important check was being accidentally skipped.

Just because you can leave out the curly braces, doesn't mean you should. Your code is not "more efficient" without them; instead, it is harder to read, less maintainable, and, potentially, more dangerous.

Pattern matching with the if statement

A feature introduced with C# 7 and later is pattern matching. The `if` statement can use the `is` keyword in combination with declaring a local variable to make your code safer. It is important to note that a single expression, for example, `o is int i`, does two things:

1. Checks the type of the data in a variable named o.
2. If the type check is `true`, then it assigns the value to the variable named `i`.

More Information: You can learn more about the `is` operator at the following link: `https://learn.microsoft.com/en-us/dotnet/csharp/language-reference/operators/is`. You will use this technique in later chapters, so keep this in mind.

Let's see it in action:

1. Add statements so that if the value stored in the variable named o is an `int`, then the value is assigned to the local variable named `i`, which can then be used inside the `if` statement. This is safer than using the variable named o because we know for sure that `i` is an `int` variable and not something else, as shown in the following code:

```
// Add and remove the "" to change between string and int.
object o = "3";
int j = 4;

if (o is int i)
{
  WriteLine($"{i} x {j} = {i * j}");
}
else
{
  WriteLine("o is not an int so it cannot multiply!");
}
```

2. Run the code and view the results, as shown in the following output:

```
o is not an int so it cannot multiply!
```

3. Delete the double-quote characters around the "3" value so that the value stored in the variable named o is an int type instead of a string type.
4. Rerun the code to view the results, as shown in the following output:

```
3 x 4 = 12
```

Branching with the switch statement

The switch statement is different from the if statement because switch compares a single expression against a list of multiple possible case statements. Every case statement is related to the single expression. Every case section must end with one of the following:

- The break keyword (like case 1 in the following code).
- The goto case keywords (like case 2 in the following code).
- They should have no statements (like case 3 in the following code).
- The goto keyword that references a named label (like case 5 in the following code).
- The return keyword to leave the current function (not shown in the code).

Let's write some code to explore the switch statements:

1. Type the following code for a switch statement. You should note that the penultimate statement is a label that can be jumped to, and the first statement generates a random number between 1 and 6 (the number 7 in the code is an exclusive upper bound). The switch statement branches are based on the value of this random number, as shown in the following code:

```
// Inclusive lower bound but exclusive upper bound.
int number = Random.Shared.Next(minValue: 1, maxValue: 7);
WriteLine($"My random number is {number}");

switch (number)
{
  case 1:
    WriteLine("One");
    break; // Jumps to end of switch statement.
  case 2:
    WriteLine("Two");
    goto case 1;
  case 3: // Multiple case section.
  case 4:
    WriteLine("Three or four");
    goto case 1;
  case 5:
```

```
    goto A_label;
  default:
    WriteLine("Default");
    break;
} // End of switch statement.
WriteLine("After end of switch");
A_label:
WriteLine($"After A_label");
```

Good Practice: You can use the `goto` keyword to jump to another case or a label. The `goto` keyword is frowned upon by most programmers but can be a good solution to code logic in some scenarios. However, you should use it sparingly, if at all. To see how often Microsoft uses `goto` in the .NET base class libraries, use the following link: `https://github.com/search?q=%22goto%20%22+repo%3Adotnet%2Fruntime+language%3AC%23&type=code&ref=advsearch`.

2. Run the code multiple times to see what happens in various cases of random numbers, as shown in the following example output:

```
// First random run.
My random number is 4
Three or four
One
After end of switch
After A_label

// Second random run.
My random number is 2
Two
One
After end of switch
After A_label

// Third random run.
My random number is 6
Default
After end of switch
After A_label

// Fourth random run.
My random number is 1
One
```

```
After end of switch
After A_label

// Fifth random run.
My random number is 5
After A_label
```

Good Practice: The `Random` class that we used to generate a random number has a `Next` method that allows you to specify an inclusive lower bound and an exclusive upper bound and will generate a pseudo-random number. Instead of creating a new instance of `Random` that is not thread-safe, since .NET 6, you can use a `Shared` instance that is thread-safe so it can be used concurrently from any thread.

Now let's take a small diversion from the `switch` keyword to review how to add a new item to a project using Visual Studio because, in the next code task, you need to add a new class file to your project and I haven't shown you how to do that yet.

Adding a new item to a project using Visual Studio

Visual Studio version 17.6 or later has an optional simplified dialog box for adding a new item to a project. After navigating to **Project** | **Add New Item...**, or right-clicking on a project in **Solution Explorer** and selecting **Add** | **New Item...**, you will see the traditional dialog box, as shown in *Figure 3.1*:

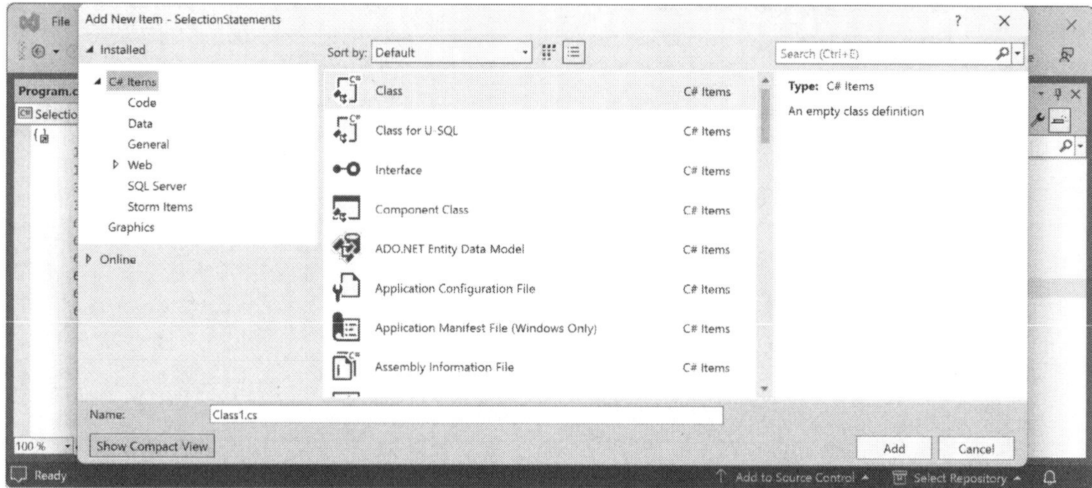

Figure 3.1: Add New Item dialog box in normal view

If you click the **Show Compact View** button, then it switches to a simplified dialog box, as shown in *Figure 3.2*:

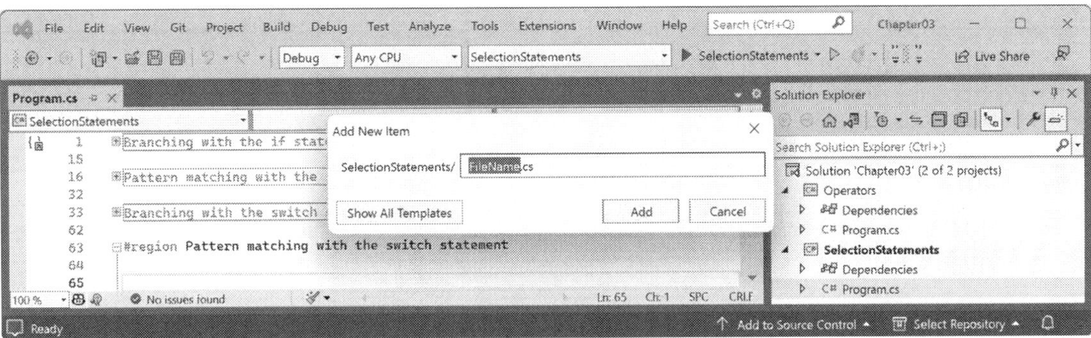

Figure 3.2: Add New Item dialog box in compact view

To revert to the normal dialog box, click the **Show All Templates** button.

Pattern matching with the switch statement

Like the `if` statement, the `switch` statement supports pattern matching in C# 7 and later. The `case` values no longer need to be literal values; they can be patterns.

In C# 7 and later, your code can more concisely branch, based on the subtype of a class, and you can declare and assign a local variable to safely use it. Additionally, `case` statements can include a `when` keyword to perform more specific pattern matching.

Let's see an example of pattern matching with the `switch` statement using a custom class hierarchy of animals with different properties:

You will learn more details about defining classes in *Chapter 5, Building Your Own Types with Object-Oriented Programming*. For now, you should be able to get the idea from reading the code.

1. In the `SelectionStatements` project, add a new class file named `Animals.cs`:
    - In Visual Studio, navigate to **Project** | **Add New Item...** or press *Ctrl* + *Shift* + *A*, type the name, and then click **Add**.
    - In VS Code, click the **New File...** button and type the name.
    - In Rider, right-click on the project and select **Add** | **Class/Interface...**.
2. In `Animals.cs`, delete any existing statements, and then define three classes, a base class, `Animal`, and two inherited classes, `Cat` and `Spider`, as shown in the following code:

```
class Animal // This is the base type for all animals.
{
  public string? Name;
  public DateTime Born;
  public byte Legs;
}
```

```
class Cat : Animal // This is a subtype of animal.
{
  public bool IsDomestic;
}

class Spider : Animal // This is another subtype of animal.
{
  public bool IsVenomous;
}
```

Since this section is about pattern matching and not the most common way to define a class with fields, it is not important how the data types are declared. The `Name` property has been declared nullable and the other two properties are not. You could make the `Name` not-nullable and make the other two nullable and the pattern-matching example is unaffected.

3. In `Program.cs`, add statements to declare an array of nullable animals, and then show a message based on what type and attributes each animal has, as shown in the following code:

```
var animals = new Animal?[]
{
  new Cat { Name = "Karen", Born = new(year: 2022, month: 8,
    day: 23), Legs = 4, IsDomestic = true },
  null,
  new Cat { Name = "Mufasa", Born = new(year: 1994, month: 6,
    day: 12) },
  new Spider { Name = "Sid Vicious", Born = DateTime.Today,
    IsVenomous = true},
  new Spider { Name = "Captain Furry", Born = DateTime.Today }
};

foreach (Animal? animal in animals)
{
  string message;

  switch (animal)
  {
    case Cat fourLeggedCat when fourLeggedCat.Legs == 4:
      message = $"The cat named {fourLeggedCat.Name} has four legs.";
      break;
    case Cat wildCat when wildCat.IsDomestic == false:
```

```
      message = $"The non-domestic cat is named {wildCat.Name}.";
      break;
    case Cat cat:
      message = $"The cat is named {cat.Name}.";
      break;
    default: // default is always evaluated last.
      message = $"{animal.Name} is a {animal.GetType().Name}.";
      break;
    case Spider spider when spider.IsVenomous:
      message = $"The {spider.Name} spider is venomous. Run!";
      break;
    case null:
      message = "The animal is null.";
      break;
  }
  WriteLine($"switch statement: {message}");
}
```

Good Practice: Always put the `default` at the end of a `switch` statement. I wrote the preceding `switch` statement with the `default` in the middle of the set of branches deliberately to show that the compiler will accept this, and will always move it to the end of the set of branches for you.

The `case` statement shown in the following code:

```
case Cat fourLeggedCat when fourLeggedCat.Legs == 4:
```

can also be written using the more concise property pattern-matching syntax, as shown in the following code:

```
case Cat { Legs: 4 } fourLeggedCat:
```

4. Run the code and note that the array named `animals` is declared to contain the `Animal?` type, so it could be any subtype of `Animal`, such as `Cat` or `Spider`, or a `null` value. In this code, we create four instances of `Animal` of different types with different properties, and one `null` one, so the result will be five messages that describe each of the animals, as shown in the following output:

```
switch statement: The cat named Karen has four legs.
switch statement: The animal is null.
switch statement: The non-domestic cat is named Mufasa.
switch statement: The Sid Vicious spider is venomous. Run!
switch statement: Captain Furry is a Spider.
```

Simplifying switch statements with switch expressions

In C# 8 or later, you can simplify `switch` statements using **switch expressions**.

Most `switch` statements are very simple, yet they require a lot of typing. `switch` expressions are designed to simplify the code you need to type while still expressing the same intent in scenarios where all cases return a value to set a single variable. `switch` expressions use a lambda, `=>`, to indicate a return value.

Let's implement the previous code that used a `switch` statement using a `switch` expression so that you can compare the two styles:

1. In `Program.cs`, at the bottom and inside the `foreach` loop, add statements to set the message based on what type and attributes the animal has, using a `switch` expression, as shown in the following code:

```
message = animal switch
{
  Cat fourLeggedCat when fourLeggedCat.Legs == 4
    => $"The cat named {fourLeggedCat.Name} has four legs.",
  Cat wildCat when wildCat.IsDomestic == false
    => $"The non-domestic cat is named {wildCat.Name}.",
  Cat cat
    => $"The cat is named {cat.Name}.",
  Spider spider when spider.IsVenomous
    => $"The {spider.Name} spider is venomous. Run!",
  null
    => "The animal is null.",
  _
    => $"{animal.Name} is a {animal.GetType().Name}."
};
WriteLine($"switch expression: {message}");
```

The main differences are the removal of the `case` and `break` keywords. The underscore character `_` is used to represent the default return value. It is known as a **discard** and you can read more about it at the following link: `https://learn.microsoft.com/en-us/dotnet/csharp/fundamentals/functional/discards`.

2. Run the code, and note that the result is the same as before, as shown in the following output:

```
switch statement: The cat named Karen has four legs.
switch expression: The cat named Karen has four legs.
switch statement: The animal is null.
switch expression: The animal is null.
switch statement: The non-domestic cat is named Mufasa.
```

```
switch expression: The non-domestic cat is named Mufasa.
switch statement: The Sid Vicious spider is venomous. Run!
switch expression: The Sid Vicious spider is venomous. Run!
switch statement: Captain Furry is a Spider.
switch expression: Captain Furry is a Spider.
```

Understanding iteration statements

Iteration statements repeat a block of statements either while a condition is `true` (`while` and `for` statements) or for each item in a collection (`foreach` statement). The choice of which statement to use is based on a combination of ease of understanding to solve the logic problem and personal preference.

Looping with the while statement

The `while` statement evaluates a Boolean expression and continues to loop while it is true. Let's explore iteration statements:

1. Use your preferred code editor to add a new **Console App** / `console` project named `IterationStatements` to the `Chapter03` solution.
2. In `Program.cs`, delete the existing statements, and then add statements to define a `while` statement that loops while an integer variable has a value less than 10, as shown in the following code:

```
int x = 0;
while (x < 10)
{
  WriteLine(x);
  x++;
}
```

3. Run the code and view the results, which should be the numbers 0 to 9, as shown in the following output:

```
0
1
2
3
4
5
6
7
8
9
```

Looping with the do statement

The `do` statement is like `while`, except the Boolean expression is checked at the bottom of the block instead of the top, which means that the block always executes at least once, as the following shows:

1. Type statements to define a `do` loop, as shown in the following code:

```
string? actualPassword = "Pa$$w0rd";
string? password;

do
{
  Write("Enter your password: ");
  password = ReadLine();
}
while (password != actualPassword);

WriteLine("Correct!");
```

2. Run the code, and note that you are prompted to enter your password repeatedly until you enter it correctly, as shown in the following output:

```
Enter your password: password
Enter your password: 12345678
Enter your password: ninja
Enter your password: correct horse battery staple
Enter your password: Pa$$w0rd
Correct!
```

3. As an optional challenge, add statements so that the user can only make three attempts before an error message is displayed.
4. At this point, you might want to comment out the code for this section so you do not have to keep entering a password every time you run the console app!

Looping with the for statement

The `for` statement is like `while`, except that it is more succinct. It combines:

- An optional **initializer expression**, which executes once at the start of the loop.
- An optional **conditional expression**, which executes on every iteration at the start of the loop to check whether the looping should continue. If the expression returns `true` or it is missing, the loop will execute again.
- An optional **iterator expression**, which executes on every loop at the bottom of the statement. This is often used to increment a counter variable.

The `for` statement is commonly used with an integer counter. Let's explore some code:

1. Type a `for` statement to output the numbers 1 to 10, as shown in the following code:

```
for (int y = 1; y <= 10; y++)
{
  WriteLine(y);
}
```

2. Run the code to view the result, which should be the numbers 1 to 10.
3. Add another `for` statement to output the numbers 0 to 10, incrementing by 3, as shown in the following code:

```
for (int y = 0; y <= 10; y += 3)
{
  WriteLine(y);
}
```

4. Run the code to view the result, which should be the numbers 0, 3, 6, and 9.
5. Optionally, experiment with changing the initializer expression, conditional expression, or iterator expression to see their effects. Only change one thing at a time so that you can clearly see the effect produced.

Looping with the foreach statement

The `foreach` statement is a bit different from the previous three iteration statements.

It is used to perform a block of statements on each item in a sequence, for example, an array or collection. Each item is usually read-only, and if the sequence structure is modified during iteration, for example, by adding or removing an item, then an exception will be thrown.

Try the following example:

1. Type statements to create an array of string variables and then output the length of each one, as shown in the following code:

```
string[] names = { "Adam", "Barry", "Charlie" };

foreach (string name in names)
{
  WriteLine($"{name} has {name.Length} characters.");
}
```

2. Run the code and view the results, as shown in the following output:

```
Adam has 4 characters.
Barry has 5 characters.
Charlie has 7 characters.
```

Understanding how foreach works internally

A developer who defines a type that represents multiple items, like an array or collection, should make sure that a programmer can use the `foreach` statement to enumerate through the type's items.

Technically, the `foreach` statement will work on any type that follows these rules:

- The type must have a method named `GetEnumerator` that returns an object.
- The returned object must have a property named `Current` and a method named `MoveNext`.
- The `MoveNext` method must change the value of `Current` and return `true` if there are more items to enumerate through or return `false` if there are no more items.

There are interfaces named `IEnumerable` and `IEnumerable<T>` that formally define these rules, but technically, the compiler does not require the type to implement these interfaces.

The compiler turns the `foreach` statement in the preceding example into something like the following pseudocode:

```
IEnumerator e = names.GetEnumerator();

while (e.MoveNext())
{
  string name = (string)e.Current; // Current is read-only!
  WriteLine($"{name} has {name.Length} characters.");
}
```

Due to the use of an iterator and its read-only `Current` property, the variable declared in a `foreach` statement cannot be used to modify the value of the current item.

Storing multiple values in an array

When you need to store multiple values of the same type, you can declare an **array**. For example, you may do this when you need to store four names in a `string` array.

Working with single-dimensional arrays

The code that you will write next will allocate memory for an array for storing four `string` values. It will then store `string` values at index positions 0 to 3 (arrays usually have a lower bound of zero, so the index of the last item is one less than the length of the array).

We could visualize the array like this:

0	1	2	3
Kate	Jack	Rebecca	Tom

Table 3.1: Visualization of an array of four string values

Good Practice: Do not assume that all arrays count from zero. The most common type of array in .NET is **szArray**, a single-dimensional zero-indexed array, and these use the normal [ ] syntax. But .NET also has **mdArray**, a multi-dimensional array, and these do not have to have a lower bound of zero. These are rarely used, but you should know they exist.

Finally, it will loop through each item in the array using a `for` statement.

Let's look at how to use an array:

1. Use your preferred code editor to add a new **Console App** / `console` project named `Arrays` to the `Chapter03` solution.
2. In `Program.cs`, delete the existing statements and then type statements to declare and use an array of `string` values, as shown in the following code:

```
string[] names; // This can reference any size array of strings.

// Allocate memory for four strings in an array.
names = new string[4];

// Store items at these index positions.
names[0] = "Kate";
names[1] = "Jack";
names[2] = "Rebecca";
names[3] = "Tom";

// Loop through the names.
for (int i = 0; i < names.Length; i++)
{
  // Output the item at index position i.
  WriteLine($"{names[i]} is at position {i}.");
}
```

3. Run the code and note the result, as shown in the following output:

```
Kate is at position 0.
Jack is at position 1.
Rebecca is at position 2.
Tom is at position 3.
```

Arrays are always of a fixed size at the time of memory allocation, so you need to decide how many items you want to store before instantiating them.

An alternative to defining the array in three steps as above is to use array initializer syntax:

1. Before the `for` loop, add a statement to declare, allocate memory, and instantiate the values of a similar array, as shown in the following code:

```
// Alternative syntax for creating and initializing an array.
string[] names2 = { "Kate", "Jack", "Rebecca", "Tom" };
```

2. Change the `for` loop to use `names2`, run the console app, and note that the results are the same.

Working with multi-dimensional arrays

Instead of a single-dimension array for storing a row of string values (or any other data type), what if we want to store a grid of values? Or a cube? Or even higher dimensions?

We could visualize a two-dimensional array, aka a grid, of `string` values like this:

	0	1	2	3
0	Alpha	Beta	Gamma	Delta
1	Anne	Ben	Charlie	Doug
2	Aardvark	Bear	Cat	Dog

Table 3.2: Visualization of a two-dimensional array

Let's look at how to use multi-dimensional arrays:

1. At the bottom of `Program.cs`, add statements to declare and instantiate a two-dimensional array of `string` values, as shown in the following code:

```
string[,] grid1 = // Two dimensional array.
{
  { "Alpha", "Beta", "Gamma", "Delta" },
  { "Anne", "Ben", "Charlie", "Doug" },
  { "Aardvark", "Bear", "Cat", "Dog" }
};
```

2. We can discover the lower and upper bounds of this array using helpful methods, as shown in the following code:

```
WriteLine($"1st dimension, lower bound: {grid1.GetLowerBound(0)}");
WriteLine($"1st dimension, upper bound: {grid1.GetUpperBound(0)}");
WriteLine($"2nd dimension, lower bound: {grid1.GetLowerBound(1)}");
WriteLine($"2nd dimension, upper bound: {grid1.GetUpperBound(1)}");
```

3. Run the code and note the result, as shown in the following output:

```
1st dimension, lower bound: 0
1st dimension, upper bound: 2
```

```
2nd dimension, lower bound: 0
2nd dimension, upper bound: 3
```

4. We can then use these values in nested `for` statements to loop through the `string` values, as shown in the following code:

```
for (int row = 0; row <= grid1.GetUpperBound(0); row++)
{
  for (int col = 0; col <= grid1.GetUpperBound(1); col++)
  {
    WriteLine($"Row {row}, Column {col}: {grid1[row, col]}");
  }
}
```

5. Run the code and note the result, as shown in the following output:

```
Row 0, Column 0: Alpha
Row 0, Column 1: Beta
Row 0, Column 2: Gamma
Row 0, Column 3: Delta
Row 1, Column 0: Anne
Row 1, Column 1: Ben
Row 1, Column 2: Charlie
Row 1, Column 3: Doug
Row 2, Column 0: Aardvark
Row 2, Column 1: Bear
Row 2, Column 2: Cat
Row 2, Column 3: Dog
```

You must supply a value for every row and every column when it is instantiated, or you will get compile errors. If you need to indicate a missing `string` value, then use `string.Empty`. Or if you declare the array to be nullable `string` values by using `string?[]`, then you can also use `null` for a missing value.

If you cannot use the array initialization syntax, perhaps because you are loading values from a file or database, then you can separate the declaration of the array dimension and the allocation of memory from the assignment of values, as shown in the following code:

```
// Alternative syntax for declaring and allocating memory
// for a multi-dimensional array.
string[,] grid2 = new string[3,4]; // Allocate memory.

grid2[0, 0] = "Alpha"; // Assign values.
grid2[0, 1] = "Beta";
// And so on.
grid2[2, 3] = "Dog";
```

When declaring the size of the dimensions, you specify the length, not the upper bound. The expression `new string[3,4]` means the array can have 3 items in its first dimension (0) with an upper bound of 2, and the array can have 4 items in its second dimension (1) with an upper bound of 3.

Working with jagged arrays

If you need a multi-dimensional array but the number of items stored in each dimension is different, then you can define an array of arrays, aka a jagged array.

We could visualize a jagged array as shown in *Figure 3.3*:

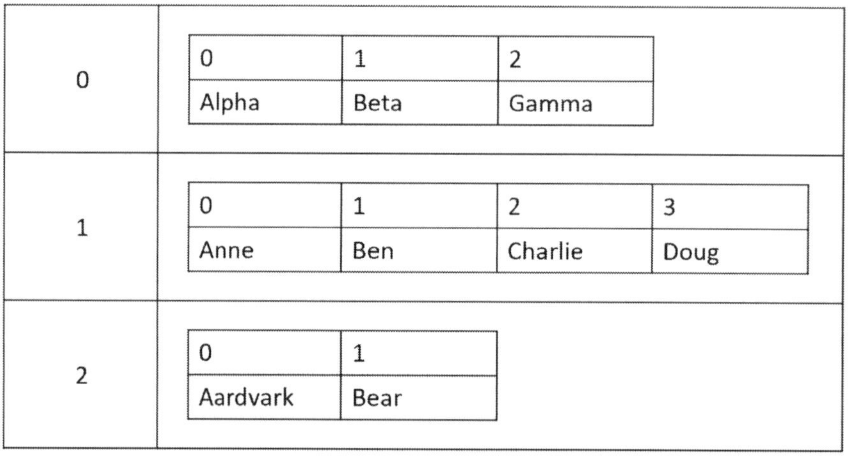

0	0	1	2	
	Alpha	Beta	Gamma	
1	0	1	2	3
	Anne	Ben	Charlie	Doug
2	0	1		
	Aardvark	Bear		

Figure 3.3: Visualization of a jagged array

Let's look at how to use a jagged array:

1. At the bottom of `Program.cs`, add statements to declare and instantiate an array of arrays of `string` values, as shown in the following code:

```
string[][] jagged = // An array of string arrays.
{
  new[] { "Alpha", "Beta", "Gamma" },
  new[] { "Anne", "Ben", "Charlie", "Doug" },
  new[] { "Aardvark", "Bear" }
};
```

2. We can discover the lower and upper bounds of the array of arrays, and then each array with it, as shown in the following code:

```
WriteLine("Upper bound of the array of arrays is: {0}",
  jagged.GetUpperBound(0));

for (int array = 0; array <= jagged.GetUpperBound(0); array++)
{
```

```
    WriteLine("Upper bound of array {0} is: {1}",
      arg0: array,
      arg1: jagged[array].GetUpperBound(0));
}
```

3. Run the code and note the result, as shown in the following output:

```
Upper bound of the array of arrays is: 2
Upper bound of array 0 is: 2
Upper bound of array 1 is: 3
Upper bound of array 2 is: 1
```

4. We can then use these values in nested `for` statements to loop through the `string` values, as shown in the following code:

```
for (int row = 0; row <= jagged.GetUpperBound(0); row++)
{
  for (int col = 0; col <= jagged[row].GetUpperBound(0); col++)
  {
    WriteLine($"Row {row}, Column {col}: {jagged[row][col]}");
  }
}
```

5. Run the code and note the result, as shown in the following output:

```
Row 0, Column 0: Alpha
Row 0, Column 1: Beta
Row 0, Column 2: Gamma
Row 1, Column 0: Anne
Row 1, Column 1: Ben
Row 1, Column 2: Charlie
Row 1, Column 3: Doug
Row 2, Column 0: Aardvark
Row 2, Column 1: Bear
```

List pattern matching with arrays

Earlier in this chapter, you saw how an individual object supports pattern matching against its type and properties. Pattern matching also works with arrays and collections.

Introduced with C# 11, list pattern matching works with any type that has a public `Length` or `Count` property and has an indexer using an `int` or `System.Index` parameter. You will learn about indexers in *Chapter 5, Building Your Own Types with Object-Oriented Programming*.

When you define multiple list patterns in the same `switch` expression, you must order them so that the more specific one comes first, or the compiler will complain because a more general pattern will match the more specific pattern too, and make the more specific one unreachable.

Table 3.3 shows examples of list pattern matching, assuming a list of `int` values:

Example	Description
`[]`	Matches an empty array or collection.
`[..]`	Matches an array or collection with any number of items, including zero, so `[..]` must come after `[]` if you need to switch on both.
`[_]`	Matches a list with any single item.
`[int item1]` or `[var item1]`	Matches a list with any single item and can use the value in the return expression by referring to `item1`.
`[7, 2]`	Matches exactly a list of two items with those values in that order.
`[_, _]`	Matches a list with any two items.
`[var item1, var item2]`	Matches a list with any two items and can use the values in the return expression by referring to `item1` and `item2`.
`[_, _, _]`	Matches a list with any three items.
`[var item1, ..]`	Matches a list with one or more items. Can refer to the value of the first item in its return expression by referring to `item1`.
`[var firstItem, .., var lastItem]`	Matches a list with two or more items. Can refer to the value of the first and last item in its return expression by referring to `firstItem` and `lastItem`.
`[.., var lastItem]`	Matches a list with one or more items. Can refer to the value of the last item in its return expression by referring to `lastItem`.

Table 3.3: Examples of list pattern matching

Let's see some examples in code:

1. At the bottom of `Program.cs`, add statements to define some arrays of `int` values, and then pass them to a method that returns descriptive text depending on the pattern that matches best, as shown in the following code:

```
int[] sequentialNumbers = { 1, 2, 3, 4, 5, 6, 7, 8, 9, 10 };
int[] oneTwoNumbers = { 1, 2 };
int[] oneTwoTenNumbers = { 1, 2, 10 };
int[] oneTwoThreeTenNumbers = { 1, 2, 3, 10 };
int[] primeNumbers = { 2, 3, 5, 7, 11, 13, 17, 19, 23, 29 };
int[] fibonacciNumbers = { 0, 1, 1, 2, 3, 5, 8, 13, 21, 34, 55, 89 };
int[] emptyNumbers = { }; // Or use Array.Empty<int>()
int[] threeNumbers = { 9, 7, 5 };
int[] sixNumbers = { 9, 7, 5, 4, 2, 10 };
```

```
WriteLine($"{nameof(sequentialNumbers)}:
{CheckSwitch(sequentialNumbers)}");
WriteLine($"{nameof(oneTwoNumbers)}: {CheckSwitch(oneTwoNumbers)}");
WriteLine($"{nameof(oneTwoTenNumbers)}:
{CheckSwitch(oneTwoTenNumbers)}");
WriteLine($"{nameof(oneTwoThreeTenNumbers)}:
{CheckSwitch(oneTwoThreeTenNumbers)}");
WriteLine($"{nameof(primeNumbers)}: {CheckSwitch(primeNumbers)}");
WriteLine($"{nameof(fibonacciNumbers)}:
{CheckSwitch(fibonacciNumbers)}");
WriteLine($"{nameof(emptyNumbers)}: {CheckSwitch(emptyNumbers)}");
WriteLine($"{nameof(threeNumbers)}: {CheckSwitch(threeNumbers)}");
WriteLine($"{nameof(sixNumbers)}: {CheckSwitch(sixNumbers)}");

static string CheckSwitch(int[] values) => values switch
{
  [] => "Empty array",
  [1, 2, _, 10] => "Contains 1, 2, any single number, 10.",
  [1, 2, .., 10] => "Contains 1, 2, any range including empty, 10.",
  [1, 2] => "Contains 1 then 2.",
  [int item1, int item2, int item3] =>
    $"Contains {item1} then {item2} then {item3}.",
  [0, _] => "Starts with 0, then one other number.",
  [0, ..] => "Starts with 0, then any range of numbers.",
  [2, .. int[] others] => $"Starts with 2, then {others.Length} more
numbers.",
  [..] => "Any items in any order.", // <-- Note the trailing comma for
easier re-ordering.
  // Use Alt + Up or Down arrow to move statements.
};
```

In C# 6, Microsoft added support for expression-bodied function members. The `CheckSwitch` function above uses this syntax. In C#, lambdas are the use of the => character to indicate a return value from a function. I will properly introduce this in *Chapter 4, Writing, Debugging, and Testing Functions.*

2. Run the code and note the result, as shown in the following output:

```
sequentialNumbers: Contains 1, 2, any range including empty, 10.
oneTwoNumbers: Contains 1 then 2.
oneTwoTenNumbers: Contains 1, 2, any range including empty, 10.
oneTwoThreeTenNumbers: Contains 1, 2, any single number, 10.
```

```
primeNumbers: Starts with 2, then 9 more numbers.
fibonacciNumbers: Starts with 0, then any range of numbers.
emptyNumbers: Empty array
threeNumbers: Contains 9 then 7 then 5.
sixNumbers: Any items in any order.
```

You can learn more about list pattern matching at the following link: `https://learn.microsoft.com/en-us/dotnet/csharp/language-reference/operators/patterns#list-patterns`.

Trailing commas

The trailing comma after the last item in the `switch` expression is optional and the compiler will not complain about it.

Most languages, including C#, allow the code style of trailing commas. When multiple items are separated by commas (for example, when declaring an anonymous object, an array, collection initializers, enums, and switch expressions), C# allows you to have the trailing comma after the last item. This makes it easy to rearrange the order of the items without having to keep adding and removing commas.

You can read a discussion about allowing trailing commas for `switch` expressions back in 2018 at the following link: `https://github.com/dotnet/csharplang/issues/2098`.

Even JSON serializers have an option to allow this because it is so common to use, as discussed at the following link: `https://learn.microsoft.com/en-us/dotnet/api/system.text.json.jsonserializeroptions.allowtrailingcommas`.

Understanding inline arrays

Inline arrays were introduced with C# 12; they are an advanced feature used by the .NET runtime team to improve performance. You are unlikely to use them yourself unless you are a public library author, but you will automatically benefit from others' use of them.

More Information: You can learn more about inline arrays at the following link: `https://learn.microsoft.com/en-us/dotnet/csharp/language-reference/proposals/csharp-12.0/inline-arrays`.

Summarizing arrays

We use slightly different syntax to declare different types of arrays, as shown in *Table 3.4*:

Type of array	Declaration syntax
One dimension	`datatype[]`, for example, `string[]`
Two dimensions	`string[,]`
Three dimensions	`string[,,]`
Ten dimensions	`string[,,,,,,,,,]`
Array of arrays, aka two-dimensional jagged array	`string[][]`
Array of arrays of arrays, aka three-dimensional jagged array	`string[][][]`

Table 3.4: Summary of array declaration syntax

Arrays are useful for temporarily storing multiple items, but collections are a more flexible option when adding and removing items dynamically. You don't need to worry about collections right now, as we will cover them in *Chapter 8, Working with Common .NET Types*.

You can convert any sequence of items into an array using the `ToArray` extension method, which we will cover in *Chapter 11, Querying and Manipulating Data Using LINQ*.

Good Practice: If you do not need to dynamically add and remove items, then you should use an array instead of a collection like `List<T>` because arrays are more efficient in memory use and the items are stored contiguously, which can improve performance.

Casting and converting between types

You will often need to convert values of variables between different types. For example, data input is often entered as text in the console, so it is initially stored in a variable of the `string` type, but it then needs to be converted into a date/time, number, or some other data type, depending on how it should be stored and processed.

Sometimes you will need to convert between number types, like between an integer and a floating point, before performing calculations.

Converting is also known as **casting**, and it has two varieties: **implicit** and **explicit**. Implicit casting happens automatically, and it is safe, meaning that you will not lose any information.

Explicit casting must be performed manually because it may lose information, for example, the precision of a number. By explicitly casting, you are telling the C# compiler that you understand and accept the risk.

Casting numbers implicitly and explicitly

Implicitly casting an `int` variable into a `double` variable is safe because no information can be lost, as the following shows:

1. Use your preferred code editor to add a new **Console App** / `console` project named `CastingConverting` to the `Chapter03` solution.

2. In `Program.cs`, delete the existing statements, then type statements to declare and assign an `int` variable and a `double` variable, and then implicitly cast the integer's value when assigning it to the `double` variable, as shown in the following code:

```
int a = 10;
double b = a; // An int can be safely cast into a double.
WriteLine($"a is {a}, b is {b}");
```

3. Type statements to declare and assign a `double` variable and an `int` variable, and then implicitly cast the `double` value when assigning it to the `int` variable, as shown in the following code:

```
double c = 9.8;
int d = c; // Compiler gives an error if you do not explicitly cast.
WriteLine($"c is {c}, d is {d}");
```

4. Run the code and note the error message, as shown in the following output:

```
Error: (6,9): error CS0266: Cannot implicitly convert type 'double' to
'int'. An explicit conversion exists (are you missing a cast?)
```

This error message will also appear in the Visual Studio **Error List**, VS Code **PROBLEMS** window, or Rider **Problems** window.

You cannot implicitly cast a `double` variable into an `int` variable because it is potentially unsafe and could lose data, like the value after the decimal point. You must explicitly cast a `double` variable into an `int` variable using a pair of round brackets around the type you want to cast the `double` type into. The pair of round brackets is the **cast operator**. Even then, you must be aware that the part after the decimal point will be trimmed off without warning because you have chosen to perform an explicit cast and therefore understand the consequences.

5. Modify the assignment statement for the `d` variable to explicitly cast the variable `c` into an `int`, and add a comment to explain what will happen, as shown highlighted in the following code:

```
double c = 9.8;
int d = (int)c; // Compiler gives an error if you do not explicitly cast.
WriteLine($"c is {c}, d is {d}"); // d loses the .8 part.
```

6. Run the code to view the results, as shown in the following output:

```
a is 10, b is 10
c is 9.8, d is 9
```

We must perform a similar operation when converting values between larger integers and smaller integers. Again, be aware that you might lose information because any value too big will have its bits copied and then be interpreted in ways that you might not expect!

7. Enter statements to declare and assign a `long` (64-bit) integer variable to an `int` (32-bit) integer variable, both using a small value that will work and a too-large value that will not, as shown in the following code:

```
long e = 10;
int f = (int)e;
WriteLine($"e is {e:N0}, f is {f:N0}");

e = long.MaxValue;
f = (int)e;
WriteLine($"e is {e:N0}, f is {f:N0}");
```

8. Run the code to view the results, as shown in the following output:

```
e is 10, f is 10
e is 9,223,372,036,854,775,807, f is -1
```

9. Modify the value of e to 5 billion, as shown in the following code:

```
e = 5_000_000_000;
```

10. Run the code to view the results, as shown in the following output:

```
e is 5,000,000,000, f is 705,032,704
```

Five billion cannot fit into a 32-bit integer, so it overflows (wraps around) to about 705 million. It is all to do with the binary representation of integer numbers. You will see more examples of integer overflow and how to handle it later in this chapter.

How negative numbers are represented in binary

You might have wondered why `f` had the value `-1` in the previous code. Negative, aka signed, numbers use the first bit to represent negativity. If the bit is `0` (zero), then it is a positive number. If the bit is `1` (one), then it is a negative number.

Let's write some code to illustrate this:

1. Enter statements to output the maximum value for an `int` in decimal and binary number formats, then output the values `8` to `-8`, decrementing by one, and finally, output the minimum value for an `int`, as shown in the following code:

```
WriteLine("{0,12} {1,34}", "Decimal", "Binary");
WriteLine("{0,12} {0,34:B32}", int.MaxValue);
for (int i = 8; i >= -8; i--)
{
  WriteLine("{0,12} {0,34:B32}", i);
}
WriteLine("{0,12} {0,34:B32}", int.MinValue);
```

Note that `,12` and `,34` mean right-align within those column widths. `:B32` means format as binary padded with leading zeros to a width of 32.

2. Run the code to view the results, as shown in the following output:

```
     Decimal                             Binary
  2147483647   01111111111111111111111111111111
           8   00000000000000000000000000001000
           7   00000000000000000000000000000111
           6   00000000000000000000000000000110
           5   00000000000000000000000000000101
           4   00000000000000000000000000000100
           3   00000000000000000000000000000011
           2   00000000000000000000000000000010
           1   00000000000000000000000000000001
           0   00000000000000000000000000000000
          -1   11111111111111111111111111111111
          -2   11111111111111111111111111111110
          -3   11111111111111111111111111111101
          -4   11111111111111111111111111111100
          -5   11111111111111111111111111111011
          -6   11111111111111111111111111111010
          -7   11111111111111111111111111111001
          -8   11111111111111111111111111111000
 -2147483648   10000000000000000000000000000000
```

3. Note that all the positive binary number representations start with `0` and all the negative binary number representations start with `1`. The decimal value `-1` is represented by all ones in binary. That is why when you have an integer too large to fit in a 32-bit integer, it becomes `-1`. But it's not always the case that the result of this type of casting is `-1`. When casting from a wider integer data type to a narrower integer data type, the most significant extra bits get truncated. For example, if you're casting from a 32-bit integer to a 16-bit integer, the 16 **most significant bits (MSBs)** of the 32-bit integer will be truncated. The **least significant bits (LSBs)** represent the result of the casting. For instance, if you're casting to a 16-bit integer, the 16 least significant bits of the original value will represent the result after casting.

4. Enter statements to show an example of a `long` integer that, when cast to an `int`, gets truncated to a non-minus-one value, as shown in the following code:

```
long r = 0b_101000101010001100100111010100101010;
int s = (int) r;
```

```
Console.WriteLine($"{r,38:B38} = {r}");
Console.WriteLine($"{s,38:B32} = {s}");
```

5. Run the code to view the results, as shown in the following output:

```
00101000101010001100100111010100101010 = 43657622826
    00101010001100100111010100101010 = 707949866
```

More Information: If you are interested in learning more about how signed numbers can be represented in computer systems, then you can read the following article: `https://en.wikipedia.org/wiki/Signed_number_representations`.

Converting with the System.Convert type

You can only cast between similar types, for example, between whole numbers like `byte`, `int`, and `long`, or between a class and its subclasses. You cannot cast a `long` to a `string` or a `byte` to a `DateTime`.

An alternative to using the cast operator is to use the `System.Convert` type. The `System.Convert` type can convert to and from all the C# number types, as well as Booleans, strings, and date and time values.

Let's write some code to see this in action:

1. At the top of `Program.cs`, statically import the `System.Convert` class, as shown in the following code:

```
using static System.Convert; // To use the ToInt32 method.
```

Alternatively, add an entry to `CastingConverting.csproj`, as shown in the following markup: `<Using Include="System.Convert" Static="true" />`.

2. At the bottom of `Program.cs`, type statements to declare and assign a value to a `double` variable, convert it into an integer, and then write both values to the console, as shown in the following code:

```
double g = 9.8;
int h = ToInt32(g); // A method of System.Convert.
WriteLine($"g is {g}, h is {h}");
```

3. Run the code and view the result, as shown in the following output:

```
g is 9.8, h is 10
```

An important difference between casting and converting is that converting rounds the `double` value `9.8` up to `10` instead of trimming the part after the decimal point. Another is that casting can allow overflows while converting will throw an exception.

Rounding numbers and the default rounding rules

You have now seen that the cast operator trims the decimal part of a real number and that the `System.Convert` method rounds up or down. However, what is the rule for rounding?

In British primary schools for children aged 5 to 11, pupils are taught to round *up* if the decimal part is .5 or higher and round *down* if the decimal part is less. Of course, these terms only make sense because, at that age, the pupils are only dealing with positive numbers. With negative numbers, these terms become confusing and should be avoided. This is why the .NET API uses the `enum` values `AwayFromZero`, `ToZero`, `ToEven`, `ToPositiveInfinity`, and `ToNegativeInfinity` for improved clarity.

Let's explore if C# follows the same primary school rule:

1. Type statements to declare and assign an array of `double` values, convert each of them into an integer, and then write the result to the console, as shown in the following code:

```
double[,] doubles = {
  { 9.49, 9.5, 9.51 },
  { 10.49, 10.5, 10.51 },
  { 11.49, 11.5, 11.51 },
  { 12.49, 12.5, 12.51 } ,
  { -12.49, -12.5, -12.51 },
  { -11.49, -11.5, -11.51 },
  { -10.49, -10.5, -10.51 },
  { -9.49, -9.5, -9.51 }
};

WriteLine($"| double | ToInt32 | double | ToInt32 | double | ToInt32 |");
for (int x = 0; x < 8; x++)
{
  for (int y = 0; y < 3; y++)
  {
    Write($"| {doubles[x, y],6} | {ToInt32(doubles[x, y]),7} ");
  }
  WriteLine("|");
}
WriteLine();
```

2. Run the code and view the result, as shown in the following output:

```
| double | ToInt32 | double | ToInt32 | double | ToInt32 |
|   9.49 |       9 |    9.5 |      10 |   9.51 |      10 |
|  10.49 |      10 |   10.5 |      10 |  10.51 |      11 |
|  11.49 |      11 |   11.5 |      12 |  11.51 |      12 |
|  12.49 |      12 |   12.5 |      12 |  12.51 |      13 |
| -12.49 |     -12 |  -12.5 |     -12 | -12.51 |     -13 |
| -11.49 |     -11 |  -11.5 |     -12 | -11.51 |     -12 |
| -10.49 |     -10 |  -10.5 |     -10 | -10.51 |     -11 |
|  -9.49 |      -9 |   -9.5 |     -10 |  -9.51 |     -10 |
```

We have shown that the rule for rounding in C# is subtly different from the primary school rule:

- It always rounds *toward zero* if the decimal part is less than the midpoint .5.
- It always rounds *away from zero* if the decimal part is more than the midpoint .5.
- It will round *away from zero* if the decimal part is the midpoint .5 and the non-decimal part is *odd*, but it will round *toward zero* if the non-decimal part is *even*.

This rule is known as **Banker's rounding**, and it is preferred because it reduces bias by alternating when it rounds toward or away from zero. Sadly, other languages such as JavaScript use the primary school rule.

Taking control of rounding rules

You can take control of the rounding rules by using the `Round` method of the `Math` class:

1. Type statements to round each of the `double` values using the "away from zero" rounding rule, also known as rounding "up," and then write the result to the console, as shown in the following code:

```
foreach (double n in doubles)
{
  WriteLine(format:
    "Math.Round({0}, 0, MidpointRounding.AwayFromZero) is {1}",
    arg0: n,
    arg1: Math.Round(value: n, digits: 0,
            mode: MidpointRounding.AwayFromZero));
}
```

You can use a `foreach` statement to enumerate all the items in a multi-dimensional array.

2. Run the code and view the result, as shown in the following partial output:

```
Math.Round(9.49, 0, MidpointRounding.AwayFromZero) is 9
Math.Round(9.5, 0, MidpointRounding.AwayFromZero) is 10
Math.Round(9.51, 0, MidpointRounding.AwayFromZero) is 10
Math.Round(10.49, 0, MidpointRounding.AwayFromZero) is 10
Math.Round(10.5, 0, MidpointRounding.AwayFromZero) is 11
Math.Round(10.51, 0, MidpointRounding.AwayFromZero) is 11
...
```

Good Practice: For every programming language that you use, check its rounding rules. They may not work the way you expect! You can read more about `Math.Round` at the following link: `https://learn.microsoft.com/en-us/dotnet/api/system.math.round`.

Converting from any type to a string

The most common conversion is from any type into a `string` variable for outputting as human-readable text, so all types have a method named `ToString` that they inherit from the `System.Object` class.

The `ToString` method converts the current value of any variable into a textual representation. Some types can't be sensibly represented as text, so they return their namespace and type name instead.

Let's convert some types into a `string`:

1. Type statements to declare some variables, convert them to their `string` representation, and write them to the console, as shown in the following code:

```
int number = 12;
WriteLine(number.ToString());
bool boolean = true;
WriteLine(boolean.ToString());
DateTime now = DateTime.Now;
WriteLine(now.ToString());
object me = new();
WriteLine(me.ToString());
```

2. Run the code and view the result, as shown in the following output:

```
12
True
08/28/2024 17:33:54
System.Object
```

Passing any object to the `WriteLine` method implicitly converts it into a `string`, so it is not necessary to explicitly call `ToString`. We are doing so here just to emphasize what is happening. Explicitly calling `ToString` does avoid a boxing operation, so if you are developing games with Unity, then that can help you avoid memory garbage collection issues.

Converting from a binary object to a string

When you have a binary object like an image or video that you want to either store or transmit, you sometimes do not want to send the raw bits because you do not know how those bits could be misinterpreted, for example, by the network protocol transmitting them or another operating system that is reading the stored binary object.

The safest thing to do is to convert the binary object into a `string` of safe characters. Programmers call this **Base64** encoding. Base64 is an encoding scheme that converts arbitrary bytes into text using a specific set of 64 characters. It's widely used for data transfer and has long been supported through various methods.

The `Convert` type has a pair of methods, `ToBase64String` and `FromBase64String`, that perform this conversion for you. Let's see them in action:

1. Type statements to create an array of bytes randomly populated with byte values, write each byte nicely formatted to the console, and then write the same bytes converted into Base64 to the console, as shown in the following code:

```
// Allocate an array of 128 bytes.
byte[] binaryObject = new byte[128];

// Populate the array with random bytes.
Random.Shared.NextBytes(binaryObject);

WriteLine("Binary Object as bytes:");
for (int index = 0; index < binaryObject.Length; index++)
{
  Write($"{binaryObject[index]:X2} ");
}
WriteLine();

// Convert the array to Base64 string and output as text.
string encoded = ToBase64String(binaryObject);
WriteLine($"Binary Object as Base64: {encoded}");
```

By default, an `int` value would output assuming decimal notation, that is, Base10. You can use format codes such as `:X2` to format the value using hexadecimal notation.

2. Run the code and view the result, as shown in the following output:

```
Binary Object as bytes:
EB 53 8B 11 9D 83 E6 4D 45 85 F4 68 F8 18 55 E5 B8 33 C9 B6 F4 00 10 7F
CB 59 23 7B 26 18 16 30 00 23 E6 8F A9 10 B0 A9 E6 EC 54 FB 4D 33 E1 68
50 46 C4 1D 5F B1 57 A1 DB D0 60 34 D2 16 93 39 3E FA 0B 08 08 E9 96 5D
64 CF E5 CD C5 64 33 DD 48 4F E8 B0 B4 19 51 CA 03 6F F4 18 E3 E5 C7 0C
11 C7 93 BE 03 35 44 D1 6F AA B0 2F A9 CE D5 03 A8 00 AC 28 8F A5 12 8B
2E BE 40 C4 31 A8 A4 1A
Binary Object as Base64: 61OLEZ2D5k1FhfRo+BhV5bgzybb0ABB/
y1kjeyYYFjAAI+aPqRCwqebsVPtNM+FoUEbEHV+xV6Hb0GA00haTOT76CwgI6ZZdZM/
lzcVkM91IT+iwtBlRygNv9Bjj5ccMEceTvgM1RNFvqrAvqc7VA6gArCiPpRKLLr5AxDGopBo=
```

Base64 for URLs

Base64 is useful, but some of the characters it uses, like + and /, are problematic for certain uses, such as query strings in URLs, where these characters have special meanings.

To address this issue, the Base64Url scheme was created. It is similar to Base64 but uses a slightly different set of characters, making it suitable for contexts like URLs.

More Information: You can learn more about the Base64Url scheme at the following link: `https://base64.guru/standards/base64url`.

.NET 9 introduces the new `Base64Url` class, which offers a range of optimized methods for encoding and decoding data using the `Base64Url` scheme. For example, you can convert some arbitrary bytes into Base64Url, as shown in the following code:

```
ReadOnlySpan<byte> bytes = ...;
string encoded = Base64Url.EncodeToString(bytes);
```

Parsing from strings to numbers or dates and times

The second most common conversion is from strings to numbers or date and time values.

The opposite of `ToString` is `Parse`. Only a few types have a `Parse` method, including all the number types and `DateTime`.

Let's see `Parse` in action:

1. At the top of `Program.cs`, import the namespace for working with cultures, as shown in the following code:

    ```
    using System.Globalization; // To use CultureInfo.
    ```

2. At the bottom of `Program.cs`, add statements to parse an integer and a date and time value from strings, and then write the result to the console, as shown in the following code:

    ```
    // Set the current culture to make sure date parsing works.
    CultureInfo.CurrentCulture = CultureInfo.GetCultureInfo("en-US");

    int friends = int.Parse("27");
    DateTime birthday = DateTime.Parse("4 June 1980");
    WriteLine($"I have {friends} friends to invite to my party.");
    WriteLine($"My birthday is {birthday}.");
    WriteLine($"My birthday is {birthday:D}.");
    ```

3. Run the code and view the result, as shown in the following output:

    ```
    I have 27 friends to invite to my party.
    My birthday is 6/4/1980 12:00:00 AM.
    My birthday is Wednesday, June 4, 1980.
    ```

    By default, a date and time value outputs with the short date and time format. You can use format codes such as `D` to output only the date part using the long date format.

Good Practice: Use the standard date and time format specifiers, as shown at the following link: `https://learn.microsoft.com/en-us/dotnet/standard/base-types/standard-date-and-time-format-strings#table-of-format-specifiers`.

Avoiding Parse exceptions by using the TryParse method

One problem with the `Parse` method is that it gives errors if the `string` cannot be converted:

1. Type a statement to attempt to parse a string containing letters into an integer variable, as shown in the following code:

    ```
    int count = int.Parse("abc");
    ```

2. Run the code and view the result, as shown in the following output:

    ```
    Unhandled Exception: System.FormatException: Input string was not in a correct format.
    ```

As well as the preceding exception message, you will see a stack trace. I have not included stack traces in this book because they take up too much space.

To avoid errors, you can use the `TryParse` method instead. `TryParse` attempts to convert the input `string` and returns `true` if it can convert it and `false` if it cannot. Exceptions are a relatively expensive operation so they should be avoided when possible.

The `out` keyword is required to allow the `TryParse` method to set the `count` variable when the conversion works.

Let's see `TryParse` in action:

1. Replace the `int count` declaration with statements to use the `TryParse` method and ask the user to input a count for the number of eggs, as shown in the following code:

```
Write("How many eggs are there? ");
string? input = ReadLine();

if (int.TryParse(input, out int count))
{
  WriteLine($"There are {count} eggs.");
}
else
{
  WriteLine("I could not parse the input.");
}
```

2. Run the code, enter `12`, and view the result, as shown in the following output:

```
How many eggs are there? 12
There are 12 eggs.
```

3. Run the code, enter `twelve`, and view the result, as shown in the following output:

```
How many eggs are there? twelve
I could not parse the input.
```

You can also use methods of the `System.Convert` type to convert `string` values into other types; however, like the `Parse` method, it gives an error if it cannot convert.

Understanding the Try method naming convention

.NET uses a standard signature for all methods that follow the `Try` naming convention. For any method named `Something` that returns a value of a specific type, its matching `TrySomething` method must return a `bool` to indicate success or failure and use an `out` parameter in place of the return value. For example:

```
// A method that might throw an exception.
int number = int.Parse("123");

// The Try equivalent of the method.
```

```
bool success = int.TryParse("123", out int number);

// Trying to create a Uri for a Web API.
bool success = Uri.TryCreate("https://localhost:5000/api/customers",
  UriKind.Absolute, out Uri serviceUrl);
```

Handling exceptions

You've seen several scenarios where errors have occurred when converting types. Some languages return error codes when something goes wrong. .NET uses exceptions that are richer and designed only for failure reporting. When this happens, we say *a runtime exception has been thrown.*

Other systems might use return values that could have multiple uses. For example, if the return value is a positive number, it might represent the count of rows in a table, or if the return value is a negative number, it might represent some error code.

Some third-party libraries make it easier to define "result" types that can indicate errors as well as successes. Many .NET developers prefer to use them instead of throwing exceptions. You can learn more about this in an optional online section at the end of this chapter.

When an exception is thrown, the thread is suspended, and if the calling code has defined a `try-catch` statement, then it is given a chance to handle the exception. If the current method does not handle it, then its calling method is given a chance, and so on up the call stack.

As you have seen, the default behavior of a console app is to output a message about the exception, including a stack trace, and then stop running the code. The application is terminated. This is better than allowing the code to continue executing in a potentially corrupt state. Your code should only catch and handle exceptions that it understands and can properly fix.

Good Practice: Avoid writing code that will throw an exception whenever possible, perhaps by performing `if` statement checks. Sometimes you can't, and sometimes it is best to allow the exception to be caught by a higher-level component that is calling your code. You will learn how to do this in *Chapter 4, Writing, Debugging, and Testing Functions*.

In .NET 9, exception handling uses a new implementation based on the NativeAOT exception handling model. This improves exception handling performance by 2 to 4 times in the .NET team's benchmarks.

Wrapping error-prone code in a try block

When you know that a statement can cause an error, you should wrap that statement in a `try` block. For example, parsing from text to a number can cause an error. Any statements in the `catch` block will be executed only if an exception is thrown by a statement in the `try` block.

We don't have to do anything inside the `catch` block. Let's see this in action:

1. Use your preferred code editor to add a new **Console App** / `console` project named `HandlingExceptions` to the `Chapter03` solution.
2. In `Program.cs`, delete any existing statements, and then type statements to prompt the user to enter their age and then write their age to the console, as shown in the following code:

```
WriteLine("Before parsing");
Write("What is your age? ");
string? input = ReadLine();

try
{
  int age = int.Parse(input);
  WriteLine($"You are {age} years old.");
}
catch
{
}
WriteLine("After parsing");
```

You will see the following compiler message: `Warning CS8604 Possible null reference argument for parameter 's' in 'int int.Parse(string s)'`.

By default, in .NET 6 or later projects, Microsoft enables nullable reference types, so you will see many more compiler warnings like this. In production code, you should add code to check for `null` and handle that possibility appropriately, as shown in the following code:

```
if (input is null)
{
  WriteLine("You did not enter a value so the app has ended.");
  return; // Exit the app.
}
```

In this book, I will not give instructions to add these `null` checks every time because the code samples are not designed to be production-quality, and having `null` checks everywhere will clutter the code and use up valuable pages.

You will probably see hundreds more examples of potentially `null` variables throughout the code samples in this book. Those warnings are safe to ignore for the book code examples. You only need to pay attention to similar warnings when you write your own production code. You will see more about null handling in *Chapter 6, Implementing Interfaces and Inheriting Classes*.

In this case, it is impossible for `input` to be `null` because the user must press *Enter* for `ReadLine` to return, and if they have not typed any characters at that point, then the `ReadLine` method will return an empty `string`. Let's tell the compiler that it does not need to show us this warning:

1. To disable the compiler warning, change `input` to `input!`, as shown highlighted in the following code:

```
int age = int.Parse(input!);
```

An exclamation mark, `!`, after an expression is called the **null-forgiving operator** and it disables the compiler warning. The **null-forgiving operator** has no effect at runtime. If the expression could evaluate to `null` at runtime, perhaps because we assigned it in another way, then an exception would be thrown.

This code includes two messages to indicate *before* parsing and *after* parsing to make the flow through the code clearer. These will be especially useful as the example code grows more complex.

2. Run the code, enter `49`, and view the result, as shown in the following output:

```
Before parsing
What is your age? 49
You are 49 years old.
After parsing
```

3. Run the code, enter `Kermit`, and view the result, as shown in the following output:

```
Before parsing
What is your age? Kermit
After parsing
```

When the code was executed, the error exception was caught, the default message and stack trace were not output, and the console app continued running. This is better than the default behavior, but it might be useful to see the type of error that occurred.

Good Practice: You should never use an empty `catch` statement like this in production code because it "swallows" exceptions and hides potential problems. You should at least log the exception if you cannot or do not want to handle it properly, or rethrow it so that higher-level code can decide instead. You will learn about logging in *Chapter 4, Writing, Debugging, and Testing Functions*.

Catching all exceptions

To get information about any type of exception that might occur, you can declare a variable of type `System.Exception` to the `catch` block:

1. Add an exception variable declaration to the `catch` block and use it to write information about the exception to the console, as shown in the following code:

```
catch (Exception ex)
{
```

```
  WriteLine($"{ex.GetType()} says {ex.Message}");
}
```

2. Run the code, enter `Kermit` again, and view the result, as shown in the following output:

```
Before parsing
What is your age? Kermit
System.FormatException says Input string was not in a correct format.
After parsing
```

Catching specific exceptions

Now that we know which specific type of exception occurred, we can improve our code by catching just that type of exception and customizing the message that we display to the user. You can think of this as a form of testing:

1. Leave the existing `catch` block and, above it, add a new `catch` block for the format exception type, as shown in the following highlighted code:

```
catch (FormatException)
{
  WriteLine("The age you entered is not a valid number format.");
}
catch (Exception ex)
{
  WriteLine($"{ex.GetType()} says {ex.Message}");
}
```

2. Run the code, enter `Kermit` again, and view the result, as shown in the following output:

```
Before parsing
What is your age? Kermit
The age you entered is not a valid number format.
After parsing
```

The reason we want to leave the more general `catch` below is that there might be other types of exceptions that can occur.

3. Run the code, enter `9876543210`, and view the result, as shown in the following output:

```
Before parsing
What is your age? 9876543210
System.OverflowException says Value was either too large or too small for
an Int32.
After parsing
```

Let's add another `catch` block for this type of exception.

4. Leave the existing `catch` blocks, and add a new `catch` block for the overflow exception type, as shown in the following highlighted code:

```
catch (OverflowException)
{
  WriteLine("Your age is a valid number format but it is either too big
or small.");
}
catch (FormatException)
{
  WriteLine("The age you entered is not a valid number format.");
}
```

5. Run the code, enter `9876543210`, and view the result, as shown in the following output:

```
Before parsing
What is your age? 9876543210
Your age is a valid number format but it is either too big or small.
After parsing
```

The order in which you catch exceptions is important. The correct order is related to the inheritance hierarchy of the exception types. You will learn about inheritance in *Chapter 5, Building Your Own Types with Object-Oriented Programming*. However, don't worry too much about this—the compiler will give you build errors if you get exceptions in the wrong order anyway.

Good Practice: Avoid over-catching exceptions. They should often be allowed to propagate up the call stack to be handled at a level where more information is known about the circumstances that could change the logic of how they should be handled. You will learn about this in *Chapter 4, Writing, Debugging, and Testing Functions*.

Catching with filters

You can also add filters to a `catch` statement using the `when` keyword, as shown in the following code:

```
Write("Enter an amount: ");
string amount = ReadLine()!;
if (string.IsNullOrEmpty(amount)) return;

try
{
  decimal amountValue = decimal.Parse(amount);
  WriteLine($"Amount formatted as currency: {amountValue:C}");
}
catch (FormatException) when (amount.Contains('$'))
{
```

```
  WriteLine("Amounts cannot use the dollar sign!");
}
catch (FormatException)
{
  WriteLine("Amounts must only contain digits!");
}
```

Good Practice: The `Contains` method on the `string` type has overloads for both a `string` value that is passed using double-quotes and a `char` value that is passed using single-quotes. When you want to check for one character, like a dollar, it is more efficient to use the `char` overload as in the preceding code.

Checking for overflow

Earlier, we saw that when casting between number types, it was possible to lose information, for example, when casting from a `long` variable to an `int` variable. If the value stored in a type is too big, it will overflow.

Throwing overflow exceptions with the checked statement

The `checked` statement tells .NET to throw an exception when an overflow happens instead of allowing it to happen silently, which is done by default for performance reasons.

We will set the initial value of an `int` variable to its maximum value minus one. Then, we will increment it several times, outputting its value each time. Once it gets above its maximum value, it overflows to its minimum value and continues incrementing from there.

Let's see this in action:

1. In `Program.cs`, type statements to declare and assign an integer to one less than its maximum possible value, and then increment it and write its value to the console three times, as shown in the following code:

   ```
   int x = int.MaxValue - 1;
   WriteLine($"Initial value: {x}");
   x++;
   WriteLine($"After incrementing: {x}");
   x++;
   WriteLine($"After incrementing: {x}");
   x++;
   WriteLine($"After incrementing: {x}");
   ```

2. Run the code and view the result that shows the value overflowing silently and wrapping around to large negative values, as shown in the following output:

```
Initial value: 2147483646
After incrementing: 2147483647
After incrementing: -2147483648
After incrementing: -2147483647
```

3. Now, let's get the compiler to warn us about the overflow by wrapping the statements using a checked statement block, as shown highlighted in the following code:

```
checked
{
  int x = int.MaxValue - 1;
  WriteLine($"Initial value: {x}");
  x++;
  WriteLine($"After incrementing: {x}");
  x++;
  WriteLine($"After incrementing: {x}");
  x++;
  WriteLine($"After incrementing: {x}");
}
```

4. Run the code and view the result that shows the overflow being checked and causing an exception to be thrown, as shown in the following output:

```
Initial value: 2147483646
After incrementing: 2147483647
Unhandled Exception: System.OverflowException: Arithmetic operation
resulted in an overflow.
```

5. Just like any other exception, we should wrap these statements in a `try` statement block and display a nicer error message for the user, as shown in the following code:

```
try
{
  // previous code goes here
}
catch (OverflowException)
{
  WriteLine("The code overflowed but I caught the exception.");
}
```

6. Run the code and view the result, as shown in the following output:

```
Initial value: 2147483646
After incrementing: 2147483647
The code overflowed but I caught the exception.
```

Disabling compiler overflow checks with the unchecked statement

The previous section was about the default overflow behavior at *runtime* and how to use the `checked` statement to change that behavior. This section is about *compile-time* overflow behavior and how to use the `unchecked` statement to change that behavior.

A related keyword is `unchecked`. This keyword switches off overflow checks performed by the compiler within a block of code. Let's see how to do this:

1. Type the following statement at the end of the previous statements. The compiler will not compile this statement because it knows it will overflow:

```
int y = int.MaxValue + 1;
```

2. Hover your mouse pointer over the error, and note that a compile-time check is shown as an error message, as shown in *Figure 3.4*:

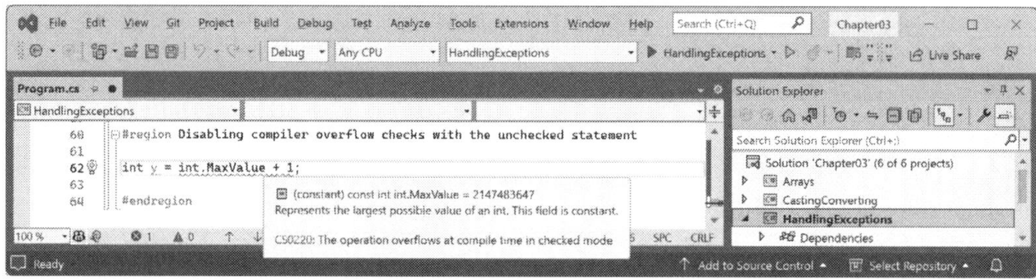

Figure 3.4: A compile-time check for integer overflow

3. To disable compile-time checks, wrap the statement in an `unchecked` block, write the value of `y` to the console, decrement it, and repeat, as shown in the following code:

```
unchecked
{
  int y = int.MaxValue + 1;
  WriteLine($"Initial value: {y}");
  y--;
  WriteLine($"After decrementing: {y}");
  y--;
  WriteLine($"After decrementing: {y}");
}
```

4. Run the code and view the results, as shown in the following output:

```
Initial value: -2147483648
After decrementing: 2147483647
After decrementing: 2147483646
```

Of course, it would be rare that you would want to explicitly switch off a check like this because it allows an overflow to occur. But perhaps you can think of a scenario where you might want that behavior.

Practicing and exploring

Test your knowledge and understanding by answering some questions, getting some hands-on practice, and exploring this chapter's topics with deeper research.

Exercise 3.1 – Online material

Online material can be extra content written by me for this book, or it can be references to content created by Microsoft or third parties.

Returning result types versus throwing exceptions

In .NET programming, handling errors often revolves around two primary mechanisms: exceptions and result types. Learn more about result types by reading the following online section:

`https://github.com/markjprice/cs13net9/blob/main/docs/ch03-result-types.md`

C# 101 notebooks

Use the links to notebooks and videos at the following link to see interactive examples of C# using Polyglot Notebooks:

`https://github.com/dotnet/csharp-notebooks#c-101`

Exercise 3.2 – Practice exercises

Practice exercises go deeper into the topics for this chapter.

Loops and overflow

What will happen if this code executes?

```
int max = 500;
for (byte i = 0; i < max; i++)
{
  WriteLine(i);
}
```

Create a console app in `Chapter03` named `Exercise_LoopsAndOverflow` and enter the preceding code. Run the console app and view the output. What happens?

What code could you add (don't change any of the preceding code) to warn us about the problem?

Practice loops and operators

FizzBuzz is a group game for children to teach them about division. Players take turns to count incrementally, replacing any number divisible by 3 with the word *fizz*, any number divisible by 5 with the word *buzz*, and any number divisible by both with *fizzbuzz*.

Create a console app in `Chapter03` named `Exercise_FizzBuzz` that outputs a simulated FizzBuzz game that counts up to 100. The output should look something like *Figure 3.5*:

```
1, 2, Fizz, 4, Buzz, Fizz, 7, 8, Fizz, Buzz,
11, Fizz, 13, 14, FizzBuzz, 16, 17, Fizz, 19, Buzz,
Fizz, 22, 23, Fizz, Buzz, 26, Fizz, 28, 29, FizzBuzz,
31, 32, Fizz, 34, Buzz, Fizz, 37, 38, Fizz, Buzz,
41, Fizz, 43, 44, FizzBuzz, 46, 47, Fizz, 49, Buzz,
Fizz, 52, 53, Fizz, Buzz, 56, Fizz, 58, 59, FizzBuzz,
61, 62, Fizz, 64, Buzz, Fizz, 67, 68, Fizz, Buzz,
71, Fizz, 73, 74, FizzBuzz, 76, 77, Fizz, 79, Buzz,
Fizz, 82, 83, Fizz, Buzz, 86, Fizz, 88, 89, FizzBuzz,
91, 92, Fizz, 94, Buzz, Fizz, 97, 98, Fizz, Buzz
```

Figure 3.5: A simulated FizzBuzz game output

Practice exception handling

Create a console app in `Chapter03` named `Exercise_Exceptions` that asks the user for two numbers in the range 0–255 and then divides the first number by the second:

```
Enter a number between 0 and 255: 100
Enter another number between 0 and 255: 8
100 divided by 8 is 12
```

Write exception handlers to catch any thrown errors, as shown in the following output:

```
Enter a number between 0 and 255: apples
Enter another number between 0 and 255: bananas
FormatException: Input string was not in a correct format.
```

Exercise 3.3 – Test your knowledge

Answer the following questions:

1. What happens when you divide an `int` variable by `0`?
2. What happens when you divide a `double` variable by `0`?
3. What happens when you overflow an `int` variable, that is, set it to a value beyond its range?
4. What is the difference between `x = y++;` and `x = ++y;`?
5. What is the difference between `break`, `continue`, and `return` when used inside a loop statement?
6. What are the three parts of a `for` statement and which of them are required?
7. What is the difference between the = and == operators?
8. Does the following statement compile?

    ```
    for ( ; ; ) ;
    ```

9. What does the underscore (_) represent in a `switch` expression?
10. What interface must an object "implement" to be enumerated over by using the `foreach` statement?

Test your knowledge of operators

What are the values of x and y after the following statements execute? Create a console app in Chapter03 named Ch03Ex03Operators to test your assumptions:

1. Increment and addition operators:

```
x = 3;
y = 2 + ++x;
```

2. Binary shift operators:

```
x = 3 << 2;
y = 10 >> 1;
```

3. Bitwise operators:

```
x = 10 & 8;
y = 10 | 7;
```

Exercise 3.4 – Explore topics

Use the links on the following page to learn about the topics covered in this chapter in more detail:

https://github.com/markjprice/cs13net9/blob/main/docs/book-links.md#chapter-3---controlling-flow-converting-types-and-handling-exceptions

Summary

In this chapter, you learned how to:

- Use operators to perform simple tasks.
- Use branch and loop statements to implement logic.
- Work with single- and multi-dimensional arrays.
- Convert between types.
- Catch exceptions and handle integer overflow.

You are now ready to learn how to reuse blocks of code by defining functions, how to pass values into them and get values back, and how to track down bugs in your code and squash them using debugging and testing tools!

4

Writing, Debugging, and Testing Functions

This chapter is about writing functions to reuse code, debugging logic errors during development, unit testing your code to remove bugs, and improving stability and reliability.

In an optional online-only section, you will learn about logging exceptions during runtime. You can find information on this at the end of the chapter.

This chapter covers the following topics:

- Writing functions
- Debugging during development
- Hot reloading during development
- Unit testing
- Throwing and catching exceptions in functions

Writing functions

A fundamental principle of programming is **don't repeat yourself** (**DRY**).

While programming, if you find yourself writing the same statements over and over again, then turn those statements into a **function**. Functions are like tiny programs that complete one small task. For example, you might write a function to calculate sales tax and then reuse that function in many places in a financial application.

Like programs, functions usually have inputs and outputs. They are sometimes described as black boxes, where you feed in some raw materials at one end and a manufactured item emerges at the other. Once created and thoroughly debugged and tested, you don't need to think about how they work.

Exploring top-level programs, functions, and namespaces

In *Chapter 1, Hello, C#! Welcome, .NET!*, we learned that since C# 10 and .NET 6, the default project template for console apps uses the top-level program feature introduced with C# 9.

Once you start writing functions, it is important to understand how they work with the automatically generated `Program` class and its `<Main>$` method.

Let's explore how the top-level program feature works when you define functions:

1. Use your preferred code editor to create a new solution and project, as defined in the following list:
    - **Project template: Console App** / `console`
    - **Project file and folder:** `TopLevelFunctions`
    - **Solution file and folder:** `Chapter04`
    - **Do not use top-level statements:** Cleared
    - **Enable native AOT** (**ahead-of-time**) **publish:** Cleared
2. In `Program.cs`, delete the existing statements, define a local function at the bottom of the file, and call it, as shown in the following code:

```
using static System.Console;

WriteLine("* Top-level functions example");

WhatsMyNamespace(); // Call the function.

void WhatsMyNamespace() // Define a local function.
{
  WriteLine("Namespace of Program class: {0}",
    arg0: typeof(Program).Namespace ?? "null");
}
```

Good Practice: Functions do not need to be at the bottom of the file, but it is good practice rather than mixing them up with other top-level statements. Types, like classes, *must* be declared at the bottom of the `Program.cs` file rather than in the middle of the file or you will see the compiler error `CS8803`, as shown at the following link: `https://learn.microsoft.com/en-us/dotnet/csharp/language-reference/compiler-messages/cs8803`. It would be better to define types like classes in a separate file.

3. Run the console app and note that the namespace for the `Program` class is `null`, as shown in the following output:

```
* Top-level functions example
Namespace of Program class: null
```

What is automatically generated for a local function?

The compiler automatically generates a `Program` class with a `<Main>$` function, then moves your statements and function inside the `<Main>$` method, which makes the function local, and renames the function, as shown highlighted in the following code:

```
using static System.Console;

partial class Program
{
  static void <Main>$(String[] args)
  {
    WriteLine("* Top-level functions example");

    <<Main>$>g__WhatsMyNamespace|0_0(); // Call the function.

    void <<Main>$>g__WhatsMyNamespace|0_0() // Define a local function.
    {
      WriteLine("Namespace of Program class: {0}",
        arg0: typeof(Program).Namespace ?? "null");
    }
  }
}
```

For the compiler to know what statements need to go where, you must follow some rules:

- Import statements (`using`) must go at the top of the `Program.cs` file.
- Statements that will go in the `<Main>$` function can be mixed with functions in the middle of the `Program.cs` file. Any functions will become **local functions** in the `<Main>$` method.

The last point is important because local functions have limitations, such as they cannot have XML comments to document them.

If the contents of the file which is code were the .NET SDK or some other tool that generated this code, then the code would need to be in a source code file that the compiler would then find in the filesystem and compile it. Because this code is generated by the compiler itself, there is no need for a source code file. The only way to discover what the compiler does is to use a decompiler on the assembly and reverse engineer the original code. You can also throw exceptions in the functions and methods to see some of the information, as I showed in *Chapter 1, Hello, C#! Welcome, .NET!*.

You are about to see some C# keywords like `static` and `partial`, which will be formally introduced in *Chapter 5, Building Your Own Types with Object-Oriented Programming*.

Defining a partial Program class with a static function

A better approach is to write any functions in a separate file and define them as `static` members of the `Program` class:

1. Add a new class file named `Program.Functions.cs`. The name of this file does not actually matter but using this naming convention is sensible. You could name the file `Gibberish.cs` and it would have the same behavior.
2. In `Program.Functions.cs`, delete any existing statements and then add statements to define a `partial Program` class. Cut and paste the `WhatsMyNamespace` function to move it from `Program.cs` into `Program.Functions.cs`, and then add the `static` keyword to the function, as shown highlighted in the following code:

```
using static System.Console;

// Do not define a namespace so this class goes in the default empty
// namespace just like the auto-generated partial Program class.

partial class Program
{
  static void WhatsMyNamespace() // Define a static function.
  {
    WriteLine("Namespace of Program class: {0}",
      arg0: typeof(Program).Namespace ?? "null");
  }
}
```

3. In `Program.cs`, confirm that its entire content is now just three statements, as shown in the following code:

```
using static System.Console;

WriteLine("* Top-level functions example");

WhatsMyNamespace(); // Call the function.
```

4. Run the console app and note that it has the same behavior as before.

What is automatically generated for a static function?

When you use a separate file to define a `partial Program` class with `static` functions, the compiler defines a `Program` class with a `<Main>$` function and merges your function as a member of the `Program` class, as shown in the following highlighted code:

```
using static System.Console;

partial class Program
{
  static void <Main>$(String[] args)
  {
    WriteLine("* Top-level functions example");

    WhatsMyNamespace(); // Call the function.
  }

  static void WhatsMyNamespace() // Define a static function.
  {
    WriteLine("Namespace of Program class: {0}",
      arg0: typeof(Program).Namespace ?? "null");
  }
}
```

Solution Explorer shows that your `Program.Functions.cs` class file merges its `partial Program` with the auto-generated `partial Program` class, as shown in *Figure 4.1*:

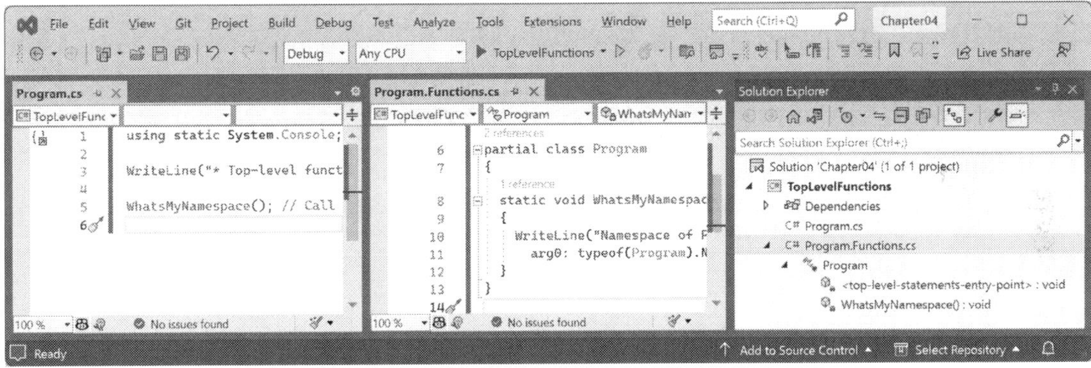

Figure 4.1: Solution Explorer showing the merged partial Program class

Good Practice: Create any functions that you will call in `Program.cs` in a separate file and manually define them inside a `partial Program` class. This will merge them into the automatically generated `Program` class *at the same level* as the `<Main>$` method, instead of as local functions *inside* the `<Main>$` method.

It is important to note the lack of namespace declarations. Both the automatically generated `Program` class and the explicitly defined `Program` class are in the default `null` namespace.

Warning! Do not define a namespace for your `partial Program` class. If you do, it will be in a different namespace and therefore will not merge with the auto-generated `partial Program` class.

Optionally, all the `static` methods in the `Program` class could be explicitly declared as `private` but this is the default anyway. Since all the functions will be called within the `Program` class itself, the access modifier is not important.

Times table example

Let's say that you want to help your child learn their times tables, so you want to make it easy to generate a times table for a number, such as the 7 times table:

```
1 x 7 = 7
2 x 7 = 14
3 x 7 = 21
...
10 x 7 = 70
11 x 7 = 77
12 x 7 = 84
```

Most times tables have 10, 12, or 20 rows, depending on how advanced the child is.

You learned about the `for` statement earlier in this book, so you know that it can be used to generate repeated lines of output when there is a regular pattern, such as a 7 times table with 12 rows, as shown in the following code:

```
for (int row = 1; row <= 12; row++)
{
  Console.WriteLine($"{row} x 7 = {row * 7}");
}
```

However, instead of always outputting the 7 times table with 12 rows, we want to make this more flexible so it can output times tables of any size for any number. We can do this by creating a function.

Let's explore functions by creating one to output any times table for numbers 0 to 255 of any size up to 255 rows (but it defaults to 12 rows):

1. Use your preferred code editor to create a new project, as defined in the following list:
    - Project template: **Console App** / `console`
    - Project file and folder: `WritingFunctions`
    - Solution file and folder: `Chapter04`
    - In Visual Studio, set the startup project for the solution to the current selection.

2. In `WritingFunctions.csproj`, after the `<PropertyGroup>` section, add a new `<ItemGroup>` section to statically import `System.Console` for all C# files using the implicit usings .NET SDK feature, as shown in the following markup:

```
<ItemGroup>
  <Using Include="System.Console" Static="true" />
</ItemGroup>
```

3. Add a new class file to the project named `Program.Functions.cs`.
4. In `Program.Functions.cs`, replace any existing code with new statements to define a function named `TimesTable` in the `partial Program` class, as shown in the following code:

```
partial class Program
{
  static void TimesTable(byte number, byte size = 12)
  {
    WriteLine($"This is the {number} times table with {size} rows:");
    WriteLine();

    for (int row = 1; row <= size; row++)
    {
      WriteLine($"{row} x {number} = {row * number}");
    }
    WriteLine();
  }
}
```

In the preceding code, note the following:

- `TimesTable` must have a `byte` value passed to it as a parameter named `number`.
- `TimesTable` can optionally have a `byte` value passed to it as a parameter named `size`. If a value is not passed, it defaults to `12`.
- `TimesTable` is a `static` method because it will be called by the `static` method `<Main>$`.
- `TimesTable` does not return a value to the caller, so it is declared with the `void` keyword before its name.
- `TimesTable` uses a `for` statement to output the times table for the `number` passed to it with its number of rows equal to `size`.

5. In `Program.cs`, delete the existing statements, and then call the function. Pass in a `byte` value for the `number` parameter, for example, `7`, as shown in the following code:

```
TimesTable(7);
```

6. Run the code and then view the result, as shown in the following output:

```
This is the 7 times table with 12 rows:

1 x 7 = 7
2 x 7 = 14
3 x 7 = 21
4 x 7 = 28
5 x 7 = 35
6 x 7 = 42
7 x 7 = 49
8 x 7 = 56
9 x 7 = 63
10 x 7 = 70
11 x 7 = 77
12 x 7 = 84
```

7. Set the `size` parameter to `20`, as shown in the following code:

```
TimesTable(7, 20);
```

8. Run the console app and confirm that the times table now has 20 rows.

Good Practice: If a function has one or more parameters where just passing the values may not provide enough meaning, then you can optionally specify the name of the parameter as well as its value, as shown in the following code: `TimesTable(number: 7, size: 10)`.

9. Change the number passed into the `TimesTable` function to other `byte` values between `0` and `255` and confirm that the output times tables are correct.
10. Note that if you try to pass a non-`byte` number, for example, an `int`, `double`, or `string`, an error is returned, as shown in the following output:

```
Error: (1,12): error CS1503: Argument 1: cannot convert from 'int' to
'byte'
```

A brief aside about arguments and parameters

In daily usage, most developers will use the terms **argument** and **parameter** interchangeably. Strictly speaking, the two terms have specific and subtly different meanings. But just like a person can be both a parent and a doctor, the two terms often apply to the same thing.

A *parameter* is a variable in a function definition. For example, `startDate` is a parameter of the `Hire` function, as shown in the following code:

```
void Hire(DateTime startDate)
```

```
{
  // Function implementation.
}
```

When a method is called, an *argument* is the data you pass into the method's parameters. For example, `when` is a variable passed as an argument to the `Hire` function, as shown in the following code:

```
DateTime when = new(year: 2024, month: 11, day: 5);
Hire(when);
```

You might prefer to specify the parameter name when passing the argument, as shown in the following code:

```
DateTime when = new(year: 2024, month: 11, day: 5);
Hire(startDate: when);
```

When talking about the call to the `Hire` function, `startDate` is the parameter, and `when` is the argument.

If you read the official Microsoft documentation, they use the phrases **named and optional arguments** and **named and optional parameters** interchangeably, as shown at the following link: `https://learn.microsoft.com/en-us/dotnet/csharp/programming-guide/classes-and-structs/named-and-optional-arguments`.

It gets complicated because a single object can act as both a parameter and an argument, depending on the context. For example, within the `Hire` function implementation, the `startDate` parameter could be passed as an argument to another function like `SaveToDatabase`, as shown in the following code:

```
void Hire(DateTime startDate)
{
  ...
  SaveToDatabase(startDate, employeeRecord);
  ...
}
```

Naming things is one of the hardest parts of computing. A classic example is the parameter to the most important function in C#, `Main`. It defines a parameter named `args`, short for arguments, as shown in the following code:

```
static void Main(String[] args)
{
  ...
}
```

To summarize, parameters define inputs to a function; arguments are passed to a function when calling the function.

Good Practice: Try to use the correct term depending on the context, but do not get pedantic with other developers if they "misuse" a term. I must have used the terms **parameter** and **argument** thousands of times in this book. I'm sure some of those times I've been imprecise. Please do not @ me about it.

Writing a function that returns a value

The previous function performed actions (looping and writing to the console), but it did not return a value. Let's say that you need to calculate sales or **value-added tax** (**VAT**). In Europe, VAT rates can range from 8% in Switzerland to 27% in Hungary. In the **United States** (**US**), state sales taxes can range from 0% in Oregon to 8.25% in California.

Tax rates change all the time, and they vary based on many factors. The values used in this example do not need to be accurate.

Let's implement a function to calculate taxes in various regions around the world:

1. In `Program.Functions.cs`, in the `Program` class, write a function named `CalculateTax`, as shown in the following code:

```
static decimal CalculateTax(
  decimal amount, string twoLetterRegionCode)
{
  decimal rate = twoLetterRegionCode switch
  {
    "CH" => 0.08M, // Switzerland
    "DK" or "NO" => 0.25M, // Denmark, Norway
    "GB" or "FR" => 0.2M, // UK, France
    "HU" => 0.27M, // Hungary
    "OR" or "AK" or "MT" => 0.0M, // Oregon, Alaska, Montana
    "ND" or "WI" or "ME" or "VA" => 0.05M,
    "CA" => 0.0825M, // California
    _ => 0.06M // Most other states.
  };

  return amount * rate;
}
```

In the preceding code, note the following:

- `CalculateTax` has two inputs: a parameter named `amount`, which will be the amount of money spent, and a parameter named `twoLetterRegionCode`, which will be the region where the amount is spent.

- `CalculateTax` will perform a calculation using a `switch` expression and then return the sales tax or VAT owed on the amount as a `decimal` value; so, before the name of the function, we have declared the data type of the return value to be `decimal`.

2. At the top of `Program.Functions.cs`, import the namespace to work with cultures, as shown in the following code:

```
using System.Globalization; // To use CultureInfo.
```

3. In `Program.Functions.cs`, in the `Program` class, write a function named `ConfigureConsole`, as shown in the following code:

```
static void ConfigureConsole(string culture = "en-US",
  bool useComputerCulture = false)
{
  // To enable Unicode characters like Euro symbol in the console.
  OutputEncoding = System.Text.Encoding.UTF8;

  if (!useComputerCulture)
  {
    CultureInfo.CurrentCulture = CultureInfo.GetCultureInfo(culture);
  }
  WriteLine($"CurrentCulture: {CultureInfo.CurrentCulture.DisplayName}");
}
```

This function enables UTF-8 encoding for the console output. This is necessary to output some special symbols like the Euro currency symbol. This function also controls the current culture used to format dates, times, and currency values.

4. In `Program.cs`, comment out any `TimesTable` method calls, and then call the `ConfigureConsole` method and the `CalculateTax` method, passing values for the amount, such as `149`, and a valid region code, such as `FR`, as shown in the following code:

```
// TimesTable(number: 7, size: 10);

ConfigureConsole();

decimal taxToPay = CalculateTax(amount: 149, twoLetterRegionCode: "FR");
WriteLine($"You must pay {taxToPay:C} in tax.");

// Alternatively, call the function in the interpolated string.
// WriteLine($"You must pay {CalculateTax(amount: 149,
//   twoLetterRegionCode: "FR"):C} in tax.");
```

5. Run the code, view the result, and note that it uses the US English culture, meaning US dollars for the currency, as shown in the following output:

```
CurrentCulture: English (United States)
You must pay $29.80 in tax.
```

6. In `Program.cs`, change the `ConfigureConsole` method to use your local computer culture, as shown in the following code:

```
ConfigureConsole(useComputerCulture: true);
```

7. Run the code, view the result, and note that the currency should now show your local currency. For example, for me in the UK, I would see `£29.80`, as shown in the following output:

```
CurrentCulture: English (United Kingdom)
You must pay £29.80 in tax.
```

8. In `Program.cs`, change the `ConfigureConsole` method to use the French culture, as shown in the following code:

```
ConfigureConsole(culture: "fr-FR");
```

9. Run the code, view the result, and note that the currency should now show Euros, as used in France, as shown in the following output:

```
CurrentCulture: French (France)
You must pay 29,80 € in tax.
```

Can you think of any problems with the `CalculateTax` function as written? What would happen if the user entered a code such as `fr` or `UK`? How could you rewrite the function to improve it? Would using a `switch` *statement* instead of a `switch` *expression* be clearer?

Converting numbers from cardinal to ordinal

Numbers that are used to count are called **cardinal** numbers, for example, 1, 2, and 3, whereas numbers used to order are **ordinal** numbers, for example, 1st, 2nd, and 3rd. Let's create a function to convert cardinals to ordinals:

1. In `Program.Functions.cs`, write a function named `CardinalToOrdinal` that converts a cardinal `uint` value into an ordinal `string` value; for example, it converts the `uint` value of `1` into the `string` value `"1st"`, the `uint` value of `2` into the `string` value of `"2nd"`, and so on, as shown in the following code:

```
static string CardinalToOrdinal(uint number)
{
  uint lastTwoDigits = number % 100;

  switch (lastTwoDigits)
  {
```

```
      case 11: // Special cases for 11th to 13th.
      case 12:
      case 13:
        return $"{number:N0}th";
      default:
        uint lastDigit = number % 10;

        string suffix = lastDigit switch
        {
          1 => "st",
          2 => "nd",
          3 => "rd",
          _ => "th"
        };

        return $"{number:N0}{suffix}";
    }
  }
```

From the preceding code, note the following:

- `CardinalToOrdinal` has one input, a parameter of the `uint` type named `number`, because we do not want to allow negative numbers, and one output: a return value of the `string` type.
- A `switch` *statement* is used to handle the special cases of `11`, `12`, and `13`.
- A `switch` *expression* then handles all other cases: if the last digit is `1`, then use `st` as the suffix; if the last digit is `2`, then use `nd` as the suffix; if the last digit is `3`, then use `rd` as the suffix; and if the last digit is anything else, then use `th` as the suffix.

2. In `Program.Functions.cs`, write a function named `RunCardinalToOrdinal` that uses a `for` statement to loop from 1 to 150, calling the `CardinalToOrdinal` function for each number and writing the returned `string` to the console, separated by a space character, as shown in the following code:

```
static void RunCardinalToOrdinal()
{
  for (uint number = 1; number <= 150; number++)
  {
    Write($"{CardinalToOrdinal(number)} ");
  }
  WriteLine();
}
```

3. In `Program.cs`, comment out the `CalculateTax` statements and call the `RunCardinalToOrdinal` method, as shown in the following code:

```
RunCardinalToOrdinal();
```

4. Run the console app and view the results, as shown in the following output:

```
1st 2nd 3rd 4th 5th 6th 7th 8th 9th 10th 11th 12th 13th 14th 15th 16th
17th 18th 19th 20th 21st 22nd 23rd 24th 25th 26th ... 139th 140th 141st
142nd 143rd 144th 145th 146th 147th 148th 149th 150th
```

5. In the `RunCardinalToOrdinal` function, change the maximum number to `1500`.
6. Run the console app and view the results, as shown in the following output:

```
... 1,480th 1,481st 1,482nd 1,483rd 1,484th 1,485th 1,486th 1,487th
1,488th 1,489th 1,490th 1,491st 1,492nd 1,493rd 1,494th 1,495th 1,496th
1,497th 1,498th 1,499th 1,500th
```

Calculating factorials with recursion

The factorial of 5 is 120 because factorials are calculated by multiplying the starting number by one less than itself, and then by one less again, and so on until the number is reduced to 1. An example can be seen here: 5 x 4 x 3 x 2 x 1 = 120.

The factorial function is defined for non-negative integers only, i.e., for 0, 1, 2, 3, and so on, and it is defined as:

```
0! = 1
n! = n × (n − 1)!, for n  { 1, 2, 3, ... }
```

We could leave it to the compiler to reject negative numbers by declaring the input parameter as `uint` as we did for the `CardinalToOrdinal` function, but this time, let's see an alternative way to handle that: throwing an argument exception.

Factorials are written like this: `5!`, where the exclamation mark is read as "bang," so `5!` = 120, or *five bang equals one hundred and twenty*. Bang is a good term to use in the context of factorials because they increase in size very rapidly, just like an explosion.

We will write a function named `Factorial`; this will calculate the factorial for an `int` passed to it as a parameter. We will use a clever technique called **recursion**, which refers to a function that calls itself within its implementation, either directly or indirectly:

1. In `Program.Functions.cs`, write a function named `Factorial`, as shown in the following code:

```
static int Factorial(int number)
{
  if (number < 0)
  {
    throw new ArgumentOutOfRangeException(message:
```

```
      $"The factorial function is defined for non-negative integers
only. Input: {number}",
      paramName: nameof(number));
  }
  else if (number == 0)
  {
    return 1;
  }
  else
  {
    return number * Factorial(number - 1);
  }
}
```

As before, there are several noteworthy elements in the preceding code, including the following:

- If the input parameter `number` is negative, `Factorial` throws an exception.
- If the input parameter `number` is `0`, `Factorial` returns `1`.
- If the input parameter `number` is more than `0` (which it will be in all other cases), `Factorial` multiplies the number by the result of calling itself and passing one less than `number`. This makes the function recursive.

More Information: Recursion is clever but it can lead to problems, such as a stack overflow due to too many function calls because memory is used to store data on every function call, and it eventually uses too much. Iteration is a more practical, if less succinct, solution in languages such as C#. You can read more about this at the following link: https://en.wikipedia.org/wiki/Recursion_(computer_science)#Recursion_versus_iteration.

2. In `Program.Functions.cs`, write a function named `RunFactorial` that uses a `for` statement to output the factorials of numbers from 1 to 15, calls the `Factorial` function inside its loop, and then outputs the result, formatted using the code `N0`, which refers to a number format using thousand separators with zero decimal places, as shown in the following code:

```
static void RunFactorial()
{
  for (int i = 1; i <= 15; i++)
  {
    WriteLine($"{i}! = {Factorial(i):N0}");
  }
}
```

3. Comment out the `RunCardinalToOrdinal` method call and call the `RunFactorial` method.

4. Run the project and view the results, as shown in the following partial output:

```
1! = 1
2! = 2
3! = 6
4! = 24
...
12! = 479,001,600
13! = 1,932,053,504
14! = 1,278,945,280
15! = 2,004,310,016
```

It is not immediately obvious in the previous output, but factorials of 13 and higher overflow the `int` type because they are so big. `12!` is 479,001,600, which is about half a billion. The maximum positive value that can be stored in an `int` variable is about two billion. `13!` is 6,227,020,800, which is about six billion, and when stored in a 32-bit integer, it overflows silently without showing any problems.

What should you do to get notified when an overflow happens? Of course, we could solve the problem for `13!` and `14!` by using a `long` (64-bit integer) instead of an `int` (32-bit integer), but we will quickly hit the overflow limit again.

The point of this section is to understand and show you that numbers can overflow, and not specifically how to calculate factorials higher than `12!`. Let's take a look:

1. Modify the `Factorial` function to check for overflows in the statement that calls itself, as shown highlighted in the following code:

```
checked // for overflow
{
  return number * Factorial(number - 1);
}
```

2. Modify the `RunFactorial` function to change the starting number to `-2` and to handle overflow and other exceptions when calling the `Factorial` function, as shown highlighted in the following code:

```
static void RunFactorial()
{
  for (int i = -2; i <= 15; i++)
  {
    try
    {
      WriteLine($"{i}! = {Factorial(i):N0}");
    }
    catch (OverflowException)
    {
```

```
      WriteLine($"{i}! is too big for a 32-bit integer.");
    }
    catch (Exception ex)
    {
      WriteLine($"{i}! throws {ex.GetType()}: {ex.Message}");
    }
  }
}
```

3. Run the code and view the results, as shown in the following partial output:

```
-2! throws System.ArgumentOutOfRangeException: The factorial function is
defined for non-negative integers only. Input: -2 (Parameter 'number')
-1! throws System.ArgumentOutOfRangeException: The factorial function is
defined for non-negative integers only. Input: -1 (Parameter 'number')
0! = 1
1! = 1
2! = 2
...
12! = 479,001,600
13! is too big for a 32-bit integer.
14! is too big for a 32-bit integer.
15! is too big for a 32-bit integer.
```

Documenting functions with XML comments

By default, when calling a function such as `CardinalToOrdinal`, code editors will show a tooltip with basic information.

Let's improve the tooltip by adding extra information:

1. If you are using VS Code with the **C#** extension, you should navigate to **View | Command Palette | Preferences: Open Settings (UI)**, and then search for `formatOnType` and make sure that it is enabled. C# XML documentation comments are a built-in feature of Visual Studio and Rider, so you do not need to do anything to use them.
2. On the line above the `CardinalToOrdinal` function, type three forward slashes, `///`, and note that they are expanded into an XML comment that recognizes that the function has a single parameter named `number`, as shown in the following code:

```
/// <summary>
///
/// </summary>
/// <param name="number"></param>
/// <returns></returns>
```

3. Enter suitable information for the XML documentation comment for the `CardinalToOrdinal` function. Add a summary and describe the input parameter and the return value, as shown highlighted in the following code:

```
/// <summary>
/// Pass a 32-bit unsigned integer and it will be converted into its
ordinal equivalent.
/// </summary>
/// <param name="number">Number as a cardinal value e.g. 1, 2, 3, and so
on.</param>
/// <returns>Number as an ordinal value e.g. 1st, 2nd, 3rd, and so on.</
returns>
```

4. Now, when calling the function, you will see more details, as shown in *Figure 4.2*:

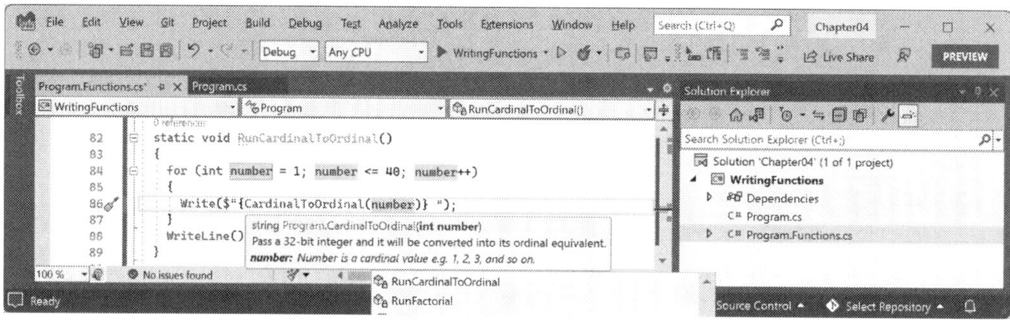

Figure 4.2: A tooltip showing the more detailed method signature

It is worth emphasizing that this feature is primarily designed to be used with a tool that converts the comments into documentation, like Sandcastle, which you can read more about at the following link: `https://github.com/EWSoftware/SHFB`. The tooltips that appear while entering code or hovering over the function name are a secondary feature.

Local functions do not support XML comments because local functions cannot be used outside the member in which they are declared, so it makes no sense to generate documentation from them. Sadly, this also means no tooltip, which would still be useful, but neither Visual Studio nor VS Code recognizes that.

Good Practice: Add XML documentation comments to all your functions except your local functions.

Using lambdas in function implementations

F# is Microsoft's strongly typed functional-first programming language that, like C#, compiles to **Intermediate Language** (IL) to be executed by .NET. Functional languages evolved from lambda calculus, a computational system based only on functions. The code looks more like mathematical functions than steps in a recipe.

Some of the important attributes of functional languages are defined in the following list:

- **Modularity**: The same benefit of defining functions in C# applies to functional languages. This breaks up a large, complex code base into smaller pieces.
- **Immutability**: Variables in the C# sense do not exist. Any data value inside a function cannot change. Instead, a new data value can be created from an existing one. This reduces bugs.
- **Maintainability**: Functional code is cleaner and clearer (for mathematically inclined programmers).

Since C# 6, Microsoft has worked to add features to the language to support a more functional approach, for example, adding **tuples** and **pattern matching** in C# 7, **non-null reference types** in C# 8, and improving pattern matching and adding records, that is, potentially **immutable objects**, in C# 9.

In C# 6, Microsoft added support for **expression-bodied function members**. We will look at an example of this now. In C#, lambdas are the use of the => character to indicate a return value from a function.

The **Fibonacci sequence** of numbers always starts with 0 and 1. Then, the rest of the sequence is generated using the rule of adding together the previous two numbers, as shown in the following sequence of numbers:

```
0 1 1 2 3 5 8 13 21 34 55 ...
```

The next term in the sequence would be 34 + 55, which is 89.

We will use the Fibonacci sequence to illustrate the difference between an imperative and declarative function implementation:

1. In `Program.Functions.cs`, write a function named `FibImperative`, which will be written in an imperative style, as shown in the following code:

```
static int FibImperative(uint term)
{
  if (term == 0)
  {
    throw new ArgumentOutOfRangeException();
  }
  else if (term == 1)
  {
    return 0;
  }
  else if (term == 2)
  {
    return 1;
  }
  else
  {
```

```
        return FibImperative(term - 1) + FibImperative(term - 2);
    }
}
```

2. In `Program.Functions.cs`, write a function named `RunFibImperative` that calls `FibImperative` inside a `for` statement that loops from 1 to 30, as shown in the following code:

```
static void RunFibImperative()
{
  for (uint i = 1; i <= 30; i++)
  {
    WriteLine("The {0} term of the Fibonacci sequence is {1:N0}.",
      arg0: CardinalToOrdinal(i),
      arg1: FibImperative(term: i));
  }
}
```

3. In `Program.cs`, comment out the other method calls and call the `RunFibImperative` method.
4. Run the console app and view the results, as shown in the following partial output:

```
The 1st term of the Fibonacci sequence is 0.
The 2nd term of the Fibonacci sequence is 1.
The 3rd term of the Fibonacci sequence is 1.
The 4th term of the Fibonacci sequence is 2.
The 5th term of the Fibonacci sequence is 3.
...
The 29th term of the Fibonacci sequence is 317,811.
The 30th term of the Fibonacci sequence is 514,229.
```

5. In `Program.Functions.cs`, write a function named `FibFunctional` written in a declarative style, as shown in the following code:

```
static int FibFunctional(uint term) => term switch
  {
    0 => throw new ArgumentOutOfRangeException(),
    1 => 0,
    2 => 1,
    _ => FibFunctional(term - 1) + FibFunctional(term - 2)
  };
```

6. In `Program.Functions.cs`, write a function to call it inside a `for` statement that loops from 1 to 30, as shown in the following code:

```
static void RunFibFunctional()
{
```

```
    for (uint i = 1; i <= 30; i++)
    {
      WriteLine("The {0} term of the Fibonacci sequence is {1:N0}.",
        arg0: CardinalToOrdinal(i),
        arg1: FibFunctional(term: i));
    }
  }
```

7. In `Program.cs`, comment out the `RunFibImperative` method call and call the `RunFibFunctional` method.
8. Run the code and view the results (which will be the same as before).

Now that you have seen some examples of functions, let's see how you can fix them when they have bugs.

Debugging during development

In this section, you will learn how to debug problems at development time. You must use a code editor that has debugging tools, such as Visual Studio or VS Code.

Creating code with a deliberate bug

Let's explore debugging by creating a console app with a deliberate bug, which we will then use the debugger tools in your code editor to track down and fix:

1. Use your preferred code editor to add a new **Console App** / `console` project named `Debugging` to the `Chapter04` solution.
2. Modify `Debugging.csproj` to statically import `System.Console` for all code files.
3. In `Program.cs`, delete any existing statements, and then, at the bottom of the file, add a function with a deliberate bug, as shown in the following code:

```
double Add(double a, double b)
{
  return a * b; // Deliberate bug!
}
```

4. Above the `Add` function, write statements to declare and set some variables and then add them together using the buggy function, as shown in the following code:

```
double a = 4.5;
double b = 2.5;
double answer = Add(a, b);

WriteLine($"{a} + {b} = {answer}");
WriteLine("Press Enter to end the app.");
ReadLine(); // Wait for user to press Enter.
```

5. Run the console application and view the result, as shown in the following output:

```
4.5 + 2.5 = 11.25
Press Enter to end the app.
```

   But wait, there's a bug! 4.5 added to 2.5 should be 7, not 11.25!

We will use the debugging tools to hunt for and squish the bug.

Setting a breakpoint and starting debugging

Breakpoints allow us to mark a line of code that we want to pause at to inspect the program state and find bugs.

Using Visual Studio

Let's set a breakpoint and then start debugging using Visual Studio:

1. Click in line 1, which is the statement that declares the variable named a.
2. Navigate to **Debug | Toggle Breakpoint** or press *F9*. A red circle will appear in the margin bar on the left-hand side and the statement will be highlighted in red to indicate that a breakpoint has been set, as shown in *Figure 4.3*:

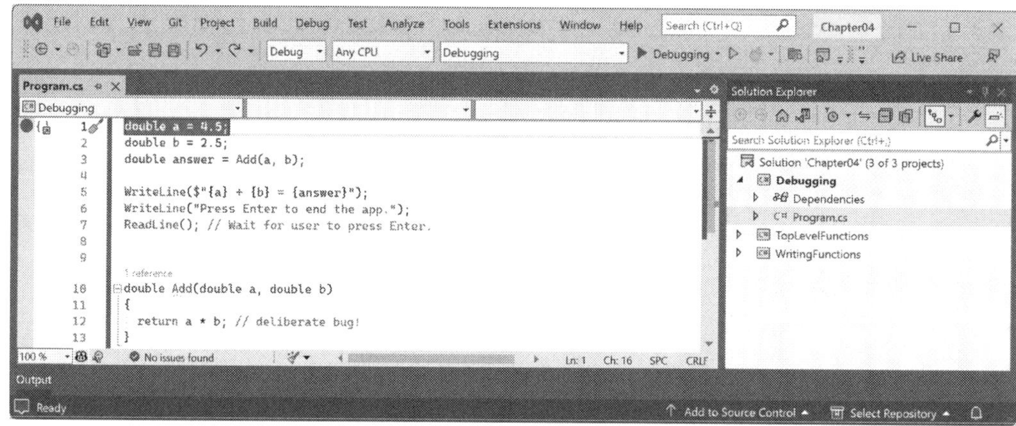

Figure 4.3: Toggling breakpoints using Visual Studio

Breakpoints can be toggled off with the same action. You can also left-click in the margin to toggle a breakpoint on and off or right-click a breakpoint to see more options, such as delete, disable, or edit conditions, or actions for an existing breakpoint.

3. Navigate to **Debug | Start Debugging** or press *F5*. Visual Studio starts the console application and then pauses when it hits the breakpoint. This is known as *break mode*. Extra windows titled **Locals** (showing the current values of local variables), **Watch 1** (showing any watch expressions you have defined), **Call Stack**, **Exception Settings**, and **Immediate Window** may appear. The **Debugging** toolbar appears. The line that will be executed next is highlighted in yellow, and a yellow arrow points at the line from the margin bar, as shown in *Figure 4.4*:

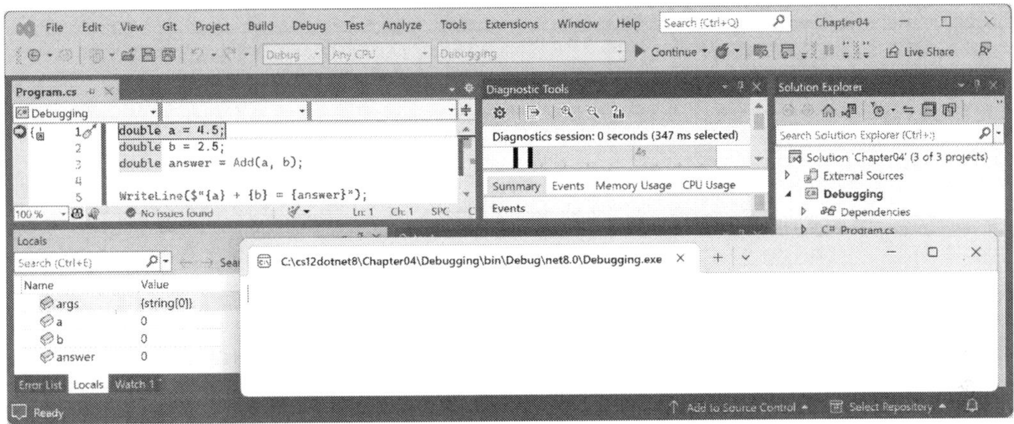

Figure 4.4: Break mode in Visual Studio

If you do not want to see how to use VS Code to start debugging, then you can skip the *Using VS Code* section and continue to the section titled *Navigating with the debugging toolbar*.

Using VS Code

Let's set a breakpoint and then start debugging using VS Code:

1. Click in line 1, which is the statement that declares the variable named a.
2. Navigate to **Run | Toggle Breakpoint** or press *F9*. A red circle will appear in the margin bar on the left-hand side to indicate that a breakpoint has been set.

Breakpoints can be toggled off with the same action. You can also left-click in the margin to toggle a breakpoint on and off; right-click to see more options, such as remove, edit, or disable an existing breakpoint; or add a breakpoint, conditional breakpoint, or Logpoint when a breakpoint does not yet exist. **Logpoints**, also known as **Tracepoints**, indicate that you want to record some information without having to stop executing the code at that point.

3. Navigate to **View | Run**, or in the left navigation bar, you can click the **Run and Debug** icon (the triangle "play" button and "bug") or press *Ctrl* + *Shift* + *D* (on Windows).

4. At the top of the **RUN AND DEBUG** window, click the **Run and Debug** button, and then select the **Debugging** project, as shown in *Figure 4.5*:

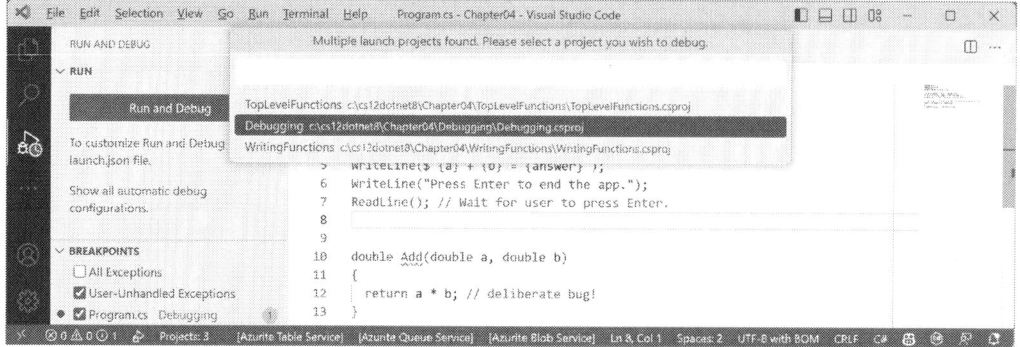

Figure 4.5: Selecting the project to debug using VS Code

If you are first prompted to choose a debugger, select **C#**, not .NET 5+ or .NET Core.

5. VS Code starts the console app and then pauses when it hits the breakpoint. This is known as **break mode**. The line that will be executed next is highlighted in yellow, and a yellow block points at the line from the margin bar, as shown in *Figure 4.6*:

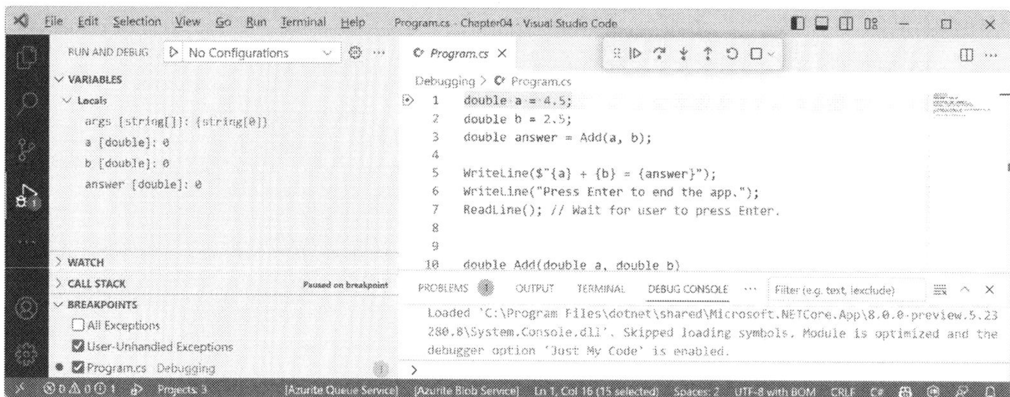

Figure 4.6: Break mode in VS Code

Navigating with the debugging toolbar

Visual Studio has two debug-related buttons in its **Standard** toolbar to start or continue debugging and to hot reload changes to the running code, and a separate **Debug** toolbar for the rest of the tools.

VS Code shows a floating toolbar with buttons to make it easy to access debugging features.

Both are shown in *Figure 4.7*:

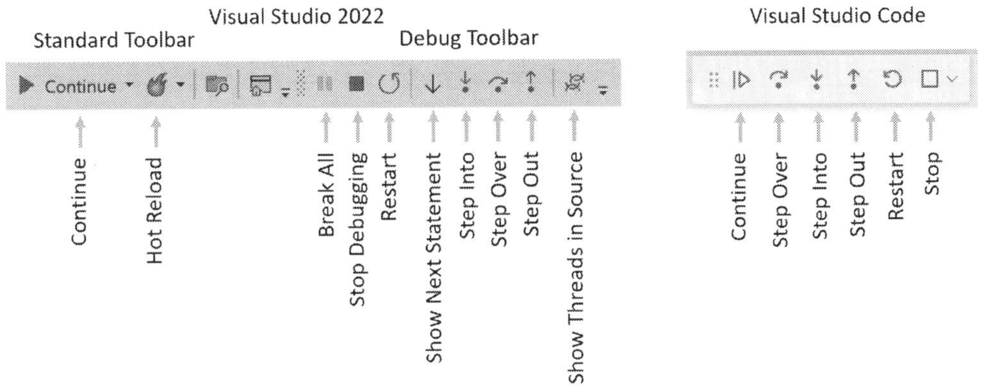

Figure 4.7: Debugging toolbars in Visual Studio and VS Code

The following list describes the most common buttons in the toolbars:

- **Start/Continue**/*F5*: This button is context-sensitive. It will either start running a project or continue running the project from the current position until it ends or hits a breakpoint.
- **Hot Reload**: This button will reload compiled code changes without needing to restart the app.
- **Break All**: This button will break into the next available line of code in a running app.
- **Stop Debugging/Stop**/*Shift* + *F5* (red square): This button will stop the debugging session.
- **Restart**/*Ctrl* or *Cmd* + *Shift* + *F5* (circular arrow): This button will stop and then immediately restart the program with the debugger attached again.
- **Show Next Statement**: This button will move the current cursor to the next statement that will execute.
- **Step Into**/*F11*, **Step Over**/*F10*, and **Step Out**/*Shift* + *F11* (blue arrows over dots): These buttons step through the code statements in various ways, as you will see in a moment.
- **Show Threads in Source**: This button allows you to examine and work with threads in the application that you're debugging.

Debugging windows

While debugging, both Visual Studio and VS Code show extra windows that allow you to monitor useful information, such as variables, while you step through your code.

The most useful windows are described in the following list:

- **VARIABLES**, including **Locals**, which shows the name, value, and type of any local variables automatically. Keep an eye on this window while you step through your code.
- **WATCH**, or **Watch 1**, which shows the value of variables and expressions that you manually enter.
- **CALL STACK**, which shows the stack of function calls.
- **BREAKPOINTS**, which shows all your breakpoints and allows finer control over them.

When in break mode, there is also a useful window at the bottom of the edit area:

- **DEBUG CONSOLE**, or **Immediate Window**, enables live interaction with your code. You can interrogate the program state, for example, by entering the name of a variable. For example, you can ask a question such as "What is 1+2?" by typing `1+2` and pressing *Enter*.

Stepping through code

Let's explore some ways to step through the code using either Visual Studio or VS Code:

The menu commands for debugging are on the **Debug** menu in Visual Studio or the **Run** menu in VS Code and Rider.

1. Navigate to **Run** or **Debug** | **Step Into**, click on the **Step Into** button in the toolbar, or press *F11*. The yellow highlight steps forward one line.
2. Navigate to **Run** or **Debug** | **Step Over**, click on the **Step Over** button in the toolbar, or press *F10*. The yellow highlight steps forward one line. At the moment, you can see that there is no difference between using **Step Into** or **Step Over** because we are executing single statements.
3. You should now be on the line that calls the `Add` method.

The difference between **Step Into** and **Step Over** can be seen when you are about to execute a method call.

If you click on **Step Into**, the debugger steps *into* the method so that you can step through every line in that method.

If you click on **Step Over**, the whole method is executed in one go; it does not skip over the method without executing it.

4. Click on **Step Into** to step inside the `Add` method.
5. Hover your mouse pointer over the `a` or `b` parameters in the code editing window and note that a tooltip appears showing their current value.
6. Select the expression `a * b`, right-click the expression, and select **Add to Watch** or **Add Watch**. The expression is added to the **WATCH** or **Watch 1** window, showing that this operator is multiplying `a` by `b` to give the result `11.25`.
7. In the **WATCH** or **Watch 1** window, right-click the expression and choose **Remove Expression** or **Delete Watch**.
8. Fix the bug by changing `*` to `+` in the `Add` function.
9. Restart debugging by clicking the circular arrow **Restart** button or pressing *Ctrl* or *Cmd* + *Shift* + *F5*.

10. Step over the function, take a minute to note how it now calculates correctly, and click the **Continue** button or press *F5*.
11. With VS Code, note that when writing to the console during debugging, the output appears in the **DEBUG CONSOLE** window instead of the **TERMINAL** window, as shown in *Figure 4.8*:

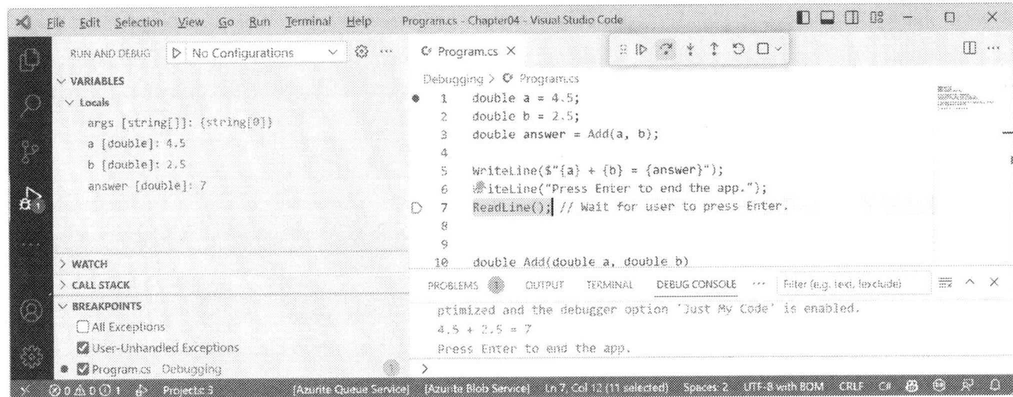

Figure 4.8: Writing to the DEBUG CONSOLE during debugging

Using the VS Code integrated terminal during debugging

By default, the console is set to use the internal **DEBUG CONSOLE** during debugging, which does not allow interactions like entering text from the `ReadLine` method.

To improve the experience, we can change the setting to use the integrated terminal instead. First, let's modify the code to require interaction with the user:

1. At the top of `Program.cs`, add statements to prompt the user to enter a number and parse that as a `double` into the variable `a`, as shown highlighted in the following code:

```
Write("Enter a number: ");
string number = ReadLine()!;

double a = double.Parse(number);
```

2. Set a breakpoint on line 1 that writes the prompt `Enter a number`.
3. At the top of the **RUN AND DEBUG** window, click the **Run and Debug** button, and then select the **Debugging** project.
4. Note that the `Enter a number` prompt is not written to either **TERMINAL** or **DEBUG CONSOLE** and neither window is waiting for the user to enter a number and press *Enter*.
5. Stop debugging.

6. At the top of the **RUN AND DEBUG** window, click the **create a launch.json file** link, and then, when prompted for the debugger, select **C#**, as shown in *Figure 4.9*:

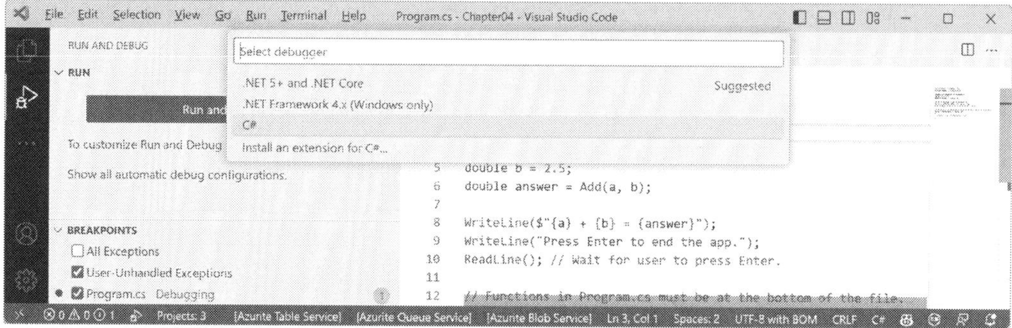

Figure 4.9: Selecting a debugger for the launch.json file

7. In the `launch.json` file editor, click the **Add Configuration...** button, and then select **.NET: Launch .NET Core Console App**, as shown in *Figure 4.10*:

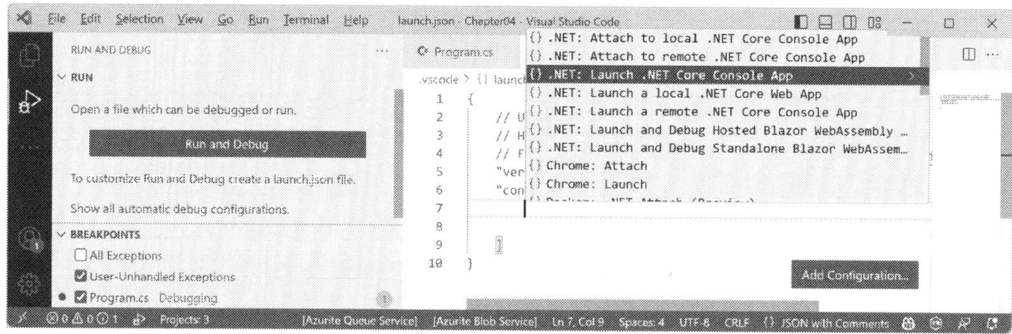

Figure 4.10: Adding a launch configuration for a .NET console app

8. In `launch.json`, make the following additions and changes, as shown highlighted in the following configuration:

   - Comment out the `preLaunchTask` setting.
   - In the `program` path, add the `Debugging` project folder after the `workspaceFolder` variable.
   - In the `program` path, change `<target-framework>` to `net9.0`.
   - In the `program` path, change `<project-name.dll>` to `Debugging.dll`.

- Change the console setting from `internalConsole` to `integratedTerminal`:

```
{
  // Use IntelliSense to learn about possible attributes.
  // Hover to view descriptions of existing attributes.
  // For more information, visit: https://go.microsoft.com/
fwlink/?linkid=830387
  "version": "0.2.0",
  "configurations": [
    {
      "name": ".NET Core Launch (console)",
      "type": "coreclr",
      "request": "launch",
      //"preLaunchTask": "build",
      "program": "${workspaceFolder}/Debugging/bin/Debug/net9.0/
Debugging.dll",
      "args": [],
      "cwd": "${workspaceFolder}",
      "stopAtEntry": false,
      "console": "integratedTerminal"
    }
  ]
}
```

Remember that with VS Code, we open the `Chapter04` folder to process the solution file, so the workspace folder is `Chapter04`, not the `Debugging` project.

9. At the top of the **RUN AND DEBUG** window, note the drop-down list of launch configurations, and click the **Start Debugging** button (green triangle), as shown in *Figure 4.11*.
10. Navigate to **View** | **Terminal** and note the **TERMINAL** window is attached to `Debugging.dll`, as shown in *Figure 4.11*.
11. Step over the statement that writes `Enter a number:` to the console.
12. Step over the statement that calls `ReadLine`.
13. Type `5.5` and press *Enter*.

14. Continue stepping through the statements or press *F5* or click **Continue**, and note the output written to the integrated terminal, as shown in *Figure 4.11*:

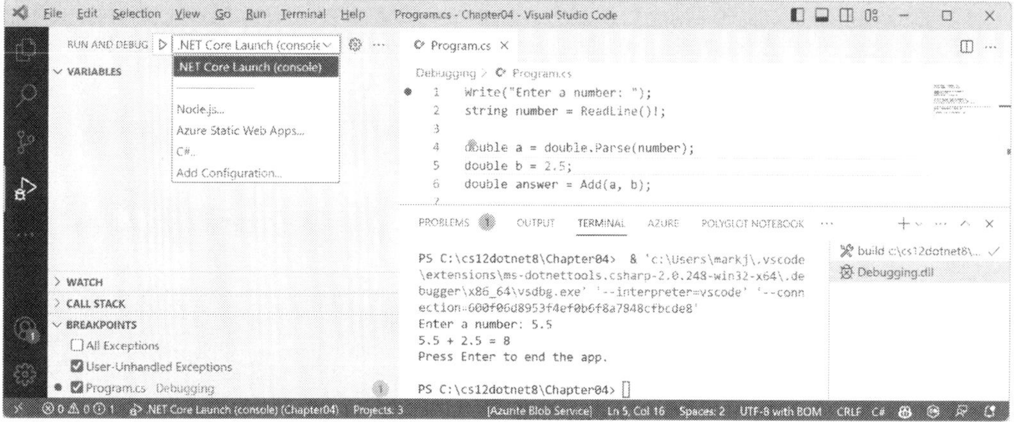

Figure 4.11: A launch configuration set to use the integrated terminal for user interaction

Customizing breakpoints

It is easy to make more complex breakpoints:

1. If you are still debugging, click the **Stop** button in the debugging toolbar, navigate to **Run** or **Debug | Stop Debugging**, or press *Shift* + *F5*.
2. Navigate to **Run | Remove All Breakpoints** or **Debug | Delete All Breakpoints.**
3. Click on the `WriteLine` statement that outputs the answer.
4. Set a breakpoint by pressing *F9* or navigating to **Run** or **Debug | Toggle Breakpoint**.
5. Right-click the breakpoint and choose the appropriate menu for your code editor:
    - In VS Code, choose **Edit Breakpoint...**.
    - In Visual Studio, choose **Conditions...**.
6. Type an expression, such as the `answer` variable must be greater than `9`, and then press *Enter* to accept it, and note the expression must evaluate to `true` for the breakpoint to activate, as shown in *Figure 4.12*:

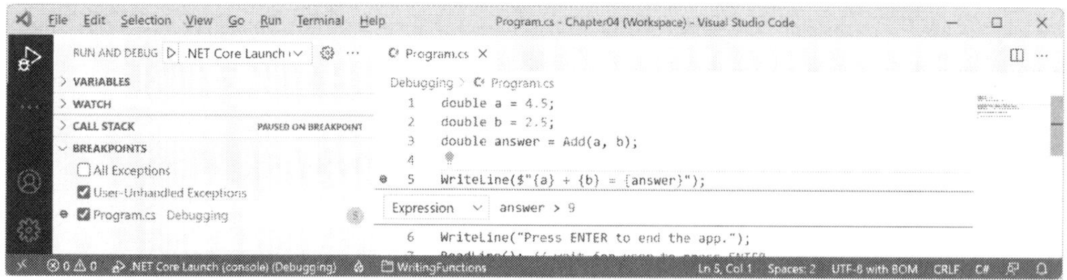

Figure 4.12: Customizing a breakpoint with an expression using VS Code

7. Start debugging and note the breakpoint is not hit.

8. Stop debugging.
9. Edit the breakpoint or its conditions and change its expression to less than 9.
10. Start debugging and note the breakpoint is hit.
11. Stop debugging.
12. Edit the breakpoint or its conditions (in Visual Studio, click **Add condition**), select **Hit Count**, then enter a number such as 3, meaning that you would have to hit the breakpoint three times before it activates, as shown in *Figure 4.13*:

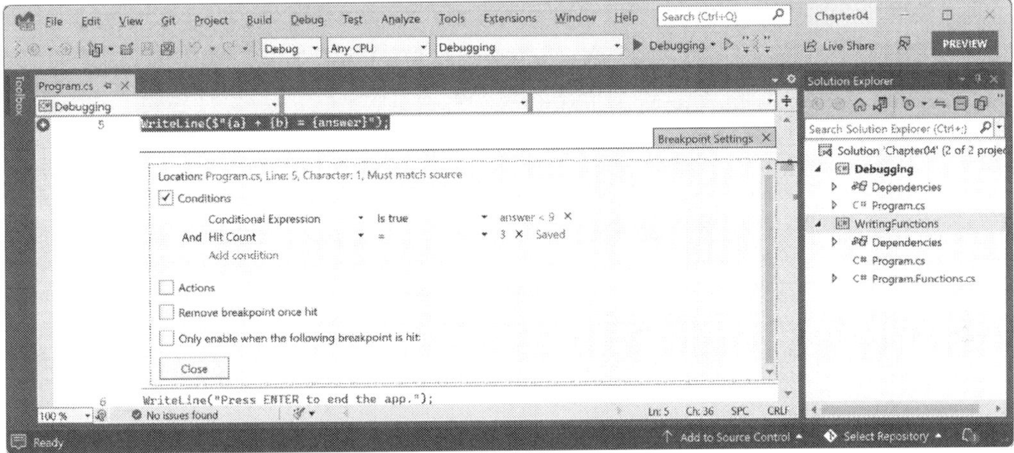

Figure 4.13: Customizing a breakpoint with an expression and hit count using Visual Studio

13. Hover your mouse over the breakpoint's red circle to see a summary, as shown in *Figure 4.14*:

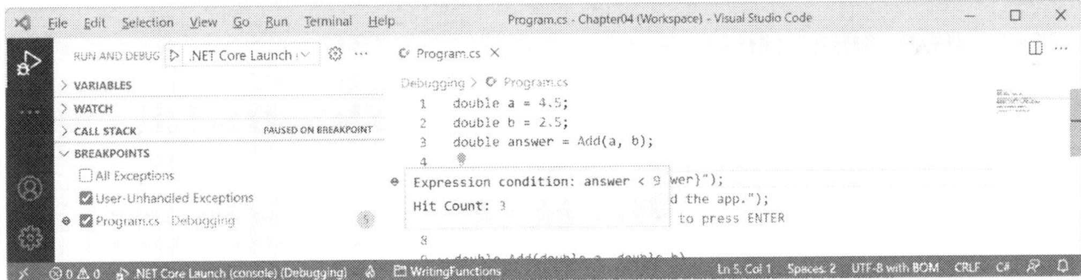

Figure 4.14: A summary of a customized breakpoint in VS Code

You have now fixed a bug using some debugging tools and seen some advanced possibilities for setting breakpoints.

Hot reloading during development

Hot Reload is a feature that allows a developer to apply changes to code while the app is running and immediately see the effect. This is great for fixing bugs quickly. Hot Reload is also known as **Edit and Continue**. A list of the types of changes you can make that support Hot Reload is found at the following link: `https://aka.ms/dotnet/hot-reload`.

Just before the release of .NET 6, a high-level Microsoft employee caused controversy by attempting to make the feature exclusive to Visual Studio. Luckily the open-source contingent within Microsoft successfully had the decision overturned. Hot Reload remains available using the command-line tool as well.

Let's see it in action:

1. Use your preferred code editor to add a new **Console App** / `console` project named `HotReloading` to the `Chapter04` solution.
2. Modify `HotReloading.csproj` to statically import `System.Console` for all code files.
3. In `Program.cs`, delete the existing statements, and then write a message to the console every two seconds, as shown in the following code:

```
/* Visual Studio: run the app, change the message, click Hot Reload.
 * VS Code: run the app using dotnet watch, change the message. */

while (true)
{
  WriteLine("Hello, Hot Reload!");
  await Task.Delay(2000);
}
```

Hot reloading using Visual Studio

If you are using Visual Studio, Hot Reload is built into the user interface:

1. In Visual Studio, start the project and note that the message is output every two seconds.
2. Leave the project running.
3. In `Program.cs`, change `Hello` to `Goodbye`.
4. Navigate to **Debug** | **Apply Code Changes** or click the **Hot Reload** button in the toolbar, as shown in *Figure 4.15*, and note the change is applied without needing to restart the console app.
5. Drop down the **Hot Reload** button menu and select **Hot Reload on File Save**, as shown in *Figure 4.15*:

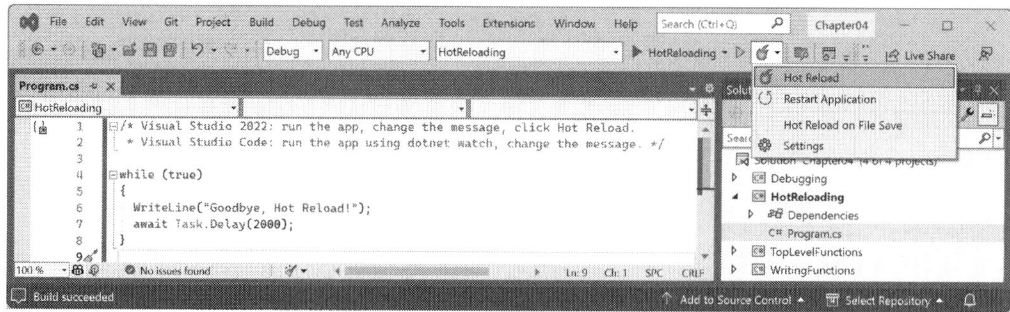

Figure 4.15: Changing Hot Reload options

6. Change the message again, save the file, and note the console app updates automatically.

Hot reloading using VS Code and dotnet watch

If you are using VS Code, you must issue a special command when starting the console app to activate Hot Reload:

1. In VS Code, in **TERMINAL**, start the console app using `dotnet watch`, and note the output that shows that Hot Reload is active, as shown in the following output:

```
dotnet watch  Hot reload enabled. For a list of supported edits, see
https://aka.ms/dotnet/hot-reload.
   Press "Ctrl + R" to restart.
dotnet watch  Building...
  Determining projects to restore...
  All projects are up-to-date for restore.
  HotReloading -> C:\cs13net9\Chapter04\HotReloading\bin\Debug\net9.0\
HotReloading.dll
dotnet watch  Started
Hello, Hot Reload!
Hello, Hot Reload!
Hello, Hot Reload!
```

2. In VS Code, change `Hello` to `Goodbye`, and note that, after a couple of seconds, the change is applied without needing to restart the console app, as shown in the following output:

```
Hello, Hot Reload!
dotnet watch  File changed: .\Program.cs.
Hello, Hot Reload!
Hello, Hot Reload!
dotnet watch  Hot reload of changes succeeded.
Goodbye, Hot Reload!
Goodbye, Hot Reload!
```

3. Press *Ctrl* + *C* to stop it running, as shown in the following output:

```
Goodbye, Hot Reload!
dotnet watch  Shutdown requested. Press Ctrl+C again to force exit.
```

Now that you've seen tools for finding and removing bugs during development, let's see how you can track down less obvious problems that might happen during development and production by writing tests.

Unit testing

Fixing bugs in code is expensive. The earlier that a bug is discovered in the development process, the less expensive it will be to fix.

Unit testing is a good way to find bugs early in the development process because they test a small unit before they are integrated together or seen by user acceptance testers. Some developers even follow the principle that programmers should create unit tests before they write code, and this is called **test-driven development** (**TDD**).

Microsoft has a proprietary unit testing framework known as **MSTest**. There is also a framework named **NUnit**. However, we will use the free and open-source third-party framework **xUnit.net**. All three do basically the same thing. xUnit was created by the same team that built NUnit, but they fixed the mistakes they felt they made previously. xUnit is more extensible and has better community support.

If you are curious about the pros and cons of the various testing systems, then there are hundreds of articles written by proponents of each. Just google them: `https://www.google.com/search?q=xunit+vs+nunit`.

Understanding types of testing

Unit testing is just one of many types of testing, as described in *Table 4.1*:

Type	Description
Unit	Tests the smallest unit of code, typically a method or function. Unit testing is performed on a unit of code isolated from its dependencies by mocking them if needed. Each unit should have multiple tests: some with typical inputs and expected outputs, some with extreme input values to test boundaries, and some with deliberately wrong inputs to test exception handling.
Integration	Tests if the smaller units and larger components work together as a single piece of software. Sometimes involves integrating with external components for which you do not have source code.
System	Tests the whole system environment in which your software will run.
Performance	Tests the performance of your software; for example, your code must return a web page full of data to a visitor in under 20 milliseconds.
Load	Tests how many requests your software can handle simultaneously while maintaining required performance, for example, 10,000 concurrent visitors to a website.
User Acceptance	Tests if users can happily complete their work using your software.

Table 4.1: Types of testing

Creating a class library that needs testing

First, we will create a function that needs testing. We will create it in a class library project separate from a console app project. A class library is a package of code that can be distributed and referenced by other .NET applications:

1. Use your preferred code editor to add a new **Class Library** / `classlib` project named `CalculatorLib` to the `Chapter04` solution.

At this point, you will have created about a dozen new console app projects and added them to a solution. The only difference when adding a **Class Library** / `classlib` project is selecting a different project template. The rest of the steps are the same as adding a **Console App** / `console` project.

If you are using Visual Studio:

   a. Navigate to **File** | **Add** | **New Project**.
   b. In the **Add a new project** dialog, search for and select **Class Library [C#]** and then click **Next**.
   c. In the **Configure your new project** dialog, for the **Project name**, enter `CalculatorLib`, leave the location as `C:\cs13net9\Chapter04`, and then click **Next**.
   d. In the **Additional information** dialog, select **.NET 9.0 (Standard Term Support)**, and then click **Create**.

If you are using VS Code:

   a. In **TERMINAL**, switch to a terminal in the `Chapter04` folder.
   b. Use the `dotnet` CLI to create a new class library project named `CalculatorLib`, as shown in the following command: `dotnet new classlib -o CalculatorLib`.
   c. Use the `dotnet` CLI to add the new project folder to the solution, as shown in the following command: `dotnet sln add CalculatorLib`.
   d. Note the results, as shown in the following output: `` Project `CalculatorLib\CalculatorLib.csproj` added to the solution. ``

2. For all code editors, in the `CalculatorLib` project, rename the file named `Class1.cs` to `Calculator.cs`.
3. In `Calculator.cs`, modify the file to define a `Calculator` class (with a deliberate bug), as shown in the following code:

```
namespace CalculatorLib;

public class Calculator
{
  public double Add(double a, double b)
  {
    return a * b;
  }
}
```

4. Compile your class library project:
   - In Visual Studio, navigate to **Build** | **Build CalculatorLib**.
   - In VS Code, in a **TERMINAL** window for the `CalculatorLib` folder, enter the command `dotnet build`. (You could also run this command in the `Chapter04` folder but that would build the whole solution, which is unnecessary in this scenario.)
5. Use your preferred code editor to add a new **xUnit Test Project [C#]** / `xunit` project named `CalculatorLibUnitTests` to the `Chapter04` solution. For example, at the command prompt or terminal in the `Chapter04` folder, enter the following commands:

```
dotnet new xunit -o CalculatorLibUnitTests
dotnet sln add CalculatorLibUnitTests
```

6. In the `CalculatorLibUnitTests` project, add a project reference to the `CalculatorLib` project:
   - If you are using Visual Studio, in **Solution Explorer**, select the `CalculatorLibUnitTests` project, navigate to **Project** | **Add Project Reference...**, check the box to select the `CalculatorLib` project, and then click **OK**.
   - If you are using VS Code, use the `dotnet add reference` command, or in the file named `CalculatorLibUnitTests.csproj`, modify the configuration to add an item group with a project reference to the `CalculatorLib` project, as shown highlighted in the following markup:

```
<ItemGroup>
  <ProjectReference
    Include="..\CalculatorLib\CalculatorLib.csproj" />
</ItemGroup>
```

The path for a project reference can use either forward (`/`) or back slashes (`\`) because the paths are processed by the .NET SDK and changed if necessary for the current operating system.

7. Build the `CalculatorLibUnitTests` project.

Writing unit tests

A well-written unit test will have three parts:

- **Arrange**: This part will declare and instantiate variables for input and output.
- **Act**: This part will execute the unit that you are testing. In our case, that means calling the method that we want to test.
- **Assert**: This part will make one or more assertions about the output. An assertion is a belief that, if not true, indicates a failed test. For example, when adding 2 and 2, we would expect the result to be 4.

Now, we will write some unit tests for the `Calculator` class:

1. Rename the file `UnitTest1.cs` to `CalculatorUnitTests.cs` and then open it.
2. In VS Code, rename the class to `CalculatorUnitTests`. (Visual Studio prompts you to rename the class when you rename the file.)
3. In `CalculatorUnitTests`, import the `CalculatorLib` namespace, and then modify the `CalculatorUnitTests` class to have two test methods, one for adding 2 and 2, and another for adding 2 and 3, as shown in the following code:

```
using CalculatorLib; // To use Calculator.

namespace CalculatorLibUnitTests;

public class CalculatorUnitTests
{
  [Fact]
  public void TestAdding2And2()
  {
    // Arrange: Set up the inputs and the unit under test.
    double a = 2;
    double b = 2;
    double expected = 4;
    Calculator calc = new();

    // Act: Execute the function to test.
    double actual = calc.Add(a, b);

    // Assert: Make assertions to compare expected to actual results.
    Assert.Equal(expected, actual);
  }

  [Fact]
  public void TestAdding2And3()
  {
    double a = 2;
    double b = 3;
    double expected = 5;
    Calculator calc = new();

    double actual = calc.Add(a, b);

    Assert.Equal(expected, actual);
  }
}
```

Visual Studio still uses an older project item template that uses a nested namespace. The preceding code shows the modern project item template used by `dotnet new` and Rider that uses a file-scoped namespace.

4. Build the `CalculatorLibUnitTests` project.

Running unit tests using Visual Studio

Now we are ready to run the unit tests and see the results:

1. In Visual Studio, navigate to **Test** | **Run All Tests**.
2. In **Test Explorer**, note that the results indicate that two tests ran, one test passed, and one test failed, as shown in *Figure 4.16*:

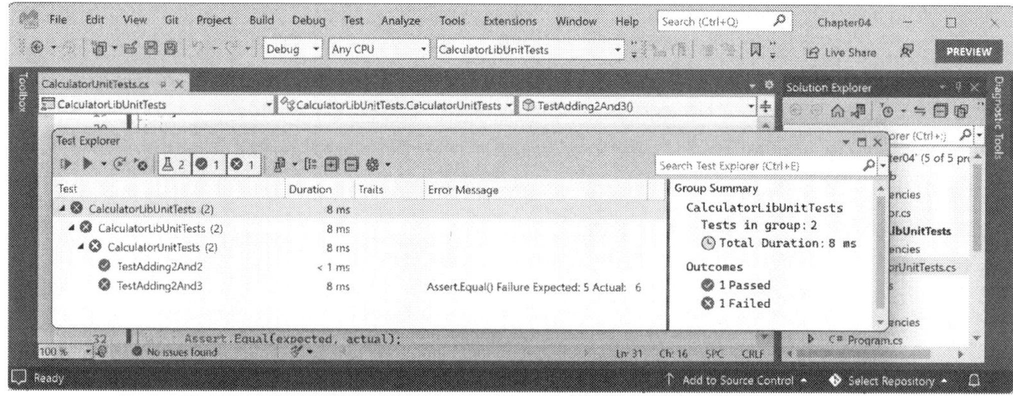

Figure 4.16: The unit test results in Visual Studio's Test Explorer

Running unit tests using VS Code

Now we are ready to run the unit tests and see the results:

1. If you have not recently built the test project, then build the `CalculatorLibUnitTests` project to make sure that the new testing feature in the C# Dev Kit extension recognizes the unit tests that you have written.
2. In VS Code, navigate to **View** | **Testing**, and note the **TESTING** window has a mini toolbar with buttons to **Refresh Tests**, **Run Tests**, **Debug Tests**, and so on.
3. In the **TESTING** window, expand the **CalculatorLibUnitTests** project to show the two tests.
4. Hover your mouse pointer over **CalculatorUnitTests** and then click the **Run Tests** button (black triangle icon) defined in that class.
5. Click the **TEST RESULTS** tab and note that the results indicate that two tests ran, one test passed, and one test failed, as shown in *Figure 4.17*:

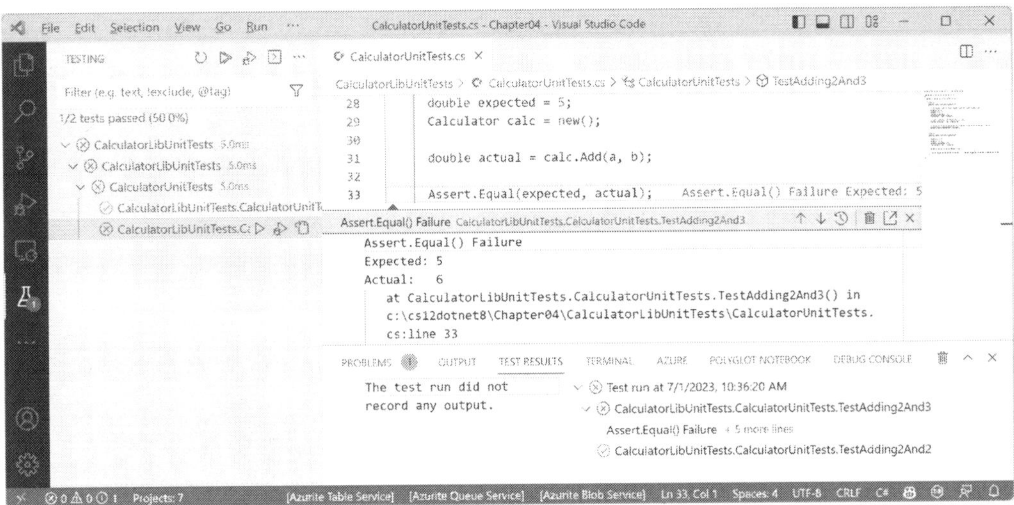

Figure 4.17: The unit test results in VS Code's TEST RESULTS

Fixing the bug

Now you can fix the bug:

1. Fix the bug in the `Add` method.
2. Run the unit tests again to see that the bug has now been fixed and both tests have passed.

Now that we've written, debugged, logged, and unit-tested functions, let's finish this chapter by looking at how to throw and catch exceptions in functions.

Specifying multiple parameter values

We created two unit test methods: the first to test adding 2 and 2, and the second to test adding 2 and 3. Instead of writing two separate methods decorated with `[Fact]`, we could have written one method decorated with `[Theory]` and passed different parameter values using `[InlineData]`, as shown in the following code:

```
using CalculatorLib; // To use Calculator.

namespace CalculatorLibUnitTests;

public class CalculatorUnitTests
{
  [Theory]
  [InlineData(2, 2, 4)]
  [InlineData(2, 3, 5)]
  public void TestAdding(double a, double b, double expected)
  {
```

```
    // Arrange: Set up the unit under test.
    Calculator calc = new();

    // Act: Execute the function to test.
    double actual = calc.Add(a, b);

    // Assert: Make assertions to compare expected to actual results.
    Assert.Equal(expected, actual);
  }
}
```

The order of parameters passed in the `[InlineData]` attribute must match the order of parameters in the test method.

Now let's see how to throw and catch exceptions when something goes wrong.

Throwing and catching exceptions in functions

In *Chapter 3, Controlling Flow, Converting Types, and Handling Exceptions*, you were introduced to exceptions and how to use a `try-catch` statement to handle them. But you should only catch and handle an exception if you have enough information to mitigate the issue. If you do not, then you should allow the exception to pass up through the call stack to a higher level.

Understanding usage errors and execution errors

Usage errors are when a programmer misuses a function, typically by passing invalid values as parameters. They could be avoided by that programmer changing their code to pass valid values. When some programmers first learn C# and .NET, they sometimes think exceptions can always be avoided because they assume all errors are usage errors. Usage errors should all be fixed before production runtime.

Execution errors are when something happens at runtime that cannot be fixed by writing "better" code. Execution errors can be split into **program errors** and **system errors**. If you attempt to access a network resource but the network is down, you need to be able to handle that system error by logging an exception, and possibly backing off for a time and trying again. However, some system errors, such as running out of memory, simply cannot be handled. If you attempt to open a file that does not exist, you might be able to catch that error and handle it programmatically by creating a new file. Program errors can be programmatically fixed by writing smart code. System errors often cannot be fixed programmatically.

Commonly thrown exceptions in functions

Very rarely should you define new types of exceptions to indicate usage errors. .NET already defines many that you should use.

When defining your own functions with parameters, your code should check the parameter values and throw exceptions if they have values that will prevent your function from properly functioning.

For example, if a parameter to a function should not be `null`, throw `ArgumentNullException`. For other problems, throw `ArgumentException`, `NotSupportedException`, or `InvalidOperationException`.

For any exception, include a message that describes the problem for whoever will have to read it (typically a developer audience for class libraries and functions, or end users if it is at the highest level of a GUI app), as shown in the following code:

```
static void Withdraw(string accountName, decimal amount)
{
  if (string.IsNullOrWhiteSpace(accountName))
  {
    throw new ArgumentException(paramName: nameof(accountName));
  }

  if (amount <= 0)
  {
    throw new ArgumentOutOfRangeException(paramName: nameof(amount),
      message: $"{nameof(amount)} cannot be negative or zero.");
  }

  // process parameters
}
```

Good Practice: If a function cannot successfully perform its operation, you should consider it a function failure and report it by throwing an exception.

Throwing exceptions using guard clauses

Instead of instantiating an exception using `new`, you can use static methods on the exception itself. When used in a function implementation to check argument values, they are known as **guard clauses**. Some were introduced with .NET 6, and more were added in .NET 8.

Common guard clauses are shown in *Table 4.2*:

Exception	Guard clause methods
`ArgumentException`	`ThrowIfNullOrEmpty`, `ThrowIfNullOrWhiteSpace`
`ArgumentNullException`	`ThrowIfNull`
`ArgumentOutOfRangeException`	`ThrowIfEqual`, `ThrowIfGreaterThan`, `ThrowIfGreaterThanOrEqual`, `ThrowIfLessThan`, `ThrowIfLessThanOrEqual`, `ThrowIfNegative`, `ThrowIfNegativeOrZero`, `ThrowIfNotEqual`, `ThrowIfZero`

Table 4.2: Common guard clauses

Instead of writing an `if` statement and then throwing a `new` exception, we can simplify the previous example, as shown in the following code:

```
static void Withdraw(string accountName, decimal amount)
{
  ArgumentException.ThrowIfNullOrWhiteSpace(accountName,
    paramName: nameof(accountName));

  ArgumentOutOfRangeException.ThrowIfNegativeOrZero(amount,
    paramName: nameof(amount));

  // process parameters
}
```

More Information: To get a sense of how often Microsoft source code uses its own guard clauses, you can read the following section of an online blog article: `https://devblogs.microsoft.com/dotnet/performance-improvements-in-net-8/#exceptions`. For example, `ThrowIfNull` is called 4,795 times from within the `src` for the core libraries in `dotnet/runtime`.

Understanding the call stack

The entry point for a .NET console application is the `Main` method (if you have explicitly defined this class) or `<Main>$` (if it was created for you by the top-level program feature) in the `Program` class.

The `Main` method will call other methods, which call other methods, and so on; these methods could be in the current project or referenced projects and NuGet packages, as shown in *Figure 4.18*:

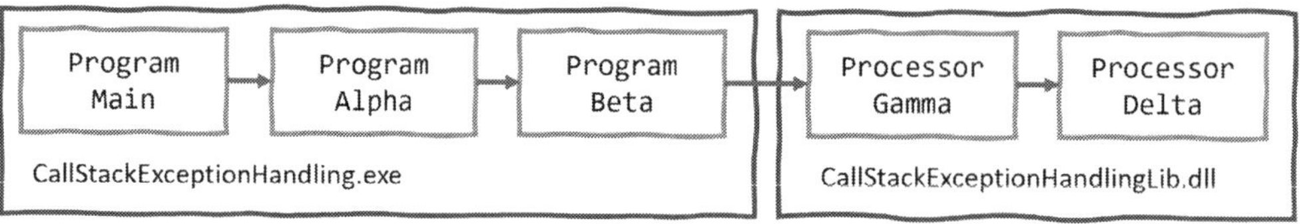

Figure 4.18: A chain of method calls that create a call stack

Let's create a similar chain of methods to explore where we could catch and handle exceptions:

1. Use your preferred code editor to add a new **Class Library** / `classlib` project named `CallStackExceptionHandlingLib` to the `Chapter04` solution.
2. Rename the `Class1.cs` file to `Processor.cs`.
3. In `Processor.cs`, modify its contents, as shown in the following code:

```
using static System.Console;

namespace CallStackExceptionHandlingLib;
```

```
public class Processor
{
  public static void Gamma() // public so it can be called from outside.
  {
    WriteLine("In Gamma");
    Delta();
  }

  private static void Delta() // private so it can only be called
internally.
  {
    WriteLine("In Delta");
    File.OpenText("bad file path");
  }
}
```

4. Use your preferred code editor to add a new **Console App** / `console` project named `CallStackExceptionHandling` to the `Chapter04` solution.
5. In the `CallStackExceptionHandling` console app project, add a reference to the `CallStackExceptionHandlingLib` class library project, as shown in the following markup:

```
<ItemGroup>
  <ProjectReference Include="..\CallStackExceptionHandlingLib\
CallStackExceptionHandlingLib.csproj" />
</ItemGroup>
```

6. Build the `CallStackExceptionHandling` console app project to make sure dependent projects are compiled and copied to the local `bin` folder.
7. In `Program.cs`, delete the existing statements, add statements to define two methods, and chain calls to them and the methods in the class library, as shown in the following code:

```
using CallStackExceptionHandlingLib; // To use Processor.
using static System.Console;

WriteLine("In Main");
Alpha();

void Alpha()
{
  WriteLine("In Alpha");
  Beta();
}
```

```
void Beta()
{
  WriteLine("In Beta");
  Processor.Gamma();
}
```

8. Run the console app *without* the debugger attached, and note the results, as shown in the following partial output:

```
In Main
In Alpha
In Beta
In Gamma
In Delta
Unhandled exception. System.IO.FileNotFoundException: Could not find file 
'C:\cs13net9\Chapter04\CallStackExceptionHandling\bin\Debug\net9.0\bad 
file path'.
File name: 'C:\cs13net9\Chapter04\CallStackExceptionHandling\bin\Debug\
net9.0\bad file path'
   at Microsoft.Win32.SafeHandles.SafeFileHandle.CreateFile(String 
fullPath, FileMode mode, FileAccess access, FileShare share, FileOptions 
options)
   at Microsoft.Win32.SafeHandles.SafeFileHandle.Open(String fullPath, 
FileMode mode, FileAccess access, FileShare share, FileOptions options, 
Int64 preallocationSize)
   at System.IO.Strategies.OSFileStreamStrategy..ctor(String path, 
FileMode mode, FileAccess access, FileShare share, FileOptions options, 
Int64 preallocationSize)
   at System.IO.Strategies.FileStreamHelpers.ChooseStrategyCore(String 
path, FileMode mode, FileAccess access, FileShare share, FileOptions 
options, Int64 preallocationSize)
   at System.IO.StreamReader.ValidateArgsAndOpenPath(String path, 
Encoding encoding, Int32 bufferSize)
   at System.IO.File.OpenText(String path)
   at CallStackExceptionHandlingLib.Calculator.Delta() in C:\cs13net9\
Chapter04\CallStackExceptionHandlingLib\Processor.cs:line 16
   at CallStackExceptionHandlingLib.Calculator.Gamma() in C:\cs13net9\
Chapter04\CallStackExceptionHandlingLib\Processor.cs:line 10
   at Program.<<Main>$>g__Beta|0_1() in C:\cs13net9\Chapter04\
CallStackExceptionHandling\Program.cs:line 16
   at Program.<<Main>$>g__Alpha|0_0() in C:\cs13net9\Chapter04\
CallStackExceptionHandling\Program.cs:line 10
```

```
   at Program.<Main>$(String[] args) in C:\cs13net9\Chapter04\
CallStackExceptionHandling\Program.cs:line 5
```

Note that the call stack is upside down. Starting from the bottom, you see:

- The first call is to the `<Main>$` entry point function in the auto-generated `Program` class. This is where arguments are passed in as a `String` array.
- The second call is to the `<<Main>$>g__Alpha|0_0` function. (The C# compiler renames it from `Alpha` when it adds it as a local function.)
- The third call is to the `Beta` function.
- The fourth call is to the `Gamma` function.
- The fifth call is to the `Delta` function. This function attempts to open a file by passing a bad file path. This causes an exception to be thrown. Any function with a `try-catch` statement could catch this exception. If it does not, the exception is automatically passed up the call stack until it reaches the top, where .NET outputs the exception (and the details of this call stack).

Good Practice: Unless you need to step through your code to debug it, you should always run your code without the debugger attached. In this case, it is especially important not to attach the debugger because, if you do, it will catch the exception and show it in a GUI dialog box instead of outputting it as shown in the book.

Where to catch exceptions

Programmers can decide if they want to catch an exception near the failure point or centralized higher up the call stack. This allows your code to be simplified and standardized. You might know that calling a function could throw one or more types of exception but you do not need to handle any of them at the current point in the call stack.

Rethrowing exceptions

Sometimes, you want to catch an exception, log it, and then rethrow it. For example, if you are writing a low-level class library that will be called from an application, your code may not have enough information to programmatically fix the error in a smart way, but the calling application might have more information and be able to. Your code should log the error in case the calling application does not, and then rethrow it up the call stack in case the calling application chooses to handle it better.

There are three ways to rethrow an exception inside a `catch` block, as shown in the following list:

- To throw the caught exception with its original call stack, call `throw`.
- To throw the caught exception as if it was thrown at the current level in the call stack, call `throw` with the caught exception, for example, `throw ex`. This is usually poor practice because you have lost some potentially useful information for debugging but it can be useful when you want to deliberately remove that information when it contains sensitive data.

- To wrap the caught exception in another exception that can include more information in a message that might help the caller understand the problem, throw a new exception, and pass the caught exception as the `innerException` parameter.

If an error occurs when we call the `Gamma` function, then we could catch the exception and perform one of the three techniques of rethrowing an exception, as shown in the following code:

This code is just illustrative. You would never use all three techniques in the same `catch` block!

```
try
{
  Gamma();
}
catch (IOException ex)
{
  LogException(ex);

  // Throw the caught exception as if it happened here
  // this will lose the original call stack.

  throw ex;

  // Rethrow the caught exception and retain its original call stack.

  throw;

  // Throw a new exception with the caught exception nested within it.

  throw new InvalidOperationException(
    message: "Calculation had invalid values. See inner exception for why.",
    innerException: ex);
}
```

Let's see this in action with our call stack example:

1. In the `CallStackExceptionHandling` project in `Program.cs`, in the `Beta` function, add a `try-catch` statement around the call to the `Gamma` function, as shown highlighted in the following code:

```
void Beta()
{
```

```
  WriteLine("In Beta");
  try
  {
    Processor.Gamma();
  }
  catch (Exception ex)
  {
    WriteLine($"Caught this: {ex.Message}");
    throw ex;
  }
}
```

Note your code editor will show a squiggle under the `throw ex` to warn you that you will lose call stack information, as described in the code analyzer message `Rethrowing caught exception changes stack information`; more details can be found at the following link: `https://learn.microsoft.com/en-us/dotnet/fundamentals/code-analysis/quality-rules/ca2200`.

2. Run the console app and note that the output excludes some details of the call stack, as shown in the following output:

```
Caught this: Could not find file 'C:\cs13net9\Chapter04\
CallStackExceptionHandling\bin\Debug\net9.0\bad file path'.
Unhandled exception. System.IO.FileNotFoundException: Could not find file
'C:\cs13net9\Chapter04\CallStackExceptionHandling\bin\Debug\net9.0\bad
file path'.
File name: 'C:\cs13net9\Chapter04\CallStackExceptionHandling\bin\Debug\
net9.0\bad file path'
   at Program.<<Main>$>g__Beta|0_1() in C:\cs13net9\Chapter04\
CallStackExceptionHandling\Program.cs:line 23
   at Program.<<Main>$>g__Alpha|0_0() in C:\cs13net9\Chapter04\
CallStackExceptionHandling\Program.cs:line 10
   at Program.<Main>$(String[] args) in C:\cs13net9\Chapter04\
CallStackExceptionHandling\Program.cs:line 5
```

3. Remove the ex by replacing the statement `throw ex;` with `throw;`.
4. Run the console app and note that the output includes all the details of the call stack.

Implementing the tester-doer and try patterns

The **tester-doer pattern** can avoid some thrown exceptions (but not eliminate them completely). This pattern uses pairs of functions: one to perform a test and the other to perform an action that would fail if the test was not passed.

.NET implements this pattern itself. For example, before adding an item to a collection by calling the `Add` method, you can test to see if it is read-only, which would cause `Add` to fail and, therefore, throw an exception.

For example, before withdrawing money from a bank account, you might test that the account is not overdrawn, as shown in the following code:

```
if (!bankAccount.IsOverdrawn())
{
  bankAccount.Withdraw(amount);
}
```

The tester-doer pattern can add performance overhead, so you can also implement the **try pattern**, which, in effect, combines the test and do parts into a single function, as we saw with `TryParse`.

Another problem with the tester-doer pattern occurs when you are using multiple threads. In this scenario, one thread calls the test function, and it returns a value that indicates that it is okay to proceed. But then another thread executes, which changes the state. Then the original thread continues executing, assuming that everything is fine, but it is not fine. This is called a *race condition*. This topic is too advanced to cover how to handle it in this book.

Good Practice: Use the try pattern in preference to the tester-doer pattern.

If you implement your own try pattern function and it fails, remember to set the `out` parameter to the default value of its type and then return `false`, as shown in the following code:

```
static bool TryParse(string? input, out Person value)
{
  if (someFailure)
  {
    value = default(Person);
    return false;
  }

  // Successfully parsed the string into a Person.
  value = new Person() { ... };
  return true;
}
```

More Information: Now that you've been introduced to the basics of exceptions, you can learn more about the details by reading the official documentation at the following link: `https://learn.microsoft.com/en-us/dotnet/standard/exceptions/`.

Practicing and exploring

Test your knowledge and understanding by answering some questions, getting some hands-on practice, and exploring, with deeper research, the topics covered in this chapter.

Exercise 4.1 – online material

Online material can be extra content written by me for this book, or it can be references to content created by Microsoft or third parties.

Logging during development and runtime

You should add code throughout your application to log what is happening, especially when exceptions occur, so that you can review the logs and use them to trace the issue and fix the problem. In an online-only section, you will learn how to implement basic logging; this is found at the following link:

`https://github.com/markjprice/cs13net9/blob/main/docs/ch04-logging.md`

Exercise 4.2 – practice exercises

Practice exercises go deeper into the topics for this chapter.

Writing functions with debugging and unit testing

Prime factors are a combination of the smallest prime numbers that, when multiplied together, will produce the original number. Consider the following examples:

- Prime factors of 4 are 2 x 2
- Prime factor of 7 is 7
- Prime factors of 30 are 5 x 3 x 2
- Prime factors of 40 are 5 x 2 x 2 x 2
- Prime factors of 50 are 5 x 5 x 2

Create three projects:

- A class library named `Exercise_PrimeFactorsLib` with a static class and static method named `PrimeFactors`, which, when passed an `int` variable as a parameter, returns a `string` showing its prime factors
- A unit test project named `Exercise_PrimeFactorsTests` with a few suitable unit tests
- A console application to use it, named `Exercise_PrimeFactorsApp`

To keep it simple, you can assume that the largest number entered will be 1,000.

Use the debugging tools and write unit tests to ensure that your function works correctly with multiple inputs and returns the correct output.

Exercise 4.3 – test your knowledge

Answer the following questions. If you get stuck, try googling the answers, if necessary, while remembering that if you get totally stuck, the answers are in the *Appendix*:

1. What does the C# keyword `void` mean?
2. What are some differences between imperative and functional programming styles?
3. In VS Code or Visual Studio, what is the difference between pressing *F5*, *Ctrl* or *Cmd* + *F5*, *Shift* + *F5*, and *Ctrl* or *Cmd* + *Shift* + *F5*?
4. Where does the `Trace.WriteLine` method write its output to?
5. What are the five trace levels?
6. What is the difference between the `Debug` and `Trace` classes?
7. When writing a unit test, what are the three "A"s?
8. When writing a unit test using xUnit, which attribute must you decorate the test methods with?
9. What `dotnet` command executes xUnit tests?
10. What statement should you use to rethrow a caught exception named `ex` without losing the stack trace?

Exercise 4.4 – explore topics

Use the links on the following page to learn more about the topics covered in this chapter:

`https://github.com/markjprice/cs13net9/blob/main/docs/book-links.md#chapter-4---writing-debugging-and-testing-functions`

Summary

In this chapter, you learned:

- How to write reusable functions with input parameters and return values, in both an imperative and functional style
- How to use the Visual Studio and VS Code debugging and diagnostic features like logging and unit tests to identify and fix any bugs in them
- How to throw and catch exceptions in functions and understand the call stack

In the next chapter, you will learn how to build your own types using object-oriented programming techniques.

Learn more on Discord

To join the Discord community for this book – where you can share feedback, ask questions to the author, and learn about new releases – follow the QR code below:

`https://packt.link/csharp13dotnet9`

Leave a Review!

Thank you for purchasing this book from Packt Publishing—we hope you enjoy it! Your feedback is invaluable and helps us improve and grow. Once you've completed reading it, please take a moment to leave an Amazon review; it will only take a minute, but it makes a big difference for readers like you.

Scan the QR or visit the link to receive a free ebook of your choice.

`https://packt.link/NzOWQ`

5

Building Your Own Types with Object-Oriented Programming

This chapter is about making your own types using **object-oriented programming** (**OOP**). You will learn about all the different categories of members that a *type* can have, including fields to store data and methods to perform actions. You will use OOP concepts such as aggregation and encapsulation. You will also learn about language features such as tuple syntax support, out variables, inferred tuple names, and default literals. Finally, you will learn about pattern matching and defining records to make the equality of variables and immutability easier to implement.

This chapter will cover the following topics:

- Talking about OOP
- Building class libraries
- Storing data in fields
- Working with methods and tuples
- Controlling access with properties and indexers
- Pattern matching with objects
- Working with record types

Talking about OOP

An object in the real world is a thing, such as a car or a person, whereas an object in programming often represents something in the real world, such as a product or bank account, but it can also be something more abstract.

In C#, we use the C# keywords `class`, `record`, and `struct` to define a type of object. You will learn about `struct` types in *Chapter 6, Implementing Interfaces and Inheriting Classes*. You can think of a type as being a blueprint or template for an object.

The concepts of OOP are briefly described here:

- **Encapsulation** is the combination of the data and actions that are related to an object. For example, a `BankAccount` type might have data, such as `Balance` and `AccountName`, as well as actions, such as `Deposit` and `Withdraw`. When encapsulating, you often want to control what can access those actions and the data, for example, restricting how the internal state of an object can be accessed or modified from the outside.
- **Composition** is about what an object is made of. For example, a `Car` is composed of different parts, such as four `Wheel` objects, several `Seat` objects, and an `Engine`.
- **Aggregation** is about what can be combined with an object. For example, a `Person` is not part of a `Car` object, but they could sit in the driver's `Seat` and then become the car's `Driver`—two separate objects that are aggregated together to form a new component.
- **Inheritance** is about reusing code by having a **subclass** derive from a **base** or **superclass**. All functionality in the base class is inherited by, and becomes available in, the **derived** class. For example, the base or super `Exception` class has some members that have the same implementation across all exceptions, and the sub or derived `SqlException` class inherits those members and has extra members that are only relevant when a SQL database exception occurs, like a property for the database connection.
- **Abstraction** is about capturing the core idea of an object and ignoring the details or specifics. C# has the `abstract` keyword that formalizes this concept but do not confuse the concept of abstraction with the use of the `abstract` keyword because it is more than that. The concept of abstraction can also be achieved using interfaces. If a class is not explicitly **abstract**, then it can be described as being **concrete**. Bases or superclasses are often abstract; for example, the superclass `Stream` is abstract, and its subclasses, like `FileStream` and `MemoryStream`, are concrete. Only concrete classes can be used to create objects; abstract classes can only be used as the base for other classes because they are missing some implementation. Abstraction is a tricky balance. If you make a class more abstract, more classes will be able to inherit from it, but at the same time, there will be less functionality to share. A real-world example of abstraction is the approach car manufacturers have taken to **electric vehicles** (**EVs**). They create a common "platform" (basically just the battery and wheels) that is an abstraction of what all EVs need, and then add on top of that to build different vehicles like cars, trucks, vans, and so on. The platform on its own is not a complete product, like an abstract class.
- **Polymorphism** is about allowing a derived class to override an inherited action to provide custom behavior.

There is a lot to cover in the next two chapters about OOP, and some parts of it are difficult to learn. At the end of *Chapter 6, Implementing Interfaces and Inheriting Classes*, I have written a summary of the categories of custom types and their capabilities with example code. This will help you review the most important facts and highlight the differences between choices, like an `abstract` class or an `interface`, and when to use them.

Building class libraries

Class library assemblies group types together into easily deployable units (DLL files). Apart from when you learned about unit testing, you have only created console apps to contain your code. To make the code that you write reusable across multiple projects, you should put it in class library assemblies, just like Microsoft does.

Creating a class library

The first task is to create a reusable .NET class library:

1. Use your preferred code editor to create a new project, as defined in the following list:
   - Project template: **Class Library** / `classlib`
   - Project file and folder: `PacktLibraryNet2`
   - Solution file and folder: `Chapter05`
2. Open the `PacktLibraryNet2.csproj` file, and note that, by default, class libraries created by the .NET 9 SDK target .NET 9 and, therefore, can only be referenced by other .NET 9-compatible assemblies, as highlighted in the following markup:

```
<Project Sdk="Microsoft.NET.Sdk">

  <PropertyGroup>
    <TargetFramework>net9.0</TargetFramework>
    <ImplicitUsings>enable</ImplicitUsings>
    <Nullable>enable</Nullable>
  </PropertyGroup>

</Project>
```

3. Modify the framework to target .NET Standard 2.0, add an entry to explicitly use the C# 13 compiler, and statically import the `System.Console` class for all C# files, as highlighted in the following markup:

```
<Project Sdk="Microsoft.NET.Sdk">

  <PropertyGroup>
    <!--.NET Standard 2.0 class library can be used by:
        .NET Framework, Xamarin, modern .NET. -->
    <TargetFramework>netstandard2.0</TargetFramework>

    <!--Compile this library using C# 13 so we can use most
        modern compiler features. -->
    <LangVersion>13</LangVersion>
```

```
    <Nullable>enable</Nullable>
    <ImplicitUsings>enable</ImplicitUsings>
  </PropertyGroup>

  <ItemGroup>
    <Using Include="System.Console" Static="true" />
  </ItemGroup>

</Project>
```

Although we can use the C# 13 compiler, some modern compiler features require a modern .NET runtime. For example, we cannot use default implementations in an interface (introduced in C# 8) because it requires .NET Standard 2.1. We cannot use the `required` keyword (introduced in C# 11) because it requires an attribute introduced in .NET 7. But many useful modern compiler features, like raw literal strings, will be available to us.

4. Save and close the file.
5. Delete the file named `Class1.cs`.
6. Compile the project so that other projects can reference it later:
   - In Visual Studio, navigate to **Build | Build PacktLibraryNet2**.
   - In VS Code, enter the following command: `dotnet build`.

Good practice: To use all the latest C# language and .NET platform features, put types in a .NET 9 class library. To support legacy .NET platforms, like .NET Core, .NET Framework, and Xamarin, put types that you might reuse in a .NET Standard 2.0 class library. By default, targeting .NET Standard 2.0 uses the C# 7 compiler, but this can be overridden so you get the benefits of the newer SDK and compiler even though you are limited to .NET Standard 2.0 APIs.

Understanding file-scoped namespaces

Traditionally, you define types like a class nested in a namespace, as shown in the following code:

```
namespace Packt.Shared
{
  public class Person
  {
  }
}
```

If you define multiple types in the same code file, then they can be in different namespaces, since the types must be explicitly inside the curly braces for each namespace.

If you use C# 10 or later, you can simplify your code by ending a namespace declaration with a semi-colon and removing the curly braces, so the type definitions do not need to be indented, as shown in the following code:

```
// All types in this file will be defined in this file-scoped namespace.
namespace Packt.Shared;

public class Person
{
}
```

This is known as a **file-scoped namespace** declaration. You can only have one file-scoped namespace per file. This feature is especially useful for book writers who have limited horizontal space.

Good practice: Put each type that you create in its own code file, or at least put types in the same namespace in the same code file so that you can use file-scoped namespace declarations.

Defining a class in a namespace

The next task is to define a class that will represent a person:

1. In the `PacktLibraryNet2` project, add a new class file named `Person.cs`.
2. In `Person.cs`, delete any existing statements and set the namespace to `Packt.Shared`, and for the `Person` class, set the access modifier to `public`, as shown in the following code:

```
// All types in this file will be defined in this file-scoped namespace.
namespace Packt.Shared;

public class Person
{
}
```

Good practice: We're doing this because it is important to put your classes in a logically named namespace. A better namespace name would be domain-specific, for example, `System.Numerics` for types related to advanced numbers. In this case, the types we will create are `Person`, `BankAccount`, and `WondersOfTheWorld`, and they do not have a typical domain, so we will use the more generic `Packt.Shared`.

Understanding type access modifiers

Note that the C# keyword `public` is applied before `class`. This keyword is an **access modifier**, and it allows for any other code to access this class even outside this class library.

If you do not explicitly apply the `public` keyword, then it will only be accessible within the assembly that defined it. This is because the implicit access modifier for a class is `internal`. We need this class to be accessible outside the assembly, so we must make sure it is `public`.

If you have nested classes, meaning a class defined in another class, then the inner class could have the `private` access modifier, which would mean it is not accessible outside its parent class.

Introduced with .NET 7, the `file` access modifier applied to a type means that type can only be used within its code file. This would only be useful if you define multiple classes in the same code file, which is rarely good practice but is used with source generators.

More information: You can learn more about the `file` access modifier at the following link: `https://learn.microsoft.com/en-us/dotnet/csharp/language-reference/keywords/file`.

Good practice: The two most common access modifiers for a class are `public` and `internal` (the default access modifier for a class if not specified). Always explicitly specify the access modifier for a class to make it clear what it is. Other access modifiers include `private` and `file`, but they are rarely used.

Understanding members

The `Person` type does not yet have any members encapsulated within it. We will create some over the following pages. Members can be fields, methods, or specialized versions of both. You'll find a description of them here:

- **Fields** are used to store data. You can think of fields as variables that belong to a type. There are also three specialized categories of field, as shown in the following bullets:
  - **Constant**: The data never changes. The compiler literally copies the data into any code that reads it. For example, `byte.MaxValue` is always `255`. Constants must be set to a literal value. Literal values can only be a `string` surrounded by `""`, a `char` surrounded with single quotes, a Boolean like `true` or `false`, or a number like `3`, `3L`, `3.0`, `3.0F`, or `3.0M`. Non-literal values are not supported, so you cannot have a constant date/time value.
  - **Read-only**: The data cannot change after the class is instantiated, but the data can be calculated or loaded from an external source at the time of instantiation. For example, `String.Empty` is `""`, `Guid.Empty` is a **GUID** (**globally unique identifier**) with all zeros, `TimeSpan.Zero` is a time interval of zero, `DateTime.MinValue` is `00:00:00.0000000 UTC, January 1, 0001` in the Gregorian calendar, and `Path.DirectorySeparatorChar` is \ on Windows.

`DateTime.UnixEpoch` is `00:00:00.0000000 UTC, January 1, 1970`, in the Gregorian calendar but this field was introduced after .NET Standard 2.0, so you won't be able to try it in your current class library.

  - **Event:** The data references one or more methods that you want to execute when something happens, such as clicking on a button or responding to a request from some other code. Events will be covered in *Chapter 6, Implementing Interfaces and Inheriting Classes*. For example, `Console.CancelKeyPress` happens when *Ctrl* + *C* or *Ctrl* + *Break* are pressed in a console app.

- **Methods** are used to execute statements. You saw some examples when you learned about functions in *Chapter 4, Writing, Debugging, and Testing Functions*. There are also four specialized categories of methods:
  - **Constructor:** The statements execute when you use the `new` keyword to allocate memory to instantiate a class. For example, to instantiate Christmas Day, 2023, you could write the following code: `new DateTime(2023, 12, 25)`.
  - **Property:** The statements execute when you get or set data. The data is commonly stored in a field but can be stored externally or calculated at runtime. Properties are the preferred way to encapsulate fields unless the memory address of the field needs to be exposed; for example, `Console.ForegroundColor` to set the current color of text in a console app.
  - **Indexer:** The statements execute when you get or set data using "array" syntax `[]`. For example, use `name[0]` to get the first character in the `name` variable, which is a `string`.
  - **Operator:** The statements execute when you apply an operator like `+` and `/` to operands of your type. For example, use `a + b` to add two variables together.

Importing a namespace to use a type

In this section, we will make an instance of the `Person` class.

Before we can instantiate a class, we need to reference the assembly that contains it from another project. We will use the class in a console app:

1. Use your preferred code editor to add a new **Console App** / `console` named `PeopleApp` to the `Chapter05` solution. Make sure you *add* the new project to the existing `Chapter05` solution because you are about to reference from the console app project to the existing class library project so both projects must be in the same solution.
2. If you use Visual Studio:
   1. Configure the startup project for the solution to the current selection.
   2. In **Solution Explorer**, select the `PeopleApp` project, navigate to **Project** | **Add Project Reference...**, check the box to select the `PacktLibraryNet2` project, and then click **OK**.

3. In `PeopleApp.csproj`, add an entry to statically import the `System.Console` class, as shown in the following markup:

```
<ItemGroup>
<Using Include="System.Console" Static="true" />
</ItemGroup>
```

4. Navigate to **Build | Build PeopleApp**.

3. If you use VS Code:

1. Edit `PeopleApp.csproj` to add a project reference to `PacktLibraryNet2`, and add an entry to statically import the `System.Console` class, as highlighted in the following markup:

```
<Project Sdk="Microsoft.NET.Sdk">

  <PropertyGroup>
    <OutputType>Exe</OutputType>
    <TargetFramework>net9.0</TargetFramework>
    <Nullable>enable</Nullable>
    <ImplicitUsings>enable</ImplicitUsings>
  </PropertyGroup>

  <ItemGroup>
  <ProjectReference Include=
  "../PacktLibraryNet2/PacktLibraryNet2.csproj" />
  </ItemGroup>

  <ItemGroup>
    <Using Include="System.Console" Static="true" />
  </ItemGroup>

</Project>
```

2. In a terminal, compile the `PeopleApp` project and its dependency `PacktLibraryNet2` project, as shown in the following command:

```
dotnet build
```

4. In the `PeopleApp` project, add a new class file named `Program.Helpers.cs`.
5. In `Program.Helpers.cs`, delete any existing statements, and define a `partial Program` class with a method to configure the console to enable special symbols, like the euro currency, and to control the current culture, as shown in the following code:

```
using System.Globalization; // To use CultureInfo.

partial class Program
{
  private static void ConfigureConsole(
    string culture = "en-US",
    bool useComputerCulture = false,
    bool showCulture = true)
  {
    OutputEncoding = System.Text.Encoding.UTF8;

    if (!useComputerCulture)
    {
      CultureInfo.CurrentCulture = CultureInfo.GetCultureInfo(culture);
    }

    if (showCulture)
    {
      WriteLine($"Current culture: {CultureInfo.CurrentCulture.
DisplayName}.");
    }
  }
}
```

By the end of this chapter, you will understand how the preceding method uses C# features like partial classes, optional parameters, and so on. If you would like to learn more about working with languages and cultures, as well as dates, times, and time zones, then there is a chapter about globalization and localization in my companion book, *Apps and Services with .NET 8*.

Instantiating a class

Now, we are ready to write statements to instantiate the `Person` class:

1. In the `PeopleApp` project, in the `Program.cs` file, delete the existing statements, then add statements to import the namespace for our `Person` class, and then call the `ConfigureConsole` method without any arguments so that it sets the current culture to US English, allowing all readers to see the same output, as shown in the following code:

```
using Packt.Shared; // To use Person.

ConfigureConsole(); // Sets current culture to US English.
```

```
// Alternatives:
// ConfigureConsole(useComputerCulture: true); // Use your culture.
// ConfigureConsole(culture: "fr-FR"); // Use French culture.
```

Although we could import the `Packt.Shared` namespace globally, it will be clearer to anyone reading this code where we import the types we use from if the `import` statement is at the top of the file, and the `PeopleApp` project will only have this one `Program.cs` file that needs the namespace imported.

2. In `Program.cs`, add statements to:
   - Create an instance of the `Person` type.
   - Output the instance using a textual description of itself.

   The `new` keyword allocates memory for the object and initializes any internal data, as shown in the following code:

```
// Person bob = new Person(); // C# 1 or later.
// var bob = new Person(); // C# 3 or later.

Person bob = new(); // C# 9 or later.
WriteLine(bob); // Implicit call to ToString().
// WriteLine(bob.ToString()); // Does the same thing.
```

3. Run the `PeopleApp` project and view the result, as shown in the following output:

```
Current culture: English (United States).
Packt.Shared.Person
```

You might be wondering, "Why does the `bob` variable have a method named `ToString`? The `Person` class is empty!" Don't worry, we're about to find out!

Inheriting from System.Object

Although our `Person` class did not explicitly choose to inherit from a type, all types ultimately inherit directly or indirectly from a special type named `System.Object`. The implementation of the `ToString` method in the `System.Object` type outputs the full namespace and type name.

Back in the original `Person` class, we could have explicitly told the compiler that `Person` inherits from the `System.Object` type, as shown in the following code:

```
public class Person : System.Object
```

When class B is inherited from class A, we say that A is the base or superclass, and B is the derived or subclass. In this case, `System.Object` is the base or superclass, and `Person` is the derived or subclass. You can also use the C# keyword `object`.

Let's make our class explicitly inherit from `object` and then review what members all objects have:

1. Modify your `Person` class to explicitly inherit from `object`, as shown in the following code:

```
public class Person : object
```

2. Click inside the `object` keyword and press *F12*, or right-click on the `object` keyword and choose **Go to Definition**.

You will see the Microsoft-defined `System.Object` type and its members. This is something you don't need to understand the details of yet, but note that the class is in a .NET Standard 2.0 class library assembly, as shown in *Figure 5.1*:

Figure 5.1: The System.Object class definition in .NET Standard 2.0

If you scroll down the class code file, note that the `Object` class has a method named `ToString`.

Good practice: Assume other programmers know that if inheritance is not specified, the class will inherit from `System.Object`.

Avoiding a namespace conflict with a using alias

We need to learn a bit more about namespaces and their types. It is possible that there are two namespaces that contain the same type name, and importing both namespaces causes ambiguity. For example, `JsonOptions` exists in multiple Microsoft-defined namespaces. If you use the wrong one to configure JSON serialization, then it will be ignored and you'll be confused as to why!

Let's review a made-up example:

```
// In the file, France.Paris.cs
namespace France
{
  public class Paris
```

```
    {
    }
}

// In the file, Texas.Paris.cs
namespace Texas
{
    public class Paris
    {
    }
}

// In the file, Program.cs
using France;
using Texas;

Paris p = new();
```

If we build this project, then the compiler would complain with the following error:

```
Error CS0104: 'Paris' is an ambiguous reference between 'France.Paris' and
'Texas.Paris'
```

We can define an alias for one of the namespaces to differentiate it, as shown in the following code:

```
using France; // To use Paris.
using Tx = Texas; // Tx becomes alias for the namespace, and it is not
imported.

Paris p1 = new(); // Creates an instance of France.Paris.
Tx.Paris p2 = new(); // Creates an instance of Texas.Paris.
```

Renaming a type with a using alias

Another situation where you might want to use an alias is if you would like to rename a type. For example, if you use the `Environment` class in the `System` namespace a lot, you could rename it with an alias to make it shorter, as shown in the following code:

```
using Env = System.Environment;

WriteLine(Env.OSVersion);
WriteLine(Env.MachineName);
WriteLine(Env.CurrentDirectory);
```

Starting with C# 12, you can alias any type. This means you can rename existing types or give a type name to unnamed types like tuples, as you will see later in this chapter.

You can learn how to refactor your code using alias any type at the following link: `https://devblogs.microsoft.com/dotnet/refactor-your-code-using-alias-any-type/`.

Storing data in fields

In this section, we will define a selection of fields in the class to store information about a person.

Defining fields

Let's say that we have decided that a person is composed of a name and a date and time of birth. We will encapsulate these two values inside a person, and the values will be visible outside it:

- Inside the `Person` class, write statements to declare two public fields to store a person's name and the date of when they were born, as highlighted in the following code:

```
public class Person : object
{
  #region Fields: Data or state for this person.

  public string? Name; // ? means it can be null.
  public DateTimeOffset Born;

  #endregion
}
```

We have multiple choices for the data type of the `Born` field. .NET 6 introduced the `DateOnly` type. This would store only the date without a time value. `DateTime` stores the date and time when the person was born, but it varies between local and **Universal Coordinated Time** (UTC). The best choice is `DateTimeOffset`, which stores the date, time, and hours offset from **UTC**, which is related to the time zone. The choice depends on how much detail you need to store.

Types for fields

Since C# 8, the compiler has had the ability to warn you if a reference type, like a `string`, could have a `null` value and, therefore, potentially throw a `NullReferenceException`. Since .NET 6, the SDK enables those warnings by default. You can suffix the `string` type with a question mark, `?`, to indicate that you accept this, and the warning disappears. You will learn more about nullability and how to handle it in *Chapter 6, Implementing Interfaces and Inheriting Classes*.

You can use any type for a field, including arrays and collections, such as lists and dictionaries. These would be used if you needed to store multiple values in one named field. In this example, a person only has one name and one date and time of birth.

Member access modifiers

Part of encapsulation is choosing how visible members are to other code.

Note that, as we did with the class, we explicitly applied the `public` keyword to these fields. If we hadn't, then they would be implicitly `private` to the class, which means they are accessible only inside the class.

There are four **member access modifier** keywords, and two combinations of access modifier keywords that you can apply to a class member, like a field or method. Member access modifiers apply to an individual member. They are similar to but separate from type access modifiers that apply to the whole type. The six possible combinations are shown in *Table 5.1*:

Member Access Modifier	Description
`private`	The member is accessible inside the type only. This is the default.
`internal`	The member is accessible inside the type and any type in the same assembly.
`protected`	The member is accessible inside the type and any type that inherits from the type.
`public`	The member is accessible everywhere.
`internal protected`	The member is accessible inside the type, any type in the same assembly, and any type that inherits from the type. Equivalent to a fictional access modifier named `internal_or_protected`.
`private protected`	The member is accessible inside the type and any type that inherits from the type and is in the same assembly. Equivalent to a fictional access modifier named `internal_and_protected`. This combination is only available with C# 7.2 or later.

Table 5.1: Six member access modifiers

Good practice: Explicitly apply one of the access modifiers to all type members, even if you want to use the implicit access modifier for members, which is `private`. Additionally, fields should usually be `private` or `protected`, and you should then create `public` properties to get or set the field values. This is because the property then controls access. You will do this later in the chapter.

Setting and outputting field values

Now, we will use those fields in your code:

1. In `Program.cs`, after instantiating `bob`, add statements to set his name and date and time of birth, and then output those fields formatted nicely, as shown in the following code:

```
bob.Name = "Bob Smith";

bob.Born = new DateTimeOffset(
  year: 1965, month: 12, day: 22,
  hour: 16, minute: 28, second: 0,
  offset: TimeSpan.FromHours(-5)); // US Eastern Standard Time.

WriteLine(format: "{0} was born on {1:D}.", // Long date.
  arg0: bob.Name, arg1: bob.Born);
```

The format code for `arg1` is one of the standard date and time formats. `D` means a long date format and `d` would mean a short date format. You can learn more about standard date and time format codes at the following link: `https://learn.microsoft.com/en-us/dotnet/standard/base-types/standard-date-and-time-format-strings`.

2. Run the `PeopleApp` project and view the result, as shown in the following output:

```
Bob Smith was born on Wednesday, December 22, 1965.
```

If you change the call to `ConfigureConsole` to use your local computer culture or a specified culture, like French in France (`"fr-FR"`), then your output will look different.

Setting field values using object initializer syntax

You can also initialize fields using a shorthand **object initializer** syntax with curly braces, which was introduced with C# 3. Let's see how:

1. Add statements underneath the existing code to create another new person named Alice. Note the different standard format code for the date and time of birth when writing her to the console, as shown in the following code:

```
Person alice = new()
{
  Name = "Alice Jones",
  Born = new(1998, 3, 7, 16, 28, 0,
    // This is an optional offset from UTC time zone.
    TimeSpan.Zero)
};

WriteLine(format: "{0} was born on {1:d}.", // Short date.
  arg0: alice.Name, arg1: alice.Born);
```

We could have used string interpolation to format the output, but for long strings, it will wrap over multiple lines, which can be harder to read in a printed book. In the code examples in this book, remember that `{0}` is a placeholder for `arg0`, and so on.

2. Run the `PeopleApp` project and view the result, as shown in the following output:

```
Alice Jones was born on 3/7/1998.
```

Good practice: Use named parameters to pass arguments, so it is clearer what the values mean, especially for types like `DateTimeOffset` where there are a bunch of numbers one after the other.

Storing a value using an enum type

Sometimes, a value needs to be one of a limited set of options. For example, there are seven ancient wonders of the world, and a person may have one favorite.

At other times, a value needs to be a combination of a limited set of options. For example, a person may have a bucket list of ancient world wonders they want to visit. We can store this data by defining an `enum` type.

An `enum` type is a very efficient way of storing one or more choices because, internally, it uses integer values in combination with a lookup table of `string` descriptions. Let's see an example:

1. Add a new file to the `PacktLibraryNet2` project named `WondersOfTheAncientWorld.cs`.
2. Modify the `WondersOfTheAncientWorld.cs` contents, as shown in the following code:

```
namespace Packt.Shared;

public enum WondersOfTheAncientWorld
{
  GreatPyramidOfGiza,
  HangingGardensOfBabylon,
  StatueOfZeusAtOlympia,
  TempleOfArtemisAtEphesus,
  MausoleumAtHalicarnassus,
  ColossusOfRhodes,
  LighthouseOfAlexandria
}
```

3. In `Person.cs`, define a field to store a person's favorite ancient world wonder, as shown in the following code:

```
public WondersOfTheAncientWorld FavoriteAncientWonder;
```

4. In Program.cs, set Bob's favorite ancient wonder of the world and output it, as shown in the following code:

```
bob.FavoriteAncientWonder = WondersOfTheAncientWorld.
StatueOfZeusAtOlympia;

WriteLine(
  format: "{0}'s favorite wonder is {1}. Its integer is {2}.",
  arg0: bob.Name,
  arg1: bob.FavoriteAncientWonder,
  arg2: (int)bob.FavoriteAncientWonder);
```

5. Run the PeopleApp project and view the result, as shown in the following output:

```
Bob Smith's favorite wonder is StatueOfZeusAtOlympia. Its integer is 2.
```

The enum value is internally stored as an int for efficiency. The int values are automatically assigned, starting at 0, so the third world wonder in our enum has a value of 2. You can assign int values that are not listed in the enum. They will output as the int value instead of a name since a match will not be found.

Storing multiple values using an enum type

For the bucket list, we could create an array or collection of instances of the enum, and collections as fields will be shown later in this chapter, but there is a better approach for this scenario. We can combine multiple choices into a single value using enum flags. Let's see how:

1. Modify the enum by decorating it with the [Flags] attribute, and explicitly set a byte value for each wonder that represents different bit columns, as highlighted in the following code:

```
namespace Packt.Shared;

[Flags]
public enum WondersOfTheAncientWorld : byte
{
  None                       = 0b_0000_0000, // i.e. 0
  GreatPyramidOfGiza         = 0b_0000_0001, // i.e. 1
  HangingGardensOfBabylon    = 0b_0000_0010, // i.e. 2
  StatueOfZeusAtOlympia      = 0b_0000_0100, // i.e. 4
  TempleOfArtemisAtEphesus   = 0b_0000_1000, // i.e. 8
  MausoleumAtHalicarnassus   = 0b_0001_0000, // i.e. 16
  ColossusOfRhodes           = 0b_0010_0000, // i.e. 32
  LighthouseOfAlexandria     = 0b_0100_0000  // i.e. 64
}
```

The integer types that an `enum` is allowed to inherit from are `Byte`, `SByte`, `Int16`, `Int32`, `Int64`, `UInt16`, `UInt32`, and `UInt64`. The new integer types `Int128` and `UInt128` are not supported.

We assign explicit values for each choice that will not overlap when looking at the bits stored in memory. We should also decorate the `enum` type with the `System.Flags` attribute so that when the value is returned, it can automatically match with multiple values as a comma-separated `string` instead of returning an `int` value.

Normally, an `enum` type uses an `int` variable internally, but since we don't need values that big, we can reduce memory requirements by 75%, that is, 1 byte per value instead of 4 bytes, by telling it to use a `byte` variable. As another example, if you wanted to define an `enum` for days of the week, there will only ever be seven of them.

If we want to indicate that our bucket list includes the *Hanging Gardens of Babylon* and the *Mausoleum at Halicarnassus* ancient world wonders, then we would want the `16` and `2` bits set to `1`. In other words, we would store the value `18`, as shown in *Table 5.2*:

64	32	16	8	4	2	1
0	0	1	0	0	1	0

Table 5.2: Storing 18 as bits in an enum

2. In `Person.cs`, leave the existing field to store a single favorite ancient world wonder and add the following statement to your list of fields to store multiple ancient world wonders:

```
public WondersOfTheAncientWorld BucketList;
```

3. In `Program.cs`, add statements to set the bucket list using the `|` operator (the bitwise logical `OR`) to combine the `enum` values. We could also set the value using the number `18` cast into the `enum` type, as shown in the comment, but we shouldn't because that would make the code harder to understand, as shown in the following code:

```
bob.BucketList =
  WondersOfTheAncientWorld.HangingGardensOfBabylon
  | WondersOfTheAncientWorld.MausoleumAtHalicarnassus;

// bob.BucketList = (WondersOfTheAncientWorld)18;

WriteLine($"{bob.Name}'s bucket list is {bob.BucketList}.");
```

4. Run the `PeopleApp` project and view the result, as shown in the following output:

```
Bob Smith's bucket list is HangingGardensOfBabylon,
MausoleumAtHalicarnassus.
```

Good practice: Use the `enum` values to store combinations of discrete options. Derive an `enum` type from `byte` if there are up to eight options, from `ushort` if there are up to 16 options, from `uint` if there are up to 32 options, and from `ulong` if there are up to 64 options.

Now that we have decorated the `enum` with the `[Flags]` attribute, combinations of values can be stored in a single variable or field. Now, a programmer could store a combination of values in the `FavoriteAncientWonder` too when it should only store one value. To enforce this, we should convert the field into a property that allows us to take control over how other programmers can get and set the value. You will see how to do this later in this chapter.

Changing an enum base type for performance

The preceding section was about storing multiple values using an `enum` type. It's about `enum` types decorated with the `[Flags]` attribute that use bitwise operations to efficiently store those multiple values. In the code example, we defined an `enum` for the Seven Wonders of the Ancient World, so it only needed seven combinable values (and `0` for `None`).

The preceding section was not about making all your `enum` types derive from `byte` to make your code faster because that would be bad advice.

On March 18, 2024, Nick Chapsas posted a YouTube video titled, *Turn All Your Enums Into Bytes Now! | Code Cop #014*, which you can watch at the following link: `https://www.youtube.com/watch?v=1gWzE9SIGkQ`. He criticized blog articles that recommend changing the default base integer type of `enum` types from `int` to `byte` to improve performance.

The original designers of the C# language spent effort on implementing the ability for `enum` types to derive from other integers than just the default `int`. For example, you can use fewer bytes by using a positive integer like `byte` or `ushort`, or the same or more bytes by using a positive integer like `uint` or `ulong`. They implemented this feature because sometimes a .NET developer will need this capability.

I think it is important that my readers know that they can do it when necessary. Microsoft's official guidance states, "Even though you can change this underlying type, it is not necessary or recommended for most scenarios. No significant performance gain is achieved by using a data type that is smaller than `Int32`.", as you can read at the following link: `https://learn.microsoft.com/en-us/dotnet/fundamentals/code-analysis/quality-rules/ca1028`.

For those developers who object to changing any `enum` from `int` to some other integer, there is a compiler code analysis warning that I linked to above. If enabled, it will trigger if you set an `enum` to anything other than `int`: "CA1028: Enum storage should be Int32." This warning is not enabled by default because Microsoft knows that there are legitimate reasons why a developer might need to use it.

Let's see some real-life examples of when you would need to change an `enum` from deriving from `int` to deriving from another integer type:

- You want to increase the size of the integer to store more than 16 options in a flag `enum`. The default `int` only allows 16 options: 0, 1, 2, 4, 8, 16, 32, 64, 128, 256, 512, 1024, 2048, 4096, 8192, and 16384. Changing to `uint` would double the number of choices to 32 without using any more space in memory. Changing to `ulong` would give 64 options. Changing to `ushort` would allow the same 16 options in half the bytes.
- You need to transfer data as a binary stream via a serial connection to an embedded device and you must carefully follow the protocol, or you are defining your own serial messaging protocol, and you want to reduce the packet size to make the best use of your available bandwidth.
- You have SQL tables with millions of records where some of the columns are `enum` values. Setting those columns to `tinyint` with a matching `enum : byte` property in the C# entity class can make indexes perform better by being smaller and reducing the number of page reads from disk. Some developers will work on systems that are 30 or more years old with spinning metal disks. Not everyone is deploying to a modern 64-bit OS with modern hardware.
- You need to reduce the size of a `struct` because it will be created 100,000 times per second on resource-constrained hardware, or you have game code that is set to use `byte` and `short` because you have millions of them in contiguous arrays for the game's data. You would gain a fair bit of performance doing this, especially from a cache point of view.

Now, let's see how to store multiple values using collections.

Storing multiple values using collections

Let's now add a field to store a person's children. This is an example of aggregation because children are instances of a class that is related to the current person, but they are not part of the person itself. We will use a generic `List<T>` collection type that can store an ordered collection of any type. You will learn more about collections in *Chapter 8, Working with Common .NET Types*. For now, just follow along:

- In `Person.cs`, declare a new field to store multiple `Person` instances that represent the children of this person, as shown in the following code:

  ```
  public List<Person> Children = new();
  ```

`List<Person>` is read aloud as "list of `Person`," for example, "the type of the property named `Children` is a list of `Person` instances."

We must ensure the collection is initialized to a new instance before we can add items to it; otherwise, the field will be `null` and throw runtime exceptions when we try to use any of its members, like `Add`.

Understanding generic collections

The angle brackets in the `List<T>` type is a feature of C# called **generics** that was introduced in 2005 with C# 2. It's a fancy term for making a collection **strongly typed**, that is, the compiler knows specifically what type of object can be stored in the collection. Generics improve the performance and correctness of your code.

Strongly typed has a different meaning than **statically typed**. The old `System.Collection` types are statically typed to contain weakly typed `System.Object` items. The newer `System.Collection.Generic` types are statically typed to contain strongly typed `<T>` instances.

Ironically, the term *generics* means we can use a more specific static type!

1. In `Program.cs`, add statements to add three children for `Bob`, and then show how many children he has and what their names are, as shown in the following code:

```
// Works with all versions of C#.
Person alfred = new Person();
alfred.Name = "Alfred";
bob.Children.Add(alfred);

// Works with C# 3 and later.
bob.Children.Add(new Person { Name = "Bella" });

// Works with C# 9 and later.
bob.Children.Add(new() { Name = "Zoe" });

WriteLine($"{bob.Name} has {bob.Children.Count} children:");

for (int childIndex = 0; childIndex < bob.Children.Count; childIndex++)
{
  WriteLine($"> {bob.Children[childIndex].Name}");
}
```

2. Run the `PeopleApp` project and view the result, as shown in the following output:

```
Bob Smith has 3 children:
> Alfred
> Bella
> Zoe
```

We could also use a `foreach` statement to enumerate over the collection. As an optional challenge, change the `for` statement to output the same information using `foreach`.

Making a field static

The fields that we have created so far have all been **instance** members, meaning that a different value of each field exists for each instance of the class that is created. The `alice` and `bob` variables have different `Name` values.

Sometimes, you want to define a field that only has one value that is shared across all instances.

These are called **static members** because fields are not the only members that can be static. Let's see what can be achieved using `static` fields using a bank account as an example. Each instance of `BankAccount` will have its own `AccountName` and `Balance` values, but all instances will share a single `InterestRate` value.

Let's do it:

1. In the `PacktLibraryNet2` project, add a new class file named `BankAccount.cs`.
2. Modify the class to give it three fields – two instance fields and one static field – as shown in the following code:

```
namespace Packt.Shared;

public class BankAccount
{
  public string? AccountName; // Instance member. It could be null.
  public decimal Balance; // Instance member. Defaults to zero.

  public static decimal InterestRate; // Shared member. Defaults to zero.
}
```

3. In `Program.cs`, add statements to set the shared interest rate, and then create two instances of the `BankAccount` type, as shown in the following code:

```
BankAccount.InterestRate = 0.012M; // Store a shared value in static
field.

BankAccount jonesAccount = new();
jonesAccount.AccountName = "Mrs. Jones";
jonesAccount.Balance = 2400;
WriteLine(format: "{0} earned {1:C} interest.",
  arg0: jonesAccount.AccountName,
  arg1: jonesAccount.Balance * BankAccount.InterestRate);

BankAccount gerrierAccount = new();
gerrierAccount.AccountName = "Ms. Gerrier";
gerrierAccount.Balance = 98;
WriteLine(format: "{0} earned {1:C} interest.",
  arg0: gerrierAccount.AccountName,
  arg1: gerrierAccount.Balance * BankAccount.InterestRate);
```

4. Run the `PeopleApp` project and view the additional output:

```
Mrs. Jones earned $28.80 interest.
Ms. Gerrier earned $1.18 interest.
```

Remember that C is a format code that tells .NET to use the current culture's currency format for the decimal numbers.

Fields are not the only members that can be static. Constructors, methods, properties, and other members can also be static.

Static methods do not need an object instance to be called. For example, `Console.WriteLine` does not need an object instance; the method is called directly from the class name. Static methods are covered in *Chapter 6, Implementing Interfaces and Inheriting Classes*.

Making a field constant

If the value of a field will never change, you can use the `const` keyword and assign a literal value at compile time. Any statement that changes the value will cause a compile-time error. Let's see a simple example:

1. In `Person.cs`, add a `string` constant for the species of a person, as shown in the following code:

```
// Constant fields: Values that are fixed at compilation.
public const string Species = "Homo Sapiens";
```

2. To get the value of a constant field, you must write the name of the class, not the name of an instance of the class. In `Program.cs`, add a statement to write Bob's name and species to the console, as shown in the following code:

```
// Constant fields are accessible via the type.
WriteLine($"{bob.Name} is a {Person.Species}.");
```

3. Run the `PeopleApp` project and view the result, as shown in the following output:

```
Bob Smith is a Homo Sapiens.
```

Examples of `const` fields in Microsoft types include `System.Int32.MaxValue` and `System.Math.PI` because neither value will ever change, as you can see in *Figure 5.2*:

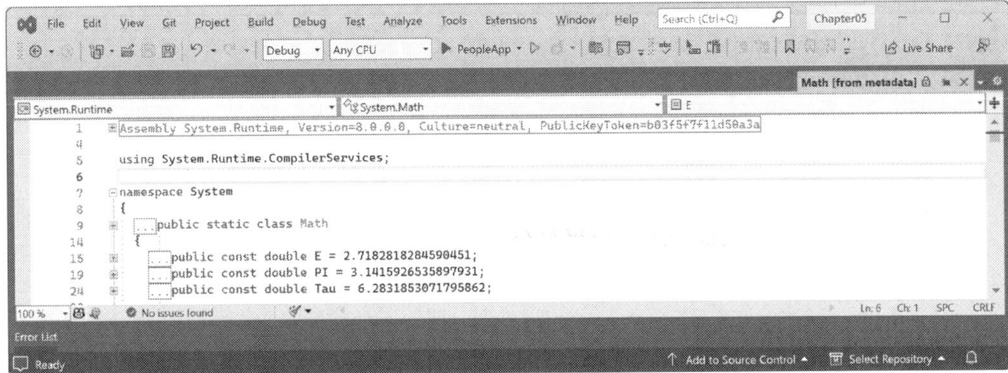

Figure 5.2: Examples of constants in the System.Math class

Good practice: Constants are not always the best choice for two important reasons: the value must be known at compile time, and it must be expressible as a literal `string`, `Boolean`, or number value. Every reference to the `const` field is replaced with the literal value at compile time, which will, therefore, not be reflected if the value changes in a future version and you do not recompile any assemblies that reference it to get the new value.

Making a field read-only

Often, a better choice for fields that should not change is to mark them as read-only:

1. In `Person.cs`, add a statement to declare an instance read-only field to store a person's home planet, as shown in the following code:

```
// Read-only fields: Values that can be set at runtime.
public readonly string HomePlanet = "Earth";
```

2. In `Program.cs`, add a statement to write Bob's name and home planet to the console, as shown in the following code:

```
// Read-only fields are accessible via the variable.
WriteLine($"{bob.Name} was born on {bob.HomePlanet}.");
```

3. Run the `PeopleApp` project and view the result, as shown in the following output:

```
Bob Smith was born on Earth.
```

Good practice: Use read-only fields over constant fields for two important reasons: the value can be calculated or loaded at runtime and can be expressed using any executable statement. So, a read-only field can be set using a constructor or a field assignment. Every reference to the read-only field is a live reference, so any future changes will be correctly reflected by the calling code.

You can also declare `static readonly` fields whose values will be shared across all instances of the type.

Requiring fields to be set during instantiation

C# 11 introduced the `required` modifier. If you use it on a field or property, the compiler will ensure that you set the field or property to a value when you instantiate it. It requires targeting .NET 7 or later, so we need to create a new class library first:

1. In the `Chapter05` solution, add a new class library project named `PacktLibraryModern` that targets .NET 9. (The oldest supported version for the `required` modifier is .NET 7.)
2. In the `PacktLibraryModern` project, rename `Class1.cs` to `Book.cs`.

3. Modify the code file contents to give the class four fields, with two set as `required`, as shown in the following code:

```
namespace Packt.Shared;

public class Book
{
  // Needs .NET 7 or later as well as C# 11 or later.
  public required string? Isbn;
  public required string? Title;

  // Works with any version of .NET.
  public string? Author;
  public int PageCount;
}
```

Note that all `three` string properties are nullable. Setting a property or field to be `required` does not mean that it cannot be `null`. It just means that it must be explicitly set to `null`.

4. In the `PeopleApp` console app project, add a reference to the `PacktLibraryModern` class library project:

    - If you use Visual Studio, then in **Solution Explorer**, select the `PeopleApp` project, navigate to **Project** | **Add Project Reference...**, check the box to select the `PacktLibraryModern` project, and then click **OK**.
    - If you use VS Code, then edit `PeopleApp.csproj` to add a project reference to `PacktLibraryModern`, as highlighted in the following markup:

    ```
    <ItemGroup>
      <ProjectReference Include=
    "..\PacktLibraryNet2\PacktLibraryNet2.csproj" />
      <ProjectReference Include=
        "..\PacktLibraryModern\PacktLibraryModern.csproj" />
    </ItemGroup>
    ```

5. Build the `PeopleApp` project. This will compile its referenced dependencies and copy the class library `.dll` to the correct subdirectory in the local `bin` folder.
6. In the `PeopleApp` project, in `Program.cs`, attempt to instantiate a `Book` without setting the `Isbn` and `Title` fields, as shown in the following code:

```
Book book = new();
```

7. Note that you will see a compiler error, as shown in the following output:

```
C:\cs13net9\Chapter05\PeopleApp\Program.cs(137,13): error CS9035:
Required member 'Book.Isbn' must be set in the object initializer or
attribute constructor. [C:\cs13net9\Chapter05\PeopleApp\PeopleApp.csproj]
C:\cs13net9\Chapter05\PeopleApp\Program.cs(137,13): error CS9035:
Required member 'Book.Title' must be set in the object initializer or
attribute constructor. [C:\cs13net9\Chapter05\PeopleApp\PeopleApp.csproj]
    0 Warning(s)
    2 Error(s)
```

8. In `Program.cs`, modify the statement to set the two required properties using object initialization syntax, as highlighted in the following code:

```
Book book = new()
{
  Isbn = "978-1803237800",
  Title = "C# 12 and .NET 8 - Modern Cross-Platform Development
Fundamentals"
};
```

9. Note that the statement now compiles without errors.
10. In `Program.cs`, add a statement to output information about the book, as shown in the following code:

```
WriteLine("{0}: {1} written by {2} has {3:N0} pages.",
  book.Isbn, book.Title, book.Author, book.PageCount);
```

Before we run the project and view the output, let's talk about an alternative way that we could initialize fields (or properties) for a type.

Initializing fields with constructors

Fields often need to be initialized at runtime. You can do this in a constructor that will be called when you make an instance of a class using the `new` keyword. Constructors execute before any fields are set by the code that uses the type:

1. In `Person.cs`, add statements after the existing read-only `HomePlanet` field to define a second read-only field, and then set the `Name` and `Instantiated` fields in a constructor, as highlighted in the following code:

```
// Read-only fields: Values that can be set at runtime.
public readonly string HomePlanet = "Earth";
public readonly DateTime Instantiated;

#endregion
```

```
#region Constructors: Called when using new to instantiate a type.

public Person()
{
  // Constructors can set default values for fields
  // including any read-only fields like Instantiated.
  Name = "Unknown";
  Instantiated = DateTime.Now;
}

#endregion
```

2. In `Program.cs`, add statements to instantiate a new person and then output its initial field values, as shown in the following code:

```
Person blankPerson = new();

WriteLine(format:
  "{0} of {1} was created at {2:hh:mm:ss} on a {2:dddd}.",
  arg0: blankPerson.Name,
  arg1: blankPerson.HomePlanet,
  arg2: blankPerson.Instantiated);
```

3. Run the `PeopleApp` project and view the result from both the code about the book as well as the blank person, as shown in the following output:

```
978-1803237800: C# 12 and .NET 8 - Modern Cross-Platform Development
Fundamentals written by  has 0 pages.
Unknown of Earth was created at 11:58:12 on a Sunday
```

Defining multiple constructors

You can have multiple constructors in a type. This is especially useful to encourage developers to set initial values for fields:

1. In `Person.cs`, add statements to define a second constructor that allows a developer to set initial values for the person's name and home planet, as shown in the following code:

```
public Person(string initialName, string homePlanet)
{
  Name = initialName;
  HomePlanet = homePlanet;
  Instantiated = DateTime.Now;
}
```

2. In `Program.cs`, add statements to create another person using the constructor with two parameters, as shown in the following code:

```
Person gunny = new(initialName: "Gunny", homePlanet: "Mars");

WriteLine(format:
  "{0} of {1} was created at {2:hh:mm:ss} on a {2:dddd}.",
  arg0: gunny.Name,
  arg1: gunny.HomePlanet,
  arg2: gunny.Instantiated);
```

3. Run the `PeopleApp` project and view the result:

```
Gunny of Mars was created at 11:59:25 on a Sunday
```

Setting required fields with a constructor

Now, let's return to the `Book` class example with its `required` fields:

1. In the `PacktLibraryModern` project, in `Book.cs`, add statements to define a pair of constructors, one that supports object initializer syntax and one to set the two required properties, as highlighted in the following code:

```
public class Book
{
  // Constructor for use with object initializer syntax.
  public Book() { }

  // Constructor with parameters to set required fields.
  public Book(string? isbn, string? title)
  {
    Isbn = isbn;
    Title = title;
  }
```

2. In `Program.cs`, comment out the statement that instantiates a book using object initializer syntax, add a statement to instantiate a book using the constructor, and then set the non-`required` properties for the book, as highlighted in the following code:

```
/*
// Instantiate a book using object initializer syntax.
Book book = new()
{
  Isbn = "978-1803237800",
  Title = "C# 12 and .NET 8 - Modern Cross-Platform Development
Fundamentals"
```

```
};
*/

Book book = new(isbn: "978-1803237800", title:
  "C# 12 and .NET 8 - Modern Cross-Platform Development Fundamentals")
{
  Author = "Mark J. Price",
  PageCount = 821
};
```

3. Note that you will see a compiler error as before because the compiler cannot automatically tell that calling the constructor will have set the two `required` properties.
4. In the `PacktLibraryModern` project, in `Book.cs`, import the namespace to perform code analysis and then decorate the constructor with the attribute to tell the compiler that it sets all the required properties and fields, as highlighted in the following code:

```
using System.Diagnostics.CodeAnalysis; // To use [SetsRequiredMembers].

namespace Packt.Shared;

public class Book
{
  public Book() { } // For use with initialization syntax.

  [SetsRequiredMembers]
  public Book(string isbn, string title)
```

5. In `Program.cs`, note the statement that calls the constructor now compiles without errors.
6. Optionally, run the `PeopleApp` project to confirm it behaves as expected, as shown in the following output:

```
978-1803237800: C# 12 and .NET 8 - Modern Cross-Platform Development
Fundamentals written by Mark J. Price has 821 pages.
```

More information: You can learn more about `required` fields and how to set them using a constructor at the following link: `https://learn.microsoft.com/en-us/dotnet/csharp/language-reference/keywords/required`.

Constructors are a special category of method. Let's look at methods in more detail.

Working with methods and tuples

Methods are members of a type that execute a block of statements. They are functions that belong to a type.

Returning values from methods

Methods can return a single value or nothing:

- A method that performs some actions but does not return a value indicates this with the `void` type before the name of the method.
- A method that performs some actions and returns a value indicates this with the type of the return value before the name of the method.

For example, in the next task, you will create two methods:

- `WriteToConsole`: This will perform an action (writing some text to the console), but it will return nothing from the method, indicated by the `void` keyword.
- `GetOrigin`: This will return a text value, indicated by the `string` keyword.

Let's write the code:

1. In `Person.cs`, add statements to define the two methods that I described earlier, as shown in the following code:

```
#region Methods: Actions the type can perform.

public void WriteToConsole()
{
  WriteLine($"{Name} was born on a {Born:dddd}.");
}

public string GetOrigin()
{
  return $"{Name} was born on {HomePlanet}.";
}

#endregion
```

2. In `Program.cs`, add statements to call the two methods, as shown in the following code:

```
bob.WriteToConsole();
WriteLine(bob.GetOrigin());
```

3. Run the `PeopleApp` project and view the result, as shown in the following output:

```
Bob Smith was born on a Wednesday.
Bob Smith was born on Earth.
```

Defining and passing parameters to methods

Methods can have parameters passed to them to change their behavior. Parameters are defined a bit like variable declarations but inside the parentheses of the method declaration, as you saw earlier in this chapter with constructors. Let's see more examples:

1. In `Person.cs`, add statements to define two methods, the first without parameters and the second with one parameter, as shown in the following code:

```
public string SayHello()
{
  return $"{Name} says 'Hello!'";
}

public string SayHelloTo(string name)
{
  return $"{Name} says 'Hello, {name}!'";
}
```

2. In `Program.cs`, add statements to call the two methods, and write the return value to the console, as shown in the following code:

```
WriteLine(bob.SayHello());
WriteLine(bob.SayHelloTo("Emily"));
```

3. Run the `PeopleApp` project and view the result:

```
Bob Smith says 'Hello!'
Bob Smith says 'Hello, Emily!'
```

When typing a statement that calls a method, IntelliSense shows a tooltip with the name, the type of any parameters, and the return type of the method.

Overloading methods

Instead of having two different method names, we could give both methods the same name. This is allowed because the methods each have a different signature.

A **method signature** is a list of parameter types that can be passed when calling the method. Overloaded methods must differ in their list of parameter types. Two overloaded methods cannot have the same list of parameter types and differ only in their return types. Let's code an example:

1. In `Person.cs`, change the name of the `SayHelloTo` method to `SayHello`.

2. In `Program.cs`, change the method call to use the `SayHello` method, and note that the quick info for the method tells you that it has an additional overload, **1 of 2**, as well as **2 of 2**, in Visual Studio, although other code editors may be different, as shown in *Figure 5.3*:

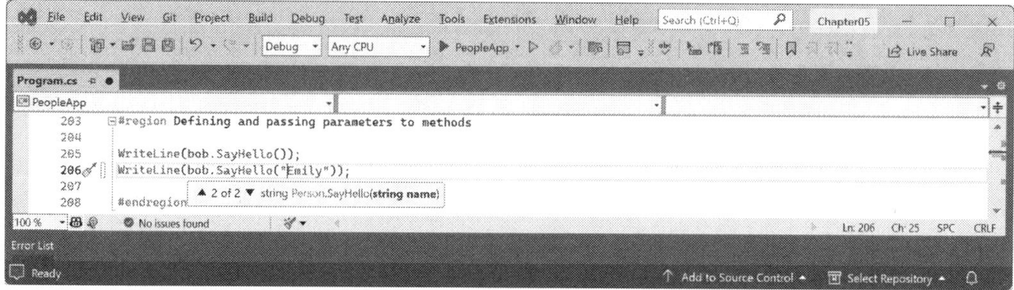

Figure 5.3: An IntelliSense tooltip for an overloaded method

Good practice: Use overloaded methods to simplify your class by making it appear to have fewer methods.

Passing optional parameters

Another way to simplify methods is to make parameters optional. You make a parameter optional by assigning a default value inside the method parameter list. Optional parameters must always come last in the list of parameters.

We will now create a method with three optional parameters:

1. In `Person.cs`, add statements to define the method, as shown in the following code:

```
public string OptionalParameters(string command = "Run!",
  double number = 0.0, bool active = true)
{
  return string.Format(
    format: "command is {0}, number is {1}, active is {2}",
    arg0: command,
    arg1: number,
    arg2: active);
}
```

2. In `Program.cs`, add a statement to call the method and write its return value to the console, as shown in the following code:

```
WriteLine(bob.OptionalParameters());
```

3. Watch IntelliSense appear as you type the code. You will see a tooltip showing the three optional parameters with their default values.

4. Run the PeopleApp project and view the result, as shown in the following output:

```
command is Run!, number is 0, active is True
```

5. In Program.cs, add a statement to pass a string value for the command parameter and a double value for the number parameter, as shown in the following code:

```
WriteLine(bob.OptionalParameters("Jump!", 98.5));
```

6. Run the PeopleApp project and see the result, as shown in the following output:

```
command is Jump!, number is 98.5, active is True
```

The default values for the command and number parameters have been replaced, but the default for active is still true.

Naming parameter values when calling methods

Optional parameters are often combined with naming parameters when you call the method because naming a parameter allows the values to be passed in a different order than how they were declared:

1. In Program.cs, add a statement to pass a string value for the command parameter and a double value for the number parameter, but using named parameters, so that the order they are passed through can be swapped around, as shown in the following code:

```
WriteLine(bob.OptionalParameters(number: 52.7, command: "Hide!"));
```

2. Run the PeopleApp project and view the result, as shown in the following output:

```
command is Hide!, number is 52.7, active is True
```

You can even use named parameters to skip over optional parameters.

3. In Program.cs, add a statement to pass a string value for the command parameter using positional order, skip the number parameter, and use the named active parameter, as shown in the following code:

```
WriteLine(bob.OptionalParameters("Poke!", active: false));
```

4. Run the PeopleApp project and view the result, as shown in the following output:

```
command is Poke!, number is 0, active is False
```

Good practice: Although you can mix named and positional parameter values, most developers prefer to read code that uses one or the other within the same method call.

Mixing optional and required parameters

At the moment, all the parameters in the `OptionalParameters` method are optional. What if one of them is required?

1. In `Person.cs`, add a fourth parameter without a default value to the `OptionalParameters` method, as highlighted in the following code:

```
public string OptionalParameters(string command = "Run!",
  double number = 0.0, bool active = true, int count)
```

2. Build the project and note the compiler error:

```
Error CS1737 Optional parameters must appear after all required
parameters.
```

3. In the `OptionalParameters` method, move the `count` parameter before the optional parameters, as shown in the following code:

```
public string OptionalParameters(int count,
  string command = "Run!",
  double number = 0.0, bool active = true)
```

4. In `Program.cs`, modify all the calls to the `OptionalParameters` method to pass an `int` value as the first argument, for example, as shown in the following code:

```
WriteLine(bob.OptionalParameters(3));
WriteLine(bob.OptionalParameters(3, "Jump!", 98.5));
WriteLine(bob.OptionalParameters(3, number: 52.7, command: "Hide!"));
WriteLine(bob.OptionalParameters(3, "Poke!", active: false));
```

Remember that if you name the arguments, then you can change their positions, for example: `bob.OptionalParameters(number: 52.7, command: "Hide!", count: 3)`.

5. As you call the `OptionalParameters` method, note the tooltip that shows the one required, three optional parameters, and their default values in Visual Studio, as shown in *Figure 5.4*:

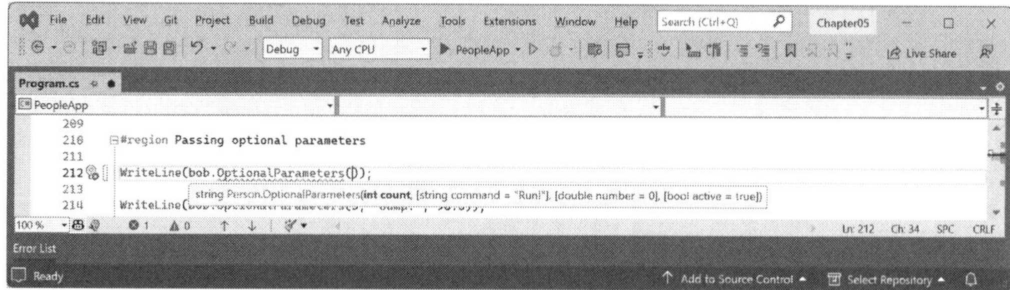

Figure 5.4: IntelliSense showing the required and optional parameters as you type code

Controlling how parameters are passed

When a parameter is passed into a method, it can be passed in one of several ways:

- By **value** (this is the default): Think of these as being *in-only*. Although the value can be changed, this only affects the parameter in the method.
- As an `out` parameter: Think of these as being *out-only*. `out` parameters cannot have a default value assigned in their declaration and cannot be left uninitialized. They must be set inside the method; otherwise, the compiler will give an error.
- By **reference** as a `ref` parameter: Think of these as being *in-and-out*. Like `out` parameters, `ref` parameters also cannot have default values, but since they can already be set outside the method, they do not need to be set inside the method.
- As an `in` parameter: Think of these as being a reference parameter that is *read-only*. `in` parameters cannot have their values changed and the compiler will show an error if you try.

Let's see some examples of passing parameters in and out of a method:

1. In `Person.cs`, add statements to define a method with three parameters, one `in` parameter, one `ref` parameter, and one `out` parameter, as shown in the following method:

```
public void PassingParameters(int w, in int x, ref int y, out int z)
{
  // out parameters cannot have a default and they
  // must be initialized inside the method.
  z = 100;

  // Increment each parameter except the read-only x.
  w++;
  // x++; // Gives a compiler error!
  y++;
  z++;

  WriteLine($"In the method: w={w}, x={x}, y={y}, z={z}");
}
```

2. In `Program.cs`, add statements to declare some `int` variables and pass them into the method, as shown in the following code:

```
int a = 10;
int b = 20;
int c = 30;
int d = 40;

WriteLine($"Before: a={a}, b={b}, c={c}, d={d}");
```

```
bob.PassingParameters(a, b, ref c, out d);

WriteLine($"After: a={a}, b={b}, c={c}, d={d}");
```

3. Run the PeopleApp project and view the result, as shown in the following output:

```
Before: a=10, b=20, c=30, d=40
In the method: w=11, x=20, y=31, z=101
After: a=10, b=20, c=31, d=101
```

Note the following about the preceding output:

- When passing a variable as a parameter by default, its current value gets passed, not the variable itself. Therefore, w has a copy of the value of the a variable. The a variable retains its original value of 10 even after w is incremented to 11.
- When passing a variable as an in parameter, a reference to the variable gets passed into the method. Therefore, x is a reference to b. If the b variable gets incremented by some other process while the method is executing, then the x parameter would show that.
- When passing a variable as a ref parameter, a reference to the variable gets passed into the method. Therefore, y is a reference to c. The c variable gets incremented when the y parameter gets incremented.
- When passing a variable as an out parameter, a reference to the variable gets passed into the method. Therefore, z is a reference to d. The value of the d variable gets replaced by whatever code executes inside the method.

We could simplify the code in the Main method by not assigning the value 40 to the d variable since it will always be replaced anyway. In C# 7 and later, we can simplify code that uses the out parameter.

4. In Program.cs, add statements to declare some more variables, including an out parameter named h declared inline, as shown in the following code:

```
int e = 50;
int f = 60;
int g = 70;
WriteLine($"Before: e={e}, f={f}, g={g}, h doesn't exist yet!");

// Simplified C# 7 or later syntax for the out parameter.
bob.PassingParameters(e, f, ref g, out int h);
WriteLine($"After: e={e}, f={f}, g={g}, h={h}");
```

5. Run the PeopleApp project and view the result, as shown in the following output:

```
Before: e=50, f=60, g=70, h doesn't exist yet!
In the method: w=51, x=60, y=71, z=101
After: e=50, f=60, g=71, h=101
```

Passing a variable number of parameters

The `params` keyword in C# is used to specify a method parameter that takes a variable number of arguments. It allows you to pass an array of arguments to a method, or to pass a list of arguments that the compiler will automatically convert to an array.

With C# 13 and later, as well as passing an array, you can pass any type that can be constructed via a collection expression, for example, a `List<T>` or `IEnumerable<T>`. Collection expressions are covered in detail in *Chapter 8, Working with Common .NET Types*, but basically, you can define a collection of items by surrounding a comma-separated list of items in square brackets `[ ]`.

The benefits of using `params` include flexibility. It allows methods to accept any number of arguments and is easy to use because it simplifies the method signature by avoiding the need to explicitly create an array before calling the method.

There are some considerations when using the `params` keyword:

- The `params` keyword can only be used for one parameter in a method, and it must be the last parameter if the method has more than one.
- You can pass individual arguments, and the compiler will automatically package them into an array. Alternatively, you can pass an array or collection expression.
- When using `params`, you can call the method without passing any arguments for that parameter, and an empty array will be passed.

Let's see some code examples:

1. In `Person.cs`, add statements to define a method with two parameters, one normal parameter and one `params` parameter, as shown in the following method:

```
public void ParamsParameters(
  string text, params int[] numbers)
{
  int total = 0;

  foreach(int number in numbers)
  {
    total += number;
  }

  WriteLine($"{text}: {total}");
}
```

With C# 12 and earlier, `params int[]` must be an array, although the items can by any data type, not just `int`. With C# 13 and later, `params int[]` could be any data type supported by collection expressions, like `params List<int>` or `IEnumerable<int>`, although the items can by any data type, not just `int`.

2. In `Program.cs`, add statements to pass a string and some numbers into the method in various ways, as shown in the following code:

```
bob.ParamsParameters("Sum using commas",
  3, 6, 1, 2);
bob.ParamsParameters("Sum using collection expression",
  [3, 6, 1, 2]);
bob.ParamsParameters("Sum using explicit array",
  new int[] { 3, 6, 1, 2 });
bob.ParamsParameters("Sum (empty)");
```

3. Run the `PeopleApp` project and view the result, as shown in the following output:

```
Sum using commas: 12
Sum using collection expression: 12
Sum using explicit array: 12
Sum (empty): 0
```

The `params` keyword is a powerful feature in C# that enhances the flexibility and readability of your code by allowing methods to accept a variable number of arguments.

Understanding ref returns

In C# 7 or later, the `ref` keyword is not just for passing parameters into a method; it can also be applied to the `return` value. This allows an external variable to reference an internal variable and modify its value after the method call. This might be useful in advanced scenarios, for example, passing placeholders into big data structures, but it's beyond the scope of this book. If you are interested in learning more, then you can read the information at the following link: `https://learn.microsoft.com/en-us/dotnet/csharp/language-reference/keywords/ref#reference-return-values`.

Now, let's return to looking at more advanced scenarios of methods that return values.

Combining multiple returned values using tuples

Each method can only return a single value that has a single type. That type could be a simple type, such as `string` in the previous example; a complex type, such as `Person`; or a collection type, such as `List<Person>`.

Imagine that we want to define a method named `GetTheData` that needs to return both a `string` value and an `int` value. We could define a new class named `TextAndNumber` with a `string` field and an `int` field, and return an instance of that complex type, as shown in the following code:

```
public class TextAndNumber
{
  public string Text;
  public int Number;
}
```

```
public class LifeTheUniverseAndEverything
{
  public TextAndNumber GetTheData()
  {
    return new TextAndNumber
    {
      Text = "What's the meaning of life?",
      Number = 42
    };
  }
}
```

But defining a class just to combine two values is unnecessary because, in modern versions of C#, we can use **tuples**. Tuples are an efficient way to combine two or more values into a single unit. I pronounce them as *tuh-ples* but I have heard other developers pronounce them as *too-ples*. To-may-toe, to-mah-toe, po-tay-toe, po-tah-toe, I guess.

Tuples have been a part of some languages, such as F#, since their first version, but .NET only added support for them with .NET 4 in 2010, using the `System.Tuple` type.

It was only with C# 7 in 2017 that C# added language syntax support for tuples using the parentheses characters `()`, and at the same time, .NET added a new `System.ValueTuple` type that is more efficient in some common scenarios than the old .NET 4 `System.Tuple` type. The C# tuple syntax uses the more efficient one.

Let's explore tuples:

1. In `Person.cs`, add statements to define a method that returns a tuple combining a `string` and `int`, as shown in the following code:

```
// Method that returns a tuple: (string, int).
public (string, int) GetFruit()
{
  return ("Apples", 5);
}
```

2. In `Program.cs`, add statements to call the `GetFruit` method, and then output the tuple's fields, which are automatically named `Item1` and `Item2`, as shown in the following code:

```
(string, int) fruit = bob.GetFruit();
WriteLine($"{fruit.Item1}, {fruit.Item2} there are.");
```

3. Run the `PeopleApp` project and view the result, as shown in the following output:

```
Apples, 5 there are.
```

Naming the fields of a tuple

To access the fields of a tuple, the default names are `Item1`, `Item2`, and so on.

You can explicitly specify the field names:

1. In `Person.cs`, add statements to define a method that returns a tuple with named fields, as shown in the following code:

```
// Method that returns a tuple with named fields.
public (string Name, int Number) GetNamedFruit()
{
  return (Name: "Apples", Number: 5);
}
```

2. In `Program.cs`, add statements to call the method and output the tuple's named fields, as shown in the following code:

```
var fruitNamed = bob.GetNamedFruit();
WriteLine($"There are {fruitNamed.Number} {fruitNamed.Name}.");
```

We use `var` to shorten the following full syntax:

```
(string Name, int Number) fruitNamed = bob.GetNamedFruit();
```

3. Run the `PeopleApp` project and view the result, as shown in the following output:

```
There are 5 Apples.
```

If you construct a tuple from another object, you can use a feature introduced in C# 7.1 called **tuple name inference.**

4. In `Program.cs`, create two tuples, each made of a `string` and `int` value, as shown in the following code:

```
var thing1 = ("Neville", 4);
WriteLine($"{thing1.Item1} has {thing1.Item2} children.");

var thing2 = (bob.Name, bob.Children.Count);
WriteLine($"{thing2.Name} has {thing2.Count} children.");
```

In C# 7, both things would use the `Item1` and `Item2` naming schemes. In C# 7.1 and later, `thing2` can infer the names `Name` and `Count`.

Aliasing tuples

The ability to alias a tuple was introduced in C# 12 so that you could name the type and use that as the type name when declaring variables and parameters, for example, as shown in the following code:

```
using UnnamedParameters = (string, int); // Aliasing a tuple type.

// Aliasing a tuple type with parameter names.
using Fruit = (string Name, int Number);
```

When aliasing tuples, use the title case naming convention for its parameters, for example, `Name`, `Number`, and `BirthDate`.

Let's see an example:

1. In `Program.cs`, at the top of the file, define a named tuple type, as shown in the following code:

```
using Fruit = (string Name, int Number); // Aliasing a tuple type.
```

2. In `Program.cs`, copy and paste the statement that calls the `GetNamedFruit` method and change `var` to `Fruit`, as shown in the following code:

```
// Without an aliased tuple type.
//var fruitNamed = bob.GetNamedFruit();

// With an aliased tuple type.
Fruit fruitNamed = bob.GetNamedFruit();
```

3. Run the `PeopleApp` project and note the result is the same.

Deconstructing tuples

You can also deconstruct tuples into separate variables. The deconstructing declaration has the same syntax as named field tuples but without a named variable for the tuple, as shown in the following code:

```
// Store return value in a tuple variable with two named fields.
(string name, int number) namedFields = bob.GetNamedFruit();

// You can then access the named fields.
WriteLine($"{namedFields.name}, {namedFields.number}");

// Deconstruct the return value into two separate variables.
(string name, int number) = bob.GetNamedFruit();

// You can then access the separate variables.
WriteLine($"{name}, {number}");
```

Deconstruction has the effect of splitting the tuple into its parts and assigning those parts to new variables. Let's see it in action:

1. In `Program.cs`, add statements to deconstruct the tuple returned from the `GetFruit` method, as shown in the following code:

```
(string fruitName, int fruitNumber) = bob.GetFruit();
WriteLine($"Deconstructed tuple: {fruitName}, {fruitNumber}");
```

2. Run the `PeopleApp` project and view the result, as shown in the following output:

```
Deconstructed tuple: Apples, 5
```

Deconstructing other types using tuples

Tuples are not the only type that can be deconstructed. Any type can have special methods, named `Deconstruct`, that break down an object into parts. You can have as many `Deconstruct` methods as you like as long as they have different signatures. Let's implement some for the `Person` class:

1. In `Person.cs`, add two `Deconstruct` methods with `out` parameters defined for the parts we want to deconstruct into, as shown in the following code:

```
// Deconstructors: Break down this object into parts.

public void Deconstruct(out string? name,
  out DateTimeOffset dob)
{
  name = Name;
  dob = Born;
}

public void Deconstruct(out string? name,
  out DateTimeOffset dob,
  out WondersOfTheAncientWorld fav)
{
  name = Name;
  dob = Born;
  fav = FavoriteAncientWonder;
}
```

2. In `Program.cs`, add statements to deconstruct `bob`, as shown in the following code:

```
var (name1, dob1) = bob; // Implicitly calls the Deconstruct method.
WriteLine($"Deconstructed person: {name1}, {dob1}");

var (name2, dob2, fav2) = bob;
WriteLine($"Deconstructed person: {name2}, {dob2}, {fav2}");
```

You do not explicitly call the `Deconstruct` method. It is called implicitly when you assign an object to a tuple variable.

3. Run the `PeopleApp` project and view the result, as shown in the following output:

```
Deconstructed person: Bob Smith, 12/22/1965 4:28:00 PM -05:00
Deconstructed person: Bob Smith, 12/22/1965 4:28:00 PM -05:00,
StatueOfZeusAtOlympia
```

Implementing functionality using local functions

A language feature introduced in C# 7 is the ability to define a **local function.**

Local functions are the method equivalent of local variables. In other words, they are methods that are only accessible from within the containing method in which they have been defined. In other languages, they are sometimes called **nested** or **inner functions.**

Local functions can be defined anywhere inside a method: the top, the bottom, or even somewhere in the middle!

We will use a local function to implement a factorial calculation:

1. In `Person.cs`, add statements to define a `Factorial` function that uses a local function inside itself to calculate the result, as shown in the following code:

```
// Method with a local function.
public static int Factorial(int number)
{
  if (number < 0)
  {
    throw new ArgumentException(
      $"{nameof(number)} cannot be less than zero.");
  }
  return localFactorial(number);

  int localFactorial(int localNumber) // Local function.
  {
    if (localNumber == 0) return 1;
    return localNumber * localFactorial(localNumber - 1);
  }
}
```

2. In `Program.cs`, add statements to call the `Factorial` function, and write the return value to the console, with exception handling, as shown in the following code:

```
// Change to -1 to make the exception handling code execute.
int number = 5;

try
{
  WriteLine($"{number}! is {Person.Factorial(number)}");
}
catch (Exception ex)
{
  WriteLine($"{ex.GetType()} says: {ex.Message} number was {number}.");
}
```

3. Run the `PeopleApp` project and view the result, as shown in the following output:

```
5! is 120
```

4. Change the number to `-1` so that we can check the exception handling.
5. Run the `PeopleApp` project and view the result, as shown in the following output:

```
System.ArgumentException says: number cannot be less than zero. number
was -1.
```

Splitting classes using partial

When working on large projects with multiple team members, or when working with especially large and complex class implementations, it is useful to be able to split the definition of a class across multiple files. You do this using the `partial` keyword.

Imagine that we want to add statements to the `Person` class that are automatically generated by a tool, like an object-relational mapper, that reads schema information from a database. If the class is defined as `partial`, then we can split the class into an autogenerated code file and a manually edited code file.

Let's write some code that simulates this example:

1. In `Person.cs`, add the `partial` keyword, as highlighted in the following code:

```
public partial class Person
```

2. In the `PacktLibraryNet2` project/folder, add a new class file named `PersonAutoGen.cs`.
3. Add statements to the new file, as shown in the following code:

```
namespace Packt.Shared;

// This file simulates an auto-generated class.
```

```
public partial class Person
{
}
```

4. Build the `PacktLibraryNet2` project. If you see the `CS0260 Missing partial modifier on declaration of type 'Person'; another partial declaration of this type exists` error, then make sure you have applied the `partial` keyword to both `Person` classes.

The rest of the code we write for this chapter will be written in the `PersonAutoGen.cs` file.

Partial methods

Partial methods were introduced with C# 3 on .NET Framework in 2007. They are a feature that allows the definition of method signatures in one part of a `partial class` or `struct`, with the actual implementation provided in another part. Partial methods are particularly useful in scenarios where code generation and manual code coexist, such as in code generated by tools like Entity Framework Core or a source code generator.

There are some key characteristics of `partial` methods, as shown in the following list:

- Partial methods are declared with the `partial` keyword. The declaration provides the method signature, while the implementation provides the method body.
- The implementation of a partial method is optional. If a partial method is declared but not implemented, the compiler removes the call to the method and no error is thrown.
- Partial methods are implicitly private and cannot have access modifiers. They must also return `void` and cannot have `out` parameters.
- Partial methods cannot be `virtual`.

Partial methods are often used in scenarios involving code generation where a base structure is provided, and custom logic can be added without modifying the generated code.

Imagine that you have a class file, as shown in the following code:

```
// MyClass1.cs
public partial class MyClass
{
  // No method body in the declaration.
  partial void OnSomethingHappened();

  public void DoWork()
  {
    // Some work here.

    // Call the partial method.
    OnSomethingHappened();
  }
}
```

Now, imagine that you have another class file, as shown in the following code:

```
// MyClass2.cs
public partial class MyClass
{
  partial void OnSomethingHappened()
  {
    Console.WriteLine("Something happened.");
  }
}
```

In the preceding examples, `OnSomethingHappened` is a partial method declared in `MyClass1.cs` and implemented in `MyClass2.cs`. The method `DoWork` calls the partial method, which prints a message if the implementation is provided.

If in the example, `OnSomethingHappened` is declared but not implemented, then the C# compiler will remove the call to `OnSomethingHappened` in `DoWork`, and no error will be thrown.

Partial methods are often used in auto-generated code where developers can hook into the process without modifying the generated code. This would be the case if the `MyClass1.cs` file was generated automatically.

Partial methods in C# provide a powerful way to extend and customize generated code without modifying it directly. They offer a clean mechanism for inserting custom behavior, ensuring that code generation and custom logic can coexist. By leveraging partial methods, developers can maintain a clear separation between generated and custom code, enhancing maintainability and readability. Now that you've seen lots of examples of fields and methods, we will look at some specialized types of methods that can be used to access fields to provide control and improve the developer's experience.

Controlling access with properties and indexers

Earlier, you created a method named `GetOrigin` that returned a `string` containing the name and origin of the person. Languages such as Java do this a lot. C# has a better way, and it is called properties.

A **property** is simply a method (or a pair of methods) that acts and looks like a field when you want to get or set a value, but it acts like a method, thereby simplifying the syntax and enabling functionality, like validation and calculation, when you set and get a value.

A fundamental difference between a field and a property is that a field provides a memory address to data. You could pass that memory address to an external component, like a Windows API C-style function call, and it could then modify the data. A property does not provide a memory address to its data, which provides more control. All you can do is ask the property to get or set the data. The property then executes statements and can decide how to respond, including refusing the request!

Defining read-only properties

A `readonly` property only has a `get` implementation:

1. In `PersonAutoGen.cs`, in the `Person` class, add statements to define three properties:
   - The first property will perform the same role as the `GetOrigin` method, using the property syntax that works with all versions of C#.
   - The second property will return a greeting message, using the lambda expression body `=>` syntax from C# 6 and later.
   - The third property will calculate the person's age.

   Here's the code:

```
#region Properties: Methods to get and/or set data or state.

// A readonly property defined using C# 1 to 5 syntax.
public string Origin
{
  get
  {
    return string.Format("{0} was born on {1}.",
      arg0: Name, arg1: HomePlanet);
  }
}

// Two readonly properties defined using C# 6 or later
// lambda expression body syntax.

public string Greeting => $"{Name} says 'Hello!'";

public int Age => DateTime.Today.Year - Born.Year;

#endregion
```

Good practice: This isn't the best way to calculate someone's age, but we aren't learning how to calculate an age from a date and time of birth. If you need to do that properly, then read the discussion at the following link: `https://stackoverflow.com/questions/9/how-do-i-calculate-someones-age-in-c`.

2. In `Program.cs`, add the statements to get the properties, as shown in the following code:

```
Person sam = new()
{
  Name = "Sam",
```

```
  Born = new(1969, 6, 25, 0, 0, 0, TimeSpan.Zero)
};

WriteLine(sam.Origin);
WriteLine(sam.Greeting);
WriteLine(sam.Age);
```

3. Run the PeopleApp project and view the result, as shown in the following output:

```
Sam was born on Earth
Sam says 'Hello!'
54
```

The output shows 54 because I ran the console app on July 5, 2023, when Sam was 54 years old.

Defining settable properties

To create a settable property, you must use the older syntax and provide a pair of methods—not just a get part, but also a set part:

1. In PersonAutoGen.cs, add statements to define a string property that has both a get and set method (also known as a **getter** and **setter**), as shown in the following code:

```
// A read-write property defined using C# 3 auto-syntax.
public string? FavoriteIceCream { get; set; }
```

Although you have not manually created a field to store the person's favorite ice cream, it is there, automatically created by the compiler for you.

Sometimes, you need more control over what happens when a property is set. In this scenario, you must use a more detailed syntax and manually create a private field to store the value of the property.

2. In PersonAutoGen.cs, add statements to define a private string field, known as a **backing field**, as shown in the following code:

```
// A private backing field to store the property value.
private string? _favoritePrimaryColor;
```

Good practice: Although there is no formal standard to name private fields, the most common is to use camel case with an underscore as a prefix.

3. In PersonAutoGen.cs, add statements to define a string property that has both get and set and validation logic in the setter, as shown in the following code:

```
// A public property to read and write to the field.
public string? FavoritePrimaryColor
{
  get
  {
    return _favoritePrimaryColor;
  }
  set
  {
    switch (value?.ToLower())
    {
      case "red":
      case "green":
      case "blue":
        _favoritePrimaryColor = value;
        break;
      default:
        throw new ArgumentException(
          $"{value} is not a primary color. " +
          "Choose from: red, green, blue.");
    }
  }
}
```

Good practice: Avoid adding too much code to your getters and setters. This could indicate a problem with your design. Consider adding private methods that you then call in the `set` and `get` methods to simplify your implementations.

4. In `Program.cs`, add statements to set Sam's favorite ice cream and color, and then write them out, as shown in the following code:

```
sam.FavoriteIceCream = "Chocolate Fudge";
WriteLine($"Sam's favorite ice-cream flavor is {sam.FavoriteIceCream}.");

string color = "Red";

try
{
  sam.FavoritePrimaryColor = color;
  WriteLine($"Sam's favorite primary color is {sam.
FavoritePrimaryColor}.");
```

```
}
catch (Exception ex)
{
  WriteLine("Tried to set {0} to '{1}': {2}",
    nameof(sam.FavoritePrimaryColor), color, ex.Message);
}
```

The print book is limited to about 820 pages. If I added exception handling code to all code examples as we have done here, then I would probably have to remove at least one chapter from the book to make enough space. In the future, I will not explicitly tell you to add exception handling code, but get into the habit of adding it yourself when needed.

5. Run the `PeopleApp` project and view the result, as shown in the following output:

```
Sam's favorite ice-cream flavor is Chocolate Fudge.
Sam's favorite primary color is Red.
```

6. Try to set the color to any value other than red, green, or blue, like black.
7. Run the `PeopleApp` project and view the result, as shown in the following output:

```
Tried to set FavoritePrimaryColor to 'Black': Black is not a primary
color. Choose from: red, green, blue.
```

Good practice: Use properties instead of fields when you want to execute statements during a read or write to a field without using a method pair, like `GetAge` and `SetAge`.

Partial properties

In the previous section, you learned about partial classes and partial methods. With C# 13 on .NET 9, you can define partial properties, and they work in a similar way. The main benefit is to support source-generated types. Partial properties make it easier for source generator designers to create natural feeling APIs, as you will see in *Chapter 8, Working with Common .NET Types*, when we see the Regex source generator.

Limiting flags enum values

Earlier in this chapter, we defined a field to store a person's favorite ancient wonder. But we then made the `enum` able to store combinations of values. Now, let's limit the favorite to one:

1. In `Person.cs`, comment out the `FavoriteAncientWonder` field and add a comment to note it has moved to the `PersonAutoGen.cs` code file, as shown in the following code:

```
// This has been moved to PersonAutoGen.cs as a property.
// public WondersOfTheAncientWorld FavoriteAncientWonder;
```

2. In `PersonAutoGen.cs`, add a `private` field and `public` property for `FavoriteAncientWonder`, as shown in the following code:

```
private WondersOfTheAncientWorld _favoriteAncientWonder;

public WondersOfTheAncientWorld FavoriteAncientWonder
{
  get { return _favoriteAncientWonder; }
  set
  {
    string wonderName = value.ToString();

    if (wonderName.Contains(','))
    {
      throw new ArgumentException(
        message: "Favorite ancient wonder can only have a single enum 
value.",
        paramName: nameof(FavoriteAncientWonder));
    }

    if (!Enum.IsDefined(typeof(WondersOfTheAncientWorld), value))
    {
      throw new ArgumentException(
        $"{value} is not a member of the WondersOfTheAncientWorld enum.",
        paramName: nameof(FavoriteAncientWonder));
    }

    _favoriteAncientWonder = value;
  }
}
```

We could simplify the validation by only checking if the value is defined in the original `enum` because `IsDefined` returns `false` for multiple values and undefined values. However, I want to show a different exception for multiple values, so I will use the fact that multiple values formatted as a string would include a comma in the list of names. This also means we must check for multiple values before we check if the value is defined. A comma-separated list is how multiple `enum` values are represented as a `string`, but you cannot use commas to set multiple `enum` values. You should use | (the bitwise `OR`).

3. In Program.cs, in the *Storing a value using an enum type* region, set Bob's favorite wonder to more than one enum value, as shown in the following code:

```
bob.FavoriteAncientWonder =
  WondersOfTheAncientWorld.StatueOfZeusAtOlympia |
  WondersOfTheAncientWorld.GreatPyramidOfGiza;
```

4. Run the PeopleApp project and note the exception, as shown in the following output:

```
Unhandled exception. System.ArgumentException: Favorite ancient wonder
can only have a single enum value. (Parameter 'FavoriteAncientWonder')
   at Packt.Shared.Person.set_
FavoriteAncientWonder(WondersOfTheAncientWorld value) in C:\cs13net9\
Chapter05\PacktLibraryNet2\PersonAutoGen.cs:line 67
   at Program.<Main>$(String[] args) in C:\cs13net9\Chapter05\PeopleApp\
Program.cs:line 57
```

5. In Program.cs, set Bob's favorite wonder to an invalid enum value like 128, as shown in the following code:

```
bob.FavoriteAncientWonder = (WondersOfTheAncientWorld)128;
```

6. Run the PeopleApp project and note the exception, as shown in the following output:

```
Unhandled exception. System.ArgumentException: 128 is not a member of the
WondersOfTheAncientWorld enum. (Parameter 'FavoriteAncientWonder')
```

7. In Program.cs, set Bob's favorite wonder back to a valid single enum value.

Defining indexers

Indexers allow the calling code to use the array syntax to access a property. For example, the string type defines an indexer so that the calling code can access individual characters in the string, as shown in the following code:

```
string alphabet = "abcdefghijklmnopqrstuvwxyz";
char letterF = alphabet[5]; // 0 is a, 1 is b, and so on.
```

You can overload indexers so that different types can be used for their parameters. For example, as well as passing an int value, you could also pass a string value.

We will define an indexer to simplify access to the children of a person:

1. In PersonAutoGen.cs, add statements to define an indexer to get and set a child using the index of the child, as shown in the following code:

```
#region Indexers: Properties that use array syntax to access them.

public Person this[int index]
{
```

```
    get
    {
      return Children[index]; // Pass on to the List<T> indexer.
    }
    set
    {
      Children[index] = value;
    }
  }

  #endregion
```

Note the use of `this[]` to define an indexer. It is not to be confused with the use of the `this` keyword to refer to the current instance of an object, or the use of `this()` to refer to the current type's constructor.

2. In `PersonAutoGen.cs`, add statements to define an indexer to get and set a child using the name of the child, as shown in the following code:

```
// A read-only string indexer.
public Person this[string name]
{
  get
  {
    return Children.Find(p => p.Name == name);
  }
}
```

You will learn more about collections like `List<T>` in *Chapter 8, Working with Common .NET Types*, and how to write lambda expressions using `=>` in *Chapter 11, Querying and Manipulating Data Using LINQ*.

3. In `Program.cs`, add statements to add two children to `Sam`, and then access the first and second children using the longer `Children` field and the shorter indexer syntax, as shown in the following code:

```
sam.Children.Add(new() { Name = "Charlie",
  Born = new(2010, 3, 18, 0, 0, 0, TimeSpan.Zero) });

sam.Children.Add(new() { Name = "Ella",
  Born = new(2020, 12, 24, 0, 0, 0, TimeSpan.Zero) });
```

```
// Get using Children List.
WriteLine($"Sam's first child is {sam.Children[0].Name}.");
WriteLine($"Sam's second child is {sam.Children[1].Name}.");

// Get using the int indexer.
WriteLine($"Sam's first child is {sam[0].Name}.");
WriteLine($"Sam's second child is {sam[1].Name}.");

// Get using the string indexer.
WriteLine($"Sam's child named Ella is {sam["Ella"].Age} years old.");
```

4. Run the `PeopleApp` project and view the result, as shown in the following output:

```
Sam's first child is Charlie.
Sam's second child is Ella.
Sam's first child is Charlie.
Sam's second child is Ella.
Sam's child named Ella is 3 years old.
```

`sam[0].Name` calls the indexer aka `this` on the `Person` class because `[]` are the symbols that call an indexer. `sam.Children[0].Name` calls the indexer on the collection named `Children`, which is a `List<Person>`. We have implemented the indexer on `Person` to do the same thing so we can use a shorter syntax.

It's similar with `string` values. If I have `string name = "Mark";` then imagine that it has a property named `Characters` that is a `List<char>`. I could use `name.Characters[0]` to return the character `M`. But since `string` has an indexer defined, I can do the same with `name[0]`.

Most types do not define an indexer. It's just a way to define a property that is called using `[]` instead of a name. It can actually do anything you want. But it should be used to define an array/collection style getter/setter. If your type doesn't need that, it doesn't need an indexer.

For any type you define, just imagine another developer using your type. What would they expect the indexer to do? For example, what if you get the first index, as shown in the following code:

```
Car car = new("Ford");
var thing1 = car[0]; // What type is thing1? What is its value?

Database db = new("Commerce");
var thing2 = db[0]; // What type is thing2? What is its value?

string name = "Mark";
var thing3 = name[0]; // thing3 is a char. Its value is "M".
```

```
List<int> numbers = new() { 27, 95, 13 };
var thing4 = numbers[0]; // thing4 is an int. Its value is 27.
```

`thing1` might be the first passenger in the car or the first wheel of the car. `thing2` might be the first table in the database or the first user. For custom types, it is up to you to decide what makes the most sense for your type.

Pattern matching with objects

In *Chapter 3, Controlling Flow, Converting Types, and Handling Exceptions*, you were introduced to basic pattern matching. In this section, we will explore pattern matching in more detail.

Pattern-matching flight passengers

In this example, we will define some classes that represent various types of passengers on a flight, and then we will use a switch expression with pattern matching to determine the cost of their flight:

1. In the `PacktLibraryNet2` project/folder, add a new file named `FlightPatterns.cs`.
2. If you use Visual Studio, in `FlightPatterns.cs`, delete the existing statements, including the class named `FlightPatterns`, because we will define multiple classes, and none match the name of the code file.
3. In `FlightPatterns.cs`, add statements to define three types of passenger with different properties, as shown in the following code:

```
// All the classes in this file will be defined in the following
namespace.
namespace Packt.Shared;

public class Passenger
{
  public string? Name { get; set; }
}

public class BusinessClassPassenger : Passenger
{
  public override string ToString()
  {
    return $"Business Class: {Name}";
  }
}

public class FirstClassPassenger : Passenger
{
  public int AirMiles { get; set; }
```

```
  public override string ToString()
  {
    return $"First Class with {AirMiles:N0} air miles: {Name}";
  }
}

public class CoachClassPassenger : Passenger
{
  public double CarryOnKG { get; set; }

  public override string ToString()
  {
    return $"Coach Class with {CarryOnKG:N2} KG carry on: {Name}";
  }
}
```

You will learn about overriding the `ToString` method in *Chapter 6, Implementing Interfaces and Inheriting Classes.*

4. In `Program.cs`, add statements to define an object array containing five passengers of various types and property values, and then enumerate them, outputting the cost of their flight, as shown in the following code:

```
// An array containing a mix of passenger types.
Passenger[] passengers = {
  new FirstClassPassenger { AirMiles = 1_419, Name = "Suman" },
  new FirstClassPassenger { AirMiles = 16_562, Name = "Lucy" },
  new BusinessClassPassenger { Name = "Janice" },
  new CoachClassPassenger { CarryOnKG = 25.7, Name = "Dave" },
  new CoachClassPassenger { CarryOnKG = 0, Name = "Amit" },
};

foreach (Passenger passenger in passengers)
{
  decimal flightCost = passenger switch
  {
    FirstClassPassenger p when p.AirMiles > 35_000 => 1_500M,
    FirstClassPassenger p when p.AirMiles > 15_000 => 1_750M,
    FirstClassPassenger _                          => 2_000M,
```

```
    BusinessClassPassenger _                        => 1_000M,
    CoachClassPassenger p when p.CarryOnKG < 10.0 => 500M,
    CoachClassPassenger _                          => 650M,
    _                                              => 800M
  };
  WriteLine($"Flight costs {flightCost:C} for {passenger}");
}
```

While reviewing the preceding code, note the following:

- Most code editors do not align the lambda symbols => as I have done above.
- To pattern match the properties of an object, you must name a local variable, like `p`, which can then be used in an expression.
- To pattern match on a type only, you can use _ to discard the local variable; for example, `FirstClassPassenger _` means that you match on the type but you don't care what values any of its properties have, so a named variable like `p` is not needed. In a moment, you will see how we can improve the code even more.
- The `switch` expression also uses _ to represent its default branch.

5. Run the `PeopleApp` project and view the result, as shown in the following output:

```
Flight costs $2,000.00 for First Class with 1,419 air miles: Suman
Flight costs $1,750.00 for First Class with 16,562 air miles: Lucy
Flight costs $1,000.00 for Business Class: Janice
Flight costs $650.00 for Coach Class with 25.70 KG carry on: Dave
Flight costs $500.00 for Coach Class with 0.00 KG carry on: Amit
```

Enhancements to pattern matching in modern C#

The previous examples worked with C# 8. Now, we will look at some enhancements in C# 9 and later. First, you no longer need to use the underscore to discard the local variable when doing type matching:

1. In `Program.cs`, comment out the C# 8 syntax, and add C# 9 and later syntax to modify the branches for first-class passengers to use a nested `switch` expression and the new support for conditionals, like `>`, as highlighted in the following code:

```
decimal flightCost = passenger switch
{
  /* C# 8 syntax
  FirstClassPassenger p when p.AirMiles > 35_000 => 1_500M,
  FirstClassPassenger p when p.AirMiles > 15_000 => 1_750M,
  FirstClassPassenger _                          => 2_000M, */

  // C# 9 or later syntax
  FirstClassPassenger p => p.AirMiles switch
```

```
    {
      > 35_000 => 1_500M,
      > 15_000 => 1_750M,
      _        => 2_000M
    },
    BusinessClassPassenger                           => 1_000M,
    CoachClassPassenger p when p.CarryOnKG < 10.0 => 500M,
    CoachClassPassenger                              => 650M,
    _                                                => 800M
};
```

2. Run the PeopleApp project to view the results, and note that they are the same as before.

You could also use the relational pattern in combination with the property pattern to avoid the nested switch expression, as shown in the following code:

```
FirstClassPassenger { AirMiles: > 35000 } => 1500M,
FirstClassPassenger { AirMiles: > 15000 } => 1750M,
FirstClassPassenger                       => 2000M,
```

Working with record types

Before we dive into the new record language feature, let us see some other related new features of C# 9 and later.

Init-only properties

You have used object initialization syntax to instantiate objects and set initial properties throughout this chapter. Those properties can also be changed after instantiation.

Sometimes, you want to treat properties like readonly fields so that they can be set during instantiation but not after. In other words, they are immutable. The init keyword enables this. It can be used in place of the set keyword in a property definition.

Since this is a language feature not supported by .NET Standard 2.0, we cannot use it in the PacktLibraryNet2 project. We must use it in the modern project:

1. In the PacktLibraryModern project, add a new file named Records.cs.
2. In Records.cs, define a person class with two immutable properties, as shown in the following code:

```
namespace Packt.Shared;

public class ImmutablePerson
{
```

```
  public string? FirstName { get; init; }
  public string? LastName { get; init; }
}
```

3. In Program.cs, add statements to instantiate a new immutable person, and then try to change one of its properties, as shown in the following code:

```
ImmutablePerson jeff = new()
{
  FirstName = "Jeff",
  LastName = "Winger"
};
jeff.FirstName = "Geoff";
```

4. Compile the console app and note the compile error, as shown in the following output:

```
C:\cs13net9\Chapter05\PeopleApp\Program.cs(404,1): error CS8852: Init-only property or indexer 'ImmutablePerson.FirstName' can only be assigned in an object initializer, or on 'this' or 'base' in an instance constructor or an 'init' accessor. [/Users/markjprice/Code/Chapter05/PeopleApp/PeopleApp.csproj]
```

5. Comment out the attempt to set the FirstName property after instantiation.

Even if you do not set FirstName in the object initializer, you still would not be able to set it post-initialization. If you need to force a property to be set, then apply the required keyword that you learned about earlier in this chapter.

Defining record types

Init-only properties provide some immutability to C#. You can take the concept further by using **record types**. These are defined by using the record keyword instead of (or as well as) the class keyword. That can make the whole object immutable, and it acts like a value when compared. We will discuss equality and comparisons of classes, records, and value types in more detail in *Chapter 6, Implementing Interfaces and Inheriting Classes*.

Immutable records should not have any state (properties and fields) that change after instantiation. Instead, the idea is that you create new records from existing ones. The new record has the changed state. This is called non-destructive mutation. To do this, C# 9 introduced the with keyword:

1. In Records.cs, add a record named ImmutableVehicle after the ImmutablePerson class, as shown in the following code:

```
public record ImmutableVehicle
{
  public int Wheels { get; init; }
```

```
  public string? Color { get; init; }
  public string? Brand { get; init; }
}
```

2. In `Program.cs`, add statements to create a `car` and then a mutated copy of it, as shown in the following code:

```
ImmutableVehicle car = new()
{
  Brand = "Mazda MX-5 RF",
  Color = "Soul Red Crystal Metallic",
  Wheels = 4
};
ImmutableVehicle repaintedCar = car
  with { Color = "Polymetal Grey Metallic" };
WriteLine($"Original car color was {car.Color}.");
WriteLine($"New car color is {repaintedCar.Color}.");
```

3. Run the `PeopleApp` project to view the results, and note the change to the car color in the mutated copy, as shown in the following output:

```
Original car color was Soul Red Crystal Metallic.
New car color is Polymetal Grey Metallic.
```

You could also release the memory for the `car` variable and the `repaintedCar` would still fully exist.

Equality of record types

One of the most important behaviors of `record` types is their equality. Two records with the same property values are considered equal. This may not sound surprising, but if you used a normal class instead of a record, then they would *not* be considered equal. Let's see:

1. In the `PacktLibraryModern` project, add a new file named `Equality.cs`.
2. In `Equality.cs`, define a `class` and a `record` type, as shown in the following code:

```
namespace Packt.Shared;

public class AnimalClass
{
  public string? Name { get; set; }
}
```

```
public record AnimalRecord
{
  public string? Name { get; set; }
}
```

3. In `Program.cs`, add statements to create two instances of `AnimalClass` and two instances of `AnimalRecord`, and then compare them for equality, as shown in the following code:

```
AnimalClass ac1 = new() { Name = "Rex" };
AnimalClass ac2 = new() { Name = "Rex" };

WriteLine($"ac1 == ac2: {ac1 == ac2}");

AnimalRecord ar1 = new() { Name = "Rex" };
AnimalRecord ar2 = new() { Name = "Rex" };

WriteLine($"ar1 == ar2: {ar1 == ar2}");
```

4. Run the `PeopleApp` project to view the results, and note that two class instances are not equal even if they have the same property values, and two record instances are equal if they have the same property values, as shown in the following output:

```
ac1 == ac2: False
ar1 == ar2: True
```

Class instances are only equal if they are literally the same object. This is true when their memory addresses are equal. You will learn more about the equality of types in *Chapter 6, Implementing Interfaces and Inheriting Classes*.

Equality of other types

It is common to compare two variables using the == and != operators. The behavior of these two operators is different for reference types and value types.

When you check the equality of two value type variables, .NET literally compares the *values* of those two variables on the stack and returns `true` if they are equal:

1. In `Program.cs`, add statements to declare two integers with equal values and then compare them, as shown in the following code:

```
int number1 = 3;
int number2 = 3;
WriteLine($"number1: {number1}, number2: {number2}");
WriteLine($"number1 == number2: {number1 == number2}");
```

2. Run the PeopleApp project and view the result, as shown in the following output:

```
number1: 3, number2: 3
number1 == number2: True
```

When you check the equality of two reference type variables, .NET compares the memory addresses of those two variables and returns `true` if they are equal.

3. In `Program.cs`, add statements to declare two `Person` instances with equal names, and then compare the variables and their names, as shown in the following code:

```
Person p1 = new() { Name = "Kevin" };
Person p2 = new() { Name = "Kevin" };
WriteLine($"p1: {p1}, p2: {p2}");
WriteLine($"p1.Name: {p1.Name}, p2.Name: {p2.Name}");
WriteLine($"p1 == p2: {p1 == p2}");
```

4. Run the PeopleApp project and view the result, as shown in the following output:

```
p1: Packt.Shared.Person, p2: Packt.Shared.Person
p1.Name: Kevin, p2.Name: Kevin
p1 == p2: False
```

This is because they are not the same object. If both variables literally pointed to the same object on the heap, then they would be equal.

5. Add statements to declare a third `Person` object and assign `p1` to it, as shown in the following code:

```
Person p3 = p1;
WriteLine($"p3: {p3}");
WriteLine($"p3.Name: {p3.Name}");
WriteLine($"p1 == p3: {p1 == p3}");
```

6. Run the PeopleApp project and view the result, as shown in the following output:

```
p3: Packt.Shared.Person
p3.Name: Kevin
p1 == p3: True
```

The one exception to this behavior of reference types is the `string` type. It is a reference type, but the equality operators have been overridden to make them behave as if they were value types.

7. Add statements to compare the `Name` properties of two `Person` instances, as shown in the following code:

```
// string is the only class reference type implemented to
// act like a value type for equality.
WriteLine($"p1.Name: {p1.Name}, p2.Name: {p2.Name}");
WriteLine($"p1.Name == p2.Name: {p1.Name == p2.Name}");
```

8. Run the `PeopleApp` project and view the result, as shown in the following output:

```
p1.Name: Kevin, p2.Name: Kevin
p1.Name == p2.Name: True
```

You can do the same as `string` with your classes to override the equality operator == to return `true`, even if the two variables are not referencing the same object (the same memory address on the heap) but, instead, their fields have the same values. However, that is beyond the scope of this book.

Good practice: Alternatively, use a `record class` because one of its benefits is that it implements this equality behavior for you.

Positional data members in records

The syntax for defining a record can be greatly simplified using positional data members. Instead of using object initialization syntax with curly braces, sometimes you might prefer to provide a constructor with positional parameters, as you saw earlier in this chapter. You can also combine this with a deconstructor to split the object into individual parts, as shown in the following code:

```
public record ImmutableAnimal
{
  public string Name { get; init; }
  public string Species { get; init; }

  public ImmutableAnimal(string name, string species)
  {
    Name = name;
    Species = species;
  }

  public void Deconstruct(out string name, out string species)
  {
    name = Name;
    species = Species;
  }
}
```

The properties, constructor, and deconstructor can be generated for you:

1. In `Records.cs`, add statements to define another record using simplified syntax, known as positional records, as shown in the following code:

```
// Simpler syntax to define a record that auto-generates the
// properties, constructor, and deconstructor.
public record ImmutableAnimal(string Name, string Species);
```

2. In `Program.cs`, add statements to construct and deconstruct immutable animals, as shown in the following code:

```
ImmutableAnimal oscar = new("Oscar", "Labrador");
var (who, what) = oscar; // Calls the Deconstruct method.
WriteLine($"{who} is a {what}.");
```

3. Run the `PeopleApp` project and view the results, as shown in the following output:

```
Oscar is a Labrador.
```

You will see records again when we look at C# 10 or later support to create `struct` records in *Chapter 6, Implementing Interfaces and Inheriting Classes.*

More information: There are many more ways to use records in your projects. I recommend that you review the official documentation at the following link: `https://learn.microsoft.com/en-us/dotnet/csharp/whats-new/tutorials/records`.

Defining a primary constructor for a class

Introduced with C# 12, you can define one constructor as part of the class definition. This is called the primary constructor. The syntax is the same as for positional data members in records, but the behavior is slightly different.

Traditionally, we separate the class definition from any constructors, as shown in the following code:

```
public class Headset // Class definition.
{
  // Constructor.
  public Headset(string manufacturer, string productName)
  {
    // You can reference manufacturer and productName parameters in
    // the constructor and the rest of the class.
  }
}
```

With class primary constructors, you combine both into a more succinct syntax, as shown in the following code:

```
public class Headset(string manufacturer, string productName);
```

Let's see an example:

1. In the PacktLibraryModern project, add a class file named Headset.cs.
2. Modify the code file contents to give the class two parameters for manufacturer and product name respectively, as shown in the following code:

```
namespace Packt.Shared;

public class Headset(string manufacturer, string productName);
```

3. In Program.cs, add statements to instantiate a headset, as shown in the following code:

```
Headset vp = new("Apple", "Vision Pro");
WriteLine($"{vp.ProductName} is made by {vp.Manufacturer}.");
```

One of the differences between a record and a class type with a primary constructor is that its parameters don't become public properties automatically, so you will see CS1061 compiler errors. Neither ProductName nor productName are accessible outside the class.

4. In Headset.cs, add statements to define two properties and set them using the parameters passed to the primary constructor, as highlighted in the following code:

```
namespace Packt.Shared;

public class Headset(string manufacturer, string productName)
{
  public string Manufacturer { get; set; } = manufacturer;
  public string ProductName { get; set; } = productName;
}
```

5. Run the PeopleApp project and view the results, as shown in the following output:

```
Vision Pro is made by Apple.
```

6. In Headset.cs, add a default parameterless constructor, as highlighted in the following code:

```
namespace Packt.Shared;

public class Headset(string manufacturer, string productName)
{
  public string Manufacturer { get; set; } = manufacturer;
```

```
    public string ProductName { get; set; } = productName;

    // Default parameterless constructor calls the primary constructor.
    public Headset() : this("Microsoft", "HoloLens") { }
}
```

Note the use of `this()` to call the constructor of the base class and pass two parameters to it when the default constructor of `Headset` is called.

7. In `Program.cs`, create an uninitialized instance of a headset and an instance for Meta Quest 3, as shown in the following code:

```
Headset holo = new();
WriteLine($"{holo.ProductName} is made by {holo.Manufacturer}.");

Headset mq = new() { Manufacturer = "Meta", ProductName = "Quest 3" };
WriteLine($"{mq.ProductName} is made by {mq.Manufacturer}.");
```

8. Run the `PeopleApp` project and view the results, as shown in the following output:

```
Vision Pro is made by Apple.
HoloLens is made by Microsoft.
Quest 3 is made by Meta.
```

More information: You can learn more about primary constructors for classes and structs at the following links: `https://learn.microsoft.com/en-us/dotnet/csharp/whats-new/tutorials/primary-constructors` and `https://devblogs.microsoft.com/dotnet/csharp-primary-constructors-refactoring/`. In particular, it is worth reading the comments at the bottom of the second link to understand why many developers do not like primary constructors in classes.

Good practice: Only use a primary constructor in a class if it will only initialize non-read-only private fields, and does not need to execute other statements.

Practicing and exploring

Test your knowledge and understanding by answering some questions, getting some hands-on practice, and exploring this chapter's topics with deeper research.

Exercise 5.1 – online material

Online material can be extra content written by me for this book, or it can be references to content created by Microsoft or third parties.

Pattern matching

There are many ways to use pattern matching in your projects. I recommend that you review the official documentation at the following link: `https://learn.microsoft.com/en-us/dotnet/csharp/fundamentals/functional/pattern-matching`.

Exercise 5.2 – practice exercises

Practice exercises go deeper into the topics for this chapter.

Practice access modifiers

Imagine that you are the compiler. What errors would you show when building the following projects? What would need to change to fix it?

In a class library project, in `Car.cs`:

```
class Car
{
  int Wheels { get; set; }
  public bool IsEV { get; set; }
  internal void Start()
  {
    Console.WriteLine("Starting...");
  }
}
```

In a console app project that references the class library project, in `Program.cs`:

```
Car fiat = new() { Wheels = 4, IsEV = true };
fiat.Start();
```

Exercise 5.3 – test your knowledge

Answer the following questions:

1. What are the seven access modifier keywords and combinations of keywords, and what do they do?
2. What is the difference between the `static`, `const`, and `readonly` keywords when applied to a type member?
3. What does a constructor do?
4. Why should you apply the `[Flags]` attribute to an `enum` type when you want to store combined values?

5. Why is the `partial` keyword useful?
6. What is a tuple?
7. What does the `record` keyword do?
8. What does overloading mean?
9. What is the difference between the following two statements? (Do not just say a > character!)

```
public List<Person> Children = new();
public List<Person> Children => new();
```

10. How do you make a method parameter optional?
11. Why can't a `DateTime` field be a constant?

Exercise 5.4 – explore topics

Use the links on the following page to learn more about the topics covered in this chapter:

`https://github.com/markjprice/cs13net9/blob/main/docs/book-links.md#chapter-5---building-your-own-types-with-object-oriented-programming`

Summary

In this chapter, you learned about:

- Making your own types using OOP.
- Some of the different categories of members that a type can have, including fields to store data and methods to perform actions.
- OOP concepts, such as aggregation and encapsulation
- How to use modern C# features, like relational and property pattern matching enhancements, `init`-only properties, and record types.

In the next chapter, you will take these concepts further by defining operators, delegates, and events, implementing interfaces, and inheriting from existing classes.

6

Implementing Interfaces and Inheriting Classes

This chapter is about deriving new types from existing ones using **object-oriented programming** (**OOP**). You will learn how to use operators as an alternative method to implement simple functionality, and you will learn how to use generics to make your code safer and more performant. You will learn about delegates and events to exchange messages between types. You will implement interfaces for common functionality. You will create a derived class to inherit from a base class to reuse functionality, override an inherited type member, and use polymorphism. Finally, you will learn how to create extension methods and cast between classes in an inheritance hierarchy.

In an online-only section, you will see the differences between reference and value types and how they are stored in memory. You will find links to online materials at the end of this chapter.

This chapter covers the following topics:

- Setting up a class library and console application
- Static methods and overloading operators
- Making types safely reusable with generics
- Raising and handling events
- Implementing interfaces
- Working with `null` values
- Inheriting from classes
- Casting within inheritance hierarchies
- Inheriting and extending .NET types
- Summarizing custom type choices

Setting up a class library and console application

We will start by defining a solution with two projects, like the one created in *Chapter 5, Building Your Own Types with Object-Oriented Programming*. Even if you completed all the exercises in that chapter, follow the instructions below so that you start this chapter with fresh working projects:

1. Use your preferred code editor to create a new project, as defined in the following list:
    - Project template: **Class Library** / `classlib`
    - Project file and folder: `PacktLibrary`
    - Solution file and folder: `Chapter06`
    - Framework: .NET 9.0 (Standard-Term Support)
2. Add a new project, as defined in the following list:
    - Project template: **Console App** / `console`
    - Project file and folder: `PeopleApp`
    - Solution file and folder: `Chapter06`
    - Framework: .NET 9.0 (Standard-Term Support)
    - Do not use top-level statements: Cleared
    - Enable native AOT publish: Cleared

In this chapter, both projects target .NET 9 and, therefore, use the C# 13 compiler by default.

3. In the `PacktLibrary` project, rename the file named `Class1.cs` to `Person.cs`.
4. In both projects, add `<ItemGroup>` to globally and statically import the `System.Console` class, as shown in the following markup:

```
<ItemGroup>
  <Using Include="System.Console" Static="true" />
</ItemGroup>
```

5. In `Person.cs`, delete any existing statements and define a `Person` class, as shown in the following code:

```
namespace Packt.Shared;

public class Person
{
  #region Properties

  public string? Name { get; set; }
```

```
  public DateTimeOffset Born { get; set; }
  public List<Person> Children { get; set; } = new();

  #endregion

  #region Methods

  public void WriteToConsole()
  {
    WriteLine($"{Name} was born on a {Born:dddd}.");
  }

  public void WriteChildrenToConsole()
  {
    string term = Children.Count == 1 ? "child" : "children";
    WriteLine($"{Name} has {Children.Count} {term}.");
  }

  #endregion
}
```

6. In the `PeopleApp` project, add a project reference to `PacktLibrary`, as shown in the following markup:

```
<ItemGroup>
  <ProjectReference
    Include="..\PacktLibrary\PacktLibrary.csproj" />
</ItemGroup>
```

7. In `Program.cs`, delete the existing statements, write statements to create an instance of `Person`, and then write information about it to the console, as shown in the following code:

```
using Packt.Shared;

Person harry = new()
{
  Name = "Harry",
  Born = new(year: 2001, month: 3, day: 25,
    hour: 0, minute: 0, second: 0,
    offset: TimeSpan.Zero)
};

harry.WriteToConsole();
```

8. If you use Visual Studio, configure the startup project for the solution as the current selection.
9. Run the `PeopleApp` project and note the result, as shown in the following output:

```
Harry was born on a Sunday.
```

Static methods and overloading operators

This section is specifically about methods that apply to two instances of the same type. It is not about the more general case of methods that apply to zero, one, or more than two instances. Remember that static methods do not require an object instance. They are called directly using the class name.

I wanted to think of some methods that would apply to two `Person` instances that could also become binary operators, like + and *. What would adding two people together represent? What would multiplying two people represent? The obvious answers are getting married and having babies.

We will design our methods to enable us to model the story of Lamech and his two wives and their children, as described at the following link:

https://www.kingjamesbibleonline.org/Genesis-4-19/

We might want two instances of `Person` to be able to marry and procreate. We can implement this by writing methods and overriding operators. Instance methods are actions that an object does to itself; static methods are actions the type does.

Which you choose depends on what makes the most sense for the action.

Good Practice: Having both static and instance methods to perform similar actions often makes sense. For example, `string` has both a `Compare` static method and a `CompareTo` instance method. This puts the choice of how to use the functionality in the hands of the programmers using your type, giving them more flexibility.

Implementing functionality using methods

Let's start by implementing some functionality by using both static and instance methods:

1. In `Person.cs`, add properties with private backing storage fields to indicate if that person is married and to whom, as shown in the following code:

```
// Allow multiple spouses to be stored for a person.
public List<Person> Spouses { get; set; } = new();

// A read-only property to show if a person is married to anyone.
public bool Married => Spouses.Count > 0;
```

2. In `Person.cs`, add one instance method and one static method that will allow two `Person` objects to marry, as shown in the following code:

```
// Static method to marry two people.
public static void Marry(Person p1, Person p2)
{
  ArgumentNullException.ThrowIfNull(p1);
  ArgumentNullException.ThrowIfNull(p2);

  if (p1.Spouses.Contains(p2) || p2.Spouses.Contains(p1))
  {
    throw new ArgumentException(
      string.Format("{0} is already married to {1}.",
      arg0: p1.Name, arg1: p2.Name));
  }

  p1.Spouses.Add(p2);
  p2.Spouses.Add(p1);
}

// Instance method to marry another person.
public void Marry(Person partner)
{
  Marry(this, partner); // "this" is the current person.
}
```

Note the following:

- In the `static` method, the `Person` objects are passed as parameters named `p1` and `p2`, and guard clauses are used to check for `null` values. If either is already married to the other, an exception is thrown; otherwise, they are each added to each other's list of spouses. You can model this differently if you want to allow two people to have multiple marriage ceremonies. In that case, you might choose to not throw an exception and instead do nothing. Their state of marriage would remain the same. Additional calls to `Marry` would not change if they are married or not. In this scenario, I want you to see that the code recognizes that they are already married by throwing an exception.
- In the instance method, a call is made to the `static` method, passing the current person (`this`) and the `partner` they want to marry.

Warning! Although instance methods can call static methods, the opposite is not true. A static method cannot call an instance method because there isn't an object instance to call it on.

3. In `Person.cs`, add an instance method to the `Person` class that will output the spouses of a person if they are married, as shown in the following code:

```
public void OutputSpouses()
{
  if (Married)
  {
    string term = Spouses.Count == 1 ? "person" : "people";

    WriteLine($"{Name} is married to {Spouses.Count} {term}:");

    foreach (Person spouse in Spouses)
    {
      WriteLine($"  {spouse.Name}");
    }
  }
  else
  {
    WriteLine($"{Name} is a singleton.");
  }
}
```

4. In `Person.cs`, add one instance method and one static method to the `Person` class that will allow two `Person` objects to procreate if they are married to each other, as shown in the following code:

```
/// <summary>
/// Static method to "multiply" aka procreate and have a child together.
/// </summary>
/// <param name="p1">Parent 1</param>
/// <param name="p2">Parent 2</param>
/// <returns>A Person object that is the child of Parent 1 and Parent
2.</returns>
/// <exception cref="ArgumentNullException">If p1 or p2 are null.</
exception>
/// <exception cref="ArgumentException">If p1 and p2 are not married.</
exception>
public static Person Procreate(Person p1, Person p2)
{
  ArgumentNullException.ThrowIfNull(p1);
  ArgumentNullException.ThrowIfNull(p2);

  if (!p1.Spouses.Contains(p2) && !p2.Spouses.Contains(p1))
  {
```

```
    throw new ArgumentException(string.Format(
      "{0} must be married to {1} to procreate with them.",
      arg0: p1.Name, arg1: p2.Name));
  }

  Person baby = new()
  {
    Name = $"Baby of {p1.Name} and {p2.Name}",
    Born = DateTimeOffset.Now
  };

  p1.Children.Add(baby);
  p2.Children.Add(baby);

  return baby;
}

// Instance method to "multiply".
public Person ProcreateWith(Person partner)
{
  return Procreate(this, partner);
}
```

Note the following:

- In the `static` method named `Procreate`, the `Person` objects that will procreate are passed as parameters named `p1` and `p2`.
- A new `Person` class named `baby` is created with a name composed of a combination of the two people who have procreated. This could be changed later by setting the returned `baby` variable's `Name` property. Although we could add a third parameter to the `Procreate` method for the baby name, we will define a binary operator later, and they cannot have third parameters. So, for consistency, we will just return the baby reference and let the calling code set the name of it.
- The `baby` object is added to the `Children` collection of both parents and then returned. Classes are reference types, meaning a reference to the `baby` object stored in memory is added, not a clone of the `baby` object.
- In the instance method named `ProcreateWith`, the `Person` object to procreate with is passed as a parameter named `partner`, and that, along with `this`, which represents the current object, is passed to the static `Procreate` method to reuse the method implementation. `this` is a keyword that references the current instance of the class. It is a convention to use different method names for related static and instance methods, but not to the extent that you would fail to match standard English meaning.

So, that could mean naming both methods the same, as in the example of `Marry(x, y)` for the static method name and `x.Marry(y)` for the instance method name. More commonly, it would mean using a different method name, for example, `Procreate(x, y)` for the static method name and `x.ProcreateWith(y)` for the instance method name, or `Compare(x, y)` for the static method name and `x.CompareTo(y)` for the instance method name.

Good Practice: A method that creates a new object, or modifies an existing object, should return a reference to that object so that the caller can access the results.

5. In `Program.cs`, create three people and have them marry and then procreate with each other, noting that to add a double-quote character into a `string`, you must prefix it with a backslash character like this, `\"`, as shown in the following code:

```
// Implementing functionality using methods.
Person lamech = new() { Name = "Lamech" };
Person adah = new() { Name = "Adah" };
Person zillah = new() { Name = "Zillah" };

// Call the instance method to marry Lamech and Adah.
lamech.Marry(adah);

// Call the static method to marry Lamech and Zillah.
Person.Marry(lamech, zillah);

lamech.OutputSpouses();
adah.OutputSpouses();
zillah.OutputSpouses();

// Call the instance method to make a baby.
Person baby1 = lamech.ProcreateWith(adah);
baby1.Name = "Jabal";
WriteLine($"{baby1.Name} was born on {baby1.Born}");

// Call the static method to make a baby.
Person baby2 = Person.Procreate(zillah, lamech);
baby2.Name = "Tubalcain";

adah.WriteChildrenToConsole();
zillah.WriteChildrenToConsole();
lamech.WriteChildrenToConsole();
```

```
for (int i = 0; i < lamech.Children.Count; i++)
{
  WriteLine(format: "  {0}'s child #{1} is named \"{2}\".",
    arg0: lamech.Name, arg1: i,
    arg2: lamech.Children[i].Name);
}
```

I used a `for` instead of a `foreach` statement so that I could use the `i` variable with the indexer to access each child.

6. Run the `PeopleApp` project and view the result, as shown in the following output:

```
Lamech is married to 2 people:
  Adah
  Zillah
Adah is married to 1 person:
  Lamech
Zillah is married to 1 person:
  Lamech
Jabal was born on 05/07/2023 15:17:03 +01:00
Adah has 1 child.
Zillah has 1 child.
Lamech has 2 children:
  Lamech's child #0 is named "Jabal".
  Lamech's child #1 is named "Tubalcain".
```

As you have just seen, for functionality that applies to two instances of an object type, it is easy to provide both static and instance methods to implement the same functionality. Neither static nor instance methods are best in all scenarios, and you cannot predict how your type might be used. It is best to provide both to allow a developer to use your types in the way that best fits their needs.

Now let's see how we can add a third way to provide the same functionality for two instances of a type.

Implementing functionality using operators

The `System.String` class has a `static` method named `Concat` that concatenates two `string` values and returns the result, as shown in the following code:

```
string s1 = "Hello ";
string s2 = "World!";
string s3 = string.Concat(s1, s2);
WriteLine(s3); // Hello World!
```

Calling a method like `Concat` works, but it might be more natural for a programmer to use the + symbol operator to "add" two `string` values together, as shown in the following code:

```
string s3 = s1 + s2;
```

A well-known biblical phrase is *Go forth and multiply*, meaning to procreate. Let's write code so that the * (multiply) symbol will allow two `Person` objects to procreate. And we will use the + operator to marry two people.

We do this by defining a `static` operator for the * symbol. The syntax is rather like a method, because in effect, an operator *is* a method, but it uses a symbol instead of a method name, which makes the syntax more concise:

1. In `Person.cs`, create a `static` operator for the + symbol, as shown in the following code:

```
#region Operators

// Define the + operator to "marry".
public static bool operator +(Person p1, Person p2)
{
  Marry(p1, p2);

  // Confirm they are both now married.
  return p1.Married && p2.Married;
}

#endregion
```

The return type for an operator does not need to match the types passed as parameters to the operator, but the return type cannot be `void`.

2. In `Person.cs`, create a `static` operator for the * symbol, as shown in the following code:

```
// Define the * operator to "multiply".
public static Person operator *(Person p1, Person p2)
{
  // Return a reference to the baby that results from multiplying.
  return Procreate(p1, p2);
}
```

Good Practice: Unlike methods, operators do not appear in IntelliSense lists for a type or a type instance when you enter a dot (`.`). For every operator that you define, make a method as well, because it may not be obvious to a programmer that the operator is available. The implementation of the operator can then call the method, reusing the code you have written. A second reason to provide a method is that operators are not supported by every language compiler; for example, although arithmetic operators like * are supported by Visual Basic and F#, there is no requirement that other languages support all operators supported by C#. You have to read the type definition or the documentation to discover whether operators are implemented.

3. In `Program.cs`, comment out the statement that calls the static `Marry` method to marry Zillah and Lamech, and replace it with an `if` statement that uses the `+` operator to marry them, as shown in the following code:

```
// Person.Marry(lamech, zillah);

if (lamech + zillah)
{
  WriteLine($"{lamech.Name} and {zillah.Name} successfully got
married.");
}
```

4. In `Program.cs`, after calling the `Procreate` method and before the statements that write the children to the console, use the `*` operator for Lamech to have two more babies with his wives, Adah and Zillah, as highlighted in the following code:

```
// Use the * operator to "multiply".
Person baby3 = lamech * adah;
baby3.Name = "Jubal";

Person baby4 = zillah * lamech;
baby4.Name = "Naamah";
```

5. Run the `PeopleApp` project and view the result, as shown in the following output:

```
Lamech and Zillah successfully got married.
Lamech is married to 2 people:
  Adah
  Zillah
Adah is married to 1 person:
  Lamech
Zillah is married to 1 person:
  Lamech
```

```
Jabal was born on 05/07/2023 15:27:30 +01:00
Adah has 2 children.
Zillah has 2 children.
Lamech has 4 children:
  Lamech's child #0 is named "Jabal".
  Lamech's child #1 is named "Tubalcain".
  Lamech's child #2 is named "Jubal".
  Lamech's child #3 is named "Naamah".
```

More Information: To learn more about operator overloading, you can read the documentation at the following link:

`https://learn.microsoft.com/en-us/dotnet/csharp/language-reference/operators/operator-overloading`

Making types safely reusable with generics

In 2005, with C# 2 and .NET Framework 2, Microsoft introduced a feature named **generics**, which enables your types to be more safely reusable and efficient. It does this by allowing a programmer to pass types as parameters, like how you can pass objects as parameters.

This topic is only about types that need to provide flexibility for the types they work with. For example, collection types need to be able to store multiple instances of any type. That flexibility can be provided either by using the `System.Object` type or generics. For other scenarios that do not need type flexibility, the use of non-generic types is good practice.

Working with non-generic types

First, let's look at an example of working with a non-generic type so that you can understand the problems that generics are designed to solve, such as weakly typed parameters and values, and performance problems caused by using `System.Object`.

`System.Collections.Hashtable` can be used to store multiple key-value pairs, each with a unique key that can later be used to quickly look up its value. Both the key and value can be any object because they are declared as `System.Object`. Although this provides flexibility, it is slow, and bugs are easier to introduce because no type checks are made when adding items.

Let's write some code:

1. In `Program.cs`, create an instance of the non-generic collection, `System.Collections.Hashtable`, and then add four items to it, as shown in the following code:

```
// Non-generic lookup collection.
System.Collections.Hashtable lookupObject = new();
lookupObject.Add(key: 1, value: "Alpha");
lookupObject.Add(key: 2, value: "Beta");
```

```
lookupObject.Add(key: 3, value: "Gamma");
lookupObject.Add(key: harry, value: "Delta");
```

Note that three items have a unique integer key to look them up. The last item has a `Person` object as its key to look it up. This is valid in a non-generic collection.

2. Add statements to define a key with the value of `2` and use it to look up its value in the *hash table*, as shown in the following code:

```
int key = 2; // Look up the value that has 2 as its key.

WriteLine(format: "Key {0} has value: {1}",
  arg0: key,
  arg1: lookupObject[key]);
```

3. Add statements to use the `harry` object to look up its value, as shown in the following code:

```
// Look up the value that has harry as its key.
WriteLine(format: "Key {0} has value: {1}",
  arg0: harry,
  arg1: lookupObject[harry]);
```

4. Run the `PeopleApp` project and note that it works, as shown in the following output:

```
Key 2 has value: Beta
Key Packt.Shared.Person has value: Delta
```

Although the code works, there is potential for mistakes because literally any type can be used for the key or value. If another developer used your variable named `lookupObject` and expected all the items to be a certain type, they might cast them to that type and get exceptions because some values might be a different type. A lookup object with lots of items would also give poor performance.

Good Practice: Avoid types in the `System.Collections` namespace. Use types in the `System.Collections.Generics` and related namespaces instead. If you need to use a library that uses non-generic types, then of course you will have to use non-generic types. This is an example of what is commonly referred to as technical debt.

Working with generic types

`System.Collections.Generic.Dictionary<TKey, TValue>` can be used to store multiple values, each with a unique key that can later be used to quickly look up its value. Both the key and value can be any object, but you must tell the compiler what the types of the key and value will be when you first instantiate the collection. You do this by specifying types for the **generic parameters** in angle brackets `<>`, `TKey`, and `TValue`.

Good Practice: When a generic type has one definable type, it should be named `T`, for example, `List<T>`, where `T` is the type stored in the list. When a generic type has multiple definable types, it should use `T` as a name prefix and have a sensible name, for example, `Dictionary<TKey, TValue>`.

Generics provides flexibility, is faster, and makes bugs easier to avoid because type checks are made when adding items at compile time. We will not need to explicitly specify the `System.Collections.Generic` namespace that contains `Dictionary<TKey, TValue>` because it is implicitly and globally imported by default.

Let's write some code to solve the problem by using generics:

1. In `Program.cs`, create an instance of the generic lookup collection `Dictionary<TKey, TValue>` and then add four items to it, as shown in the following code:

```
// Define a generic lookup collection.
Dictionary<int, string> lookupIntString = new();
lookupIntString.Add(key: 1, value: "Alpha");
lookupIntString.Add(key: 2, value: "Beta");
lookupIntString.Add(key: 3, value: "Gamma");
lookupIntString.Add(key: harry, value: "Delta");
```

2. Note the compile error when using `harry` as a key, as shown in the following output:

```
/Users/markjprice/Code/Chapter06/PeopleApp/Program.cs(98,32): error
CS1503: Argument 1: cannot convert from 'Packt.Shared.Person' to 'int' [/
Users/markjprice/Code/Chapter06/PeopleApp/PeopleApp.csproj]
```

3. Replace `harry` with `4`.
4. Add statements to set the `key` to `3`, and use it to look up its value in the dictionary, as shown in the following code:

```
key = 3;

WriteLine(format: "Key {0} has value: {1}",
  arg0: key,
  arg1: lookupIntString[key]);
```

5. Run the `PeopleApp` project and note that it works, as shown in the following output:

```
Key 3 has value: Gamma
```

You have now seen the difference between non-generic and generic types that need the flexibility to store any type. You know to always use generic collection types if possible. Unless you are unlucky enough to be forced to use a legacy non-generic library, you never need to write code that uses non-generic types that can store any type again.

Just because it is good practice to use generic collection types in preference to non-generic collection types does not mean the more general case is also true. Non-generic non-collection types and other types that do not need the flexibility to work with any type are used all the time. Collection types just happen to be the most common type that benefits from generics.

Raising and handling events

Methods are often described as *actions that an object can perform, either on itself or on related objects.* For example, `List<T>` can add an item to itself or clear itself, and `File` can create or delete a file in the filesystem.

Events are often described as *actions that happen to an object*. For example, in a user interface, `Button` has a `Click` event, a click being something that happens to a button. Also, `FileSystemWatcher` listens to the filesystem for change notifications and raises events like `Created` and `Deleted`, which are triggered when a directory or file changes.

Another way to think of events is that they provide a way of exchanging messages between objects.

Events are built on **delegates**, so let's start by having a look at what delegates are and how they work.

Calling methods using delegates

You have already seen the most common way to call or execute a method: using the `.` operator to access the method using its name. For example, `Console.WriteLine` tells the `Console` type to call its `WriteLine` method.

The other way to call or execute a method is to use a delegate. If you have used languages that support **function pointers**, then think of a delegate as being a **type-safe method pointer.**

In other words, a delegate contains the memory address of a method that must match the same signature as the delegate, enabling it to be called safely with the correct parameter types.

The code in this section is illustrative and not meant to be typed into a project. You will explore code like this in the next section, so for now, just read the code and try to understand its meaning.

For example, imagine there is a method in the `Person` class that must have a `string` type passed as its only parameter, and it returns an `int` type, as shown in the following code:

```
public class Person
{
  public int MethodIWantToCall(string input)
  {
    return input.Length; // It doesn't matter what the method does.
  }
```

I can call this method on an instance of Person named p1 like this:

```
Person p1 = new();
int answer = p1.MethodIWantToCall("Frog");
```

Alternatively, I can define a delegate with a matching signature to call the method indirectly. Note that the names of the parameters do not have to match. Only the types of parameters and return values must match, as shown in the following code:

```
delegate int DelegateWithMatchingSignature(string s);
```

Good Practice: A `delegate` is a reference type like a `class`, so if you define one in `Program.cs`, then it must be at the bottom of the file. It would be best to define it in its own class file, for example, `Program.Delegates.cs`. If you defined a delegate in the middle of `Program.cs`, then you would see the following compiler error: `CS8803: Top-level statements must precede namespace and type declarations.`

Now, I can create an instance of the delegate, point it at the method, and finally, call the delegate (which calls the method), as shown in the following code:

```
// Create a delegate instance that points to the method.
DelegateWithMatchingSignature d = new(p1.MethodIWantToCall);

// Call the delegate, which then calls the method.
int answer2 = d("Frog");
```

Examples of delegate use

You are probably thinking, *"What's the point of that?"*

It provides flexibility. For example, we could use delegates to create a queue of methods that need to be called in order. Queuing actions that need to be performed is common in services to provide improved scalability.

Another example is to allow multiple actions to execute in parallel. Delegates have built-in support for asynchronous operations that run on a different thread, which can provide improved responsiveness.

The most important example is that delegates allow us to implement events to send messages between different objects that do not need to know about each other. Events are an example of loose coupling between components because they do not need to know about each other; they just need to know the event signature.

Status: It's complicated

Delegates and events are two of the most confusing features of C# and can take a few attempts to understand, so don't worry if you feel lost as we walk through how they work! Move on to other topics and come back again another day when your brain has had the opportunity to process the concepts while you sleep.

Defining and handling delegates

Microsoft has two predefined delegates for use as events. They both have two parameters:

- `object? sender`: This parameter is a reference to the object raising the event or sending the message. The `?` indicates that this reference could be `null`.
- `EventArgs e` or `TEventArgs e`: This parameter contains additional relevant information about the event. For example, in a GUI app, you might define `MouseMoveEventArgs`, which has properties for the `X` and `Y` coordinates for the mouse pointer. A bank account might have a `WithdrawEventArgs` with a property for the `Amount` to withdraw.

Their signatures are simple, yet flexible, as shown in the following code:

```
// For methods that do not need additional argument values passed in.
public delegate void EventHandler(object? sender, EventArgs e);

// For methods that need additional argument values passed in as
// defined by the generic type TEventArgs.
public delegate void EventHandler<TEventArgs>(object? sender, TEventArgs e);
```

Good Practice: When you want to define an event in your own type, you should use one of these two predefined delegates.

Some types provide "empty" values for when you need an instance, but it doesn't need to have any particular value. For example:

- `string.Empty` is an empty string value `""`. It is better to use `string.Empty` than `""` because every time you use `""` the compiler must instantiate a new empty string instance. You can think of `string.Empty` as a global single instance of an empty string. Sharing it is more efficient than creating your own new instances every time you need to reference an empty string value.
- `EventArgs.Empty` is an empty `EventArgs` value. Use it when you must conform to the built-in event delegates that require an `EventArgs` instance to be passed as a parameter, but it doesn't need any particular value because it won't be read or used in the method anyway.

You might have a delegate defined, as shown in the following code:

```
public EventHandler? Shout; // This field could be null.
```

In that case, there are multiple ways to call the delegate, as shown in the following code examples:

- Use its variable name, `Shout`:

  ```
  Shout(this, EventArgs.Empty);
  ```

- Use its `Invoke` method to call it synchronously:

  ```
  Shout.Invoke(this, EventArgs.Empty);
  ```

- Use its `BeginInvoke` method to call it asynchronously, without a callback function or any state:

```
IAsyncResult result = Shout.BeginInvoke(
  this, EventArgs.Empty, null, null);
```

The `BeginInvoke` method is beyond the scope of this book, but I have included it so that you know it exists.

Delegates and their method handlers have a potentially many-to-many relationship. One delegate can have one method handler. But one delegate can also have many method handlers (you will do this when you hook up `Shout` to both `Harry_Shout` and `Harry_Shout_2`). And many delegates can reference one method handler, or any combination of these. Let's explore delegates and events:

1. Add statements to the `Person` class and note the following points, as shown in the following code:
    - It defines an `EventHandler` delegate field named `Shout`.
    - It defines an `int` field to store `AngerLevel`.
    - It defines a method named `Poke`.
    - Each time a person is poked, their `AngerLevel` increments. Once their `AngerLevel` reaches three, they raise the `Shout` event, but only if there is at least one event delegate pointing at a method defined somewhere else in the code; that is, it is not `null`:

```
#region Events

// Delegate field to define the event.
public EventHandler? Shout; // null initially.

// Data field related to the event.
public int AngerLevel;

// Method to trigger the event in certain conditions.
public void Poke()
{
  AngerLevel++;

  if (AngerLevel < 3) return;

  // If something is listening to the event...
  if (Shout is not null)
  {
    // ...then call the delegate to "raise" the event.
    Shout(this, EventArgs.Empty);
```

```
    }
  }

  #endregion
```

Checking whether an object is not `null` before calling one of its methods is very common. C# 6 and later allow `null` checks to be simplified inline using a ? symbol before the `.` operator, as shown in the following code:

```
Shout?.Invoke(this, EventArgs.Empty);
```

2. In the `PeopleApp` project, add a new class file named `Program.EventHandlers.cs`.
3. In `Program.EventHandlers.cs`, delete any existing statements, and then add a method with a matching signature that gets a reference to the `Person` object from the `sender` parameter and outputs some information about them, as shown in the following code:

```
using Packt.Shared; // To use Person.

// No namespace declaration so this extends the Program class
// in the null namespace.

partial class Program
{
  // A method to handle the Shout event received by the harry object.
  private static void Harry_Shout(object? sender, EventArgs e)
  {
    // If no sender, then do nothing.
    if (sender is null) return;

    // If sender is not a Person, then do nothing and return; else assign
sender to p.
    if (sender is not Person p) return;

    WriteLine($"{p.Name} is this angry: {p.AngerLevel}.");
  }
}
```

Good Practice: Microsoft's convention for method names that handle events is `ObjectName_EventName`. In this project, `sender` will always be a `Person` instance, so the `null` checks are not necessary, and the event handler could be much simpler with just the `WriteLine` statement. However, it is important to know that these types of `null` checks make your code more robust in cases of event misuse.

You can have as many methods as you like to be event handlers, named whatever you like, as long as the method signature matches the delegate signature. This means you could have 50 `Person` instances, each with their own method, or have one method that they all share. The methods can be declared at any level that makes sense for the scenario and matches the access levels set (like `protected`, `private`, `public`, and so on). One of the key benefits of delegates and events is loose binding between components so maximum flexibility is desired.

4. In `Program.cs`, add a statement to assign the method to the delegate field, and then add statements to call the `Poke` method four times, as shown in the following code:

```
// Assign the method to the Shout delegate.
harry.Shout = Harry_Shout;

// Call the Poke method that eventually raises the Shout event.
harry.Poke();
harry.Poke();
harry.Poke();
harry.Poke();
```

5. Run the `PeopleApp` project and view the result. Note that Harry says nothing the first two times he is poked, and only gets angry enough to shout once he's been poked at least three times, as shown in the following output:

```
Harry is this angry: 3.
Harry is this angry: 4.
```

In Step 3, note that the `sender` is checked to make sure it is a `Person` instance, and if it is, then it is assigned to a local variable named `p`, as shown in the following code:

```
// If sender is not a Person, then do nothing and return; else assign sender to
p.
if (sender is not Person p) return;
WriteLine($"{p.Name} is this angry: {p.AngerLevel}.");
```

The first statement does two things at once, which needs more explanation. The parameter named `sender` is declared to be of type `object`. This means we cannot just say `sender.Name` or `sender.AngerLevel`. We need to cast `sender` to a local variable that is explicitly defined as `Person`. We also need to check that `sender` actually is a `Person`.

We can do both things at once in a single expression: `sender is not Person p`. This expression will return `true` if `sender` is not a `Person`, and hence the statement executes `return` so the method immediately returns. Or the expression returns `false` if `sender` is a `Person`, and sender will be stored in the local variable named `p`, which is of type `Person`. After that, we can use expressions like `p.Name` and `p.AngerLevel`.

Defining and handling events

You've now seen how delegates implement the most important functionality of events: the ability to define a signature for a method that can be implemented by a completely different piece of code, calling that method and any others that are hooked up to the delegate field.

But what about events? There is less to them than you might think.

When assigning a method to a delegate field, you should not use the simple assignment operator as we did in the preceding example.

Delegates are multicast, meaning that you can assign multiple delegates to a single delegate field. Instead of the = assignment, we could have used the += operator so that we could add more methods to the same delegate field. When the delegate is called, all the assigned methods are called, although you have no control over the order in which they are called. Do not use events to implement a queuing system to buy concert tickets; otherwise, the wrath of millions of Swifties will fall upon you.

If the `Shout` delegate field already referenced one or more methods, by assigning another method, that method would replace all the others. With delegates that are used for events, we usually want to make sure that a programmer only ever uses either the += operator or the -= operator to assign and remove methods:

1. To enforce this, in `Person.cs`, add the `event` keyword to the delegate field declaration, as highlighted in the following code:

    ```
    public event EventHandler? Shout;
    ```

2. Build the `PeopleApp` project and note the compiler error message, as shown in the following output:

    ```
    Program.cs(41,13): error CS0079: The event 'Person.Shout' can only appear
    on the left hand side of += or -=
    ```

    This is (almost) all that the `event` keyword does! If you will never have more than one method assigned to a delegate field, then technically you do not need events, but it is still good practice to indicate your meaning and that you expect a delegate field to be used as an event.

3. In `Program.cs`, modify the comment and the method assignment to use += instead of just =, as highlighted in the following code:

    ```
    // Assign the method to the Shout event delegate.
    harry.Shout += Harry_Shout;
    ```

4. Run the `PeopleApp` project and note that it has the same behavior as before.
5. In `Program.EventHandlers.cs`, create a second event handler for Harry's `Shout` event, as shown in the following code:

    ```
    // Another method to handle the event received by the harry object.
    private static void Harry_Shout_2(object? sender, EventArgs e)
    {
    ```

```
    WriteLine("Stop it!");
}
```

6. In `Program.cs`, after the statement that assigns the `Harry_Shout` method to the `Shout` event, add a statement to attach the new event handler to the `Shout` event too, as shown highlighted in the following code:

```
// Assign the method(s) to the Shout event delegate.
harry.Shout += Harry_Shout;
harry.Shout += Harry_Shout_2;
```

7. Run the `PeopleApp` project, view the result. Note that both event handlers execute whenever an event is raised, which only happens once the anger level is three or more, as shown in the following output:

```
Harry is this angry: 3.
Stop it!
Harry is this angry: 4.
Stop it!
```

In Windows desktop development, imagine that you have three buttons: `AddButton`, `SaveButton`, and `DeleteButton`. Each button has very different functionality. Good practice would be to create three methods to handle their `Click` events, named `AddButton_Click`, `SaveButton_Click`, and `DeleteButton_Click`. Each would have a different implementation code.

But now, imagine you have 26 buttons: `AButton`, `BButton`, `CButton`, and so on, up to `ZButton`. Each button has the same functionality: to filter a list of people by the first letter of their name. Good practice would be to create one method to handle their `Click` events, perhaps named `AtoZButtons_Click`. This method would have an implementation code that would use the `sender` parameter to know which button was clicked, and therefore how to apply the filtering, but otherwise be the same for all the buttons.

That's it for events. Now, let's look at interfaces.

Implementing interfaces

Interfaces are a way to implement standard functionality and connect different types to make new things. Think of them like the studs on top of LEGO™ bricks, which allow them to "stick" together, or electrical standards for plugs and sockets.

If a type implements an interface, then it makes a promise to the rest of .NET that it supports specific functionality. Therefore, they are sometimes described as contracts.

Common interfaces

Table 6.1 shows some common interfaces that your types might implement:

Interface	Method(s)	Description
`IComparable`	`CompareTo(other)`	This defines a comparison method that a type implements to order or sort its instances.
`IComparer`	`Compare(first, second)`	This defines a comparison method that a secondary type implements to order or sort instances of a primary type.
`IDisposable`	`Dispose()`	This defines a disposal method to release unmanaged resources more efficiently than waiting for a finalizer. See the *Releasing unmanaged resources* section later in this chapter for more details.
`IFormattable`	`ToString(format, culture)`	This defines a culture-aware method to format the value of an object into a string representation.
`IFormatter`	`Serialize(stream, object)` `Deserialize(stream)`	This defines methods to convert an object to and from a stream of bytes for storage or transfer.
`IFormatProvider`	`GetFormat(type)`	This defines a method to format inputs based on a language and region.

Table 6.1: Some common interfaces that your types might implement

Comparing objects when sorting

One of the most common interfaces that you will want to implement in your types that represent data is `IComparable`. If a type implements one of the `IComparable` interfaces, then arrays and collections containing instances of that type can be sorted.

This is an example of an abstraction for the concept of sorting. To sort any type, the minimum functionality would be the ability to compare two items and decide which goes before the other. If a type implements that minimum functionality, then a sorting algorithm can use it to sort instances of that type in any way the sorting algorithm wants.

The `IComparable` interface has one method named `CompareTo`. This has two variations, one that works with a nullable `object` type and one that works with a nullable generic type `T`, as shown in the following code:

```
namespace System
{
  public interface IComparable
  {
    int CompareTo(object? obj);
  }
```

```
  public interface IComparable<in T>
  {
    int CompareTo(T? other);
  }
}
```

The in keyword specifies that the type parameter T is contravariant, which means that you can use a less derived type than that specified. For example, if Employee derives from Person, then both can be compared to each other.

For example, the string type implements IComparable by returning -1 if the string should be sorted before the string is compared to, 1 if it should be sorted after, and 0 if they are equal. The int type implements IComparable by returning -1 if the int is less than the int being compared to, 1 if it is greater, and 0 if they are equal.

CompareTo return values can be summarized as shown in *Table 6.2*:

this before other	this is equal to other	this after other
-1	0	1

Table 6.2: Summary of the CompareTo return values

Before we implement the IComparable interface and its CompareTo method for the Person class, let's see what happens when we try to sort an array of Person instances without implementing this interface, including some that are null or have a null value for their Name property:

1. In the PeopleApp project, add a new class file named Program.Helpers.cs.
2. In Program.Helpers.cs, delete any existing statements. Then define a method for the partial Program class that will output all the names of a collection of people passed as a parameter, with a title beforehand, as shown in the following code:

```
using Packt.Shared;

partial class Program
{
  private static void OutputPeopleNames(
    IEnumerable<Person?> people, string title)
  {
    WriteLine(title);
    foreach (Person? p in people)
    {
      WriteLine("  {0}",
        p is null ? "<null> Person" : p.Name ?? "<null> Name");
```

```
        /* if p is null then output: <null> Person
           else output: p.Name
           unless p.Name is null then output: <null> Name */
      }
    }
  }
```

3. In `Program.cs`, add statements that create an array of `Person` instances, call the `OutputPeopleNames` method to write the items to the console, and then attempt to sort the array and write the items to the console again, as shown in the following code:

```
Person?[] people =
{
  null,
  new() { Name = "Simon" },
  new() { Name = "Jenny" },
  new() { Name = "Adam" },
  new() { Name = null },
  new() { Name = "Richard" }
};

OutputPeopleNames(people, "Initial list of people:");

Array.Sort(people);

OutputPeopleNames(people,
  "After sorting using Person's IComparable implementation:");
```

4. Run the `PeopleApp` project and an exception will be thrown. As the message explains, to fix the problem, our type must implement `IComparable`, as shown in the following output:

```
Unhandled Exception: System.InvalidOperationException: Failed to compare
two elements in the array. ---> System.ArgumentException: At least one
object must implement IComparable.
```

5. In `Person.cs`, after inheriting from `object`, add a comma and enter `IComparable<Person?>`, as highlighted in the following code:

```
public class Person : IComparable<Person?>
```

Your code editor will draw a red squiggle under the new code to warn you that you have not yet implemented the method you promised to. Your code editor can write the skeleton implementation for you.

6. Click on the light bulb and then click **Implement interface**.
7. Scroll down to the bottom of the `Person` class to find the method that was written for you, as shown in the following code:

```
public int CompareTo(Person? other)
{
  throw new NotImplementedException();
}
```

8. Delete the statement that throws the `NotImplementedException` error.
9. Add statements to handle variations of input values, including `null`. Call the `CompareTo` method of the `Name` field, which uses the `string` type's implementation of `CompareTo`. Return the result, as shown in the following code:

```
int position;

if (other is not null)
{
  if ((Name is not null) && (other.Name is not null))
  {
    // If both Name values are not null, then
    // use the string implementation of CompareTo.
    position = Name.CompareTo(other.Name);
  }
  else if ((Name is not null) && (other.Name is null))
  {
    position = -1; // this Person precedes other Person.
  }
  else if ((Name is null) && (other.Name is not null))
  {
    position = 1; // this Person follows other Person.
  }
  else // Name and other.Name are both null.
  {
    position = 0; // this and other are at same position.
  }
}
else if (other is null)
{
  position = -1; // this Person precedes other Person.
}
else // this and other are both null.
{
```

```
    position = 0; // this and other are at same position.
  }
  return position;
```

We have chosen to compare two `Person` instances by comparing their `Name` fields. `Person` instances will, therefore, be sorted alphabetically by their name. `null` values will be sorted to the bottom of the collection. Storing the calculated `position` before returning it is useful when debugging. I've also used more round brackets than the compiler needs to make the code easier for me to read. If you prefer fewer brackets, then feel free to remove them.

Also, note that the final `else` statement will never execute because the logic of the `if` and `else if` clauses means it will only execute when `this` (the current object instance) is `null`. In that scenario, the method could not execute anyway since the object wouldn't exist! I wrote the `if` statement to exhaustively cover all combinations of `null` and not `null` for `other` and `this`, but the last of those combinations could, in practice, never happen.

10. Run the `PeopleApp` project. Note that this time it works as it should, sorted alphabetically by name, as shown in the following output:

```
Initial list of people:
  Simon
  <null> Person
  Jenny
  Adam
  <null> Name
  Richard
After sorting using Person's IComparable implementation:
  Adam
  Jenny
  Richard
  Simon
  <null> Name
  <null> Person
```

Good Practice: If you want to sort an array or collection of instances of your type, then implement the `IComparable` interface.

Comparing objects using a separate class

Sometimes, you won't have access to the source code for a type, and it might not implement the `IComparable` interface. Luckily, there is another way to sort instances of a type. You can create a separate type that implements a slightly different interface, named `IComparer`:

1. In the `PacktLibrary` project, add a new class file named `PersonComparer.cs`, containing a class implementing the `IComparer` interface that will compare two people, that is, two `Person` instances. Implement it by comparing the length of their `Name` fields, or if the names are the same length, then compare the names alphabetically, as shown in the following code:

```
namespace Packt.Shared;

public class PersonComparer : IComparer<Person?>
{
  public int Compare(Person? x, Person? y)
  {
    int position;

    if ((x is not null) && (y is not null))
    {
      if ((x.Name is not null) && (y.Name is not null))
      {
        // If both Name values are not null...

        // ...then compare the Name lengths...
        int result = x.Name.Length.CompareTo(y.Name.Length);

        // ...and if they are equal...
        if (result == 0)
        {
          // ...then compare by the Names...
          return x.Name.CompareTo(y.Name);
        }
        else
        {
          // ...otherwise compare by the Lengths.
          position = result;
        }
      }
      else if ((x.Name is not null) && (y.Name is null))
      {
```

```
          position = -1; // x Person precedes y Person.
        }
        else if ((x.Name is null) && (y.Name is not null))
        {
          position = 1; // x Person follows y Person.
        }
        else // x.Name and y.Name are both null.
        {
          position = 0; // x and y are at same position.
        }
      }
      else if ((x is not null) && (y is null))
      {
        position = -1; // x Person precedes y Person.
      }
      else if ((x is null) && (y is not null))
      {
        position = 1; // x Person follows y Person.
      }
      else // x and y are both null.
      {
        position = 0; // x and y are at same position.
      }
      return position;
    }
  }
```

2. In `Program.cs`, add statements to sort the array using an alternative implementation, as shown in the following code:

```
Array.Sort(people, new PersonComparer());

OutputPeopleNames(people,
  "After sorting using PersonComparer's IComparer implementation:");
```

3. Run the `PeopleApp` project, and view the result of sorting the people by the length of their names and then alphabetically, as shown in the following output:

```
After sorting using PersonComparer's IComparer implementation:
  Adam
  Jenny
  Simon
  Richard
```

```
<null> Name
<null> Person
```

This time, when we sort the `people` array, we explicitly ask the sorting algorithm to use the `PersonComparer` type instead so that the people are sorted with the shortest names first, like `Adam`, and the longest names last, like `Richard`. When the lengths of two or more names are equal, they are sorted alphabetically, like `Jenny` and `Simon`.

Implicit and explicit interface implementations

Interfaces can be implemented implicitly and explicitly. Implicit implementations are simpler and more common. Explicit implementations are only necessary if a type must have multiple methods with the same name and signature. Personally, the only time I can remember ever having to explicitly implement an interface is when writing the code example for this book.

For example, both `IGamePlayer` and `IKeyHolder` might have a method called `Lose` with the same parameters because both a game and a key can be lost. The members of an interface are always and automatically `public` because they have to be accessible for another type to implement them!

In a type that must implement both interfaces, only one implementation of `Lose` can be the implicit method. If both interfaces can share the same implementation, there is no problem, but if not, then the other `Lose` method will have to be implemented differently and called explicitly, as shown in the following code:

```
public interface IGamePlayer // Defaults to internal.
{
  void Lose(); // Defaults to public. Could be set to internal.
}

public interface IKeyHolder
{
  void Lose();
}

public class Human : IGamePlayer, IKeyHolder
{
  // Implicit implementation must be public.
  public void Lose() // Implicit implementation.
  {
    // Implement losing a key.
    WriteLine("Implementation for losing a key.");
  }
```

```
    // Explicit implementation can be any access modifier.
    void IGamePlayer.Lose() // Defaults to private.
    {
      // Implement losing a game.
      WriteLine("Implementation for losing a game.");
    }
}

Human human = new();

human.Lose(); // Calls implicit implementation of losing a key.
// Outputs: Implementation for losing a key.

((IGamePlayer)human).Lose(); // Calls explicit implementation of losing a game.
// Outputs: Implementation for losing a game.

// Alternative way to do the same.
IGamePlayer player = human as IGamePlayer;
player.Lose(); // Calls explicit implementation of losing a game.
// Outputs: Implementation for losing a game.
```

Although the implementation of `IGamePlayer.Lose` in `Human` is `private`, the `IGamePlayer.Lose` member itself has an access modifier of `public`, so if we cast the `Human` instance into the interface type, then that `Lose` implementation is accessible.

Warning! Method access modifiers in an implementation type must match the method definition in the interface. For example, the `Lose` method in the interface is `public`, so the method implementation in the class must also be `public`.

Defining interfaces with default implementations

A language feature introduced in C# 8 is the **default implementation** for an interface. This allows an interface to contain implementation. This breaks the clean separation between interfaces that define a contract and classes and other types that implement them. It is considered by some .NET developers to be a perversion of the language.

Let's see it in action:

1. In the `PacktLibrary` project, add a new file named `IPlayable.cs`, and modify the statements to define a public `IPlayable` interface with two methods to `Play` and `Pause`, as shown in the following code:

```
namespace Packt.Shared;

public interface IPlayable
{
  void Play();
  void Pause();
}
```

2. In the `PacktLibrary` project, add a new class file named `DvdPlayer.cs`, and modify the statements in the file to implement the `IPlayable` interface, as shown in the following code:

```
namespace Packt.Shared;

public class DvdPlayer : IPlayable
{
  public void Pause()
  {
    WriteLine("DVD player is pausing.");
  }

  public void Play()
  {
    WriteLine("DVD player is playing.");
  }
}
```

This is useful, but what if we decide to add a third method named `Stop`? Before C# 8, this would be impossible once at least one type is implemented in the original interface. One of the main traits of an interface is that it is a fixed contract.

C# 8 allows you to add new members to an interface after release if those new members have a default implementation. C# purists do not like the idea, but for practical reasons, such as avoiding breaking changes or having to define a whole new interface, it is useful, and other languages such as Java and Swift enable similar techniques.

Support for default interface implementations requires some fundamental changes to the underlying platform, so they are only supported with C# if the target framework is .NET 5 or later, .NET Core 3 or later, or .NET Standard 2.1. They are, therefore, not supported by .NET Framework.

Let's add a default implementation to the interface:

1. Modify the `IPlayable` interface to add a `Stop` method with a default implementation, as highlighted in the following code:

```
namespace Packt.Shared;

public interface IPlayable
{
  void Play();
  void Pause();
  void Stop() // Default interface implementation.
  {
    WriteLine("Default implementation of Stop.");
  }
}
```

2. Build the `PeopleApp` project, and note that the projects compile successfully despite the `DvdPlayer` class not implementing `Stop`. In the future, we could override the default implementation of `Stop` by implementing it in the `DvdPlayer` class.

Although controversial, default implementations in interfaces might be useful in scenarios where the most common implementation is known at the time of defining the interface. Therefore, it is best if the interface defines that implementation once, and then most types that implement that interface can inherit it without needing to implement their own. However, if the interface definer does not know how the member should or even could be implemented, then it is a waste of effort to add a default implementation because it will always be replaced.

Think about the `IComparable` interface that you saw earlier in this chapter. It defines a `CompareTo` method. What might a default implementation of that method be? Personally, I think it's obvious that there is no default implementation that would make any practical sense. The least-worst implementation that I can think of would be to compare the `string` values returned from calling `ToString` on the two objects. However, every type really should implement its own `CompareTo` method. You are likely to find the same with 99.9% of the interfaces you use.

Now let's look at how to work with null values.

Working with null values

What if a variable does not yet have a value? How can we indicate that? C# has the concept of a `null` value, which can be used to indicate that a variable has not been set.

If you are unsure of the difference between reference types and value types in .NET, then I recommend that you read the following online-only section before you continue with this section: `https://github.com/markjprice/cs13net9/blob/main/docs/ch06-memory.md`.

Making a value type nullable

By default, **value types** like `int` and `DateTime` must always have a *value*, hence their name. Sometimes, for example, when reading values stored in a database that allows empty, missing, or `null` values, it is convenient to allow a value type to be `null`. We call this a **nullable value type**.

You can enable this by adding a question mark as a suffix to the type when declaring a variable.

Let's see an example. We will create a new project because some of the null handling options are set at the project level:

1. Use your preferred code editor to add a new **Console App**/`console` project named `NullHandling` to the `Chapter06` solution.
2. In `NullHandling.csproj`, add an `<ItemGroup>` to globally and statically import the `System.Console` class.
3. In `Program.cs`, delete the existing statements, and then add statements to declare and assign values, including `null`, two `int` variables, one suffixed with `?` and one not, as shown in the following code:

```
int thisCannotBeNull  = 4;
thisCannotBeNull = null; // CS0037 compiler error!
WriteLine(thisCannotBeNull);

int? thisCouldBeNull = null;

WriteLine(thisCouldBeNull);
WriteLine(thisCouldBeNull.GetValueOrDefault());

thisCouldBeNull = 7;

WriteLine(thisCouldBeNull);
WriteLine(thisCouldBeNull.GetValueOrDefault());
```

4. Build the project and note the compile error, as shown in the following output:

```
Cannot convert null to 'int' because it is a non-nullable value type
```

5. Comment out the statement that gives the compile error, as shown in the following code:

```
//thisCannotBeNull = null; // CS0037 compiler error!
```

6. Run the project and view the result, as shown in the following output:

```
4

0
```

```
7
7
```

The second line is blank because it outputs the `null` value.

7. Add statements to use alternative syntax, as shown in the following code:

```
// The actual type of int? is Nullable<int>.
Nullable<int> thisCouldAlsoBeNull = null;
thisCouldAlsoBeNull = 9;
WriteLine(thisCouldAlsoBeNull);
```

8. Click on `Nullable<int>` and press *F12*, or right-click and choose **Go To Definition**.
9. Note that the generic value type, `Nullable<T>`, must have a type `T`, which is a `struct`, or a value type. It has useful members like `HasValue`, `Value`, and `GetValueOrDefault`, as shown in *Figure 6.1*:

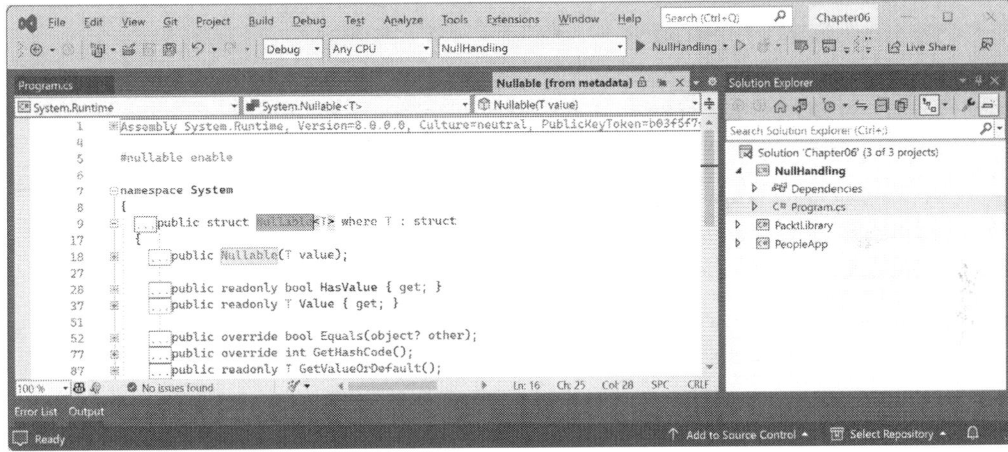

Figure 6.1: Revealing Nullable<T> members

Good Practice: When you append a `?` after a `struct` type, you change it to a different type. For example, `DateTime?` becomes `Nullable<DateTime>`.

Understanding null-related initialisms

Before we see some code, let's review some commonly used initialisms in *Table 6.3*:

Initialism	Meaning	Description
NRT	Nullable Reference Type	This is a compiler feature introduced with C# 8 and enabled by default in new projects with C# 10, which performs static analysis of your code at design time and shows warnings of potential misuse of `null` values for reference types.
NRE	`NullReferenceException`	This is an exception thrown at runtime when **dereferencing** a `null` value, also known as accessing a variable or member on an object that is `null`.
ANE	`ArgumentNullException`	This is an exception thrown at runtime by a method, property, or indexer invocation when an argument or value is `null`, and when the business logic determines that it is not valid.

Table 6.3: Commonly used initialisms

Understanding nullable reference types

The use of the `null` value is so common, in so many languages, that many experienced programmers never question the need for its existence. However, there are many scenarios where we could write better, simpler code if a variable is not allowed to have a `null` value.

The most significant change to the C# 8 language compiler was the introduction of checks and warnings for nullable and non-nullable reference types. *"But wait!"*, you are probably thinking. *"Reference types are already nullable!"*

And you would be right, but in C# 8 and later, reference types can be configured to warn you about `null` values by setting a file- or project-level option, enabling this useful new feature. Since this is a big change for C#, Microsoft decided to make the feature an opt-in.

It will take several years for this new C# language compiler feature to make an impact, since thousands of existing library packages and apps will expect the old behavior. Even Microsoft did not have time to fully implement this new feature in all the main .NET packages until .NET 6. Important libraries like `Microsoft.Extensions` for logging, dependency injections, and configuration were not annotated until .NET 7.

During the transition, you can choose between several approaches for your own projects:

- **Default**: For projects created using .NET 5 or earlier, no changes are needed. Non-nullable reference types are not checked. For projects created using .NET 6 or later, nullability checks are enabled by default, but this can be disabled by either deleting the `<Nullable>` entry in the project file or setting it to `disable`.

- **Opt-in project and opt-out files:** Enable the feature at the project level, and for any files that need to remain compatible with old behavior, opt out. This was the approach Microsoft used internally while it updated its own packages to use this new feature.
- **Opt-in files:** Only enable the NRT feature for individual files.

Warning! This NRT feature does not *prevent* `null` values – it just *warns* you about them, and the warnings can be disabled, so you still need to be careful! For example, using the ? character to declare a variable as nullable does not allow `null` values, it just disables the warning, as shown in the following code:

```
string firstName; // Allows null but gives warning when
potentially null.
string? lastName; // Allows null and does not give warning
if
null.
```

Controlling the nullability warning check feature

To enable the nullability warning check feature at the project level, have the `<Nullable>` element set to `enable` in your project file, as highlighted in the following markup:

```
<PropertyGroup>
  ...
  <Nullable>enable</Nullable>
</PropertyGroup>
```

To disable the nullability warning check feature at the project level, have the `<Nullable>` element set to `disable` in your project file, as highlighted in the following markup:

```
<PropertyGroup>
  ...
  <Nullable>disable</Nullable>
</PropertyGroup>
```

You could also remove the `<Nullable>` element completely because the default, if not explicitly set, is disabled.

To disable the feature at the file level, add the following at the top of a code file:

```
#nullable disable
```

To enable the feature at the file level, add the following at the top of a code file:

```
#nullable enable
```

Disabling null and other compiler warnings

You could decide to enable the nullability feature at the project or file level but then disable some of the 50+ warnings related to it. Some common nullability warnings are shown in *Table 6.4*:

Code	Description
CS8600	Converting a null literal or a possible null value to a non-nullable type
CS8601	A possible null reference assignment
CS8602	A dereference of a possibly null reference
CS8603	A possible null reference return
CS8604	A possible null reference argument for a parameter
CS8618	A non-nullable field `<field_name>` must contain a non-null value when exiting a constructor. Consider declaring the field as nullable
CS8625	Cannot convert a null literal to a non-nullable reference type
CS8655	The switch expression does not handle some null inputs (it is not exhaustive)

Table 6.4: Common nullability warnings

You can disable compiler warnings for a whole project. To do so, add a `NoWarn` element with a semi-colon-separated list of compiler warning codes, as shown in the following markup:

```
<NoWarn>CS8600;CS8602</NoWarn>
```

To disable compiler warnings at the statement level, you can disable and then restore a specified compiler warning to temporarily suppress it for a block of statements, as shown in the following code:

```
#pragma warning disable CS8602
WriteLine(firstName.Length);
WriteLine(lastName.Length);
#pragma warning restore CS8602
```

These techniques can be used for any compiler warnings, not just those related to nullability.

Declaring non-nullable variables and parameters

If you enable NRTs and you want a reference type to be assigned the `null` value, then you will have to use the same syntax to make a value type nullable, that is, adding a ? symbol after the type declaration.

So, how do NRTs work? Let's look at an example. When storing information about an address, you might want to force a value for the street, city, and region, but the building can be left blank, that is, `null`:

1. In the `NullHandling` project, add a class file named `Address.cs`.
2. In `Address.cs`, delete any existing statements and then add statements to declare an `Address` class with four fields, as shown in the following code:

```
namespace Packt.Shared;

public class Address
{
  public string? Building;
  public string Street;
  public string City;
  public string Region;
}
```

3. After a few seconds, note the warnings about non-nullable fields, like `Street` not being initialized, as shown in *Figure 6.2*:

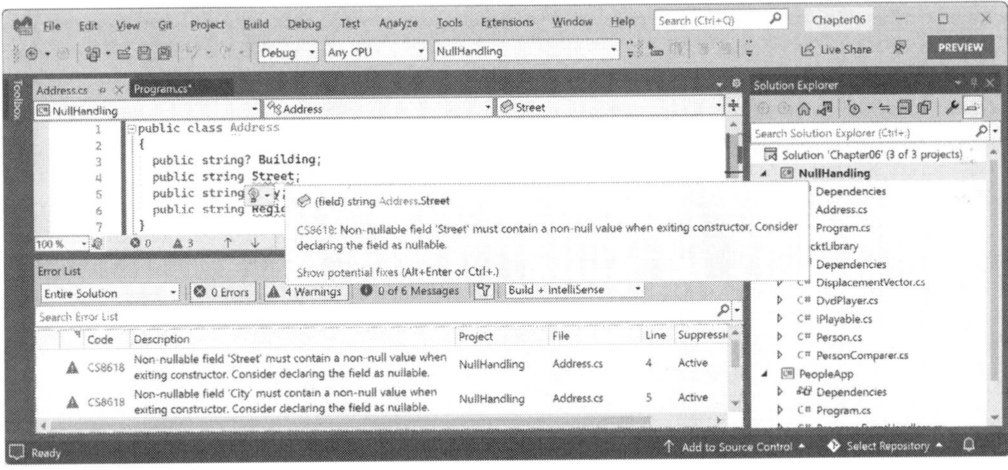

Figure 6.2: Warning messages about non-nullable fields in the Error List window

4. Assign the empty `string` value to the `Street` field, and define constructors to set the other fields that are non-nullable, as highlighted in the following code:

```
public string Street = string.Empty;
public string City;
public string Region;

public Address()
{
  City = string.Empty;
  Region = string.Empty;
}

// Call the default parameterless constructor
// to ensure that Region is also set.
public Address(string city) : this()
```

```
{
  City = city;
}
```

5. In `Program.cs`, import the namespace to use `Address`, as shown in the following code:

```
using Packt.Shared; // To use Address.
```

6. In `Program.cs`, add statements to instantiate an `Address` and set its properties, as shown in the following code:

```
Address address = new(city: "London")
{
  Building = null,
  Street = null,
  Region = "UK"
};
```

7. Note the `Warning CS8625` on setting the `Street` but not the `Building`, as shown in the following output:

```
CS8625 Cannot convert null literal to non-nullable reference type.
```

8. Append an exclamation mark to `null` when setting `Street`, as highlighted in the following code:

```
Street = null!, // null-forgiving operator.
```

9. Note that the warning disappears.
10. Add statements that will dereference the `Building` and `Street` properties, as shown in the following code:

```
WriteLine(address.Building.Length);
WriteLine(address.Street.Length);
```

11. Note the `Warning CS8602` on setting the `Building` but not the `Street`, as shown in the following output:

```
CS8602 Dereference of a possibly null reference.
```

At runtime it is still possible for an exception to be thrown when working with `Street`, but the compiler should continue to warn you of potential exceptions when working with `Building` so that you can change your code to avoid them.

12. Use the null-conditional operator to return `null` instead of accessing the `Length`, as shown in the following code:

```
WriteLine(address.Building?.Length);
```

13. Run the console app, and note that the statement that accesses the `Length` of the `Building` outputs a `null` value (blank line), but a runtime exception occurs when we access the `Length` of the `Street`, as shown in the following output:

```
Unhandled exception. System.NullReferenceException: Object reference not
set to an instance of an object.
```

14. Wrap the statement that accesses the `Street` length in a null check, as shown in the following code:

```
if (address.Street is not null)
{
  WriteLine(address.Street.Length);
}
```

It is worth reminding yourself that an NRT is only about asking the compiler to provide warnings about potential `null` values that might cause problems. It does not actually change the behavior of your code. It performs a static analysis of your code at compile time.

This explains why the new language feature is named **nullable reference types** (NRTs). Starting with C# 8, unadorned reference types can become non-nullable, and the same syntax is used to make a reference type nullable, as it is used for value types.

Suffixing a reference type with ? does not change the type. This is different from suffixing a value type with ?, which changes its type to `Nullable<T>`. Reference types can already have `null` values. All you do with **NRTs** is tell the compiler that you expect it to be `null`, so the compiler does not need to warn you. However, this does not remove the need to perform `null` checks throughout your code.

Now let's look at language features to work with `null` values that change the behavior of your code and work well as a complement to NRTs.

Checking for null

Checking whether a nullable reference type or value type variable currently contains `null` is important because if you do not, a `NullReferenceException` can be thrown, which results in an error. You should check for a `null` value before using a nullable variable, as shown in the following code:

```
// Check that the variable is not null before using it.
if (thisCouldBeNull != null)
{
  // Access a member of thisCouldBeNull.
  int length = thisCouldBeNull.Length;
  ...
}
```

C# 7 introduced `is` combined with the `!` (not) operator as an alternative to `!=`, as shown in the following code:

```
if (!(thisCouldBeNull is null))
{
```

C# 9 introduced `is not` as an even clearer alternative, as shown in the following code:

```
if (thisCouldBeNull is not null)
{
```

Good Practice: Although you would traditionally use the expression `(thisCouldBeNull != null)`, this is no longer considered good practice because the developer could have overloaded the `!=` operator to change how it works. Using pattern matching with `is null` and `is not null` is the only guaranteed way to check for `null`. For many developers, it is still instinctual to use `!=`, so I apologize in advance if you catch me still using it!

Since the `is` operator cannot be overloaded by a developer, its behavior is guaranteed. Both the `==` and `!=` operators could be overloaded by a developer so their behavior could change. The .NET team has now consistently implemented their null checks using `is`, so I recommend that you do the same, as you can see at the following link:

`https://twitter.com/Dave_DotNet/status/1764995927868076428`

The term *overriding* is about altering or extending the behavior of an inherited method in the context of object-oriented inheritance. *Overloading* occurs when you provide a new implementation for an existing operator or method.

If you try to use a member of a variable that might be `null`, use the **null-conditional operator**, `?.`, as shown in the following code:

```
string authorName = null;
int? authorNameLength;

// The following throws a NullReferenceException.
authorNameLength = authorName.Length;

// Instead of throwing an exception, null is assigned.
authorNameLength = authorName?.Length;
```

Sometimes, you want to either assign a variable to a result or use an alternative value, such as `3`, if the variable is `null`. You do this using the **null-coalescing operator**, `??`, as shown in the following code:

```
// Result will be 25 if authorName?.Length is null.
authorNameLength = authorName?.Length ?? 25;
```

Checking for null in method parameters

Even if you enable NRTs, when defining methods with parameters, it is good practice to check for `null` values.

In earlier versions of C#, you would have to write `if` statements to check for `null` parameter values and then throw an `ArgumentNullException` for any parameter that is `null`, as shown in the following code:

```
public void Hire(Person manager, Person employee)
{
  if (manager is null)
  {
    throw new ArgumentNullException(paramName: nameof(manager));
  }

  if (employee is null)
  {
    throw new ArgumentNullException(paramName: nameof(employee));
  }
  ...
}
```

C# 10 introduced a convenience method to throw an exception if an argument is `null`, as shown in the following code:

```
public void Hire(Person manager, Person employee)
{
  ArgumentNullException.ThrowIfNull(manager);
  ArgumentNullException.ThrowIfNull(employee);
  ...
}
```

C# 11 previews proposed and introduced a new `!!` operator that does this for you, as shown in the following code:

```
public void Hire(Person manager!!, Person employee!!)
{
  ...
}
```

The `if` statement and throwing of the exception would be done for you. The code is injected and executed before any statements that you write.

This proposal was controversial within the C# developer community. Some would prefer the use of attributes to decorate parameters instead of a pair of characters. The .NET product team said they reduced the .NET libraries by more than 10,000 lines of code by using this feature.

That sounds like a good reason to use it to me! And no one must use it if they choose not to. Unfortunately, the team eventually decided to remove the feature, so now we all have to write the null checks manually. If you're interested in this story, then you can read more about it at the following link:

https://devblogs.microsoft.com/dotnet/csharp-11-preview-updates/#remove-parameter-null-checking-from-c-11

I include this story in this book because I think it's an interesting example of Microsoft being transparent, by developing .NET in the open and listening to and responding to feedback from the community.

Good Practice: Always remember that nullable is a warning check, not an enforcement. You can read more about the compiler warnings relating to `null` at the following link:

https://learn.microsoft.com/en-us/dotnet/csharp/language-reference/compiler-messages/nullable-warnings

That's more than enough talk about "nothing"! Let's look at the meat of this chapter, inheritance.

Inheriting from classes

The `Person` type we created earlier derived (inherited) from `System.Object`. Now, we will create a subclass that inherits from `Person`:

1. In the `PacktLibrary` project, add a new class file named `Employee.cs`.
2. Modify its contents to define a class named `Employee` that derives from `Person`, as shown in the following code:

```
namespace Packt.Shared;

public class Employee : Person
{
}
```

3. In the `PeopleApp` project, in `Program.cs`, add statements to create an instance of the `Employee` class, as shown in the following code:

```
Employee john = new()
{
  Name = "John Jones",
  Born = new(year: 1990, month: 7, day: 28,
    hour: 0, minute: 0, second: 0, offset: TimeSpan.Zero)
};

john.WriteToConsole();
```

4. Run the PeopleApp project and view the result, as shown in the following output:

```
John Jones was born on a Saturday.
```

Note that the Employee class has inherited all the members of Person.

Extending classes to add functionality

Now, we will add some employee-specific members to extend the class:

1. In Employee.cs, add statements to define two properties, for an employee code and the date they were hired (we do not need to know a start time, so we can use the DateOnly type), as shown in the following code:

```
public string? EmployeeCode { get; set; }
public DateOnly HireDate { get; set; }
```

2. In Program.cs, add statements to set John's employee code and hire date, as shown in the following code:

```
john.EmployeeCode = "JJ001";
john.HireDate = new(year: 2014, month: 11, day: 23);
WriteLine($"{john.Name} was hired on {john.HireDate:yyyy-MM-dd}.");
```

3. Run the PeopleApp project and view the result, as shown in the following output:

```
John Jones was hired on 2014-11-23.
```

Hiding members

So far, the WriteToConsole method is inherited from Person, and it only outputs the employee's name and date and time of birth. We might want to change what this method does for an employee:

1. In Employee.cs, add statements to redefine the WriteToConsole method, as highlighted in the following code:

```
namespace Packt.Shared;

public class Employee : Person
{
  public string? EmployeeCode { get; set; }
  public DateOnly HireDate { get; set; }

  public void WriteToConsole()
  {
    WriteLine(format:
      "{0} was born on {1:dd/MM/yy} and hired on {2:dd/MM/yy}.",
```

```
        arg0: Name, arg1: Born, arg2: HireDate);
    }
}
```

2. Run the PeopleApp project, view the result, and note that the first line of output is before the employees were hired; hence, it has a default date, as shown in the following output:

```
John Jones was born on 28/07/90 and hired on 01/01/01.
John Jones was hired on 2014-11-23.
```

Your code editor warns you that your method now hides the method from Person by drawing a squiggle under the method name, the **PROBLEMS/Error List** window includes more details, and the compiler will output a warning when you build and run the console application, as shown in *Figure 6.3*:

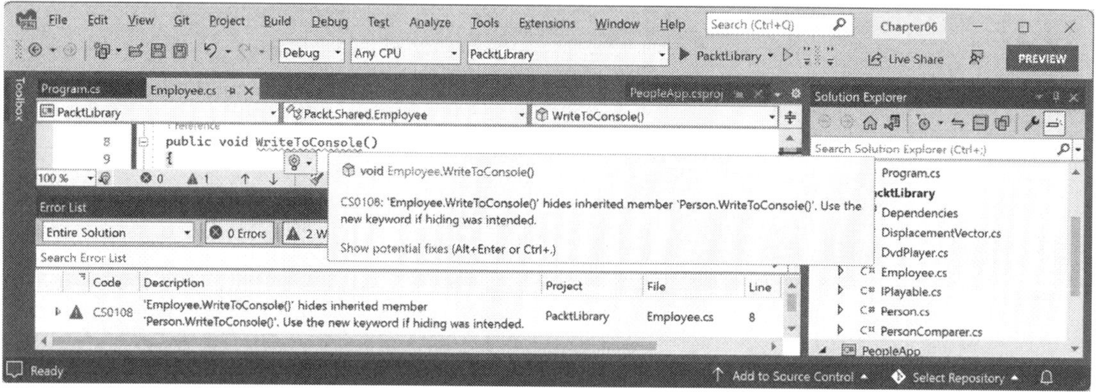

Figure 6.3: Hidden method warning

As the warning describes, you should hide this message by applying the new keyword to the method, indicating that you are deliberately replacing the old method, as highlighted in the following code:

```
public new void WriteToConsole()
```

Make this fix now.

Understanding the this and base keywords

There are two special C# keywords that can be used to refer to the current object instance or the base class that it inherits from:

- this: It represents the current object instance. For example, in the Person class instance members (but not in static members), you could use the expression this.Born to access the Born field of the current object instance. You rarely need to use it, since the expression Born would also work. It is only when there is a local variable also named Born that you would need to use this.Born to explicitly say you are referring to the field, not the local variable.
- base: It represents the base class that the current object inherits from. For example, anywhere in the Person class, you could use the expression base.ToString() to call the base class implementation of that method.

You will (hopefully) remember from *Chapter 5, Building Your Own Types with Object-Oriented Programming*, that to access static members, you must use the type name.

Overriding members

Rather than hiding a method, it is usually better to **override** it. You can only override it if the base class chooses to allow overriding, by applying the `virtual` keyword to any methods that should allow overriding.

Let's see an example:

1. In `Program.cs`, add a statement to write the value of the `john` variable to the console using its `string` representation, as shown in the following code:

```
WriteLine(john.ToString());
```

2. Run the `PeopleApp` project and note that the `ToString` method is inherited from `System.Object`, so the implementation returns the namespace and type name, as shown in the following output:

```
Packt.Shared.Employee
```

3. In `Person.cs` (not in the `Employee` class!), override this behavior by adding a `ToString` method to output the name of the person as well as the type name, as shown in the following code:

```
#region Overridden methods

public override string ToString()
{
  return $"{Name} is a {base.ToString()}.";
}

#endregion
```

The `base` keyword allows a subclass to access members of its superclass, that is, the **base class** that it inherits or derives from.

Good Practice: Some code editors like Visual Studio will automatically add a call to the base class member when you override a member. To decide if you need to keep the call or not, view the tooltip or definition of the member. For example, in *Chapter 10, Working with Data Using Entity Framework Core*, you will override a method of the `DbContext` class named `OnConfiguring`. If you hover over the method name, its tooltip tells you "The base implementation does nothing." If you **Go To Definition** (*F12*) of the method, you will see that it does nothing, as shown in the following code:

```
protected internal virtual void OnConfiguring(
DbContextOptionsBuilder optionsBuilder)
{
}
```

The method only exists in the base class so that subclasses can override it and then EF Core can call your code when it needs to configure the data context. It would be a waste for your overridden method implementation to call the base implementation.

4. Run the `PeopleApp` project and view the result. Now, when the `ToString` method is called, it outputs the person's name, as well as returning the base class's implementation of `ToString`, as shown in the following output:

```
John Jones is a Packt.Shared.Employee.
```

Good Practice: Many real-world APIs, for example, Microsoft's Entity Framework Core, Castle's DynamicProxy, and Optimizely CMS's content models, require the properties that you define in your classes to be marked as `virtual` so that they can be overridden. Carefully decide which of your methods and property members should be marked as `virtual`.

Inheriting from abstract classes

Earlier in this chapter, you learned about interfaces that can define a set of members that a type must have to meet a basic level of functionality. These are very useful, but their main limitation is that until C# 8 they could not provide any implementation of their own.

This is a particular problem if you still need to create class libraries that will work with .NET Framework and other platforms that do not support .NET Standard 2.1.

In those earlier platforms, you could use an **abstract class** as a sort of halfway house between a pure interface and a fully implemented class.

When a class is marked as `abstract`, this means that it cannot be instantiated because you have indicated that the class is not complete. It needs more implementation before it can be instantiated.

For example, the `System.IO.Stream` class is `abstract` because it implements common functionality that all streams would need but is not complete. Therefore, it is useless without more implementation that is specific to the type of stream, so you cannot instantiate it using `new Stream()`.

Let's compare the two types of interface and the two types of class, as shown in the following code:

```
public interface INoImplementation // C# 1 and later.
{
  void Alpha(); // Must be implemented by derived type.
}

public interface ISomeImplementation // C# 8 and later.
{
  void Alpha(); // Must be implemented by derived type.

  void Beta()
  {
    // Default implementation; can be overridden.
  }
}

public abstract class PartiallyImplemented // C# 1 and later.
{
  public abstract void Gamma(); // Must be implemented by derived type.

  public virtual void Delta() // Can be overridden.
  {
    // Implementation.
  }
}

public class FullyImplemented : PartiallyImplemented, ISomeImplementation
{
  public void Alpha()
  {
    // Implementation.
  }

  public override void Gamma()
  {
    // Implementation.
  }
```

```
}

// You can only instantiate the fully implemented class.
FullyImplemented a = new();

// All the other types give compile errors.
PartiallyImplemented b = new(); // Compile error!
ISomeImplementation c = new(); // Compile error!
INoImplementation d = new(); // Compile error!
```

Choosing between an interface and an abstract class

You have now seen examples of implementing the concept of abstraction using either an interface or an `abstract` class. Which should you pick? Now that an interface can have default implementations for its members, is the `abstract` keyword for a class obsolete?

Well, let's think about a real example. `Stream` is an `abstract` class. Would or could the .NET team use an interface for that today?

Every member of an interface must be `public` (or at least match the interface's access level, which could be `internal` if it should only be used in the class library that it's defined in). An `abstract` class has more flexibility in its members' access modifiers.

Another advantage of an `abstract` class over an interface is that serialization often does not work for an interface. So, no, we still need to be able to define abstract classes.

Preventing inheritance and overriding

You can prevent another developer from inheriting from your class by applying the `sealed` keyword to its definition. For example, no one can inherit from Scrooge McDuck, as shown in the following code:

```
public sealed class ScroogeMcDuck
{
}
```

An example of `sealed` in .NET is the `string` class. Microsoft has implemented some extreme optimizations inside the `string` class that could be negatively affected by your inheritance, so Microsoft prevents that.

You can prevent someone from further overriding a `virtual` method in your class by applying the `sealed` keyword to the method. For example, no one can change the way Lady Gaga sings, as shown in the following code:

```
namespace Packt.Shared;

public class Singer
{
```

```
    // Virtual allows this method to be overridden.
    public virtual void Sing()
    {
      WriteLine("Singing...");
    }
  }

  public class LadyGaga : Singer
  {
    // The sealed keyword prevents overriding the method in subclasses.
    public sealed override void Sing()
    {
      WriteLine("Singing with style...");
    }
  }
```

You can only seal an overridden method.

Understanding polymorphism

You have now seen two ways to change the behavior of an inherited method. We can *hide* it using the `new` keyword (known as **non-polymorphic inheritance**), or we can *override* it (known as **polymorphic inheritance**).

Both ways can access members of the base or superclass by using the `base` keyword, so what is the difference?

It all depends on the type of variable holding a reference to the object. For example, a variable of the `Person` type can hold a reference to a `Person` class, or any type that derives from `Person`.

Let's see how this could affect your code:

1. In `Employee.cs`, add statements to override the `ToString` method so that it writes the employee's name and code to the console, as shown in the following code:

   ```
   public override string ToString()
   {
     return $"{Name}'s code is {EmployeeCode}.";
   }
   ```

2. In `Program.cs`, write statements to create a new employee named Alice stored in a variable of type `Employee`. Also, store Alice in a second variable of type `Person`, and then call both variables' `WriteToConsole` and `ToString` methods, as shown in the following code:

   ```
   Employee aliceInEmployee = new()
     { Name = "Alice", EmployeeCode = "AA123" };
   ```

```
Person aliceInPerson = aliceInEmployee;
aliceInEmployee.WriteToConsole();
aliceInPerson.WriteToConsole();
WriteLine(aliceInEmployee.ToString());
WriteLine(aliceInPerson.ToString());
```

3. Run the `PeopleApp` project and view the result, as shown in the following output:

```
Alice was born on 01/01/01 and hired on 01/01/01
Alice was born on a Monday
Alice's code is AA123
Alice's code is AA123
```

When a method is hidden with `new`, the compiler is not smart enough to know that the object is an `Employee`, so it calls the `WriteToConsole` method in `Person`.

When a method is overridden with `virtual` and `override`, the compiler is smart enough to know that although the variable is declared as a `Person` class and the object itself is an `Employee` class, and therefore, the `Employee` implementation of `ToString` is called.

The member modifiers and the effect they have are summarized in *Table 6.5*:

Variable type	Member modifier	Method executed	In class
`Person`		`WriteToConsole`	`Person`
`Employee`	`new`	`WriteToConsole`	`Employee`
`Person`	`virtual`	`ToString`	`Employee`
`Employee`	`override`	`ToString`	`Employee`

Table 6.5: Member modifiers and the effect they have

In my opinion, polymorphism is academic to most programmers. If you get the concept, that's cool; but if not, I suggest that you don't worry about it. Some people like to make others feel inferior by saying understanding polymorphism is important for all C# programmers, but in my opinion, it's not. There are thousands of other topics that your time and effort will be better spent on.

You can have a successful career with C# and never need to be able to explain polymorphism, just as a racing car driver doesn't need to explain the engineering behind fuel injection.

Good Practice: You should use `virtual` and `override` rather than `new` to change the implementation of an inherited method whenever possible.

Casting within inheritance hierarchies

Casting between types is subtly different from converting between types. Casting is between similar types, like between a 16-bit integer and a 32-bit integer, or between a superclass and one of its subclasses. **Converting** is between dissimilar types, such as between text and a number.

For example, if you need to work with multiple types of `stream`, then instead of declaring specific types of stream like `MemoryStream` or `FileStream`, you could declare an array of `Stream`, the supertype of `MemoryStream`, and `FileStream`.

Implicit casting

In the previous example, you saw how an instance of a derived type can be stored in a variable of its base type (or its base's base type, and so on). When we do this, it is called **implicit casting.**

Explicit casting

The opposite of implicit casting is explicit casting, and you must use parentheses around the type you want to cast into as a prefix to do it:

1. In `Program.cs`, add a statement to assign the `aliceInPerson` variable to a new `Employee` variable, as shown in the following code:

```
Employee explicitAlice = aliceInPerson;
```

2. Your code editor displays a red squiggle and a compile error, as shown in *Figure 6.4*:

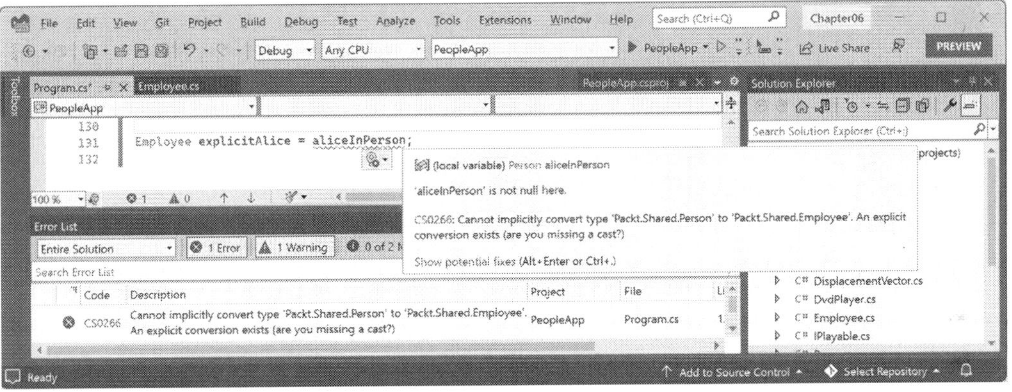

Figure 6.4: A missing explicit cast compile error

3. Change the statement to prefix the assigned variable name with a cast to the `Employee` type, as shown highlighted in the following code:

```
Employee explicitAlice = (Employee)aliceInPerson;
```

Avoiding casting exceptions

The compiler is now happy; however, because `aliceInPerson` might be a different derived type, like `Student` instead of `Employee`, we need to be careful. In a real application with more complex code, the current value of this variable could have been set to a `Student` instance, and then this statement would throw an `InvalidCastException` error at runtime.

Using is to check a type

We can handle this by writing a `try` statement, but there is a better way. We can check the type of an object using the `is` keyword:

1. Wrap the explicit cast statement in an `if` statement, as highlighted in the following code:

```
if (aliceInPerson is Employee)
{
  WriteLine($"{nameof(aliceInPerson)} is an Employee.");

  Employee explicitAlice = (Employee)aliceInPerson;

  // Safely do something with explicitAlice.
}
```

2. Run the `PeopleApp` project and view the result, as shown in the following output:

```
aliceInPerson is an Employee.
```

Good Practice: You could simplify the code further using a declaration pattern, and this will avoid the need to perform an explicit cast, as shown in the following code:

```
if (aliceInPerson is Employee explicitAlice)
{
  WriteLine($"{nameof(aliceInPerson)} is an Employee.");

  // Safely do something with explicitAlice.
}
```

This technique of both checking and casting was covered in *Chapter 3, Controlling Flow, Converting Types, and Handling Exceptions*.

What if you want to execute a block of statements when Alice is *not* an employee?

In the past, you would have had to use the `!` (not) operator, as shown in the following code:

```
if (!(aliceInPerson is Employee))
```

With C# 9 and later, you can use the `not` keyword, as shown in the following code:

```
if (aliceInPerson is not Employee)
```

Using as to cast a type

Alternatively, you can use the `as` keyword to cast a type. Instead of throwing an exception, the `as` keyword returns `null` if the type cannot be cast:

1. In `Program.cs`, add statements to cast Alice using the `as` keyword, and then check whether the return value is not null, as shown in the following code:

```
Employee? aliceAsEmployee = aliceInPerson as Employee;

if (aliceAsEmployee is not null)
{
  WriteLine($"{nameof(aliceInPerson)} as an Employee.");

  // Safely do something with aliceAsEmployee.
}
```

   Since accessing a member of a `null` variable will throw a `NullReferenceException` error, you should always check for `null` before using the result.

2. Run the `PeopleApp` project and view the result, as shown in the following output:

```
aliceInPerson as an Employee.
```

Good Practice: Use the `is` and `as` keywords to prevent throwing exceptions when casting between derived types. If you don't do this, you must write `try-catch` statements for `InvalidCastException`.

Inheriting and extending .NET types

.NET has pre-built class libraries containing hundreds of thousands of types. Rather than creating your own completely new types, you can often get a head start by deriving from one of Microsoft's types to inherit some or all its behavior, and then overriding or extending it.

Inheriting exceptions

As an example of inheritance, we will derive a new type of exception:

1. In the `PacktLibrary` project, add a new class file named `PersonException.cs`.
2. Modify the contents of the file to define a class named `PersonException` with three constructors, as shown in the following code:

```
namespace Packt.Shared;

public class PersonException : Exception
{
```

```
  public PersonException() : base() { }

  public PersonException(string message) : base(message) { }

  public PersonException(string message, Exception innerException)
    : base(message, innerException) { }
}
```

Unlike ordinary methods, constructors are not inherited, so we must explicitly declare and explicitly call the `base` constructor implementations in `System.Exception` (or whichever exception class you derived from) to make them available to programmers who might want to use those constructors with our custom exception.

3. In `Person.cs`, add statements to define a method that throws an exception if a date/time parameter is earlier than a person's date and time of birth, as shown in the following code:

```
public void TimeTravel(DateTime when)
{
  if (when <= Born)
  {
    throw new PersonException("If you travel back in time to a date
earlier than your own birth, then the universe will explode!");
  }
  else
  {
    WriteLine($"Welcome to {when:yyyy}!");
  }
}
```

4. In `Program.cs`, add statements to test what happens when employee John Jones tries to time-travel too far back, as shown in the following code:

```
try
{
  john.TimeTravel(when: new(1999, 12, 31));
  john.TimeTravel(when: new(1950, 12, 25));
}
catch (PersonException ex)
{
  WriteLine(ex.Message);
}
```

5. Run the PeopleApp project and view the result, as shown in the following output:

```
Welcome to 1999!
If you travel back in time to a date earlier than your own birth, then
the universe will explode!
```

Good Practice: When defining your own exceptions, give them the same three constructors that explicitly call the built-in ones in `System.Exception`. Other exceptions that you might inherit from may have more.

Extending types when you can't inherit

Earlier, we saw how the `sealed` modifier can be used to prevent inheritance.

Microsoft has applied the `sealed` keyword to the `System.String` class so that no one can inherit and potentially break the behavior of strings.

Can we still add new methods to strings? Yes, if we use a language feature named **extension methods**, which was introduced with C# 3.0. To properly understand extension methods, we need to review static methods first.

Using static methods to reuse functionality

Since the first version of C#, we've been able to create `static` methods to reuse functionality, such as the ability to validate that a `string` contains an email address. The implementation will use a regular expression that you will learn more about in *Chapter 8, Working with Common .NET Types*.

Let's write some code:

1. In the `PacktLibrary` project, add a new class file named `StringExtensions.cs`.
2. Modify `StringExtensions.cs`, as shown in the following code, and note the following:
    - The class imports a namespace to handle regular expressions.
    - The `IsValidEmail` method is `static`, and it uses the `Regex` type to check for matches against a simple email pattern that looks for valid characters before and after the @ symbol:

```
using System.Text.RegularExpressions; // To use Regex.

namespace Packt.Shared;

public class StringExtensions
{
  public static bool IsValidEmail(string input)
  {
```

```
      // Use a simple regular expression to check
      // that the input string is a valid email.

      return Regex.IsMatch(input,
        @"[a-zA-Z0-9\.-_]+@[a-zA-Z0-9\.-_]+");
    }
  }
```

3. In `Program.cs`, add statements to validate two examples of email addresses, as shown in the following code:

```
string email1 = "pamela@test.com";
string email2 = "ian&test.com";

WriteLine("{0} is a valid e-mail address: {1}",
  arg0: email1,
  arg1: StringExtensions.IsValidEmail(email1));

WriteLine("{0} is a valid e-mail address: {1}",
  arg0: email2,
  arg1: StringExtensions.IsValidEmail(email2));
```

4. Run the `PeopleApp` project and view the result, as shown in the following output:

```
pamela@test.com is a valid e-mail address: True
ian&test.com is a valid e-mail address: False
```

This works, but extension methods can reduce the amount of code we must type and simplify the usage of this function.

Using extension methods to reuse functionality

It is easy to turn `static` methods into extension methods:

1. In `StringExtensions.cs`, add the `static` modifier before the class, and then add the `this` modifier before the `string` type, as highlighted in the following code:

```
public static class StringExtensions
{
  public static bool IsValidEmail(this string input)
  {
```

Good Practice: These two changes tell the compiler that it should treat the method as one that extends the `string` type.

2. In `Program.cs`, add statements to use the extension method for `string` values that need to be checked for valid email addresses, as shown in the following code:

```
WriteLine("{0} is a valid e-mail address: {1}",
  arg0: email1,
  arg1: email1.IsValidEmail());

WriteLine("{0} is a valid e-mail address: {1}",
  arg0: email2,
  arg1: email2.IsValidEmail());
```

Note the subtle simplification in the syntax to call the `IsValidEmail` method. The older, longer syntax still works too.

3. The `IsValidEmail` extension method now appears to be a method just like all the actual instance methods of the `string` type, such as `IsNormalized`, except with a small down arrow on the method icon to indicate an extension method, as shown in *Figure 6.5*:

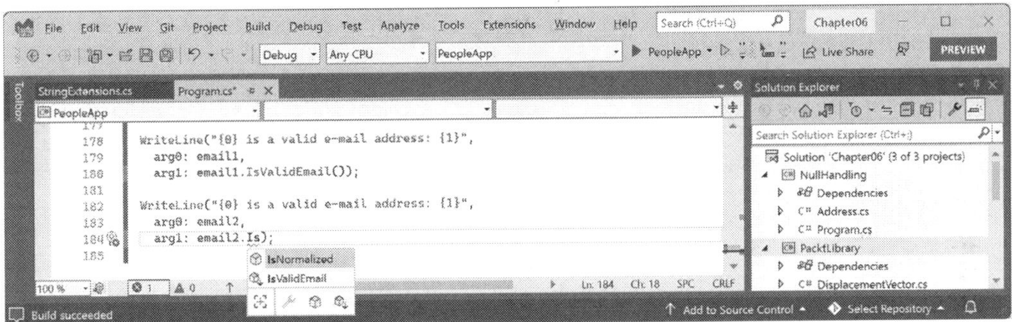

Figure 6.5: Extension methods appear in IntelliSense alongside instance methods

4. Run the `PeopleApp` project and view the result, which will be the same as before.

Good Practice: Extension methods cannot replace or override existing instance methods. You cannot, for example, redefine the `Insert` method. The extension method will appear as an overload in IntelliSense, but an instance method will be called in preference to an extension method with the same name and signature.

Although extension methods might not seem to give a big benefit, in *Chapter 11, Querying and Manipulating Data Using LINQ*, you will see some extremely powerful uses of extension methods.

Method chaining or fluent style

Method chaining, also known as fluent style, is a programming technique where multiple method calls are chained together in a single statement. This is achieved by having each method return an instance of the same object (often the object the method was called on), enabling a sequence of method calls on the same object. This style is popular in C# for creating readable and concise code, especially for configuring objects or building complex queries.

For example, consider this `Person` class:

```
public class Person
{
    public string Name { get; set; }
    public int Age { get; set; }

    public Person SetName(string name)
    {
        Name = name;
        return this;
    }

    public Person SetAge(int age)
    {
        Age = age;
        return this;
    }
}
```

To construct a `Person` instance, you can chain method calls, as shown in the following code:

```
Person person = new()
  .SetName("John Doe")
  .SetAge(30);
```

Extension methods allow adding new methods to existing types without modifying their source code or creating a new derived type. When combined with method chaining, this can enhance the readability and functionality of existing classes.

Suppose you have a `Car` class, as shown in the following code:

```
public class Car
{
  public string Model { get; set; }
  public string Color { get; set; }
}
```

You can create extension methods to enable method chaining for the `Car` class:

```
public static class CarExtensions
{
  public static Car SetModel(this Car car, string model)
  {
    car.Model = model;
    return car;
  }

  public static Car SetColor(this Car car, string color)
  {
    car.Color = color;
    return car;
  }
}
```

Now you can use these extension methods to chain method calls on a Car object:

```
Car car = new()
  .SetModel("Tesla Model S")
  .SetColor("Red");
```

The benefits of method chaining include:

- **Readability:** The code is more readable and resembles natural language
- **Conciseness:** Reduces the need for repetitive code
- **Configurability:** Allows the flexible and easy configuration of objects

But the potential drawbacks of method chaining include debugging difficulty because long chains can be harder to debug, and misleading code because if not used carefully, it can lead to less intuitive code, especially when dealing with nullable objects or complex logic.

Good practice: When method chaining, use clear method names to ensure they clearly indicate their purpose, return `this` or the object itself to maintain the chain, and carefully consider how to handle potential errors or exceptions within chained methods.

Method chaining and the fluent style, especially when combined with extension methods in C#, provide clean, readable, and maintainable code.

Summarizing custom type choices

Now that we have covered OOP and the C# features that enable you to define your own types, let's summarize what you've learned.

Categories of custom types and their capabilities

Categories of custom types and their capabilities are summarized in *Table 6.6*:

Type	Instantiation	Inheritance	Equality	Memory
`class`	Yes	Single	Reference	Heap
`sealed class`	Yes	None	Reference	Heap
`abstract class`	No	Single	Reference	Heap
`record` or `record class`	Yes	Single	Value	Heap
`struct` or `record struct`	Yes	None	Value	Stack
`interface`	No	Multiple	Reference	Heap

Table 6.6: Categories of custom types and their capabilities

It is best to think about these differences by starting with the "normal" case and then spotting the differences in other cases. For example, a "normal" `class` can be instantiated with `new`, it supports single inheritance, it uses memory reference equality, and its state is stored in heap memory.

Now let's highlight what is different about the more specialized types of classes:

- A `sealed` class does not support inheritance.
- An `abstract` class does not allow instantiation with `new`.
- A `record` class uses value equality instead of reference equality.

We can do the same for other types compared to a "normal" class:

- A `struct` or `record struct` does not support inheritance, it uses value equality instead of reference equality, and its state is stored in stack memory.
- An `interface` does not allow instantiation with `new` and supports multiple inheritance.

Mutability and records

A common misconception is that `record` types are immutable, meaning their instance property and field values cannot be changed after initialization. However, the mutability of a `record` type actually depends on how the `record` is defined. Let's explore mutability:

1. In the `PacktLibrary` project, add a new class file named `Mutability.cs`.
2. Modify `Mutability.cs`, as shown in the following code, and note the following:

```
namespace Packt.Shared;

// A mutable record class.
public record class C1
{
  public string? Name { get; set; }
}
```

```
// An immutable record class.
public record class C2(string? Name);

// A mutable record struct.
public record struct S1
{
  public string? Name { get; set; }
}

// Another mutable record struct.
public record struct S2(string? Name);

// An immutable record struct.
public readonly record struct S3(string? Name);
```

3. In the `PeopleApp` project, in `Program.cs`, create an instance of each type, setting the initial `Name` value to `Bob`, and then modify the `Name` property to `Bill`. You will see the two types that are immutable after initialization because they will give the compiler error `CS8852`, as shown in the following code:

```
C1 c1 = new() { Name = "Bob" };
c1.Name = "Bill";

C2 c2 = new(Name: "Bob");
c2.Name = "Bill"; // CS8852: Init-only property.

S1 s1 = new() { Name = "Bob" };
s1.Name = "Bill";

S2 s2 = new(Name: "Bob");
s2.Name = "Bill";

S3 s3 = new(Name: "Bob");
s3.Name = "Bill"; // CS8852: Init-only property.
```

4. Note that record `C1` is mutable and `C2` is immutable. Note that `S1` and `S2` are mutable and `S3` is immutable.
5. Comment out the two statements that cause compiler errors.

Microsoft made some interesting design choices with records. Make sure you remember the subtle differences in behavior when combining record, class, and struct, and use different types of declaration of each.

Comparing inheritance and implementation

For me, the terms *inherit* and *implement* are different, and in the early days of C# and .NET you could strictly apply them to classes and interfaces, respectively. For example, the `FileStream` class inherits from the `Stream` class, and the `Int32` `struct` implements the `IComparable` interface.

Inherit implies some functionality that a subclass gets "for free" by inheriting from its **base**, or **superclass**. *Implement* implies some functionality that is NOT inherited but instead MUST be provided by the subclass. This is why I chose to title this chapter *Implementing Interfaces and Inheriting Classes*.

Before C# 8, interfaces were always purely contracts. There was no functionality in an interface that you could inherit. In those days, you could strictly use the term *implement* for interfaces that represent a list of members that your type must implement, and *inherit* for classes with functionality that your type can inherit and potentially override.

With C# 8, interfaces can now include default implementations, making them more like abstract classes, and the term *inherit* for an interface that has default implementations does make sense. But I feel uncomfortable with this capability, as do many other .NET developers, because it messes up what used to be a clean language design. Default interfaces also require changes to the underlying .NET runtime, so they cannot be used with legacy platforms like .NET Standard 2.0 class libraries and .NET Framework.

Classes can also have abstract members, for example, methods or properties without any implementation, just like an interface could have. When a subclass inherits from this class, it MUST provide an implementation of those abstract members, and the base class must be decorated with the `abstract` keyword to prevent it from being instantiated using `new` because it is missing some functionality.

Reviewing illustrative code

Let's review some example code that illustrates some of the important differences between types.

Note the following:

- To simplify the code, I have left out access modifiers like `private` and `public`.
- Instead of normal brace formatting, to save vertical space I have put all the method implementations in one statement, for example:

  ```
  void M1() { /* implementation */ }
  ```

- Using "I" as a prefix for interfaces is a convention, not a requirement. It is useful to highlight interfaces using this prefix, since only interfaces support multiple inheritance.

Here's the code:

```
// These are both "classic" interfaces in that they are pure contracts.
// They have no functionality, just the signatures of members that
// must be implemented.

interface IAlpha
```

```
{
  // A method that must be implemented in any type that implements
  // this interface.
  void M1();
}

interface IBeta
{
  void M2(); // Another method.
}

// A type (a struct in this case) implementing an interface.
// ": IAlpha" means Gamma promises to implement all members of IAlpha.

struct Gamma : IAlpha
{
  void M1() { /* implementation */ }
}

// A type (a class in this case) implementing two interfaces.

class Delta : IAlpha, IBeta
{
  void M1() { /* implementation */ }
  void M2() { /* implementation */ }
}

// A sub class inheriting from a base aka super class.
// ": Delta" means inherit all members from Delta.

class Episilon : Delta
{
  // This can be empty because this inherits M1 and M2 from Delta.
  // You could also add new members here.
}

// A class with one inheritable method and one abstract method
// that must be implemented in sub classes. A class with at least
// one abstract member must be decorated with the abstract keyword
// to prevent instantiation.
```

```
abstract class Zeta
{
  // An implemented method would be inherited.
  void M3() { /* implementation */ }

  // A method that must be implemented in any type that inherits
  // this abstract class.
  abstract void M4();
}

// A class inheriting the M3 method from Zeta but it must provide
// an implementation for M4.

class Eta : Zeta
{
  void M4() { /* implementation */ }
}

// In C# 8 and later, interfaces can have default implementations
// as well as members that must be implemented.
// Requires: .NET Standard 2.1, .NET Core 3.0 or later.

interface ITheta
{
  void M3() { /* implementation */ }
  void M4();
}

// A class inheriting the default implementation from an interface
// and must provide an implementation for M4.

class Iota : ITheta
{
  void M4() { /* implementation */ }
}
```

Practicing and exploring

Test your knowledge and understanding by answering some questions, getting some hands-on practice, and exploring this chapter's topics with more in-depth research.

Exercise 6.1 – Online material

Online material can be extra content written by me for this book, or it can be references to content created by Microsoft or third parties.

Managing memory with reference and value types

Read the following online-only section to learn how to manage memory with reference and value types:

`https://github.com/markjprice/cs13net9/blob/main/docs/ch06-memory.md`

Writing better code

Read the following online-only section to learn how to use analyzers to write better code:

`https://github.com/markjprice/cs13net9/blob/main/docs/ch06-writing-better-code.md`

Exercise 6.2 – Practice creating an inheritance hierarchy

Explore inheritance hierarchies by following these steps:

1. Add a new console app named `Exercise_Inheritance` to your `Chapter06` solution.
2. Create a class named `Shape` with properties named `Height`, `Width`, and `Area`.
3. Add three classes that derive from it—`Rectangle`, `Square`, and `Circle`—with any additional members you feel are appropriate and that override and implement the `Area` property correctly.
4. In `Program.cs`, add statements to create one instance of each shape, as shown in the following code:

```
Rectangle r = new(height: 3, width: 4.5);
WriteLine($"Rectangle H: {r.Height}, W: {r.Width}, Area: {r.Area}");

Square s = new(5);
WriteLine($"Square H: {s.Height}, W: {s.Width}, Area: {s.Area}");

Circle c = new(radius: 2.5);
WriteLine($"Circle H: {c.Height}, W: {c.Width}, Area: {c.Area}");
```

5. Run the console app and ensure that the result looks like the following output:

```
Rectangle H: 3, W: 4.5, Area: 13.5
Square H: 5, W: 5, Area: 25
Circle H: 5, W: 5, Area: 19.6349540849362
```

Exercise 6.3 – Test your knowledge

Answer the following questions:

1. What is a delegate?
2. What is an event?

3. How are a base class and a derived class related, and how can the derived class access the base class?
4. What is the difference between the `is` and `as` operators?
5. Which keyword is used to prevent a class from being derived from or a method from being further overridden?
6. Which keyword is used to prevent a class from being instantiated with the `new` keyword?
7. Which keyword is used to allow a member to be overridden?
8. What's the difference between a destructor and a deconstruct method?
9. What are the signatures of the constructors that all exceptions should have?
10. What is an extension method, and how do you define one?

Exercise 6.4 – Explore topics

Use the links on the following page to learn more about the topics covered in this chapter:

```
https://github.com/markjprice/cs13net9/blob/main/docs/book-links.md#chapter-6---implementing-interfaces-and-inheriting-classes
```

Summary

In this chapter, you learned about:

- Operators
- Generic types
- Delegates and events
- Implementing interfaces
- Memory usage differences between reference and value types
- Working with null values
- Deriving and casting types using inheritance
- Base and derived classes, how to override a type member, and using polymorphism

In the next chapter, you will learn how .NET is packaged and deployed, and in subsequent chapters, the types that it provides you with to implement common functionality, such as file handling and database access.

7

Packaging and Distributing .NET Types

This chapter is about how C# keywords are related to .NET types and the relationship between namespaces and assemblies. You'll become familiar with how to package and publish your .NET apps and libraries for cross-platform use.

There are several online sections for this chapter, and you will find links to them all at the end of the chapter.

In an online-only section, *Decompiling .NET assemblies,* we cover how to decompile .NET assemblies for learning purposes and why you cannot prevent others from decompiling your code. In another online-only section, *Porting from .NET Framework to modern .NET,* you can learn how to use legacy .NET Framework libraries in .NET libraries and how it is possible to port legacy .NET Framework code bases to modern .NET. Finally, in a third online-only section, *Creating source generators,* you will learn how to create source generators that can dynamically add source code to your projects—a very powerful feature.

This chapter covers the following topics:

- The road to .NET 9
- Understanding .NET components
- Publishing your applications for deployment
- Native ahead-of-time compilation
- Packaging your libraries for NuGet distribution
- Working with preview features

The road to .NET 9

This part of the book is about the functionality in the **Base Class Library** (**BCL**) APIs provided by .NET and how to reuse functionality across all the different .NET platforms, using .NET Standard.

.NET Standard-compatible frameworks

From .NET Core 2.0 onward, the support for a minimum of .NET Standard 2.0 is important because it provides many of the APIs that were missing from the first version of .NET Core. The 15 years' worth of libraries and applications that .NET Framework developers had available to them that are relevant for modern development have now been migrated to .NET, and they can run cross-platform on macOS and Linux variants, as well as on Windows.

.NET Standard 2.1 added about 3,000 new APIs. Some of those APIs need runtime changes that would break backward compatibility. To summarize:

- .NET Framework 4.8 only implements .NET Standard 2.0.
- .NET Core 3.0, Xamarin, Mono, and Unity implement .NET Standard 2.1.

.NET 5 removed the need for .NET Standard because all project types can now target a single version of .NET. The same applies to .NET 6 and later. Each version from .NET 5 onward is backward compatible with previous versions. This means a class library that targets .NET 5 can be used by any .NET 5 or later projects of any type. Now that .NET versions have been released with full support for mobile and desktop apps built using .NET MAUI, the need for .NET Standard has been further reduced.

Since you might still need to create class libraries for legacy .NET Framework projects or legacy Xamarin mobile apps, there is still a need to create .NET Standard 2.0 class libraries.

Officially, you must also use a .NET Standard 2.0 class library to create a source generator even though it's a rather modern feature of .NET.

To summarize the progress that .NET has made since the first version of .NET Core in 2016, I have compared the major .NET Core and modern .NET versions with the equivalent .NET Framework versions in the following list:

- **.NET Core 1.x**: Much smaller API compared to .NET Framework 4.6.1, which was the current version in March 2016.
- **.NET Core 2.x**: Reached API parity with .NET Framework 4.7.1 for modern APIs because they both implement .NET Standard 2.0.
- **.NET Core 3.x**: Larger API compared to .NET Framework for modern APIs because .NET Framework 4.8 does not implement .NET Standard 2.1.
- **.NET 5**: Even larger API compared to .NET Framework 4.8 for modern APIs, with much-improved performance.
- **.NET 6**: Continued improvements to performance and expanded APIs, with optional support for mobile apps in .NET MAUI, which was added in May 2022.
- **.NET 7**: Final unification with support for mobile apps, with .NET MAUI available as an optional workload. This book does not cover .NET MAUI development. Packt has multiple books that specialize in .NET MAUI, and you can find them by searching their website.

- **.NET 8:** Continued to improve the platform and should be used for all new development if you need long-term support.
- **.NET 9:** Continues to improve the platform, especially with performance, and should be used if you are comfortable with standard-term support. You can see the .NET 9 release notes for each preview at the following link: `https://github.com/dotnet/core/discussions/9234`.

You can read more details in the GitHub repository at the following link: `https://github.com/markjprice/cs13net9/blob/main/docs/ch07-features.md`.

More Information: A useful way to compare two versions of .NET is the APIs of the .NET website. For example, you can compare what's new in .NET 9 compared to .NET 8 at the following link: `https://apisof.net/catalog?diff=net8.0-vs-net9.0`.

Checking your .NET SDKs for updates

Microsoft introduced a command with .NET 6 to check the versions of .NET SDKs and runtimes that you have installed, and it warns you if any need updating. For example, enter the following command:

```
dotnet sdk check
```

You will see results, including the status of available updates, as shown in the following partial output:

```
.NET SDKs:
Version             Status
------------------------------------------------------------
8.0.400             Patch 8.0.401 is available.
9.0.100             Up to date.
```

Good Practice: To maintain support from Microsoft, you must keep your .NET SDKs and .NET runtimes up to date with the latest patches.

A related third-party tool is `dotnet-outdated`, which allows you to quickly report on any outdated NuGet packages in your .NET projects. You can learn more about it at the following link: `https://github.com/dotnet-outdated/dotnet-outdated`.

Understanding .NET components

.NET is made up of several pieces, which are shown in the following list:

- **Language compilers:** These turn your source code, written with languages such as C#, F#, and Visual Basic, into **intermediate language (IL)** code stored in assemblies. With C# 6 and later, Microsoft switched to an open-source rewritten compiler known as **Roslyn**, which is also used by Visual Basic.

- **Common Language Runtime** (**CLR**): This runtime loads assemblies, compiles the IL code stored in them into native code instructions for your computer's CPU, and executes the code within an environment that manages resources, such as threads and memory.
- **Base Class Libraries** (**BCL**): These are prebuilt assemblies of types packaged and distributed using NuGet to perform common tasks when building applications. You can use them to quickly build anything you want, rather like combining LEGO™ pieces.

Assemblies, NuGet packages, and namespaces

An **assembly** is where a type is stored in the filesystem. Assemblies are a mechanism for deploying code. For example, the `System.Data.dll` assembly contains types to manage data. To use types in other assemblies, they must be referenced. Assemblies can be static (pre-created) or dynamic (generated at runtime). Dynamic assemblies are an advanced feature that we will not cover in this book. Assemblies can be compiled into a single file as a DLL (class library) or an EXE (console app).

Assemblies are distributed as **NuGet packages**, which are files that are downloadable from public online feeds and can contain multiple assemblies and other resources. You will also hear about **project SDKs**, **workloads**, and **platforms**, which are combinations of NuGet packages.

Microsoft's NuGet feed is found here: `https://www.nuget.org/`.

What is a namespace?

A namespace is the address of a type. Namespaces are a mechanism to uniquely identify a type by requiring a full address, rather than just a short name. In the real world, *Bob of 34 Sycamore Street* is different from *Bob of 12 Willow Drive*.

In .NET, the `IActionFilter` interface of the `System.Web.Mvc` namespace is different from the `IActionFilter` interface of the `System.Web.Http.Filters` namespace.

Dependent assemblies

If an assembly is compiled as a class library and provides types for other assemblies to use, then it has the file extension `.dll` (**dynamic link library**), and it cannot be executed standalone.

Likewise, if an assembly is compiled as an application, then it has the file extension `.exe` (**executable**) and can be executed standalone. Before .NET Core 3, console apps were compiled to `.dll` files and had to be executed by the `dotnet run` command or a host executable.

Any assembly can reference one or more class library assemblies as dependencies, but you cannot have circular references. So assembly *B* cannot reference assembly *A* if assembly *A* already references assembly *B*. The compiler will warn you if you attempt to add a dependency reference that would cause a circular reference. Circular references are often a warning sign of poor code design. If you are sure that you need a circular reference, then use an interface to solve it.

Microsoft .NET project SDKs

By default, console applications have a dependency reference on the Microsoft .NET project SDK. This platform contains thousands of types in NuGet packages that almost all applications would need, such as the `System.Int32` and `System.String` types.

When using .NET, you reference the dependency assemblies, NuGet packages, and platforms that your application needs in a project file.

Let's explore the relationship between assemblies and namespaces:

1. Use your preferred code editor to create a new project, as defined in the following list:
    - Project template: **Console App** / `console`
    - Project file and folder: `AssembliesAndNamespaces`
    - Solution file and folder: `Chapter07`
2. Open `AssembliesAndNamespaces.csproj`, and note that it is a typical project file for a .NET application, as shown in the following markup:

```
<Project Sdk="Microsoft.NET.Sdk">

  <PropertyGroup>
    <OutputType>Exe</OutputType>
    <TargetFramework>net9.0</TargetFramework>
    <Nullable>enable</Nullable>
    <ImplicitUsings>enable</ImplicitUsings>
  </PropertyGroup>

</Project>
```

3. After the `<PropertyGroup>` section, add a new `<ItemGroup>` section to statically import `System.Console` for all C# files, using the implicit usings .NET SDK feature, as shown in the following markup:

```
<ItemGroup Label="To simplify use of Console methods.">
  <Using Include="System.Console" Static="true" />
</ItemGroup>
```

PropertyGroup element

The `PropertyGroup` element is used to define configuration properties that control the build process. These properties can include anything, from the output path of the compiled binaries to versioning information. Each property within a `PropertyGroup` is defined as a simple name-value pair, as shown in the following markup:

```
<PropertyGroup>
<OutputType>Exe</OutputType>
```

```
<TargetFramework>net9.0</TargetFramework>
<ImplicitUsings>enable</ImplicitUsings>
<Nullable>enable</Nullable>
<RootNamespace>Packt.Shared</RootNamespace>
</PropertyGroup>
```

In the preceding example:

- `OutputType` specifies the type of output, such as an executable (`Exe`) or a library (`Library`).
- `TargetFramework` specifies the target framework for the project, for example, `net9.0`.
- `RootNamespace` sets the default namespace for the project.

ItemGroup element

The `ItemGroup` element is used to define collections of items that are included in the build process. Items can be source files, references to other assemblies, package references, and other resources.

You often want to define multiple `ItemGroup` elements for different purposes. They will all be merged at build time. For example, you might have one `ItemGroup` for all your project references and a separate `ItemGroup` for all your package references.

The `ItemGroup` element can have a `Label` attribute so that you can document what each section should be used for, as shown in the following markup:

```
<ItemGroup Label="Additional files to include during build.">
  <Compile Include="Utils.cs" />
  <None Include="readme.txt" />
</ItemGroup>

<ItemGroup Label="NuGet packages for this project.">
  <PackageReference Include="Newtonsoft.Json" Version="13.0.3" />
</ItemGroup>
```

Label and Condition attributes

Both `PropertyGroup` and `ItemGroup` can have a `Label` attribute that can be used to document the purpose of the element, and a `Condition` attribute that can be used to define when they apply. For example, a typical condition might be when building a `DEBUG` release, as shown in the following markup:

```
<PropertyGroup Condition="'$(CompileConfig)' == 'DEBUG'" >
  <Optimization>false</Optimization>
  <Obfuscate>false</Obfuscate>
  <OutputPath>$(OutputPath)\debug</OutputPath>
</PropertyGroup>
```

Warning! PropertyGroup and ItemGroup elements are both at the same level within a root Project element. Do not nest ItemGroup elements within PropertyGroup elements, or vice versa!

Namespaces and types in assemblies

Many common .NET types are in the System.Runtime.dll assembly. There is not always a one-to-one mapping between assemblies and namespaces. A single assembly can contain many namespaces, and a namespace can be defined in many assemblies. You can see the relationship between some assemblies and the namespaces that they supply types for in *Table 7.1*:

Assembly	Example namespaces	Example types
System.Runtime.dll	System, System.Collections, System.Collections.Generic	Int32, String, IEnumerable<T>
System.Console.dll	System	Console
System.Threading.dll	System.Threading	Interlocked, Monitor, Mutex
System.Xml.XDocument.dll	System.Xml.Linq	XDocument, XElement, XNode

Table 7.1: Examples of assemblies and their namespaces

NuGet packages

.NET is split into a set of packages, distributed using a Microsoft-supported package management technology named NuGet. Each of these packages represents a single assembly of the same name. For example, the System.Collections package contains the System.Collections.dll assembly.

The following are the benefits of packages:

- Packages can be easily distributed on public feeds.
- Packages can be reused.
- Packages can ship on their own schedule.
- Packages can be tested independently of other packages.
- Packages can support different OSes and CPUs by including multiple versions of the same assembly, built for different OSes and CPUs.
- Packages can have dependencies specific to only one library.
- Apps are smaller because unreferenced packages aren't part of the distribution. *Table 7.2* lists some of the more important packages and their important types:

Package	Important types
System.Runtime	Object, String, Int32, Array
System.Collections	List<T>, Dictionary<TKey, TValue>
System.Net.Http	HttpClient, HttpResponseMessage

System.IO.FileSystem	File, Directory
System.Reflection	Assembly, TypeInfo, MethodInfo

Table 7.2: Some important packages and their important types

Package sources

Package sources are locations where NuGet can look for packages, such as `https://www.nuget.org`, local directories, or other package repositories. The `nuget.config` file allows you to specify, prioritize, and configure these sources, as well as other related settings.

The `nuget.config` file uses XML to define package sources, as shown in the following markup:

```
<?xml version="1.0" encoding="utf-8"?>
<configuration>
  <packageSources>
    <add key="nuget.org"
         value="https://api.nuget.org/v3/index.json" />
    <add key="MyCustomSource"
         value="https://mycustomsource.com/nuget" />
  </packageSources>
</configuration>
```

In the preceding configuration:

- `<packageSources>`: This section lists all the sources NuGet will use to look for packages.
- `<add key="sourceName" value="sourceUrl"/>`: Each `<add>` element defines a package source.

NuGet can use multiple `nuget.config` files located in different directories. The settings are applied in a specific order, with the most specific settings taking precedence:

1. Project directory: `nuget.config` in the project directory.
2. Solution directory: `nuget.config` in the solution directory.
3. User directory: `nuget.config` in the user's profile directory—for example, `%AppData%\nuget\nuget.config` on Windows.
4. Machine-wide configuration: `nuget.config` in the machine-wide settings—for example, `%ProgramFiles(x86)%\nuget\config` on Windows.

NuGet merges these configurations, with settings from more specific files overriding those from less specific files.

If no `nuget.config` file is found, NuGet uses the default package source, which is usually the official NuGet.org repository. This means that, by default, tools like Visual Studio, `dotnet`, and `nuget.exe` will look for packages on NuGet.org unless otherwise configured.

Understanding frameworks

There is a two-way relationship between frameworks and packages. Packages define the APIs, while frameworks group packages. A framework without any packages would not define any APIs.

.NET packages each support a set of frameworks. For example, the `System.IO.FileSystem` package version 4.3.0 supports the following frameworks:

- .NET Standard, version 1.3 or later
- .NET Framework, version 4.6 or later
- Six Mono and Xamarin platforms (for example, Xamarin.iOS)

More Information: You can read the details at the following link: `https://www.nuget.org/packages/System.IO.FileSystem/#supportedframeworks-body-tab`.

Importing a namespace to use a type

Let's explore how namespaces are related to assemblies and types:

1. In the `AssembliesAndNamespaces` project, in `Program.cs`, delete the existing statements, and then enter the following code:

```
XDocument doc = new();
```

Recent versions of code editors will often automatically add a namespace import statement to fix the problem I want you to see. Please delete the `using` statement that your code editor writes for you.

2. Build the project and note the compiler error message, as shown in the following output:

```
CS0246 The type or namespace name 'XDocument' could not be found (are you missing a using directive or an assembly reference?)
```

The `XDocument` type is not recognized because we have not told the compiler what the namespace of the type is. Although this project already has a reference to the assembly that contains the type, we also need to either prefix the type name with its namespace (for example, `System.Xml.Linq.XDocument`) or import the namespace.

3. Click inside the `XDocument` class name. Your code editor displays a light bulb, showing that it recognizes the type and can automatically fix the problem for you.
4. Click the light bulb, and select `using System.Xml.Linq;` from the menu.

This will *import the namespace* by adding a `using` statement to the top of the file. Once a namespace is imported at the top of a code file, all the types within the namespace are available for use in that code file by just typing their name, without the type name needing to be fully qualified by prefixing it with its namespace.

I like to add a comment after importing a namespace to remind me why I need to import it, as shown in the following code:

```
using System.Xml.Linq; // To use XDocument.
```

If you do not comment on your namespaces, you or other developers will not know why they are imported and might delete them, breaking the code. Conversely, they might never delete imported namespaces "just in case" they are needed, potentially cluttering the code unnecessarily. This is why most modern code editors have features to remove unused namespaces. This technique also subconsciously trains you, while you are learning, to remember which namespace you need to import to use a particular type or extension method.

Relating C# keywords to .NET types

One of the common questions I get from new C# programmers is, "What is the difference between `string` with a lowercase s and `String` with an uppercase S?"

The short answer is easy: none. The long answer is that all C# keywords that represent types like `string` or `int` are aliases for a .NET type in a class library assembly.

When you use the `string` keyword, the compiler recognizes it as a `System.String` type. When you use the `int` type, the compiler recognizes it as a `System.Int32` type.

Let's see this in action with some code:

1. In `Program.cs`, declare two variables to hold `string` values, one using lowercase `string` and one using uppercase `String`, as shown in the following code:

   ```
   string s1 = "Hello";
   String s2 = "World";
   WriteLine($"{s1} {s2}");
   ```

2. Run the `AssembliesAndNamespaces` project, and note that `string` and `String` both work and literally mean the same thing.
3. In `AssembliesAndNamespaces.csproj`, add an entry to prevent the `System` namespace from being globally imported, as shown in the following markup:

   ```
   <ItemGroup>
     <Using Remove="System" />
   </ItemGroup>
   ```

4. In `Program.cs`, and the **Error List** or **PROBLEMS** window, note the compiler error message, as shown in the following output:

```
CS0246 The type or namespace name 'String' could not be found (are you missing a using directive or an assembly reference?)
```

5. At the top of `Program.cs`, import the `System` namespace with a `using` statement that will fix the error, as shown in the following code:

```
using System; // To use String.
```

Good Practice: When you have a choice, use the C# keyword instead of the actual type because the keywords do not need a namespace to be imported.

Mapping C# aliases to .NET types

Table 7.3 shows the 18 C# type keywords along with their actual .NET types:

Keyword	.NET type		Keyword	.NET type
`string`	`System.String`		`char`	`System.Char`
`sbyte`	`System.SByte`		`byte`	`System.Byte`
`short`	`System.Int16`		`ushort`	`System.UInt16`
`int`	`System.Int32`		`uint`	`System.UInt32`
`long`	`System.Int64`		`ulong`	`System.UInt64`
`nint`	`System.IntPtr`		`nuint`	`System.UIntPtr`
`float`	`System.Single`		`double`	`System.Double`
`decimal`	`System.Decimal`		`bool`	`System.Boolean`
`object`	`System.Object`		`dynamic`	`System.Dynamic.DynamicObject`

Table 7.3: C# type keywords and their actual .NET types

Other .NET programming language compilers can do the same thing. For example, the Visual Basic .NET language has a type named `Integer`, which is its alias for `System.Int32`.

Understanding native-sized integers

C# 9 introduced the `nint` and `nuint` keyword aliases for **native-sized integers**, meaning that the storage size for the integer value is platform-specific. They store a 32-bit integer in a 32-bit process and `sizeof()` returns 4 bytes; they store a 64-bit integer in a 64-bit process and `sizeof()` returns 8 bytes. The aliases represent pointers to the integer value in memory, which is why their .NET names are `IntPtr` and `UIntPtr`. The actual storage type will be either `System.Int32` or `System.Int64`, depending on the process.

In a 64-bit process, the following code:

```
WriteLine($"Environment.Is64BitProcess = {Environment.Is64BitProcess}");
WriteLine($"int.MaxValue = {int.MaxValue:N0}");
WriteLine($"nint.MaxValue = {nint.MaxValue:N0}");
```

produces this output:

```
Environment.Is64BitProcess = True
int.MaxValue = 2,147,483,647
nint.MaxValue = 9,223,372,036,854,775,807
```

Revealing the location of a type

Most code editors provide built-in documentation for .NET types. Let's start by making sure that you have the expected experience with Visual Studio by checking that you have the relevant option set correctly, and then explore:

1. If you are using Visual Studio, then make sure you have disabled the Source Link feature:
    a. Navigate to **Tools** | **Options**.
    b. In the search box, enter `navigation to source`.
    c. Select **Advanced** in the **Text Editor** | **C#** section.
    d. Clear the **Enable navigation to Source Link and Embedded sources** checkbox, and then click **OK**.
2. Right-click on `XDocument` and choose **Go to Definition**.
3. Navigate to the top of the code file, expand the collapsed region, and note that the assembly filename is `System.Xml.XDocument.dll` but the class is in the `System.Xml.Linq` namespace, as shown in the following code and *Figure 7.1*:

```
#region Assembly System.Runtime, Version=9.0.0.0, Culture=neutral,
PublicKeyToken=b03f5f7f11d50a3a
// C:\Program Files\dotnet\packs\Microsoft.NETCore.App.Ref\9.0.0\ref\
net9.0\System.Runtime.dll
#endregion
```

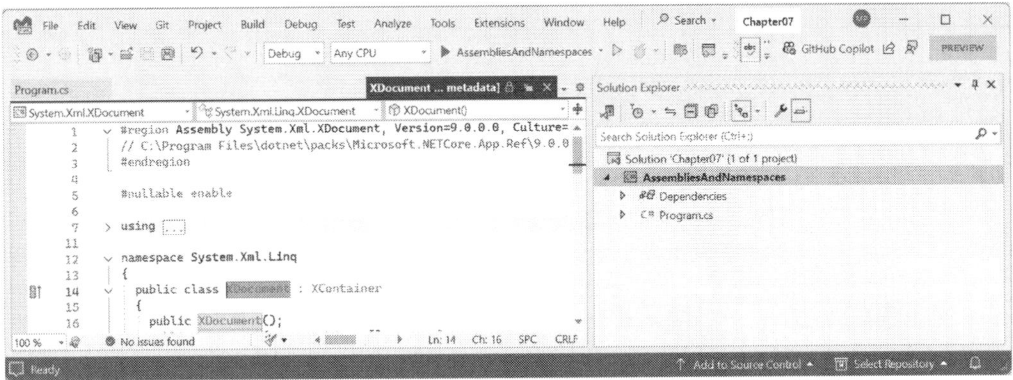

Figure 7.1: Assembly and namespace that contains the XDocument type

4. Close the **XDocument [from metadata]** tab.
5. Right-click inside `string` or `String` and choose **Go to Definition.**
6. Navigate to the top of the code file, expand the collapsed region, and note that the assembly filename is `System.Runtime.dll` but the class is in the `System` namespace.

Your code editor is technically lying to you. If you remember when we wrote code in *Chapter 2, Speaking C#*, when we revealed the extent of the C# vocabulary, we discovered that the `System.Runtime.dll` assembly contains zero types.

What the `System.Runtime.dll` assembly does contain are type-forwarders. These are special types that appear to exist in an assembly but are implemented elsewhere. In this case, they are implemented deep inside the .NET runtime using highly optimized code.

You might want to use a type-forwarder if you refactor a type to move it from its original assembly to a different one. Without defining a type-forwarder, any projects that reference the original assembly will not find the type in it, and a runtime exception will be thrown. You can read more about this contrived example at the following link: `https://learn.microsoft.com/en-us/dotnet/standard/assembly/type-forwarding`.

Sharing code with legacy platforms using .NET Standard

Before .NET Standard, there were **Portable Class Libraries** (**PCLs**). With PCLs, you could create a library of code and explicitly specify which platforms you want the library to support, such as Xamarin, Silverlight, and Windows 8. Your library could then use the intersection of APIs that are supported by the specified platforms.

Microsoft realized that this was unsustainable, so they created .NET Standard—a single API that all future .NET platforms would support. There are older versions of .NET Standard, but .NET Standard 2.0 was an attempt to unify all important recent .NET platforms. .NET Standard 2.1 was released in late 2019, but only .NET Core 3.0 and that year's version of Xamarin support its new features. For the rest of this book, I will use the term .NET Standard to mean .NET Standard 2.0.

.NET Standard is like HTML5 in that they are both standards that a platform should support. Just as Google's Chrome browser and Microsoft's Edge browser implement the HTML5 standard, .NET Core, .NET Framework, and Xamarin all implement .NET Standard. If you want to create a library of types that will work across variants of legacy .NET, you can do so most easily with .NET Standard.

Good Practice: Since many of the API additions in .NET Standard 2.1 required runtime changes, and .NET Framework is Microsoft's legacy platform, which needs to remain as unchanged as possible, .NET Framework 4.8 remained on .NET Standard 2.0 rather than implementing .NET Standard 2.1. If you need to support .NET Framework customers, then you should create class libraries on .NET Standard 2.0, even though it is not the latest and does not support all the recent language and BCL new features.

Your choice of which .NET Standard version to target comes down to a balance between maximizing platform support and available functionality. An older version supports more platforms but has a smaller set of APIs. A newer version supports fewer platforms but has a larger set of APIs. Generally, you should choose the lowest version that supports all the APIs that you need.

Understanding defaults for class libraries with different SDKs

When using the `dotnet` SDK tool to create a class library, it might be useful to know which target framework will be used by default, as shown in *Table 7.4*:

SDK	Default target framework for new class libraries
.NET Core 3.1	`netstandard2.0`
.NET 6	`net6.0`
.NET 7	`net7.0`
.NET 8	`net8.0`
.NET 9	`net9.0`

Table 7.4: .NET SDKs and their default target framework for new class libraries

Of course, just because a class library targets a specific version of .NET by default, it does not mean that you cannot change it after creating a class library project using the default template.

You can manually set the target framework to a value that supports the projects that need to reference that library, as shown in *Table 7.5*:

Class library target framework	Can be used by projects that target the following
`netstandard2.0`	.NET Framework 4.6.1 or later, .NET Core 2 or later, .NET 5 or later, Mono 5.4 or later, Xamarin.Android 8 or later, and Xamarin.iOS 10.14 or later
`netstandard2.1`	.NET Core 3 or later, .NET 5 or later, Mono 6.4 or later, Xamarin.Android 10 or later, and Xamarin.iOS 12.16 or later

net6.0	.NET 6 or later
net7.0	.NET 7 or later
net8.0	.NET 8 or later
net9.0	.NET 9 or later

Table 7.5: Class library target frameworks and the projects that can use them

Good Practice: Always check the target framework of a class library, and then manually change it to something more appropriate if necessary. Make a conscious decision about what it should be rather than accepting the default.

Creating a .NET Standard class library

We will create a class library using .NET Standard 2.0 so that it can be used across all important .NET legacy platforms and cross-platform on Windows, macOS, and Linux OSes, while also having access to a wide set of .NET APIs:

1. Use your preferred code editor to add a new **Class Library** / `classlib` project, named `SharedLibrary`, that targets .NET Standard 2.0 for the `Chapter07` solution:
   - If you use Visual Studio, when prompted for the **Target Framework**, select **.NET Standard 2.0**, and then configure the startup project for the solution to the current selection.
   - If you use VS Code, include a switch to target .NET Standard 2.0, as shown in the following command:

   ```
   dotnet new classlib -f netstandard2.0
   ```

   Good Practice: If you need to create types that use new features in .NET 9, as well as types that only use .NET Standard 2.0 features, then you can create two separate class libraries: one targeting .NET Standard 2.0 and one targeting .NET 0.

2. Build the `SharedLibrary` project.

An alternative to manually creating two class libraries is to create one that supports **multi-targeting**. If you would like me to add a section about multi-targeting to the next edition, please let me know. You can read about multi-targeting here: `https://learn.microsoft.com/en-us/dotnet/standard/library-guidance/cross-platform-targeting#multi-targeting`.

Controlling the .NET SDK

By default, executing `dotnet` commands uses the highest version of the .NET SDK installed. There may be times when you want to control which SDK is used.

For example, once .NET 10 becomes available in preview, starting in February 2025, or the final version becomes available in November 2025, you might install it. But you would probably want your experience to match the book steps, which use the .NET 9 SDK. But once you install a .NET 10 SDK, it will be used by default.

You can control the .NET SDK used by default by using a `global.json` file, which contains the version to use. The `dotnet` command searches the current folder and then each ancestor folder in turn for a `global.json` file, seeing if it should use a different .NET SDK version.

You do not need to complete the following steps, but if you want to try and do not already have .NET 8 SDK installed, then you can install it from the following link:

`https://dotnet.microsoft.com/download/dotnet/8.0`

1. Create a subdirectory/folder in the `Chapter07` folder named `ControlSDK`.
2. On Windows, start **Command Prompt** or **Windows Terminal.** On macOS, start **Terminal.** If you are using VS Code, then you can use the integrated terminal.
3. In the `ControlSDK` folder, at the command prompt or terminal, enter a command to list the installed .NET SDKs, as shown in the following command:

   ```
   dotnet --list-sdks
   ```

4. Note the results and the version number of the latest .NET 8 SDK installed, as shown highlighted in the following output:

   ```
   8.0.400 [C:\Program Files\dotnet\sdk]
   9.0.100 [C:\Program Files\dotnet\sdk]
   ```

5. Create a `global.json` file that forces the use of the latest .NET 8 SDK that you have installed (which might be later than mine), as shown in the following command:

   ```
   dotnet new globaljson --sdk-version 8.0.400
   ```

6. Note the result, as shown in the following output:

   ```
   The template "global.json file" was created successfully.
   ```

7. Use your preferred code editor to open the `global.json` file and review its contents, as shown in the following markup:

   ```
   {
     "sdk": {
       "version": "8.0.400"
     }
   }
   ```

For example, to open it with VS Code, enter the command `code global.json`.

8. In the `ControlSDK` folder, at the command prompt or terminal, enter a command to create a class library project, as shown in the following command:

```
dotnet new classlib
```

9. If you do not have the .NET 8 SDK installed, then you will see an error, as shown in the following output:

```
Could not execute because the application was not found or a compatible
.NET SDK is not installed.
```

10. If you do have the .NET 8 SDK installed, then a class library project will be created that targets .NET 8 by default, as shown highlighted in the following markup:

```
<Project Sdk="Microsoft.NET.Sdk">

  <PropertyGroup>
    <TargetFramework>net8.0</TargetFramework>
    <ImplicitUsings>enable</ImplicitUsings>
    <Nullable>enable</Nullable>
  </PropertyGroup>

</Project>
```

Mixing SDKs and framework targets

Many organizations decide to target a long-term support version of .NET to get up to three years of support from Microsoft. Doing this does not mean you lose the benefits of improvements to the C# language during the lifetime of the .NET runtime that you need to target.

You can easily continue to target the .NET 9 runtime while installing and using future C# compilers, as shown in *Figure 7.2* and illustrated in the following list:

1. **November 2024:** Install .NET SDK 9.0.100, and use it to build projects that target .NET 9 and use the C# 13 compiler by default. Every month, update to .NET 9 SDK patches on the development computer and update to .NET 9 runtime patches on any deployment computers.
2. **February 2025:** Optionally, install .NET SDK 10 Preview 1 to explore the new C# 14 language and .NET 10 library features. Note that you won't be able to use new library features while targeting .NET 9. Previews are released monthly between February and October each year. Read the monthly announcements to find out about the new features in that preview at the following link: `https://github.com/dotnet/Announcements/issues`.

3. **November 2025:** Install .NET SDK 10.0.100, and use it to build projects that continue to target .NET 9 and use the C# 14 compiler for its new features. You will use a fully supported SDK and fully supported runtime.
4. **February 2026:** Optionally, install .NET 11 previews to explore new C# 15 language and .NET 11 library features. Start planning for any new library and ASP.NET Core features in .NET 10 and .NET 11 that can be applied to your .NET 9 projects when you are ready to migrate.
5. **November 2026:** Install .NET 11.0.100 SDK, and use it to build projects that target .NET 10 and use the C# 15 compiler.

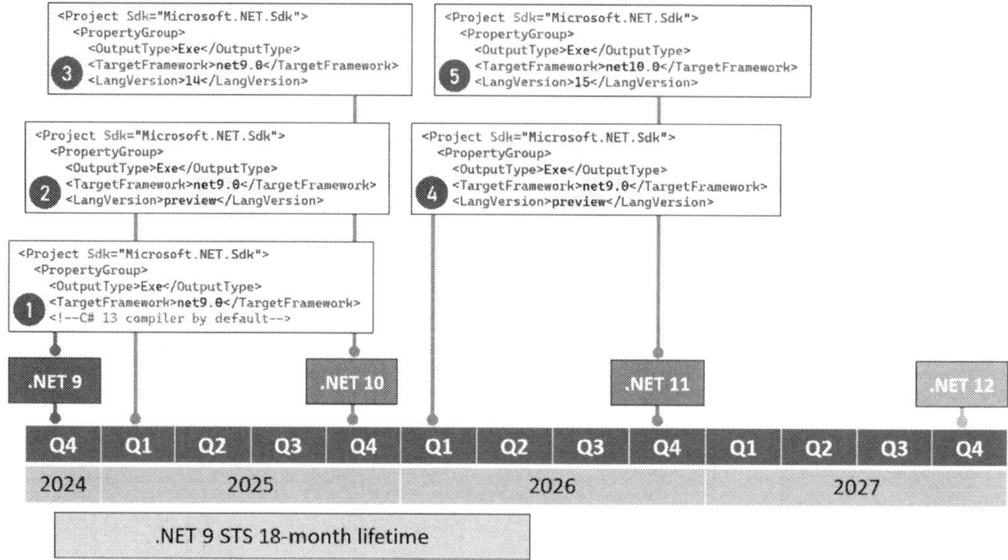

Figure 7.2: Targeting .NET 8 for long-term support while using the latest C# compilers

When deciding to install a .NET SDK, remember that the latest is used by default to build any .NET projects. Once you've installed a .NET 9 SDK preview, it will be used by default for all projects, unless you force the use of an older, fully supported SDK version like 8.0.100 or a later patch.

Publishing your code for deployment

If you write a novel and you want other people to read it, you must publish it.

Most developers write code for other developers to use in their own projects, or for users to run as an app. To do so, you must publish your code as packaged class libraries or executable applications.

There are three ways to publish and deploy a .NET application. They are:

- **Framework-dependent deployment (FDD)**
- **Framework-dependent executable (FDE)**
- **Self-contained**

If you choose to deploy your application and its package dependencies, but not .NET itself, then you rely on .NET already being on the target computer. This works well for web applications deployed to a server because .NET and lots of other web applications are likely already on the server.

FDD means you deploy a DLL that must be executed by the `dotnet` command-line tool. FDE means you deploy an EXE that can be run directly from the command line. Both require the appropriate version of the .NET runtime to be installed on a system.

Sometimes, you want to be able to give someone a USB stick containing your application, built for their OS, and know that it can execute on their computer. You would want to perform a self-contained deployment. While the size of the deployment files will be larger, you'll know that it will work.

Creating a console app to publish

Let's explore how to publish a console app:

1. Use your preferred code editor to add a new **Console App** / `console` project named `DotNetEverywhere` to the `Chapter07` solution. Make sure you target .NET 9.
2. Modify the project file to statically import the `System.Console` class into all C# files.
3. In `Program.cs`, delete the existing statements, and then add a statement to output a message, saying that the console app can run everywhere, and some information about the OS, as shown in the following code:

```
WriteLine("I can run everywhere!");
WriteLine($"OS Version is {Environment.OSVersion}.");

if (OperatingSystem.IsMacOS())
{
  WriteLine("I am macOS.");
}
else if (OperatingSystem.IsWindowsVersionAtLeast(
  major: 10, build: 22000))
{
  WriteLine("I am Windows 11.");
}
else if (OperatingSystem.IsWindowsVersionAtLeast(major: 10))
{
  WriteLine("I am Windows 10.");
}
else
{
  WriteLine("I am some other mysterious OS.");
}
WriteLine("Press any key to stop me.");
ReadKey(intercept: true); // Do not output the key that was pressed.
```

4. Run the `DotNetEverywhere` project, and note the results when run on Windows 11, as shown in the following output:

```
I can run everywhere!
OS Version is Microsoft Windows NT 10.0.22000.0.
I am Windows 11.
Press any key to stop me.
```

5. In `DotNetEverywhere.csproj`, add the **runtime identifiers** (**RIDs**) to target three OSes inside the `<PropertyGroup>` element, as shown highlighted in the following markup:

```
<Project Sdk="Microsoft.NET.Sdk">

  <PropertyGroup>
    <OutputType>Exe</OutputType>
    <TargetFramework>net9.0</TargetFramework>
    <Nullable>enable</Nullable>
    <ImplicitUsings>enable</ImplicitUsings>
    <RuntimeIdentifiers>
      win-x64;osx-arm64;linux-x64
    </RuntimeIdentifiers>
  </PropertyGroup>

</Project>
```

- The `win-x64` RID value means Windows on an x64-compatible CPU. Use the `win-arm64` RID value to deploy to a Windows ARM device.
- The `osx-arm64` RID value means macOS on Apple Silicon.
- The `linux-x64` RID value means most desktop distributions of Linux, like Ubuntu, CentOS, Debian, or Fedora on an x64-compatible CPU. Use `linux-arm` for Raspbian or Raspberry Pi OS 32-bit. Use `linux-arm64` for a Raspberry Pi running Ubuntu 64-bit.

You can check the latest allowed RID values at the following link: `https://learn.microsoft.com/en-us/dotnet/core/rid-catalog#known-rids`.

Warning! There are two elements that you can use to specify runtime identifiers. Use `<RuntimeIdentifier>` if you only need to specify one. Use `<RuntimeIdentifiers>` if you need to specify multiple, as we did in the preceding example. If you use the wrong one, then the compiler will give an error, and it can be difficult to understand why with only one character difference!

Understanding dotnet commands

When you install the .NET SDK, it includes a **command-line interface** (**CLI**) named `dotnet`.

The .NET CLI has commands that work on the current folder to create a new project, using templates:

1. On Windows, start **Command Prompt** or **Windows Terminal**. On macOS, start **Terminal**. If you prefer to use Visual Studio or VS Code, then you can use the integrated terminal.
2. Enter the `dotnet new list` (or `dotnet new -l` or `dotnet new --list` with older SDKs) command to list your currently installed templates, the most common of which are shown in *Table 7.6*:

Template Name	Short Name	Language
`.NET MAUI App`	`maui`	C#
`.NET MAUI Blazor App`	`maui-blazor`	C#
`ASP.NET Core Empty`	`web`	C# and F#
`ASP.NET Core gRPC Service`	`grpc`	C#
`ASP.NET Core Web API`	`webapi`	C# and F#
`ASP.NET Core Web API (native AOT)`	`webapiaot`	C#
`ASP.NET Core Web App (Model-View-Controller)`	`mvc`	C# and F#
`Blazor Web App`	`blazor`	C#
`Class Library`	`classlib`	C#, F#, and VB
`Console App`	`console`	C#, F#, and VB
`EditorConfig File`	`editorconfig`	
`global.json File`	`globaljson`	
`Solution File`	`sln`	
`xUnit Test Project`	`xunit`	

Table 7.6: Project template full and short names

.NET MAUI projects are not supported for Linux. The team has said they have left that work to the open source community. If you need to create a truly cross-platform graphical app, then take a look at Avalonia at the following link: `https://avaloniaui.net/`.

Getting information about .NET and its environment

It is useful to see what .NET SDKs and runtimes are currently installed, alongside information about the OS, as shown in the following command:

```
dotnet --info
```

Note the results, as shown in the following partial output:

```
.NET SDK:
 Version:   9.0.100
 Commit:    81a714c6d3
 Workload version:  9.0.100-manifests.bbb3781c
 MSBuild version:   17.12.0

Runtime Environment:
 OS Name:     Windows
 OS Version:  10.0.26100
 OS Platform: Windows
 RID:         win-arm64
 Base Path:   C:\Program Files\dotnet\sdk\9.0.100\

.NET workloads installed:
Configured to use loose manifests when installing new manifests.
 [aspire]
   Installation Source: VS 17.12.35309.182, VS 17.11.35303.130
   Manifest Version:    9.0.0/0.0.100
   Manifest Path:       C:\Program Files\dotnet\sdk-manifests\9.0.100\
microsoft.net.sdk.aspire\9.0.0\WorkloadManifest.json
   Install Type:        FileBased

Host (useful for support):
  Version: 9.0.0
  Architecture: arm64
  Commit:  static

.NET SDKs installed:
  8.0.400 [C:\Program Files\dotnet\sdk]
  9.0.100 [C:\Program Files\dotnet\sdk]

.NET runtimes installed:
  Microsoft.AspNetCore.App 8.0.8 [...\dotnet\shared\Microsoft.AspNetCore.App]
...
```

Managing projects using the dotnet CLI

The .NET CLI has the following commands that work on the project in the current folder, enabling you to manage the project:

- `dotnet help`: This shows the command-line help.

- `dotnet new`: This creates a new .NET project or file.
- `dotnet tool`: This installs or manages tools that extend the .NET experience.
- `dotnet workload`: This manages optional workloads like .NET MAUI.
- `dotnet restore`: This downloads dependencies for the project.
- `dotnet build`: This builds, aka compiles, a .NET project. A new switch introduced with .NET 8 is `--tl` (meaning terminal logger), which provides a modern output. For example, it provides real-time information about what the build is doing. You can learn more at the following link: `https://learn.microsoft.com/en-us/dotnet/core/tools/dotnet-build#options`.
- `dotnet build-server`: This interacts with servers started by a build.
- `dotnet msbuild`: This runs MS Build Engine commands.
- `dotnet clean`: This removes the temporary outputs from a build.
- `dotnet test`: This builds and then runs unit tests for the project.
- `dotnet run`: This builds and then runs the project.
- `dotnet pack`: This creates a NuGet package for the project.
- `dotnet publish`: This builds and then publishes the project, either with dependencies or as a self-contained application. In .NET 7 and earlier, this published the `Debug` configuration by default. In .NET 8 and later, it now publishes the `Release` configuration by default.
- `dotnet add`: This adds a reference to a package or class library to the project.
- `dotnet remove`: This removes a reference to a package or class library from the project.
- `dotnet list`: This lists the package or class library references for the project.
- `dotnet package search`: This allows you to search one or more package sources for packages that match a search term. The command format is `dotnet package search [search term] [options]`. You can learn more at the following link: `https://devblogs.microsoft.com/nuget/announcing-nuget-6-9/#support-for-dotnet-search-command`.

Publishing a self-contained app

Now that you have seen some example `dotnet` tool commands, we can publish our cross-platform console app:

1. At the command prompt or terminal, make sure that you are in the `DotNetEverywhere` folder.
2. Enter a command to build and publish the self-contained release version of the console application for Windows, as shown in the following command:

```
dotnet publish -c Release -r win-x64 --self-contained
```

The publishing process can take a while, so be patient.

3. Note that the build engine restores any needed packages, compiles the project source code into an assembly DLL, and creates a `publish` folder, as shown in the following output:

```
MSBuild version 17.11.0+14c24b2d3 for .NET
  Determining projects to restore...
  All projects are up-to-date for restore.
  DotNetEverywhere -> C:\cs13net9\Chapter07\DotNetEverywhere\bin\Release\
net9.0\win-x64\DotNetEverywhere.dll
  DotNetEverywhere -> C:\cs13net9\Chapter07\DotNetEverywhere\bin\Release\
net9.0\win-x64\publish\
```

4. Enter the following commands to build and publish the release versions for the macOS and Linux variants, as shown in the following commands:

```
dotnet publish -c Release -r osx-arm64 --self-contained
dotnet publish -c Release -r linux-x64 --self-contained
```

Good Practice: You could automate these commands by using a scripting language like PowerShell and execute the script file on any OS, using the cross-platform PowerShell Core. I have done this for you at the following link: `https://github.com/markjprice/cs13net9/tree/main/scripts/publish-scripts`.

5. Open Windows **File Explorer** or a macOS **Finder** window, navigate to `DotNetEverywhere\bin\Release\net9.0`, and note the output folders for the five OSes.
6. In the `win-x64` folder, open the `publish` folder, and note all the supporting assemblies, like `Microsoft.CSharp.dll`.
7. Select the `DotNetEverywhere` executable file, and note that it is 154 KB, as shown in *Figure 7.3*:

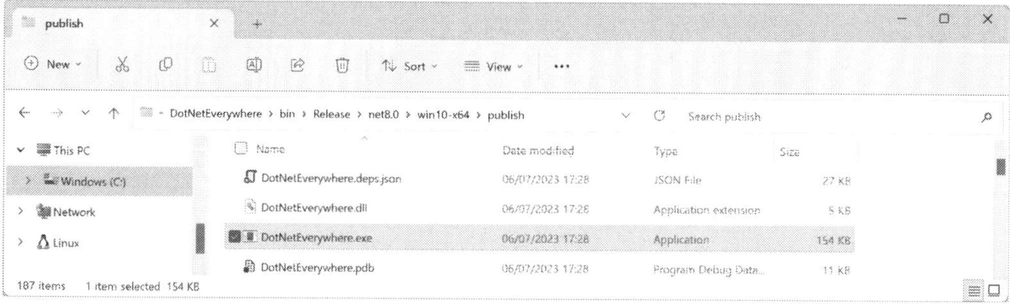

Figure 7.3: The DotNetEverywhere executable file for Windows 64-bit

8. If you are on Windows with an x64-compatible chip, then double-click to execute the program and note the result, as shown in the following output:

```
I can run everywhere!
OS Version is Microsoft Windows NT 10.0.26100.0.
I am Windows 11.
Press any key to stop me.
```

I executed the program on my Surface Laptop 7, so I targeted and built for the `win-arm64` architecture. The preceding output shows the results on my computer.

9. Press any key to close the console app and its window.
10. Note that the total size of the `publish` folder and all its files is 68.3 MB.
11. In the `osx-arm64` folder, select the `publish` folder, note all the supporting assemblies, and then select the `DotNetEverywhere` executable file. Note that the executable is 125 KB and the `publish` folder is about 73.9 MB. There is no `.exe` file extension for published applications on macOS, so the filename will not have an extension.

If you copy any of those `publish` folders to the appropriate **operating system** (**OS**), the console app will run; this is because it is a self-contained, deployable .NET application. For example, here it is on macOS:

```
I can run everywhere!
OS Version is Unix 13.5.2
I am macOS.
Press any key to stop me.
```

This example used a console app, but you could just as easily create an ASP.NET Core website or web service, or a Windows Forms or WPF app. Of course, you can only deploy Windows desktop apps to Windows computers, not Linux or macOS.

Publishing a single-file app

If you know that .NET is already installed on the computer on which you want to run your app, then you can use the extra flags when you publish your app for release to say that it does not need to be self-contained and that you want to publish it as a single file (if possible), as shown in the following command (which must be entered on a single line):

```
dotnet publish -r win-x64 -c Release --no-self-contained
/p:PublishSingleFile=true
```

This will generate two files: `DotNetEverywhere.exe` and `DotNetEverywhere.pdb`. The `.exe` file is the executable. The `.pdb` file is a **program debug database** file that stores debugging information.

If you prefer the `.pdb` file to be embedded in the `.exe` file (for example, to ensure that it is deployed with its assembly), then add a `<DebugType>` element to the `<PropertyGroup>` element in your `.csproj` file and set it to `embedded`, as shown highlighted in the following markup:

```
<PropertyGroup>

<OutputType>Exe</OutputType>
<TargetFramework>net9.0</TargetFramework>
<Nullable>enable</Nullable>
```

```
<ImplicitUsings>enable</ImplicitUsings>

<RuntimeIdentifiers>
win-x64;osx-arm64;linux-x64
</RuntimeIdentifiers>

<DebugType>embedded</DebugType>

</PropertyGroup>
```

If you cannot assume that .NET is already installed on a computer, then although Linux also only generates the two files, expect the following additional files for Windows: `coreclr.dll`, `clrjit.dll`, `clrcompression.dll`, and `mscordaccore.dll`.

Let's see an example for Windows:

1. At the command prompt or terminal, in the `DotNetEverywhere` folder, enter the command to build the self-contained release version of the console app for Windows, as shown in the following command:

   ```
   dotnet publish -c Release -r win-x64 --self-contained
   /p:PublishSingleFile=true
   ```

2. Navigate to the `DotNetEverywhere\bin\Release\net9.0\win-x64\publish` folder and select the `DotNetEverywhere` executable file. Note that the executable is now 62.6 MB, and there is also a `.pdb` file that is 11 KB. The sizes of these files on your system will vary.

Reducing the size of apps using app trimming

One of the problems with deploying a .NET app as a self-contained app is that the .NET libraries take up a lot of space. One of the biggest needs is to reduce the size of Blazor WebAssembly components because all the .NET libraries need to be downloaded to the browser.

Luckily, you can reduce this size by not packaging unused assemblies with your deployments. Introduced with .NET Core 3, the app trimming system can identify the assemblies needed by your code and remove those that are not needed. This was known as `copyused` trim mode.

With .NET 5, the trimming went further by removing individual types, and even members, like methods from within an assembly if they are not used. For example, with a **Hello World** console app, the `System.Console.dll` assembly is trimmed from 61.5 KB to 31.5 KB. This was known as `link` trim mode, but it was not enabled by default.

With .NET 6, Microsoft added annotations to their libraries to indicate how they can be safely trimmed, so the trimming of types and members was made the default.

With .NET 7, Microsoft renamed `link` to `full` and `copyused` to `partial`.

The catch is how well the trimming identifies unused assemblies, types, and members. If your code is dynamic, perhaps using reflection, then it might not work correctly, so Microsoft also allows manual control.

There are two ways to enable type-level and member-level, aka `full`, trimming. Since this level of trimming is the default with .NET 6 or later, all we need to do is enable trimming without setting a trim level or mode.

The first way is to add an element to the project file, as shown in the following markup:

```
<PublishTrimmed>true</PublishTrimmed> <!--Enable trimming.-->
```

The second way is to add a flag when publishing, as shown highlighted in the following command:

```
dotnet publish ... -p:PublishTrimmed=True
```

There are two ways to enable assembly-level, aka `partial`, trimming.

The first way is to add two elements to the project file, as shown in the following markup:

```
<PublishTrimmed>true</PublishTrimmed> <!--Enable trimming.-->
<TrimMode>partial</TrimMode> <!--Set assembly-level trimming.-->
```

The second way is to add two flags when publishing, as shown highlighted in the following command:

```
dotnet publish ... -p:PublishTrimmed=True -p:TrimMode=partial
```

Controlling where build artifacts are created

Traditionally, each project has its own `bin` and `obj` subfolders where temporary files are created during the build process. When you publish, the files are created in the `bin` folder.

You might prefer to put all these temporary files and folders somewhere else. Introduced with .NET 8 is the ability to control where build artifacts are created. Let's see how:

1. At the command prompt or terminal for the `Chapter07` folder, enter the following command:

   ```
   dotnet new buildprops --use-artifacts
   ```

2. Note the success message, as shown in the following output:

   ```
   The template "MSBuild Directory.Build.props file" was created
   successfully.
   ```

   We could have created this file in the `cs13net9` folder, and it would then affect all projects in all chapters.

3. In the `Chapter07` folder, open the `Directory.Build.props` file, as shown in the following markup:

```
<Project>
<!-- See https://aka.ms/dotnet/msbuild/customize for more details on
customizing your build -->
<PropertyGroup>

    <ArtifactsPath>$(MSBuildThisFileDirectory)artifacts</ArtifactsPath>

</PropertyGroup>
</Project>
```

4. Build any project or the whole solution.
5. In the `Chapter07` folder, note that there is now an `artifacts` folder that contains subfolders for any recently built projects.
6. You can delete this file, or rename it to something like `Directory.Build.props.disabled` so that it does not affect the rest of this chapter by building these artifacts in a folder that you do not expect based on default behavior. This is optional but recommended.

Warning! If you leave this build configuration enabled, then remember that your build artifacts are now created in this new folder structure.

Native ahead-of-time compilation

Native AOT produces assemblies that are:

- **Self-contained**, meaning they can run on systems that do not have the .NET runtime installed.
- **Ahead-of-time (AOT) compiled for native code**, meaning a faster startup time and a potentially smaller memory footprint.

Native AOT compiles IL code to native code at the time of writing, rather than at runtime using the **Just in Time (JIT)** compiler. But native AOT assemblies must target a specific runtime environment like Windows x64 or Linux Arm.

Since native AOT happens at publish time, you should remember that while you are debugging and working live on a project in your code editor, it still uses the runtime JIT compiler, not native AOT, even if you have AOT enabled in the project!

However, some features that are incompatible with native AOT will be disabled or throw exceptions, and a source analyzer is enabled to show warnings about potential code incompatibilities.

Limitations of native AOT

Native AOT has limitations that apply to projects that enable it, some of which are shown in the following list:

- No dynamic loading of assemblies.
- No runtime code generation, for example, using `System.Reflection.Emit`.
- It requires trimming, which has its own limitations, as we covered in the previous section.
- The projects must be self-contained, so they must embed any libraries they call, which increases their size.

Although your own assemblies might not use the features listed above, major parts of .NET itself do. For example, ASP.NET Core MVC (including web API services that use controllers) and EF Core do runtime code generation to implement their functionality.

The .NET teams are hard at work making as much of .NET compatible with native AOT as possible, as soon as possible. But .NET 9 only includes basic support for ASP.NET Core if you use Minimal APIs, and no support for EF Core.

My guess is that .NET 10 will include support for ASP.NET Core MVC and some parts of EF Core, but it could take until .NET 11 or .NET 12 before we can all confidently use most of .NET and know we can build our assemblies with native AOT to gain the benefits.

The native AOT publishing process includes code analyzers to warn you if you use any features that are not supported, but not all packages have been annotated to work well with these yet.

The most common annotation used to indicate that a type or member does not support AOT is the `[RequiresDynamicCode]` attribute.

More Information: You can learn more about AOT warnings at the following link: `https://learn.microsoft.com/en-us/dotnet/core/deploying/native-aot/fixing-warnings`.

Reflection and native AOT

Reflection is frequently used for runtime inspection of type `metadata`, dynamic invocation of members, and code generation.

Native AOT does allow some reflection features, but the trimming performed during the native AOT compilation process cannot statically determine when a type has members that might be only accessed via reflection. These members would be removed by AOT, which would then cause a runtime exception.

Good Practice: Developers must annotate their types with `[DynamicallyAccessedMembers]` to indicate a member that is only dynamically accessed via reflection and should, therefore, be left untrimmed.

Requirements for native AOT

There are additional requirements for different OSes:

- On Windows, you must install the Visual Studio **Desktop development with C++** workload with all default components.
- On Linux, you must install the compiler toolchain and developer packages for libraries that the .NET runtime depends on. For example, for Ubuntu 18.04 or later: `sudo apt-get install clang zlib1g-dev`.
- **Warning!** Cross-platform native AOT publishing is not supported. This means that you must run the publish action on the OS that you will deploy to. For example, you cannot publish a native AOT project on Linux to later run on Windows, and vice versa.

Enabling native AOT for a project

To enable native AOT publishing in a project, add the `<PublishAot>` element to the project file, as shown highlighted in the following markup:

```
<PropertyGroup>
  <TargetFramework>net9.0</TargetFramework>
  <PublishAot>true</PublishAot>
```

Building a native AOT project

Now, let's see a practical example using the new AOT option for a console app:

1. In the solution named `Chapter07`, add a native AOT-compatible console app project, as defined in the following list:

   - Project template: **Console App** / `console --aot`
   - Solution file and folder: `Chapter07`
   - Project file and folder: `AotConsole`
   - Do not use top-level statements: Cleared
   - Enable native AOT publish: Selected

If your code editor does not yet provide the option for AOT, create a traditional console app, and then you will need to manually enable AOT, as shown in step 2, or use the `dotnet` CLI.

2. In the project file, note that native AOT publishing is enabled, as well as invariant globalization, as shown highlighted in the following markup:

```
<Project Sdk="Microsoft.NET.Sdk.Web">
```

```
  <PropertyGroup>
    <TargetFramework>net9.0</TargetFramework>
    <Nullable>enable</Nullable>
    <ImplicitUsings>enable</ImplicitUsings>
    <PublishAot>true</PublishAot>
    <InvariantGlobalization>true</InvariantGlobalization>
  </PropertyGroup>

</Project>
```

Explicitly setting invariant globalization to `true` is new in the **Console App** project template with .NET 8. It is designed to make a console app non-culture-specific so that it can be deployed anywhere in the world and have the same behavior. If you set this property to `false`, or if the element is missing, then the console app will default to the culture of the current computer it is hosted on. You can read more about invariant globalization mode at the following link: `https://github.com/dotnet/runtime/blob/main/docs/design/features/globalization-invariant-mode.md`.

3. Modify the project file to statically import the `System.Console` class in all C# files.
4. In `Program.cs`, delete any existing statements, and then add statements to output the current culture and OS version, as shown in the following code:

```
using System.Globalization; // To use CultureInfo.

WriteLine("This is an ahead-of-time (AOT) compiled console app.");
WriteLine("Current culture: {0}", CultureInfo.CurrentCulture.
DisplayName);
WriteLine("OS version: {0}", Environment.OSVersion);

Write("Press any key to exit.");
ReadKey(intercept: true); // Do not output the key that was pressed.
```

5. Run the console app project and note that the culture is invariant, as shown in the following output:

```
This is an ahead-of-time (AOT) compiled console app.
Current culture: Invariant Language (Invariant Country)
OS version: Microsoft Windows NT 10.0.22621.0
```

Warning! Actually, the console app is not AOT-compiled; it is still currently JIT-compiled because we have not yet published it.

Publishing a native AOT project

A console app that functions correctly during development when the code is untrimmed and JIT-compiled could still fail once you publish it using native AOT, as the code is then trimmed and JIT-compiled, and therefore, it is different code with different behavior. You should, therefore, perform a publish action before assuming your project will work.

If your project does not produce any AOT warnings at publish time, you can then be confident that your service will work after publishing for AOT.

Let's publish our console app:

1. At the command prompt or terminal for the `AotConsole` project, publish the console app using native AOT, as shown in the following command:

```
dotnet publish
```

2. Note the message about generating native code, as shown in the following output:

```
MSBuild version 17.8.0+4ce2ff1f8 for .NET
  Determining projects to restore...
  Restored C:\cs13net9\Chapter07\AotConsole\AotConsole.csproj (in 173
ms).
  AotConsole -> C:\cs13net9\Chapter07\AotConsole\bin\Release\net9.0\
win-x64\AotConsole.dll
  Generating native code
  AotConsole -> C:\cs13net9\Chapter07\AotConsole\bin\Release\net9.0\
win-x64\publish\
```

3. Start **File Explorer**, open the `bin\Release\net9.0\win-x64\publish` folder, and note that the `AotConsole.exe` file is about 1.2 MB. The `AotConsole.pdb` file is only needed for debugging.
4. Run the `AotConsole.exe` and note that the console app has the same behavior as before.
5. In `Program.cs`, import namespaces to work with dynamic code assemblies, as shown in the following code:

```
using System.Reflection; // To use AssemblyName.
using System.Reflection.Emit; // To use AssemblyBuilder.
```

6. In `Program.cs`, create a dynamic assembly builder, as shown in the following code:

```
AssemblyBuilder ab = AssemblyBuilder.DefineDynamicAssembly(
  new AssemblyName("MyAssembly"), AssemblyBuilderAccess.Run);
```

7. At the command prompt or terminal for the `AotConsole` project, publish the console app using native AOT, as shown in the following command:

```
dotnet publish
```

8. Note the warning about calling the `DefineDynamicAssembly` method, which the .NET team has decorated with the `[RequiresDynamicCode]` attribute, as shown in the following output:

```
C:\cs13net9\Chapter07\AotConsole\Program.cs(9,22): warning
IL3050: Using member 'System.Reflection.Emit.AssemblyBuilder.
DefineDynamicAssembly(AssemblyName, AssemblyBuilderAccess)' which
has 'RequiresDynamicCodeAttribute' can break functionality when AOT
compiling. Defining a dynamic assembly requires dynamic code. [C:\
cs13net9\Chapter07\AotConsole\AotConsole.csproj]
```

9. Comment out the statement that we cannot use in an AOT project.

More Information: You can learn more about native AOT at the following link: `https://learn.microsoft.com/en-us/dotnet/core/deploying/native-aot/`.

Packaging your libraries for NuGet distribution

Before we learn how to create and package our own libraries, we will review how a project can use an existing package.

Referencing a NuGet package

Let's say that you want to add a package created by a third-party developer, for example, `Newtonsoft.Json`, a popular package for working with the **JavaScript Object Notation** (**JSON**) serialization format:

1. In the `AssembliesAndNamespaces` project, add a reference to the `Newtonsoft.Json` NuGet package, either using the GUI for Visual Studio or the `dotnet add package` command using the CLI.

With the April release of C# Dev Kit, you can now manage your NuGet packages directly from VS Code using some commands in the command palette, as described at the following link: `https://devblogs.microsoft.com/nuget/announcing-nuget-commands-in-c-dev-kit/`.

2. Open the `AssembliesAndNamespaces.csproj` file and note that a package reference has been added, as shown in the following markup:

```
<ItemGroup>
  <PackageReference Include="Newtonsoft.Json"
                    Version="13.0.3" />
</ItemGroup>
```

If you have a more recent version of the `Newtonsoft.Json` package, then it has been updated since this chapter was written.

Fixing dependencies

To consistently restore packages and write reliable code, it's important that you **fix dependencies.** Fixing dependencies means you are using the same family of packages released for a specific version of .NET, for example, SQLite for .NET 9, as shown highlighted in the following markup:

```
<Project Sdk="Microsoft.NET.Sdk">

  <PropertyGroup>
    <OutputType>Exe</OutputType>
    <TargetFramework>net9.0</TargetFramework>
    <Nullable>enable</Nullable>
    <ImplicitUsings>enable</ImplicitUsings>
  </PropertyGroup>

  <ItemGroup>
    <PackageReference Version="9.0.0"
      Include="Microsoft.EntityFrameworkCore.Sqlite" />
  </ItemGroup>

</Project>
```

To fix dependencies, every package should have a single version with no additional qualifiers. Additional qualifiers include betas (`beta1`), release candidates (`rc4`), and wildcards (`*`).

Wildcards allow future versions to be automatically referenced and used because they always represent the most recent release. Wildcards are, therefore, dangerous because they could result in the use of future incompatible packages that break your code.

This can be worth the risk while writing a book where new preview versions are released every month and you do not want to keep updating the preview package references, as I did during 2024, and as shown in the following markup:

```
<PackageReference Version="9.0.0-preview.*"
  Include="Microsoft.EntityFrameworkCore.Sqlite" />
```

To also automatically use the release candidates that arrive in September and October each year, you can make the pattern even more flexible, as shown in the following markup:

```
<PackageReference Version="9.0-*"
  Include="Microsoft.EntityFrameworkCore.Sqlite" />
```

If you use the `dotnet add package` command or Visual Studio's **Manage NuGet Packages**, then it will by default use the latest specific version of a package. But if you copy and paste configuration from a blog article or manually add a reference yourself, you might include wildcard qualifiers.

The following dependencies are examples of NuGet package references that are *not* fixed and, therefore, should be avoided unless you know the implications:

```
<PackageReference Include="System.Net.Http" Version="4.1.0-*" />
<PackageReference Include="Newtonsoft.Json" Version="13.0.2-beta1" />
```

Good Practice: Microsoft guarantees that if you fix your dependencies to what ships with a specific version of .NET, for example, `9.0.0`, those packages will all work together. Almost always fix your dependencies, especially in production deployments.

Packaging a library for NuGet

Now, let's package the `SharedLibrary` project that you created earlier:

1. In the `SharedLibrary` project, note that the class library targets .NET Standard 2.0 and, therefore, by default, uses the C# 7.3 compiler. Explicitly specify the C# 12 compiler, as shown in the following markup:

```
<Project Sdk="Microsoft.NET.Sdk">

  <PropertyGroup>
    <TargetFramework>netstandard2.0</TargetFramework>
    <LangVersion>12</LangVersion>
  </PropertyGroup>

</Project>
```

2. In the `SharedLibrary` project, rename the `Class1.cs` file `StringExtensions.cs`.
3. Modify its contents to provide some useful extension methods to validate various text values, using regular expressions, as shown in the following code:

```
using System.Text.RegularExpressions; // To use Regex.

namespace Packt.Shared;

public static class StringExtensions
{
  public static bool IsValidXmlTag(this string input)
  {
    return Regex.IsMatch(input,
      @"^<([a-z]+)([^<]+)*(?:>(.*)<\/\1>|\s+\/>)$");
  }

  public static bool IsValidPassword(this string input)
```

```
  {
    // Minimum of eight valid characters.
    return Regex.IsMatch(input, "^[a-zA-Z0-9_-]{8,}$");
  }

  public static bool IsValidHex(this string input)
  {
    // Three or six valid hex number characters.
    return Regex.IsMatch(input,
      "^#?([a-fA-F0-9]{3}|[a-fA-F0-9]{6})$");
  }
}
```

You will learn how to write regular expressions in *Chapter 8, Working with Common .NET Types*.

4. In `SharedLibrary.csproj`, modify its contents, as shown highlighted in the following markup, and note the following:
    - `PackageId` must be globally unique, so you must use a different value if you want to publish this NuGet package to the `https://www.nuget.org/` public feed for others to reference and download.
    - `PackageLicenseExpression` must be a value from `https://spdx.org/licenses/`, or you could specify a custom license.

Warning! If you rely on IntelliSense to edit the file, then it could mislead you to use deprecated tag names. For example, `<PackageIconUrl>` is deprecated in favor of `<PackageIcon>`. Sometimes, you cannot trust automated tools to help you correctly! The recommended tag names are documented in the **MSBuild Property** column in the table found at the following link: `https://learn.microsoft.com/en-us/nuget/reference/msbuild-targets#pack-target`.

    - All the other elements are self-explanatory:

```
<Project Sdk="Microsoft.NET.Sdk">

  <PropertyGroup>
    <TargetFramework>netstandard2.0</TargetFramework>
    <LangVersion>12</LangVersion>
```

```
    <GeneratePackageOnBuild>true</GeneratePackageOnBuild>
    <PackageId>Packt.CSdotnet.SharedLibrary</PackageId>
    <PackageVersion>9.0.0.0</PackageVersion>
    <Title>C# 13 and .NET 9 Shared Library</Title>
    <Authors>Mark J Price</Authors>
    <PackageLicenseExpression>
      MS-PL
    </PackageLicenseExpression>
    <PackageProjectUrl>
      https://github.com/markjprice/cs13net9
    </PackageProjectUrl>
    <PackageReadmeFile>readme.md</PackageReadmeFile>
    <PackageIcon>packt-csdotnet-sharedlibrary.png</PackageIcon>
    <PackageRequireLicenseAcceptance>true</
PackageRequireLicenseAcceptance>
    <PackageReleaseNotes>
      Example shared library packaged for NuGet.
    </PackageReleaseNotes>
    <Description>
      Three extension methods to validate a string value.
    </Description>
    <Copyright>
      Copyright © 2016-2023 Packt Publishing Limited
    </Copyright>
    <PackageTags>string extensions packt csharp dotnet</
PackageTags>
  </PropertyGroup>

  <ItemGroup>
    <None Include="packt-csdotnet-sharedlibrary.png"
          PackagePath="\" Pack="true" />
    <None Include="readme.md"
          PackagePath="\" Pack="true" />
  </ItemGroup>

</Project>
```

`<None>` represents a file that does not participate in the build process. `Pack="true"` means the file will be included in the NuGet package created in the specified package path location. You can learn more at the following link: `https://learn.microsoft.com/en-us/nuget/reference/msbuild-targets#packing-an-icon-image-file`.

Good Practice: Configuration property values that are `true` or `false` values cannot have any whitespace.

5. Download the icon file and save it in the `SharedLibrary` project folder from the following link: `https://github.com/markjprice/cs13net9/blob/main/code/Chapter07/SharedLibrary/packt-csdotnet-sharedlibrary.png`.
6. In the `SharedLibrary` project folder, create a file named `readme.md`, with some basic information about the package, as shown in the following markup:

```
# README for C# 13 and .NET 9 Shared Library

This is a shared library that readers build in the book,
*C# 13 and .NET 9 - Modern Cross-Platform Development Fundamentals*.
```

7. Build the release assembly:
   - In Visual Studio, select **Release** in the toolbar, and then navigate to **Build** | **Build SharedLibrary**.
   - In VS Code, in **Terminal**, enter `dotnet build -c Release`.

   If we had not set `<GeneratePackageOnBuild>` to `true` in the project file, then we would have had to create a NuGet package manually using the following additional steps:

   - In Visual Studio, navigate to **Build** | **Pack SharedLibrary**.
   - In VS Code, in **Terminal**, enter `dotnet pack -c Release`.

Publishing a package to a public NuGet feed

If you want everyone to be able to download and use your NuGet package, then you must upload it to a public NuGet feed like Microsoft's:

1. Start your favorite browser and navigate to the following link: `https://www.nuget.org/packages/manage/upload`.
2. You will need to sign up for, and then sign in with, a Microsoft account at `https://www.nuget.org/` if you want to upload a NuGet package for other developers to reference as a dependency package.

3. Click the **Browse...** button and select the `.nupkg` file that was created by generating the NuGet package. The folder path should be `cs13net9\Chapter07\SharedLibrary\bin\Release` and the file is named `Packt.CSdotnet.SharedLibrary.9.0.0.nupkg`.
4. Verify that the information you entered in the `SharedLibrary.csproj` file has been correctly filled in, and then click **Submit**.
5. Wait a few seconds, and then you will see a success message, showing that your package has been uploaded, as shown in *Figure 7.4*:

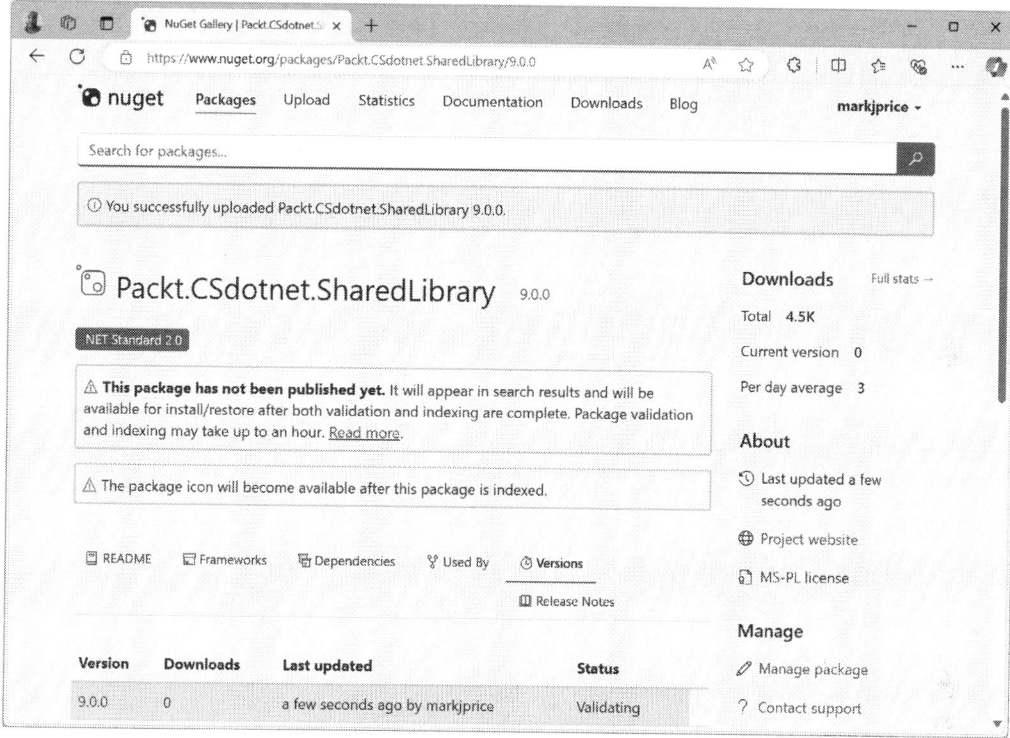

Figure 7.4: A NuGet package upload message

Good Practice: If you get an error, then review the project file for mistakes, or read more information about the `PackageReference` format at `https://learn.microsoft.com/en-us/nuget/reference/msbuild-targets`.

6. Click the **Frameworks** tab, and note that because we targeted .NET Standard 2.0, our class library can be used by every .NET platform, as shown in *Figure 7.5*:

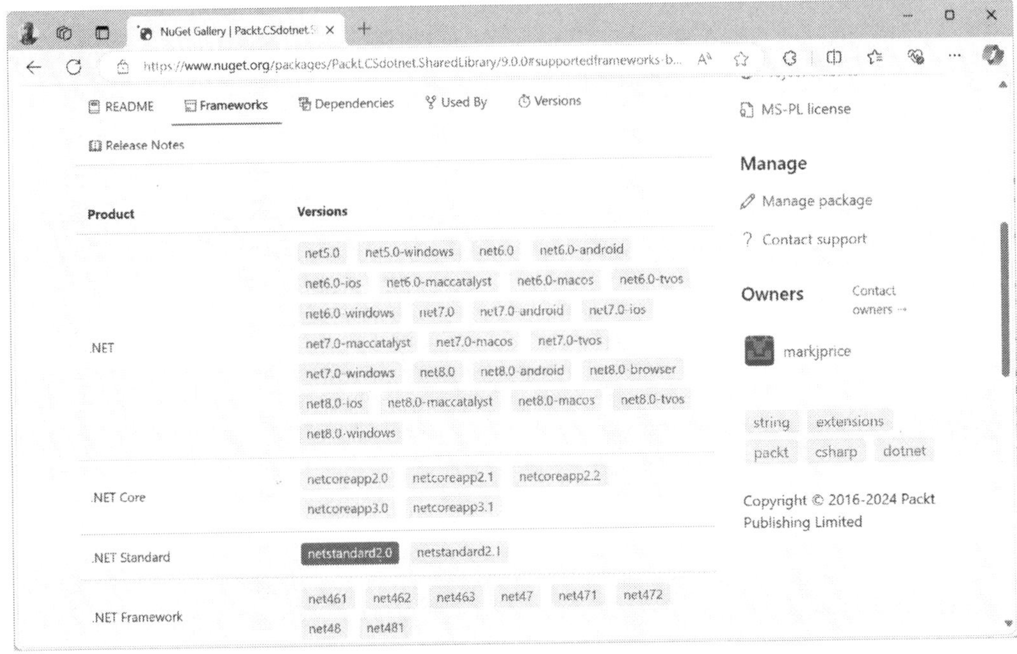

Figure 7.5: .NET Standard 2.0 class library package can be used by all .NET platforms

Publishing a package to a private NuGet feed

Organizations can host their own private NuGet feeds. This can be a handy way for many developer teams to share work. You can read more at the following link: `https://learn.microsoft.com/en-us/nuget/hosting-packages/overview`.

Exploring NuGet packages with a tool

A handy tool named **NuGet Package Explorer**, for opening and reviewing more details about a NuGet package, was created by Uno Platform. As well as being a website, it can be installed as a cross-platform app. Let's see what it can do:

1. Start your favorite browser and navigate to the following link: `https://nuget.info`.
2. In the search box, enter `Packt.CSdotnet.SharedLibrary`.
3. Select the package **v9.0.0** published by **Mark J Price**, and then click the **Open** button.
4. In the **Contents** section, expand the `lib` folder and the `netstandard2.0` folder.
5. Select `SharedLibrary.dll`, and note the details, as shown in *Figure 7.6*:

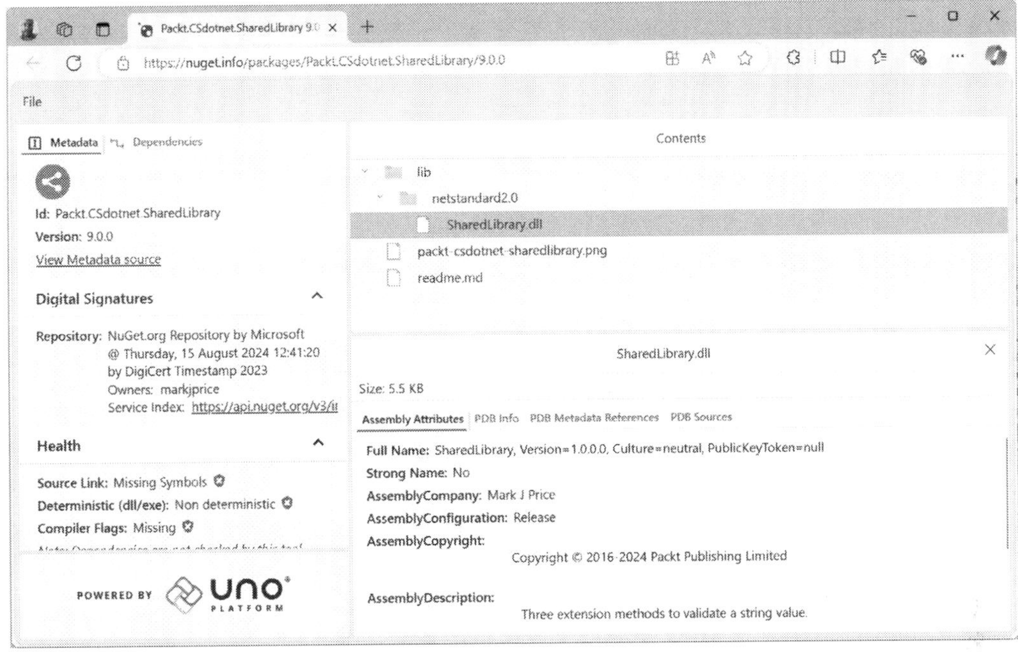

Figure 7.6: Exploring my package using NuGet Package Explorer from Uno Platform

6. If you want to use this tool locally in the future, click the **Install** button in your browser.
7. Close your browser.

Not all browsers support installing web apps like this. I recommend Chrome for testing and development.

Testing your class library package

You will now test your uploaded package by referencing it in the `AssembliesAndNamespaces` project:

1. In the `AssembliesAndNamespaces` project, add a reference to your (or my) package, as shown highlighted in the following markup:

```
<ItemGroup>
  <PackageReference Include="Newtonsoft.Json" Version="13.0.3" />
  <PackageReference Include="Packt.CSdotnet.SharedLibrary"
                    Version="9.0.0" />
</ItemGroup>
```

2. Build the `AssembliesAndNamespaces` project.
3. In `Program.cs`, import the `Packt.Shared` namespace.

4. In `Program.cs`, prompt the user to enter some `string` values, and then validate them using the extension methods in the package, as shown in the following code:

```
Write("Enter a color value in hex: ");
string? hex = ReadLine();
WriteLine("Is {0} a valid color value? {1}",
  arg0: hex, arg1: hex.IsValidHex());

Write("Enter a XML element: ");
string? xmlTag = ReadLine();
WriteLine("Is {0} a valid XML element? {1}",
  arg0: xmlTag, arg1: xmlTag.IsValidXmlTag());

Write("Enter a password: ");
string? password = ReadLine();
WriteLine("Is {0} a valid password? {1}",
  arg0: password, arg1: password.IsValidPassword());
```

5. Run the `AssembliesAndNamespaces` project, enter some values as prompted, and view the results, as shown in the following output:

```
Enter a color value in hex: 00ffc8
Is 00ffc8 a valid color value? True
Enter an XML element: <h1 class="<" />
Is <h1 class="<" /> a valid XML element? False
Enter a password: secretsauce
Is secretsauce a valid password? True
```

Working with preview features

It is a challenge for Microsoft to deliver some new features that have cross-cutting effects across many parts of .NET, like the runtime, language compilers, and API libraries. It is the classic chicken and egg problem. What do you do first?

From a practical perspective, it means that although Microsoft might have completed most of the work needed for a feature, the whole thing might not be ready until very late in their now annual cycle of .NET releases, which is too late for proper testing in "the wild."

So from .NET 6 onward, Microsoft will include preview features in **general availability** (**GA**) releases. Developers can opt into these preview features and provide Microsoft with feedback. In a later GA release, they can be enabled for everyone.

It is important to note that this topic is about *preview features*. This is different from a preview version of .NET or Visual Studio. Microsoft releases preview versions of Visual Studio and .NET while developing them to get feedback from developers, and then they do a final GA release. With GA, the feature is available for everyone. Before GA, the only way to get the new functionality was to install a preview version. *Preview features* are different because they are installed with GA releases and must be optionally enabled.

For example, when Microsoft released .NET SDK 6.0.200 in February 2022, it included the C# 11 compiler as a preview feature. This meant that .NET 6 developers could optionally set the language version to `preview` and then start exploring C# 11 features, like raw string literals and the `required` keyword.

Once .NET SDK 7.0.100 was released in November 2022, any .NET 6 developer who wanted to continue to use the C# 11 compiler would then need to use the .NET 7 SDK for their .NET 6 projects and set the target framework to `net6.0`, with a `<LangVersion>` set to `11`. This way, they use the supported .NET 7 SDK with the supported C# 11 compiler to build .NET 6 projects.

In November 2025, Microsoft is likely to release .NET 10 SDK with a C# 14 compiler. You can then install and use the .NET 10 SDK to gain the benefits of whatever new features are available in C# 14, while still targeting .NET 9, as shown highlighted in the following `Project` file:

```
<Project Sdk="Microsoft.NET.Sdk">

  <PropertyGroup>
    <OutputType>Exe</OutputType>
    <TargetFramework>net9.0</TargetFramework>
    <LangVersion>14</LangVersion> <!--Requires .NET 10 SDK GA-->
    <ImplicitUsings>enable</ImplicitUsings>
    <Nullable>enable</Nullable>
  </PropertyGroup>

</Project>
```

Good Practice: Preview features are not supported in production code. Preview features are likely to have breaking changes before the final release. Enable preview features at your own risk. Switch to a GA-release future SDK like .NET 11 to use new compiler features, while still targeting older but longer-supported versions of .NET like .NET 8 or 10.

Requiring preview features

The `[RequiresPreviewFeatures]` attribute is used to indicate assemblies, types, or members that use, and, therefore, require warnings about, preview features. A code analyzer can scan for this attribute and then generate warnings if needed. If your code does not use any preview features, you will not see any warnings. If your code does use any preview features, then you will see warnings. Your code should also be decorated with this attribute to warn other developers that your code uses preview features.

Enabling preview features

In the `Project` file, add an element to enable preview features and an element to enable preview language features, as shown highlighted in the following markup:

```
<Project Sdk="Microsoft.NET.Sdk">

  <PropertyGroup>
    <OutputType>Exe</OutputType>
    <TargetFramework>net9.0</TargetFramework>
    <Nullable>enable</Nullable>
    <ImplicitUsings>enable</ImplicitUsings>
    <EnablePreviewFeatures>true</EnablePreviewFeatures>
    <LangVersion>preview</LangVersion>
  </PropertyGroup>

</Project>
```

Method interceptors

An interceptor is a method that substitutes a call to an interceptable method with a call to itself. This is an advanced feature most commonly used in source generators. If you are interested, then I might add a section about them to the ninth edition.

More Information: You can learn more about interceptors at the following link: `https://learn.microsoft.com/en-us/dotnet/csharp/whats-new/csharp-12#interceptors`.

Practicing and exploring

Test your knowledge and understanding by answering some questions, getting some hands-on practice, and researching the topics of this chapter further.

Exercise 7.1 – Online material

Online material can be extra content written by me for this book, or it can be references to content created by Microsoft or third parties.

Experimental MSBuild editor

MSBuild is an essential component of the .NET development experience, yet it can be difficult for both new and experienced developers to navigate. To help developers better understand their project files and utilize advanced features that require direct project file editing, the Microsoft team is introducing an experimental MSBuild editor. This new editor offers a deeper understanding of MSBuild files beyond what the current XML schema provides.

If you want to try out the extension, search for **MSBuild Editor** in the Visual Studio Extension Manager. Currently, the extension is only available for Visual Studio, but efforts are underway to port it to VS Code.

You can learn more about the new MSBuild editor experience at the following link:

`https://devblogs.microsoft.com/visualstudio/experimental-msbuild-editor/`

Improving performance in .NET

Microsoft has made significant improvements to .NET performance in the past few years. You should review the blog posts written by Stephen Toub to learn what the team changed and why. His posts are famously long, detailed, and brilliant!

You can find the posts about the improvements at the following links:

- `https://devblogs.microsoft.com/dotnet/performance-improvements-in-net-core/` - 25 pages
- `https://devblogs.microsoft.com/dotnet/performance-improvements-in-net-core-2-1/` - 20 pages
- `https://devblogs.microsoft.com/dotnet/performance-improvements-in-net-core-3-0/` - 41 pages
- `https://devblogs.microsoft.com/dotnet/performance-improvements-in-net-5/` - 43 pages
- `https://devblogs.microsoft.com/dotnet/performance-improvements-in-net-6/` - 100 pages
- `https://devblogs.microsoft.com/dotnet/performance_improvements_in_net_7/` - 156 pages
- `https://devblogs.microsoft.com/dotnet/performance-improvements-in-net-8/` - 218 pages
- `https://devblogs.microsoft.com/dotnet/performance-improvements-in-net-9/` - 307 pages

Decompiling .NET assemblies

To learn about decompiling .NET assemblies, you can read the following link:

`https://github.com/markjprice/cs13net9/blob/main/docs/ch07-decompiling.md`

Porting from .NET Framework to modern .NET

If you are interested in porting legacy projects from .NET Framework to modern .NET, then I have written an online-only section at the following link:

`https://github.com/markjprice/cs13net9/blob/main/docs/ch07-porting.md`

Exercise 7.2 – Practice exercises

Practice exercises to dive deeper into the topics of this chapter.

Creating source generators

If you are interested in creating source generators, then I have written an online-only section at the following link:

`https://github.com/markjprice/cs13net9/blob/main/docs/ch07-source-generators.md`

You can find examples of source generators at the following link:

`https://github.com/amis92/csharp-source-generators`

Explore PowerShell

PowerShell is Microsoft's scripting language for automating tasks on every OS. Microsoft recommends VS Code with the PowerShell extension to write PowerShell scripts.

Since PowerShell is its own extensive language, there is not enough space in this book to cover it. You can learn about some key concepts from a Microsoft training module at the following link: `https://learn.microsoft.com/en-us/training/modules/introduction-to-powershell/`.

You can read the official documentation at the following link: `https://learn.microsoft.com/en-us/powershell/`.

Exercise 7.3 – Test your knowledge

Answer the following questions:

1. What is the difference between a namespace and an assembly?
2. How do you reference another project in a `.csproj` file?
3. What is the benefit of a tool like ILSpy?
4. Which .NET type does the C# `float` alias represent?
5. What are the `<PropertyGroup>` and `<ItemGroup>` elements used for in a .NET project file?
6. What is the difference between framework-dependent and self-contained deployments of .NET applications?
7. What is a RID?
8. What is the difference between the `dotnet pack` and `dotnet publish` commands?
9. How can you control the sources that a build process uses to download NuGet packages?
10. What are some limitations of using AOT compilation?

Exercise 7.4 – Explore topics

Use the links on the following page to learn more details about the topics covered in this chapter:

`https://github.com/markjprice/cs13net9/blob/main/docs/book-links.md#chapter-7---packaging-and-distributing-net-types`

Summary

In this chapter, we:

- Reviewed the journey of .NET for BCL functionality.
- Explored the relationship between assemblies and namespaces.
- Saw options to publish an app for distribution to multiple OSes.
- Learned how to publish to native AOT for faster startup and a smaller memory footprint.
- Packaged and distributed a class library.
- Learned how to activate preview features.

In the next chapter, you will learn about some common BCL types that are included with modern .NET.

8

Working with Common .NET Types

This chapter is about some common types that are included with .NET. These include types for manipulating numbers, text, and collections; improving working with spans, indexes, and ranges; and, in an optional online-only section, working with network resources.

This chapter covers the following topics:

- Working with numbers
- Working with text
- Pattern matching with regular expressions
- Storing multiple objects in collections
- Working with spans, indexes, and ranges

Working with numbers

One of the most common types of data is numbers. The most common types in .NET for working with numbers are shown in *Table 8.1*:

Namespace	Example types	Description
`System`	`SByte`, `Int16`, `Int32`, `Int64`, `Int128`	Integers; that is, zero, and positive and negative whole numbers.
`System`	`Byte`, `UInt16`, `UInt32`, `UInt64`, `UInt128`	Cardinals; that is, zero and positive whole numbers.
`System`	`Half`, `Single`, `Double`	Reals; that is, floating-point numbers.
`System`	`Decimal`	Accurate reals; that is, for use in science, engineering, or financial scenarios.

System. Numerics	BigInteger, Complex, Quaternion	Arbitrarily large integers, complex numbers, and quaternion numbers.

Table 8.1: Common .NET number types

.NET has had the 32-bit `float` and 64-bit `double` types since .NET Framework 1.0 was released in 2002. The IEEE 754 specification also defines a 16-bit floating-point standard. Machine learning and other algorithms would benefit from this smaller, lower-precision number type; so, Microsoft introduced the `System.Half` type with .NET 5 and later. Currently, the C# language does not define a `half` alias, so you must use the .NET type `System.Half`. This might change in the future.

`System.Int128` and `System.UInt128` were introduced with .NET 7, and they too do not yet have a C# alias keyword.

Working with big integers

The largest whole number that can be stored in .NET types that have a C# alias is about eighteen and a half quintillion, stored in an unsigned 64-bit integer using `ulong`. But what if you need to store numbers larger than that?

Let's explore numerics:

1. Use your preferred code editor to create a new project, as defined in the following list:
   - Project template: **Console App** / `console`
   - Project file and folder: `WorkingWithNumbers`
   - Solution file and folder: `Chapter08`
2. In the project file, add an element to statically and globally import the `System.Console` class.
3. In `Program.cs`, delete the existing statements and then add a statement to import `System.Numerics`, as shown in the following code:

```
using System.Numerics; // To use BigInteger.
```

4. Add statements to output the maximum value of the `ulong` type, and a number with 30 digits using `BigInteger`, as shown in the following code:

```
const int width = 40;

WriteLine("ulong.MaxValue vs a 30-digit BigInteger");
WriteLine(new string('-', width));

ulong big = ulong.MaxValue;
WriteLine($"{big,width:N0}");

BigInteger bigger =
  BigInteger.Parse("123456789012345678901234567890");
WriteLine($"{bigger,width:N0}");
```

The `width` constant with the value `40` in the format code means "right-align 40 characters," so both numbers are lined up to the right-hand edge. The `N0` means "use a thousand separators and zero decimal places."

5. Run the code and view the result, as shown in the following output:

```
ulong.MaxValue vs a 30-digit BigInteger
----------------------------------------
              18,446,744,073,709,551,615
 123,456,789,012,345,678,901,234,567,890
```

`BigInteger` can represent integer values of nearly unlimited length. However, in reality, this is constrained by factors like available memory and computation time. Some APIs can also fail if the value produced is excessively large. Therefore, in .NET 9 and later, a maximum length for `BigInteger` has been enforced, limiting it to `(2^31) - 1` bits (approximately 2.14 billion bits). This translates to about 256 MB and approximately 646.5 million digits. This limit ensures APIs remain reliable and consistent while still supporting numbers far larger than most practical needs.

Multiplying big integers

When you multiply two 32-bit integers, the result can exceed the range of a 32-bit integer. For instance, the maximum value of a 32-bit integer (`int`) is 2,147,483,647. If you multiply this by another large integer, the result can exceed this limit, causing an overflow and potentially resulting in incorrect calculations if only 32 bits are used to store the result.

Since the earliest versions of .NET, the `System.Math` class has a `BigMul` method that performs a multiplication of two 32-bit integers and returns the full 64-bit result as a `long`. This method is necessary because multiplying two 32-bit integers using the * operator can result in a 64-bit integer, which might overflow the typical 32-bit integer type.

Since .NET 5, `BigMul` has gained overloads to multiply two signed or unsigned `long` values. In .NET 9 and later, the number types themselves, like `int` and `long` and `uint` and `ulong`, now gain `BigMul` methods to multiply their values and return the results in the next largest type. For example, `long.BigMul` returns a `System.Int128`.

Now, let's look at how to multiply big integers:

1. In `Program.cs`, add statements to multiply some big integers using old and new methods, as shown in the following code:

```
WriteLine("Multiplying big integers");
int number1 = 2_000_000_000;
int number2 = 2;
WriteLine($"number1: {number1:N0}");
WriteLine($"number2: {number2:N0}");
WriteLine($"number1 * number2: {number1 * number2:N0}");
```

```
WriteLine($"Math.BigMul(number1, number2): {
  Math.BigMul(number1, number2):N0}");
WriteLine($"int.BigMul(number1, number2): {
  int.BigMul(number1, number2):N0}");
```

2. Run the code and view the result, as shown in the following output:

```
Multiplying big integers
number1: 2,000,000,000
number2: 2
number1 * number2: -294,967,296
Math.BigMul(number1, number2): 4,000,000,000
int.BigMul(number1, number2): 4,000,000,000
```

Working with complex numbers

A complex number can be expressed as $a + bi$, where a and b are real numbers and i is an imaginary unit, where $i^2 = -1$. If the real part a is zero, it is a purely imaginary number. If the imaginary part b is zero, it is a real number.

Complex numbers have practical applications in many **science, technology, engineering, and mathematics** (**STEM**) fields of study. They are added by separately adding the real and imaginary parts of the summands; consider this:

```
(a + bi) + (c + di) = (a + c) + (b + d)i
```

Let's explore complex numbers:

1. In `Program.cs`, add statements to add two complex numbers, as shown in the following code:

```
Complex c1 = new(real: 4, imaginary: 2);
Complex c2 = new(real: 3, imaginary: 7);
Complex c3 = c1 + c2;

// Output using the default ToString implementation.
WriteLine($"{c1} added to {c2} is {c3}");

// Output using a custom format.
WriteLine("{0} + {1}i added to {2} + {3}i is {4} + {5}i",
  c1.Real, c1.Imaginary,
  c2.Real, c2.Imaginary,
  c3.Real, c3.Imaginary);
```

2. Run the code and view the result, as shown in the following output:

```
<4; 2> added to <3; 7> is <7; 9>
4 + 2i added to 3 + 7i is 7 + 9i
```

.NET 6 and earlier used a different default format for complex numbers: `(4, 2) added to (3, 7) is (7, 9)`. In .NET 7 and later, the default format was changed to use angle brackets and semi-colons because some cultures use round brackets to indicate negative numbers and use commas for decimal numbers. At the time of writing, the official documentation has not been updated to use the new format, as shown at the following link: `https://learn.microsoft.com/en-us/dotnet/api/system.numerics.complex.tostring`.

Working with tensors

Tensors are a fundamental data structure in **artificial intelligence** (**AI**). You can think of them as specialized multidimensional arrays.

Tensors are used to represent and encode various forms of data, including text sequences (tokens), images, videos, and audio. Tensors perform computations on high-dimensional data with optimal efficiency and store weight information and intermediate calculations within neural networks.

.NET 9 introduced a new `Tensor<T>` type that enables efficient interoperation with AI libraries, such as ML.NET, TorchSharp, and ONNX Runtime, and minimizes data copying where possible. It simplifies data manipulation through intuitive indexing and slicing operations, ensuring ease of use and efficiency.

Generating random numbers for games and similar apps

In scenarios that don't need truly random numbers like games, you can create an instance of the `Random` class, as shown in the following code example:

```
Random r = new();
```

`Random` has a constructor with a parameter for specifying a seed value used to initialize its pseudo-random number generator, as shown in the following code:

```
Random r = new(Seed: 46378);
```

As you learned in *Chapter 2*, *Speaking C#*, parameter names should use *camel case*. The developer who defined the constructor for the `Random` class broke this convention. The parameter name should be `seed`, not `Seed`.

Good practice: Shared seed values act as a secret key; so, if you use the same random number generation algorithm with the same seed value in two applications, then they can generate the same "random" sequences of numbers. Sometimes this is necessary, for example, when synchronizing a GPS receiver with a satellite, or when a game needs to randomly generate the same level. But usually, you want to keep your seed secret.

Generating random numbers with the Random class

To avoid allocating more memory than necessary, .NET 6 introduced a shared static instance of `Random` that you can access instead of creating your own.

The `Random` class has commonly used methods for generating random numbers, as described in the following list:

- `Next`: This method returns a random `int` (whole number) and it takes two parameters, `minValue` and `maxValue`, but `maxValue` is not the maximum value that the method returns! It is an *exclusive upper bound*, meaning `maxValue` is one more than the maximum value returned. Use the `NextInt64` method to return a `long` integer.
- `NextDouble`: This method returns a number that is greater than or equal to `0.0` and less than and never equal to `1.0`. Use the `NextSingle` method to return a `float`.
- `NextBytes`: This method populates an array of any size with random `byte` (`0` to `255`) values. It is common to format `byte` values as hexadecimal, for example, `00` to `FF`.

Let's see some examples of generating pseudo-random numbers:

1. In `Program.cs`, add statements to access the shared `Random` instance, and then call its methods to generate random numbers, as shown in the following code:

```
Random r = Random.Shared;

// minValue is an inclusive lower bound i.e. 1 is a possible value.
// maxValue is an exclusive upper bound i.e. 7 is not a possible value.
int dieRoll = r.Next(minValue: 1, maxValue: 7); // Returns 1 to 6.
WriteLine($"Random die roll: {dieRoll}");

double randomReal = r.NextDouble(); // Returns 0.0 to less than 1.0.
WriteLine($"Random double: {randomReal}");

byte[] arrayOfBytes = new byte[256];
r.NextBytes(arrayOfBytes); // Fills array with 256 random bytes.
Write("Random bytes: ");
for (int i = 0; i < arrayOfBytes.Length; i++)
{
  Write($"{arrayOfBytes[i]:X2} ");
}
WriteLine();
```

2. Run the code and view the result, as shown in the following output:

```
Random die roll: 1
Random double: 0.06735275453092382
```

```
Random bytes: D9 38 CD F3 5B 40 2D F4 5B D0 48 DF F7 B6 67 C1 95 A1 2C 58
42 CF 70 6C C3 BE 82 D7 EC 61 0D D2 2D C4 49 7B C7 0F EA CC B3 41 F3 04
5D 29 25 B7 F7 99 8A 0F 56 20 A6 B3 57 C4 48 DA 94 2B 07 F1 15 64 EA 8D
FF 79 E6 E4 9A C8 65 C5 D8 55 3D 3C C0 2B 0B 4C 3A 0E E6 A5 91 B7 59 6C
9A 94 97 43 B7 90 EE D8 9A C6 CA A1 8F DD 0A 23 3C 01 48 E0 45 E1 D6 BD
7C 41 C8 22 8A 81 82 DC 1F 2E AD 3F 93 68 0F B5 40 7B 2B 31 FC A6 BF BA
05 C0 76 EE 58 B3 41 63 88 E5 5C 8B B5 08 5C C3 52 FF 73 69 B0 97 78 B5
3B 87 2C 12 F3 C3 AE 96 43 7D 67 2F F8 C9 31 70 BD AD B3 9B 44 53 39 5F
19 73 C8 43 0E A5 5B 6B 5A 9D 2F DF DC A3 EE C5 CF AF A4 8C 0F F2 9C 78
19 48 CE 49 A8 28 06 A3 4E 7D F7 75 AA 49 E7 4E 20 AF B1 77 0A 90 CF C1
E0 62 BC 4F 79 76 64 98 BF 63 76 B4 F9 1D A4 C4 74 03 63 02
```

In scenarios that do need truly random numbers, like cryptography, there are specialized types for that, like `RandomNumberGenerator`. I cover this and other cryptographic types in the companion book, *Tools and Skills for .NET 8*, in a chapter titled *Protecting Data and Apps Using Cryptography*, published in July 2024.

New random methods in .NET 8 and later

.NET 8 introduced two new `Random` methods, as described in the following list:

- `GetItems<T>`: This method is passed an array or read-only span of any type `T` of choices and the number of items you want to generate, and then it returns that number of items randomly selected from the choices.
- `Shuffle<T>`: This method is passed an array or span of any type `T` and the order of items is randomized.

Let's see an example of each:

1. In `Program.cs`, add statements to access the shared `Random` instance, and then call its methods to generate random numbers, as shown in the following code:

```
string[] beatles = r.GetItems(
  choices: new[] { "John", "Paul", "George", "Ringo" },
  length: 10);

Write("Random ten beatles:");
foreach (string beatle in beatles)
{
  Write($" {beatle}");
}
WriteLine();

r.Shuffle(beatles);
```

```
Write("Shuffled beatles:");
foreach (string beatle in beatles)
{
  Write($" {beatle}");
}
WriteLine();
```

2. Run the code and view the result, as shown in the following output:

```
Random ten beatles: Paul Paul John John John John Paul John George Ringo
Shuffled beatles: George John Paul Paul John John John Ringo Paul John
```

Generating GUIDs

A **globally unique identifier (GUID)** is a 128-bit text string that represents a unique value for identification. As a developer, you will need to generate GUIDs when a unique reference is needed to identify information. Traditionally, database and computer systems may have used an incrementing integer value, but a GUID is more likely to avoid conflicts in multi-tasking systems.

The `System.Guid` type is a value type (`struct`) that represents a GUID value. It has `Parse` and `TryParse` methods to take an existing GUID value represented as a `string` and convert it into the `Guid` type. It has a `NewGuid` method to generate a new value.

The `NewGuid` method implements version 4 of the **universally unique identifiers (UUIDs)** specification. You can learn about the UUID specification at the following link: `https://datatracker.ietf.org/doc/rfc9562/`.

.NET 9 introduced support for version 7 of the UUID specification, which allows you to control the first 48-bit timestamp millisecond ticks since Unix Epoch so that you can create sortable GUIDs. This is much better for unique IDs for databases.

Let's break down an example value: `0191a398-12ec-7d5e-a8d3-8a629eae8be1`:

- The first 48 bits are a timestamp in hexadecimal: `0191a39812ec`.
    - `0191a398`: The first 32 bits represent the high part of the timestamp.
    - `12ec`: The next 16 bits continue the timestamp.
- The next 12 bits are a version and variant in hexadecimal: `7d5e`.
    - The `7` indicates that this is a version 7 UUID.
    - The remaining three digits (`d5e`) combine parts of the timestamp and the variant, which is used to identify the layout of the UUID.
- The final 68 bits are a random section in hexadecimal: `a8d3-8a629eae8be1`.
    - This ensures the uniqueness of the GUID.

Let's see how we can generate GUID values and output them:

1. In `Program.cs`, add statements to output the value of an empty `Guid`, generate a new random `Guid` and then output its value, and finally output each individual byte of the random `Guid` value, as shown in the following code:

```
WriteLine($"Empty GUID: {Guid.Empty}.");
Guid g = Guid.NewGuid();
WriteLine($"Random GUID: {g}.");

byte[] guidAsBytes = g.ToByteArray();
Write("GUID as byte array: ");
for (int i = 0; i < guidAsBytes.Length; i++)
{
  Write($"{guidAsBytes[i]:X2} ");
}
WriteLine();

WriteLine("Generating three v7 GUIDs:");
for (int i = 0; i < 3; i++)
{
  Guid g7 = Guid.CreateVersion7(DateTimeOffset.UtcNow);
  WriteLine($"  {g7}.");
}
```

2. Run the code, view the result, and note that the first parts of the three GUID values (`0191a398-12ec` and so on) are sortable because they are based on the same `DateTimeOffset` value, as shown in the following output:

```
Empty GUID: 00000000-0000-0000-0000-000000000000.
Random GUID: c7a11eea-45a5-4619-964a-a9cce1e4220c.
GUID as byte array: EA 1E A1 C7 A5 45 19 46 96 4A A9 CC E1 E4 22 0C
Generating three v7 GUIDs:
  0191a398-12ec-7d5e-a8d3-8a629eae8be1.
  0191a398-12ed-7913-8096-c6f70b5edd8f.
  0191a398-12ed-7475-8284-0588b573080e.
```

The EF Core team has a proposal to enable the use of version 7 GUIDs in your data models. You can learn more about that idea at the following link: `https://github.com/dotnet/efcore/issues/34158`.

Working with text

One of the other most common types of data for variables is text. The most common types in .NET for working with text are shown in *Table 8.2*:

Namespace	Type	Description
`System`	`Char`	Storage for a single text character
`System`	`String`	Storage for multiple text characters
`System.Text`	`StringBuilder`	Efficiently manipulates strings
`System.Text.RegularExpressions`	`Regex`	Efficiently pattern-matches strings

Table 8.2: Common .NET types for working with text

Getting the length of a string

Let's explore some common tasks when working with text; for example, sometimes you need to find out the length of a piece of text stored in a `string` variable:

1. Use your preferred code editor to add a new **Console App** / `console` project named `WorkingWithText` to the `Chapter08` solution.
2. In the `WorkingWithText` project, in `Program.cs`, delete the existing statements and then add statements to define a variable to store the name of the city, London, and then write its name and length to the console, as shown in the following code:

```
string city = "London";
WriteLine($"{city} is {city.Length} characters long.");
```

3. Run the code and view the result, as shown in the following output:

```
London is 6 characters long.
```

Getting the characters of a string

The `string` class uses an array of `char` internally to store the text. It also has an indexer, which means that we can use the array syntax to read its characters. Array indexes start at 0, so the third character will be at index 2.

Let's see this in action:

1. Add a statement to write the characters at the first and fourth positions in the `string` variable, as shown in the following code:

```
WriteLine($"First char is {city[0]} and fourth is {city[3]}.");
```

2. Run the code and view the result, as shown in the following output:

```
First char is L and fourth is d.
```

Splitting a string

Sometimes, you need to split some text wherever there is a character, such as a comma:

1. Add statements to define a single `string` variable containing comma-separated city names, then use the `Split` method and specify that you want to treat commas as the separator, and then enumerate the returned array of `string` values, as shown in the following code:

```
string cities = "Paris,Tehran,Chennai,Sydney,New York,Medellín";

string[] citiesArray = cities.Split(',');

WriteLine($"There are {citiesArray.Length} items in the array:");

foreach (string item in citiesArray)
{
  WriteLine($"  {item}");
}
```

2. Run the code and view the result, as shown in the following output:

```
There are 6 items in the array:
  Paris
  Tehran
  Chennai
  Sydney
  New York
  Medellín
```

Later in this chapter, you will learn how to handle more complex string-splitting scenarios using a regular expression.

Getting part of a string

Sometimes, you need to get part of some text. The `IndexOf` method has nine overloads that return the index position of a specified `char` or `string` within a `string`. The `Substring` method has two overloads, as shown in the following list:

- `Substring(startIndex, length)`: This returns part of a string starting at `startIndex` and containing the next `length` characters.
- `Substring(startIndex)`: This returns part of a string starting at `startIndex` and containing all characters up to the end of the string.

Let's explore a simple example:

1. Add statements to store a person's full name in a `string` variable with a space character between the first and last names, find the position of the space, and then extract the first name and last name as two parts so that they can be recombined in a different order, as shown in the following code:

```
string fullName = "Alan Shore";

int indexOfTheSpace = fullName.IndexOf(' ');

string firstName = fullName.Substring(
  startIndex: 0, length: indexOfTheSpace);

string lastName = fullName.Substring(
  startIndex: indexOfTheSpace + 1);

WriteLine($"Original: {fullName}");
WriteLine($"Swapped: {lastName}, {firstName}");
```

2. Run the code and view the result, as shown in the following output:

```
Original: Alan Shore
Swapped: Shore, Alan
```

If the format of the initial full name was different, for example, `"LastName, FirstName"`, then the code would need to be different. As an optional exercise, try writing some statements that would change the input `"Shore, Alan"` into `"Alan Shore"`.

Checking a string for content

Sometimes, you need to check whether a piece of text starts or ends with some characters or contains some characters. You can achieve this with methods named `StartsWith`, `EndsWith`, and `Contains`:

1. Add statements to store a `string` value and then check if it starts with or contains a couple of different `char` values, as shown in the following code:

```
string company = "Microsoft";
WriteLine($"Text: {company}");
WriteLine("Starts with M: {0}, contains an N: {1}",
  arg0: company.StartsWith('M'),
  arg1: company.Contains('N'));
```

Good practice: `StartsWith`, `EndsWith`, and `Contains` can be passed a `string` or `char` value. The single quotes indicate a `char` value in the preceding code. Searching for a `char` like `'M'` is more efficient than searching for a `string` value like `"M"`.

2. Run the code and view the result, as shown in the following output:

```
Text: Microsoft
Starts with M: True, contains an N: False
```

Comparing string values

Two common tasks with string values are sorting (aka collating) and comparing. For example, when a user enters their username or password, you need to compare what they entered with what is stored.

The `string` class implements the `IComparable` interface, meaning that you can easily compare two string values using the `CompareTo` instance method and it will return `-1`, `0`, or `1` depending on if the value is "less than," "equal to," or "greater than" the other. You saw an example of this when you implemented the `IComparable` interface for the `Person` class in *Chapter 6, Implementing Interfaces and Inheriting Classes*.

However, the lower or upper casing of characters can affect ordering, and the ordering rules for text are culture-dependent. For example, double-L is treated as a single character in traditional Spanish, as shown in *Table 8.3*:

Culture	Description	Example string values
Spanish	In 1994, the Royal Spanish Academy issued a new alphabetization rule to treat LL and CH as Latin alphabetic characters instead of separate individual characters.	Modern: `llegar` comes before `lugar`. Traditional: `llegar` comes after `lugar`.
Swedish	In 2006, the Swedish Academy issued a new rule. Before 2006, V and W were the same character. Since 2006, they have been treated as separate characters.	Swedish words mostly only use V. Loanwords (words taken from other languages) that contain W can now keep those Ws instead of replacing the Ws with Vs.
German	Phonebook ordering is different than dictionary ordering, for example, umlauts are sorted as combinations of letters.	`Müller` and `Mueller` in phonebook ordering are the same name.
German	The character ß is sorted as `SS`. This is a common issue with addresses since the word for `street` is `Straße`.	`Straße` and `Strasse` have the same meaning.

Table 8.3: Examples of ordering rules in European languages

For consistency and performance, you sometimes want to make comparisons in a culture-invariant way. It is therefore better to use the `static` method `Compare`.

Let's see some examples:

1. At the top of `Program.cs`, import the namespace for working with cultures and enable special characters like the Euro currency symbol, as shown in the following code:

```
using System.Globalization; // To use CultureInfo.

OutputEncoding = System.Text.Encoding.UTF8; // Enable Euro symbol.
```

2. In `Program.cs`, define some text variables and compare them in different cultures, as shown in the following code:

```
CultureInfo.CurrentCulture = CultureInfo.GetCultureInfo("en-US");

string text1 = "Mark";
string text2 = "MARK";

WriteLine($"text1: {text1}, text2: {text2}");

WriteLine("Compare: {0}.", string.Compare(text1, text2));

WriteLine("Compare (ignoreCase): {0}.",
  string.Compare(text1, text2, ignoreCase: true));

WriteLine("Compare (InvariantCultureIgnoreCase): {0}.",
  string.Compare(text1, text2,
  StringComparison.InvariantCultureIgnoreCase));

// German string comparisons

CultureInfo.CurrentCulture = CultureInfo.GetCultureInfo("de-DE");

text1 = "Strasse";
text2 = "Straße";

WriteLine($"text1: {text1}, text2: {text2}");

WriteLine("Compare: {0}.", string.Compare(text1, text2,
  CultureInfo.CurrentCulture, CompareOptions.IgnoreNonSpace));

WriteLine("Compare (IgnoreCase, IgnoreNonSpace): {0}.",
  string.Compare(text1, text2, CultureInfo.CurrentCulture,
  CompareOptions.IgnoreNonSpace | CompareOptions.IgnoreCase));
```

```
WriteLine("Compare (InvariantCultureIgnoreCase): {0}.",
  string.Compare(text1, text2,
  StringComparison.InvariantCultureIgnoreCase));
```

3. Run the code, view the result, and note that a lowercase "a" is "less than" (`-1`) an uppercase "A," so the comparison returns -1. But we can either set an option to ignore case, or even better, do a culture- and case-invariant comparison to treat the two string values as equal (`0`), as shown in the following output:

```
text1: Mark, text2: MARK
Compare: -1.
Compare (ignoreCase): 0.
Compare (InvariantCultureIgnoreCase): 0.
text1: Strasse, text2: Straße
Compare: 0.
Compare (IgnoreCase, IgnoreNonSpace): 0.
Compare (InvariantCultureIgnoreCase): -1.
```

More Information: You can learn more about `string` comparisons at the following link: `https://learn.microsoft.com/en-us/globalization/locale/sorting-and-string-comparison`.

Joining, formatting, and other string members

There are many other `string` members, as shown in *Table 8.4*:

Member	Description
`Trim`, `TrimStart`, `TrimEnd`	These methods trim whitespace characters such as space, tab, and carriage return from the start and/or end.
`ToUpper`, `ToLower`	These convert all the characters into uppercase or lowercase.
`Insert`, `Remove`	These methods insert or remove some text.
`Replace`	This replaces some text with other text.
`string.Empty`	This can be used instead of allocating memory each time you use a literal string value using an empty pair of double quotes (`""`).
`string.Concat`	This concatenates two string variables. The `+` operator does the equivalent when used between string operands.
`string.Join`	This concatenates one or more string variables with a character in between each one.
`string.IsNullOrEmpty`	This checks whether a string variable is null or empty.

`string.IsNullOrWhiteSpace`	This checks whether a string variable is null or whitespace; that is, a mix of any number of horizontal and vertical spacing characters, for example, tab, space, carriage return, line feed, and so on.
`string.Format`	An alternative method to string interpolation for outputting formatted string values, which uses positioned instead of named parameters.

Table 8.4: Joining, formatting, and other string members

Some of the preceding methods are `static` methods. This means that the method can only be called from the type, not from a variable instance. In the preceding table, I indicated the static methods by prefixing them with `string.`, as in `string.Format`.

Let's explore some of these methods:

1. Add statements to take an array of `string` values and combine them back together into a single `string` variable with separators using the `Join` method, as shown in the following code:

```
string recombined = string.Join(" => ", citiesArray);
WriteLine(recombined);
```

2. Run the code and view the result, as shown in the following output:

```
Paris => Tehran => Chennai => Sydney => New York => Medellín
```

3. Add statements to use positioned parameters and interpolated `string` formatting syntax to output the same three variables twice, as shown in the following code:

```
string fruit = "Apples";
decimal price =  0.39M;
DateTime when = DateTime.Today;

WriteLine($"Interpolated:  {fruit} cost {price:C} on {when:dddd}.");
WriteLine(string.Format("string.Format: {0} cost {1:C} on {2:dddd}.",
  arg0: fruit, arg1: price, arg2: when));
```

Some code editors, like Rider, will warn you about boxing operations. These are slow but not a problem in this scenario. To avoid boxing, call `ToString` on `price` and `when`.

4. Run the code and view the result, as shown in the following output:

```
Interpolated:  Apples cost $0.39 on Friday.
string.Format: Apples cost $0.39 on Friday.
```

Note that we could have simplified the second statement because `Console.WriteLine` supports the same format code as `string.Format`, as shown in the following code:

```
WriteLine("WriteLine: {0} cost {1:C} on {2:dddd}.",
  arg0: fruit, arg1: price, arg2: when);
```

Building strings efficiently

You can concatenate two strings to make a new `string` using the `String.Concat` method or simply by using the + operator. But both choices are bad practices when combining more than a few values because .NET must create a completely new `string` in memory.

This might not be noticeable if you are only adding two `string` values, but if you concatenate inside a loop with many iterations, it can have a significant negative impact on performance and memory use. You can concatenate `string` variables more efficiently using the `StringBuilder` type.

I have written an online-only section for the companion book, *Apps and Services with .NET 8*, about performance benchmarking using `string` concatenations as the main example. You can optionally complete the section and its practical coding tasks at the following link: `https://github.com/markjprice/apps-services-net8/blob/main/docs/ch01-benchmarking.md`.

More information: You can see examples of using `StringBuilder` at the following link: `https://learn.microsoft.com/en-us/dotnet/api/system.text.stringbuilder#examples`.

Working with characters

Sometimes, you need to work with individual characters. The `System.Char` type (aka `char`) has some useful methods, as shown in *Table 8.5*:

Method	Description
`char.IsDigit(char)`, `char.IsDigit(string, int)`	Returns `true` if the character is a decimal digit.
`char.IsLetter(char)`, `char.IsLetter(string, int)`	Returns `true` if the character is a letter.
`char.IsLower(char)`, `char.IsLower(string, int)`	Returns `true` if the character is a lowercase letter.
`char.IsUpper(char)`, `char.IsUpper(string, int)`	Returns `true` if the character is an uppercase letter.

`char.IsSeparator(char)`, `char.IsSeparator(string, int)`	Returns `true` if the character is a separator.
`char.IsSymbol(char)`, `char.IsSymbol(string, int)`	Returns `true` if the character is a symbol.
`char.IsWhiteSpace(char)`, `char.IsWhiteSpace(string, int)`	Returns `true` if the character is whitespace like a space or tab.

Table 8.5: Methods for working with characters

Searching in strings

.NET 8 introduced the `SearchValues` type, which implements an optimized method for searching for sets of characters or bytes within spans. For example, `SearchValues` helps to find the first instance of a vowel in a text value, as shown in the following code:

```
string vowels = "AEIOUaeiou";

// .NET 8 and later.
SearchValues<char> vowelsSearchValues = SearchValues.Create(vowels);

ReadOnlySpan<char> text = "Fred";

WriteLine($"vowels: {vowels}");
WriteLine($"text: {text}");
WriteLine($"text.IndexOfAny(vowelsSearchValues): {
  text.IndexOfAny(vowelsSearchValues)}");
```

The first vowel in the text `Fred` is at index position 2, as shown in the following output:

```
vowels: AEIOUaeiou
text: Fred
text.IndexOfAny(vowelsSearchValues): 2
```

In .NET 9 or later, `SearchValues` now supports searching for substrings as well as searching for characters within a larger string, as shown in the following code:

```
string[] names = [ "Cassian", "Luthen", "Mon Mothma",
  "Dedra", "Syril", "Kino" ];

// .NET 9 and later.
SearchValues<string> namesSearchValues = SearchValues.Create(
  names, StringComparison.OrdinalIgnoreCase);
```

```
ReadOnlySpan<char> sentence = "In Andor, Diego Luna returns as the titular
character, Cassian Andor, to whom audiences were first introduced in Rogue
One.";

WriteLine($"names: {string.Join(' ', names)}");
WriteLine($"sentence: {sentence}");
WriteLine($"sentence.IndexOfAny(vowelsSearchValues): {
  sentence.IndexOfAny(namesSearchValues)}");
```

The first name from the list of Andor characters starts at index position 55, as shown in the following output:

```
names: Cassian Luthen Mon Mothma Dedra Syril Kino
sentence: In Andor, Diego Luna returns as the titular character, Cassian Andor,
to whom audiences were first introduced in Rogue One.
sentence.IndexOfAny(vowelsSearchValues): 55
```

Pattern matching with regular expressions

Regular expressions are useful for validating input from the user. They are very powerful and can get very complicated. Almost all programming languages have support for regular expressions and use a common set of special characters to define them.

Let's try out some example regular expressions:

1. Use your preferred code editor to add a new **Console App** / `console` project named `WorkingWithRegularExpressions` to the `Chapter08` solution.
2. In `Program.cs`, delete the existing statements and then import the following namespace:

   ```
   using System.Text.RegularExpressions; // To use Regex.
   ```

Checking for digits entered as text

We will start by implementing the common example of validating number input:

1. In `Program.cs`, add statements to prompt the user to enter their age and then check that it is valid using a regular expression that looks for a digit character, as shown in the following code:

   ```
   Write("Enter your age: ");
   string input = ReadLine()!; // Null-forgiving operator.
   Regex ageChecker = new(@"\d");
   WriteLine(ageChecker.IsMatch(input) ? "Thank you!" :
     $"This is not a valid age: {input}");
   ```

Note the following about the code:

- The `@` character switches off the ability to use escape characters in the `string`. Escape characters are prefixed with a backslash. For example, `\t` means a tab and `\n` means a new line. When writing regular expressions, we need to disable this feature. To paraphrase the television show *The West Wing*, "Let backslash be backslash."
- Once escape characters are disabled with `@`, then they can be interpreted by a regular expression. For example, `\d` means digit. You will learn about more regular expression symbols that are prefixed with a backslash later in this topic.

2. Run the code, enter a whole number, such as `34`, for the age, and view the result, as shown in the following output:

```
Enter your age: 34
Thank you!
```

3. Run the code again, enter `carrots`, and view the result, as shown in the following output:

```
Enter your age: carrots
This is not a valid age: carrots
```

4. Run the code again, enter `bob30smith`, and view the result, as shown in the following output:

```
Enter your age: bob30smith
Thank you!
```

The regular expression we used is `\d`, which means *one digit*. However, it does not specify what can be entered before and after that one digit. This regular expression could be described in English as "Enter any characters you want as long as you enter at least one digit character."

In regular expressions, you indicate the start of some input with the caret `^` symbol and the end of some input with the dollar `$` symbol. Let's use these symbols to indicate that we expect nothing else between the start and end of the input except for a digit.

5. Add a `^` and a `$` to change the regular expression to `^\d$`, as shown highlighted in the following code:

```
Regex ageChecker = new(@"^\d$");
```

6. Run the code again and note that it rejects any input except a single digit.
7. Add a `+` after the `\d` expression to modify the meaning to one or more digits, as shown highlighted in the following code:

```
Regex ageChecker = new(@"^\d+$");
```

8. Run the code again and note the regular expression only allows zero or positive whole numbers of any length.

Regular expression performance improvements

The .NET types for working with regular expressions are used throughout the .NET platform and many of the apps built with it. As such, they have a significant impact on performance. However, until .NET 5, they had not received much optimization attention from Microsoft.

With .NET 5 and later, the types in the `System.Text.RegularExpressions` namespace have rewritten implementations to squeeze out maximum performance. Common regular expression benchmarks using methods like `IsMatch` are now five times faster. The best thing is you do not have to change your code to get the benefits!

With .NET 7 and later, the `IsMatch` method of the `Regex` class now has an overload for a `ReadOnlySpan<char>` as its input, which gives even better performance.

Understanding the syntax of a regular expression

Some common symbols that you can use in regular expressions are shown in *Table 8.6*:

Symbol	Meaning	Symbol	Meaning
`^`	Start of input	`$`	End of input
`\d`	A single digit	`\D`	A single *non*-digit
`\s`	Whitespace	`\S`	*Non*-whitespace
`\w`	Word characters	`\W`	*Non*-word characters
`[A-Za-z0-9]`	Range(s) of characters	`\^`	^ (caret) character
`[aeiou]`	Set of characters	`[^aeiou]`	*Not* in a set of characters
`.`	Any single character	`\.`	. (dot) character

Table 8.6: Common regular expression symbols

In addition, some common regular expression quantifiers that affect the previous symbols in a regular expression are shown in *Table 8.7*:

Symbol	Meaning	Symbol	Meaning
`+`	One or more	`?`	One or none
`{3}`	Exactly three	`{3,5}`	Three to five
`{3,}`	At least three	`{,3}`	Up to three

Table 8.7: Common regular expression quantifiers

Examples of regular expressions

Some examples of regular expressions with a description of their meaning are shown in *Table 8.8*:

Expression	Meaning
`\d`	A single digit somewhere in the input
`a`	The character "a" somewhere in the input
`Bob`	The word "Bob" somewhere in the input
`^Bob`	The word "Bob" at the start of the input
`Bob$`	The word "Bob" at the end of the input
`^\d{2}$`	Exactly two digits
`^[0-9]{2}$`	Exactly two digits
`^[A-Z]{4,}$`	At least four uppercase English letters in the ASCII character set only
`^[A-Za-z]{4,}$`	At least four upper or lowercase English letters in the ASCII character set only
`^[A-Z]{2}\d{3}$`	Two uppercase English letters in the ASCII character set and three digits only
`^[A-Za-z\u00c0-\u017e]+$`	At least one uppercase or lowercase English letter in the ASCII character set or European letters in the Unicode character set, as shown in the following list: ÀÁÂÃÄÅÆÇÈÉÊËÌÍÎÏÐÑÒÓÔÕÖ×ØÙÚÛÜÝ Þßàáâãäåæçèéêëìíîïðñòóôõö÷øùúûüýþÿıŒœŠšŸŽž
`^d.g$`	The letter `d`, then any character, and then the letter `g`, so it would match both `dig` and `dog` or any single character between the `d` and `g`
`^d\.g$`	The letter `d`, then a dot `.`, and then the letter `g`, so it would match `d.g` only

Table 8.8: Examples of regular expressions with descriptions of their meaning

Good practice: Use regular expressions to validate input from the user. The same regular expressions can be reused in other languages, such as JavaScript and Python.

A handy website for building and testing regular expressions can be found at the following link: `https://regex101.com/`.

Splitting a complex comma-separated string

Earlier in this chapter, you learned how to split a simple comma-separated string variable. But what about the following example of film titles?

```
"Monsters, Inc.","I, Tonya","Lock, Stock and Two Smoking Barrels"
```

The `string` value uses double quotes around each film title. We can use these to identify whether we need to split on a comma (or not). The `Split` method is not powerful enough, so we can use a regular expression instead.

Good practice: You can read a fuller explanation in the Stack Overflow article that inspired this task at the following link: `https://stackoverflow.com/questions/18144431/regex-to-split-a-csv`.

To include double quotes inside a `string` value, we prefix them with a backslash, or we could use the raw string literal feature in C# 11 or later:

1. Add statements to store a complex comma-separated `string` variable, and then split it in a dumb way using the `Split` method, as shown in the following code:

```
// C# 1 to 10: Use escaped double-quote characters \"
// string films = "\"Monsters, Inc.\",\"I, Tonya\",\"Lock, Stock and Two
Smoking Barrels\"";

// C# 11 or later: Use """ to start and end a raw string literal
string films = """
"Monsters, Inc.","I, Tonya","Lock, Stock and Two Smoking Barrels"
""";

WriteLine($"Films to split: {films}");

string[] filmsDumb = films.Split(',');

WriteLine("Splitting with string.Split method:");
foreach (string film in filmsDumb)
{
  WriteLine($"  {film}");
}
```

2. Add statements to define a regular expression to split and write the film titles in a smart way, as shown in the following code:

```
Regex csv = new(
  "(?:^|,)(?=[^\"]|(\")?)\"?((?(1)[^\"]*|[^,\"]*))\"?(?=,|$)");

MatchCollection filmsSmart = csv.Matches(films);

WriteLine("Splitting with regular expression:");
```

```
foreach (Match film in filmsSmart)
{
  WriteLine($"  {film.Groups[2].Value}");
}
```

In a later section, you will see how you can get a source generator to auto-generate XML comments for a regular expression to explain how it works. This is really useful for regular expressions that you might have copied from a website.

3. Run the code and view the result, as shown in the following output:

```
Splitting with string.Split method:
  "Monsters
   Inc."
  "I
   Tonya"
  "Lock
   Stock and Two Smoking Barrels"
Splitting with regular expression:
  Monsters, Inc.
  I, Tonya
  Lock, Stock and Two Smoking Barrels
```

Activating regular expression syntax coloring

If you use Visual Studio as your code editor, then you probably noticed that when passing a `string` value to the `Regex` constructor, you see color syntax highlighting, as shown in *Figure 8.1*:

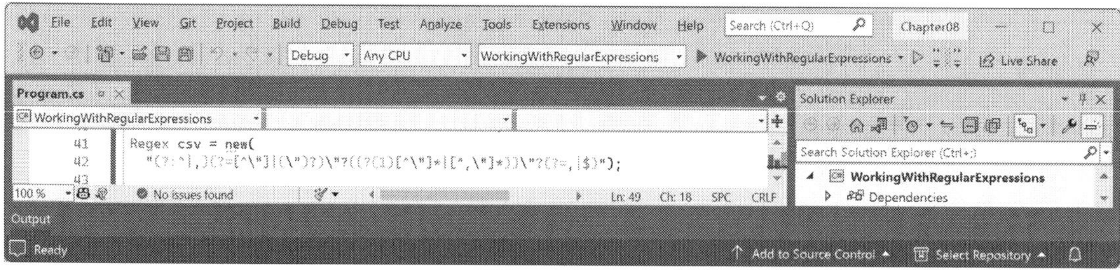

Figure 8.1: Regular expression color syntax highlighting when using the Regex constructor

This would be a good time to remind print book readers, who will only see the preceding figure in grayscale, that they can see all figures in full color as a PDF at the following link: `https://packt.link/gbp/9781837635870`.

Why does this string get syntax coloring for regular expressions when most string values do not? Let's find out:

1. Right-click on the new constructor, select **Go To Implementation**, and note the string parameter named pattern is decorated with an attribute named StringSyntax that has the string constant Regex value passed to it, as shown highlighted in the following code:

```
public Regex([StringSyntax(StringSyntaxAttribute.Regex)] string pattern)
:
  this(pattern, culture: null)
{
}
```

2. Right-click on the StringSyntax attribute, select **Go To Implementation**, and note there are 12 recognized string syntax formats that you can choose from as well as Regex, as shown in the following partial code:

```
[AttributeUsage(AttributeTargets.Property | AttributeTargets.Field |
AttributeTargets.Parameter, AllowMultiple = false, Inherited = false)]
public sealed class StringSyntaxAttribute : Attribute
{
  public const string CompositeFormat = "CompositeFormat";
  public const string DateOnlyFormat = "DateOnlyFormat";
  public const string DateTimeFormat = "DateTimeFormat";
  public const string EnumFormat = "EnumFormat";
  public const string GuidFormat = "GuidFormat";
  public const string Json = "Json";
  public const string NumericFormat = "NumericFormat";
  public const string Regex = "Regex";
  public const string TimeOnlyFormat = "TimeOnlyFormat";
  public const string TimeSpanFormat = "TimeSpanFormat";
  public const string Uri = "Uri";
  public const string Xml = "Xml";
  …
}
```

3. In the WorkingWithRegularExpressions project, add a new class file named Program.Strings.cs, delete any existing statements, and then define some string constants in a partial Program class, as shown in the following code:

```
partial class Program
{
  private const string DigitsOnlyText = @"^\d+$";
```

```
  private const string CommaSeparatorText =
    "(?:^|,)(?=[^\"]|(\")?)\"?((?(1)[^\"]*|[^,\"]*))\"?(?=,|$)";
}
```

Note that the two `string` constants do not have any color syntax highlighting yet.

4. In `Program.cs`, replace the literal `string` with the `string` constant for the digits-only regular expression, as shown highlighted in the following code:

```
Regex ageChecker = new(DigitsOnlyText);
```

5. In `Program.cs`, replace the literal `string` with the `string` constant for the comma-separator regular expression, as shown highlighted in the following code:

```
Regex csv = new(CommaSeparatorText);
```

6. Run the `WorkingWithRegularExpressions` project and confirm that the regular expression behavior is as before.
7. In `Program.Strings.cs`, import the namespace for the `[StringSyntax]` attribute and then decorate both `string` constants with it, as shown highlighted in the following code:

```
using System.Diagnostics.CodeAnalysis; // To use [StringSyntax].

partial class Program
{
  [StringSyntax(StringSyntaxAttribute.Regex)]
  private const string DigitsOnlyText = @"^\d+$";

  [StringSyntax(StringSyntaxAttribute.Regex)]
  private const string CommaSeparatorText =
    "(?:^|,)(?=[^\"]|(\")?)\"?((?(1)[^\"]*|[^,\"]*))\"?(?=,|$)";
}
```

8. In `Program.Strings.cs`, add another `string` constant for formatting a date, as shown in the following code:

```
[StringSyntax(StringSyntaxAttribute.DateTimeFormat)]
private const string FullDateTime = "";
```

9. Click inside the empty string, type the letter `d`, and note the IntelliSense explanation, as shown in *Figure 8.2*:

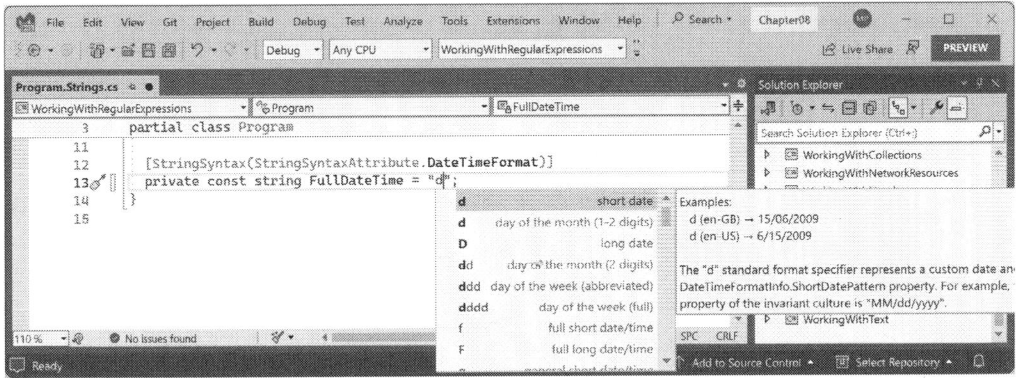

Figure 8.2: IntelliSense activated due to the StringSyntax attribute

10. Finish entering the date format and as you type, note the IntelliSense value: `dddd, d MMMM yyyy`.
11. Inside, at the end of the `DigitsOnlyText` string literal, enter `\`, and note the IntelliSense to help you write a valid regular expression, as shown in *Figure 8.3*:

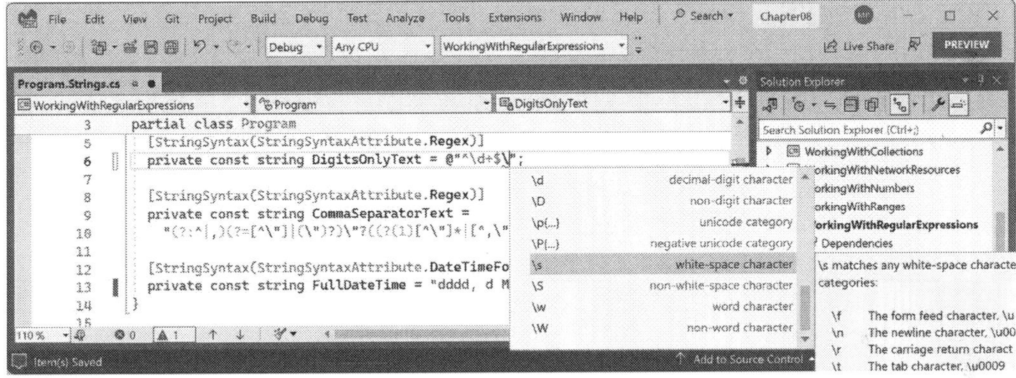

Figure 8.3: IntelliSense for writing a regular expression

12. Delete the `\` that you entered to trigger IntelliSense.

The `[StringSyntax]` attribute is a feature introduced in .NET 7. It depends on your code editor whether it is recognized. The .NET BCL has more than 350 parameters, properties, and fields that are now decorated with this attribute.

Improving regular expression performance with source generators

When you pass a string literal or string constant to the constructor of `Regex`, the class parses the string and transforms it into an internal tree structure that represents the expression in an optimized way that can be executed efficiently by a regular expression interpreter.

You can also compile regular expressions by specifying RegexOptions, as in the following code:

```
Regex ageChecker = new(DigitsOnlyText, RegexOptions.Compiled);
```

Unfortunately, compiling has the negative effect of slowing down the initial creation of the regular expression. After creating the tree structure that would then be executed by the interpreter, the compiler then must convert the tree into IL code, and then that IL code needs to be JIT compiled into native code. If you're only running the regular expression a few times, it is not worth compiling it, which is why it is not the default behavior. If you're running the regular expression more than a few times, for example, because it will be used to validate the URL for every incoming HTTP request to a website, then it is worth compiling. But even then, you should only use compilation if you must use .NET 6 or earlier.

.NET 7 introduced a source generator for regular expressions that recognizes if you decorate a partial method that returns Regex with the [GeneratedRegex] attribute. It generates an implementation of that method that implements the logic for the regular expression. .NET 9 allows you to use partial properties, which have a more natural syntax.

Let's see this in action:

1. In the WorkingWithRegularExpressions project, add a new class file named Program.Regexs.cs and modify its content to define some partial properties, as shown in the following code:

```
using System.Text.RegularExpressions; // To use [GeneratedRegex].

partial class Program
{
  [GeneratedRegex(DigitsOnlyText, RegexOptions.IgnoreCase)]
  private static partial Regex DigitsOnly { get; }

  [GeneratedRegex(CommaSeparatorText, RegexOptions.IgnoreCase)]
  private static partial Regex CommaSeparator { get; }
}
```

2. In Program.cs, replace the new constructor with a call to the partial method that returns the digits-only regular expression, as shown highlighted in the following code:

```
Regex ageChecker = DigitsOnly;
```

3. In Program.cs, replace the new constructor with a call to the partial method that returns the comma-separator regular expression, as shown highlighted in the following code:

```
Regex csv = CommaSeparator;
```

4. Hover your mouse pointer over the partial methods and note the tooltip describes the behavior of the regular expression, as shown in *Figure 8.4*:

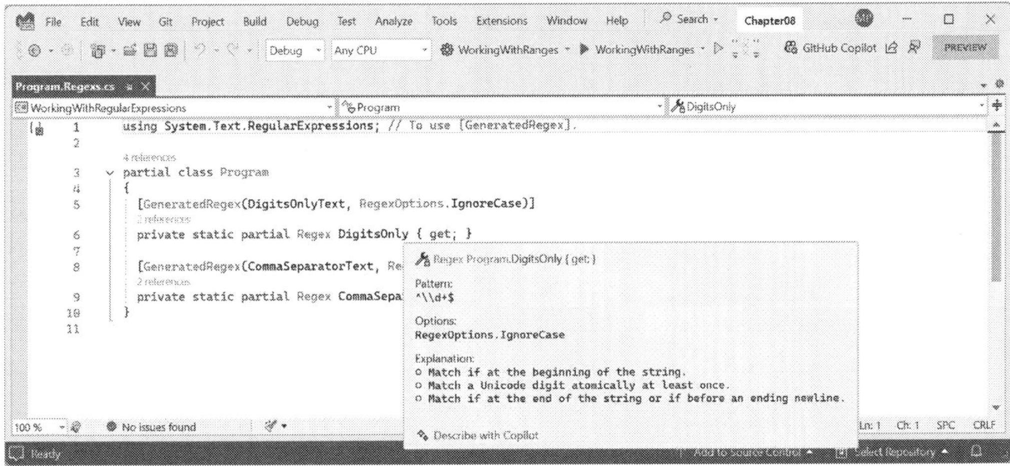

Figure 8.4: Tooltip for a partial method shows a description of the regular expression

5. Right-click the `DigitsOnly` `partial` method, select **Go To Definition**, and note that you can review the implementation of the auto-generated partial methods, as shown in *Figure 8.5*:

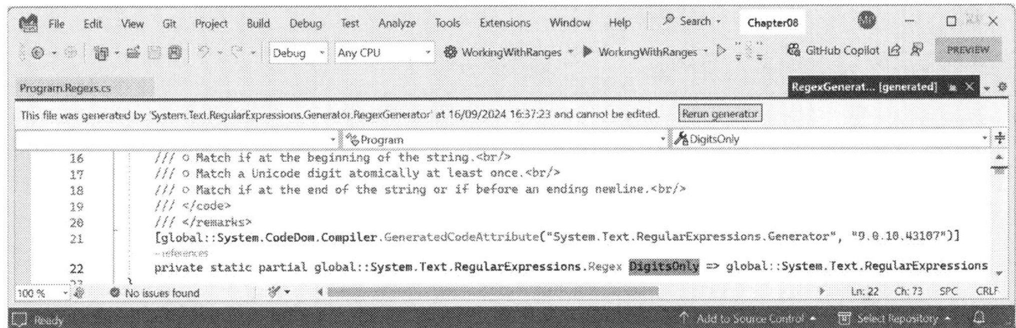

Figure 8.5: The auto-generated source code for the regular expression

6. Run the project and confirm that the functionality is the same as before.

You can learn more about the improvements to regular expressions with .NET 7 at the following link: `https://devblogs.microsoft.com/dotnet/regular-expression-improvements-in-dotnet-7`.

Storing multiple objects in collections

Another of the most common types of data is collections. If you need to store multiple values in a variable, then you can use a collection.

A collection is a data structure in memory that can manage multiple items in different ways, although all collections have some shared functionality.

The most common types in .NET for working with collections are shown in *Table 8.9*:

Namespace	Example types	Description
System.Collections	IEnumerable, IEnumerable<T>	Interfaces and base classes used by collections.
System.Collections.Generic	List<T>, Dictionary<T>, Queue<T>, Stack<T>	Introduced in C# 2.0 with .NET Framework 2.0. These collections allow you to specify the type you want to store using a generic type parameter (which is safer, faster, and more efficient).
System.Collections.Concurrent	BlockingCollection, ConcurrentDictionary, ConcurrentQueue	These collections are safe to use in multithreaded scenarios.
System.Collections.Immutable	ImmutableArray, ImmutableDictionary, ImmutableList, ImmutableQueue	Designed for scenarios where the contents of the original collection will never change, although they can create modified collections as a new instance.

Table 8.9: Common .NET collection types

Common features of all collections

All collections implement the `ICollection` interface; this means that they must have a `Count` property to tell you how many objects are in them, and three other members, as shown in the following code:

```
namespace System.Collections;

public interface ICollection : IEnumerable
{
  int Count { get; }
  bool IsSynchronized { get; }
  object SyncRoot { get; }
  void CopyTo(Array array, int index);
}
```

For example, if we had a collection named `passengers`, we could do this:

```
int howMany = passengers.Count;
```

As you have probably surmised, `CopyTo` copies the collection to an array. `IsSynchronized` and `SyncRoot` are used in multithreading scenarios, so I do not cover them in this book.

All collections implement the `IEnumerable` interface, which means that they can be iterated using the `foreach` statement. They must have a `GetEnumerator` method that returns an object that implements `IEnumerator`; this means that the returned `object` must have `MoveNext` and `Reset` methods for navigating through the collection and a `Current` property containing the current item in the collection, as shown in the following code:

```
namespace System.Collections;

public interface IEnumerable
{
  IEnumerator GetEnumerator();
}

public interface IEnumerator
{
  object Current { get; }
  bool MoveNext();
  void Reset();
}
```

For example, to perform an action on each object in the `passengers` collection, we could write the following code:

```
foreach (Passenger p in passengers)
{
  // Perform an action on each passenger.
}
```

As well as the `object`-based collection interface, there is also a generic collection interface, where the generic type defines the type stored in the collection. It has additional members like `IsReadOnly`, `Add`, `Clear`, `Contains`, and `Remove`, as shown in the following code:

```
namespace System.Collections.Generic;

public interface ICollection<T> : IEnumerable<T>, IEnumerable
{
  int Count { get; }
  bool IsReadOnly { get; }
  void Add(T item);
  void Clear();
  bool Contains(T item);
  void CopyTo(T[] array, int index);
  bool Remove(T item);
}
```

Working with lists

Lists, that is, a type that implements `IList<T>`, are **ordered collections**, with an `int` index to show the position of an item within the list, as shown in the following code:

```
namespace System.Collections.Generic;

[DefaultMember("Item")] // aka "this" indexer.
public interface IList<T> : ICollection<T>, IEnumerable<T>, IEnumerable
{
  T this[int index] { get; set; }
  int IndexOf(T item);
  void Insert(int index, T item);
  void RemoveAt(int index);
}
```

The `[DefaultMember]` attribute allows you to specify which member is accessed by default when no member name is specified. To make `IndexOf` the default member, you would use `[DefaultMember("IndexOf")]`. To specify the indexer, you use `[DefaultMember("Item")]`.

`IList<T>` derives from `ICollection<T>`, so it has a `Count` property, and an `Add` method to put an item at the end of the collection, as well as an `Insert` method to put an item in the list at a specified position, and `RemoveAt` to remove an item at a specified position.

Lists are a good choice when you want to manually control the order of items in a collection. Each item in a list has a unique index (or position) that is automatically assigned. Items can be any type defined by `T` and items can be duplicated. Indexes are `int` types and start from `0`, so the first item in a list is at index `0`, as shown in *Table 8.10*:

Index	Item
0	London
1	Paris
2	London
3	Sydney

Table 8.10: Cities in a list with indexes

If a new item (for example, Santiago) is inserted between London and Sydney, then the index of Sydney is automatically incremented. Therefore, you must be aware that an item's index can change after inserting or removing items, as shown in *Table 8.11*:

Index	Item
0	London
1	Paris
2	London
3	Santiago
4	Sydney

Table 8.11: Cities list after an item is inserted

Good practice: Some developers can get into the poor habit of using `List<T>` and other collections when an array would be better. Use arrays instead of collections if the data will not change size after instantiation. You should also use lists initially while you are adding and removing items, but then convert them into an array once you are done with manipulating the items.

Let's explore lists:

1. Use your preferred code editor to add a new **Console App** / `console` project named `WorkingWithCollections` to the `Chapter08` solution.
2. Add a new class file named `Program.Helpers.cs`.
3. In `Program.Helpers.cs`, define a partial `Program` class with a generic method to output a collection of `T` values with a title, as shown in the following code:

```
partial class Program
{
  private static void OutputCollection<T>(
    string title, IEnumerable<T> collection)
  {
    WriteLine($"{title}:");
    foreach (T item in collection)
    {
      WriteLine($"  {item}");
    }
  }
}
```

4. In `Program.cs`, delete the existing statements and then add some statements to illustrate some of the common ways of defining and working with lists, as shown in the following code:

```
// Simple syntax for creating a list and adding three items.
List<string> cities = new();
cities.Add("London");
cities.Add("Paris");
```

```
cities.Add("Milan");

/* Alternative syntax that is converted by the compiler into
   the three Add method calls above.
List<string> cities = new()
  { "London", "Paris", "Milan" }; */

/* Alternative syntax that passes an array
   of string values to AddRange method.
List<string> cities = new();
cities.AddRange(new[] { "London", "Paris", "Milan" }); */

OutputCollection("Initial list", cities);
WriteLine($"The first city is {cities[0]}.");
WriteLine($"The last city is {cities[cities.Count - 1]}.");
WriteLine($"Are all cities longer than four characters? {
  cities.TrueForAll(city => city.Length > 4)}.");
WriteLine($"Do all cities contain the character 'e'? {
  cities.TrueForAll(city => city.Contains('e'))}.");

cities.Insert(0, "Sydney");
OutputCollection("After inserting Sydney at index 0", cities);

cities.RemoveAt(1);
cities.Remove("Milan");
OutputCollection("After removing two cities", cities);
```

5. Run the code and view the result, as shown in the following output:

```
Initial list:
  London
  Paris
  Milan
The first city is London.
The last city is Milan.
Are all cities longer than four characters? True.
Do all cities contain the character 'e'? False.
After inserting Sydney at index 0:
  Sydney
  London
  Paris
  Milan
After removing two cities:
```

```
Sydney
Paris
```

Good practice: LINQ has extension methods named `All()` and `Count()` that work like the `List<T>` class `TrueForAll()` method and `Count` property respectively. You will generally achieve better performance by using methods provided by a class rather than the more general LINQ methods.

Working with dictionaries

Dictionaries are a good choice when each **value** (or object) has a unique sub-value (or a made-up value) that can be used as a **key** to quickly find a value in the collection later. The key must be unique. For example, if you are storing a list of people, you could choose to use a government-issued identity number as the key. Dictionaries are called **hashmaps** in other languages like Python and Java.

Think of the key as being like an index entry in a real-world dictionary. It allows you to quickly find the definition of a word because the words (in other words, keys) are kept sorted; if we know we're looking for the definition of *manatee*, we will jump to the middle of the dictionary to start looking, because the letter *m* is in the middle of the alphabet.

Dictionaries in programming are similarly smart when looking something up. They must implement the interface `IDictionary<TKey, TValue>`, as shown in the following code:

```
namespace System.Collections.Generic;

[DefaultMember("Item")] // aka "this" indexer.
public interface IDictionary<TKey, TValue>
  : ICollection<KeyValuePair<TKey, TValue>>,
    IEnumerable<KeyValuePair<TKey, TValue>>, IEnumerable
{
  TValue this[TKey key] { get; set; }
  ICollection<TKey> Keys { get; }
  ICollection<TValue> Values { get; }
  void Add(TKey key, TValue value);
  bool ContainsKey(TKey key);
  bool Remove(TKey key);
  bool TryGetValue(TKey key, [MaybeNullWhen(false)] out TValue value);
}
```

Items in a dictionary are instances of the `struct`, aka the value type, `KeyValuePair<TKey, TValue>`, where `TKey` is the type of the key and `TValue` is the type of the value, as shown in the following code:

```
namespace System.Collections.Generic;

public readonly struct KeyValuePair<TKey, TValue>
```

```
{
  public KeyValuePair(TKey key, TValue value);
  public TKey Key { get; }
  public TValue Value { get; }
  [EditorBrowsable(EditorBrowsableState.Never)]
  public void Deconstruct(out TKey key, out TValue value);
  public override string ToString();
}
```

An example `Dictionary<string, Person>` uses a `string` as the key and a `Person` instance as the value. `Dictionary<string, string>` uses `string` values for both, as shown in *Table 8.12*:

Key	Value
BSA	Bob Smith
MW	Max Williams
BSB	Bob Smith
AM	Amir Mohammed

Table 8.12: Dictionary with keys and values

Let's explore dictionaries:

1. At the top of `Program.cs`, define an alias for the `Dictionary<TKey, TValue>` class, where `TKey` and `TValue` are both `string`, as shown in the following code:

```
// Define an alias for a dictionary with string key and string value.
using StringDictionary = System.Collections.Generic.Dictionary<string,
string>;
```

2. In `Program.cs`, add some statements to illustrate some of the common ways of working with dictionaries, for example, looking up word definitions, as shown in the following code:

```
// Declare a dictionary without the alias.
// Dictionary<string, string> keywords = new();

// Use the alias to declare the dictionary.
StringDictionary keywords = new();

// Add using named parameters.
keywords.Add(key: "int", value: "32-bit integer data type");

// Add using positional parameters.
keywords.Add("long", "64-bit integer data type");
keywords.Add("float", "Single precision floating point number");
```

```
/* Alternative syntax; compiler converts this to calls to Add method.
Dictionary<string, string> keywords = new()
{
  { "int", "32-bit integer data type" },
  { "long", "64-bit integer data type" },
  { "float", "Single precision floating point number" },
}; */

/* Alternative syntax; compiler converts this to calls to Add method.
Dictionary<string, string> keywords = new()
{
  ["int"] = "32-bit integer data type",
  ["long"] = "64-bit integer data type",
  ["float"] = "Single precision floating point number",
}; */

OutputCollection("Dictionary keys", keywords.Keys);
OutputCollection("Dictionary values", keywords.Values);

WriteLine("Keywords and their definitions:");
foreach (KeyValuePair<string, string> item in keywords)
{
  WriteLine($"  {item.Key}: {item.Value}");
}

// Look up a value using a key.
string key = "long";
WriteLine($"The definition of {key} is {keywords[key]}.");
```

The trailing commas after the third item is added to the dictionary are optional and the compiler will not complain about them. This is convenient so that you can change the order of the three items without having to delete and add commas in the right places.

3. Run the code and view the result, as shown in the following output:

```
Dictionary keys:
  int
  long
  float
Dictionary values:
  32-bit integer data type
```

```
  64-bit integer data type
  Single precision floating point number
Keywords and their definitions:
  int: 32-bit integer data type
  long: 64-bit integer data type
  float: Single precision floating point number
The definition of long is 64-bit integer data type
```

In *Chapter 11*, *Querying and Manipulating Data Using LINQ*, you will learn how to create dictionaries and lookups from existing data sources, like tables in a database, using LINQ methods, like `ToDictionary` and `ToLookup`. This is much more common than manually adding items to a dictionary, as shown in this section.

Sets, stacks, and queues

Sets are a good choice when you want to perform set operations between two collections. For example, you may have two collections of city names, and you want to know which names appear in both sets (known as the *intersect* between the sets). Items in a set must be unique.

Common set methods are shown in *Table 8.13*:

Method	Description
`Add`	If the item does not already exist in the set, then it is added. This returns `true` if the item was added, and `false` if it was already in the set.
`ExceptWith`	This removes the items in the set passed as the parameter from the set.
`IntersectWith`	This removes the items not in the set passed as the parameter and in the set.
`IsProperSubsetOf`, `IsProperSupersetOf`, `IsSubsetOf`, `IsSupersetOf`	A subset is a set whose items are all in the other set. A proper subset is a set whose items are all in the other set but there is at least one item in the other set that is not in the set. A superset is a set that contains all the items in the other set. A proper superset is a set that contains all the items in the other set and at least one more not in the other set.
`Overlaps`	The set and the other set share at least one common item.
`SetEquals`	The set and the other set contain exactly the same items.
`SymmetricExceptWith`	This removes the items not in the set passed as the parameter from the set and adds any that are missing.
`UnionWith`	This adds any items in the set passed as the parameter to the set that are not already in the set.

Table 8.13: Set methods

Let's explore example code for sets:

1. In `Program.cs`, add some statements to add items to a set, as shown in the following code:

```
HashSet<string> names = new();

foreach (string name in
  new[] { "Adam", "Barry", "Charlie", "Barry" })
{
  bool added = names.Add(name);
  WriteLine($"{name} was added: {added}.");
}

WriteLine($"names set: {string.Join(',', names)}.");
```

2. Run the code and view the result, as shown in the following output:

```
Adam was added: True.
Barry was added: True.
Charlie was added: True.
Barry was added: False.
names set: Adam,Barry,Charlie.
```

You will see more set operations in *Chapter 11, Querying and Manipulating Data Using LINQ*.

Stacks are a good choice when you want to implement **last-in, first-out** (**LIFO**) behavior. With a stack, you can only directly access or remove the one item at the top of the stack, although you can enumerate to read through the whole stack of items. You cannot, for example, directly access the second item in a stack.

For example, word processors use a stack to remember the sequence of actions you have recently performed, and then, when you press *Ctrl* + *Z*, it will undo the last action in the stack, and then the next-to-last action, and so on.

Queues are a good choice when you want to implement **first-in, first-out** (**FIFO**) behavior. With a queue, you can only directly access or remove the item at the front of the queue, although you can enumerate to read through the whole queue of items. You cannot, for example, directly access the second item in a queue.

For example, background processes use a queue to process work items in the order that they arrive, just like people standing in line at the post office.

.NET 6 introduced the `PriorityQueue`, where each item in the queue has a priority value assigned, as well as its position in the queue.

Let's explore example code for queues:

1. In `Program.cs`, add some statements to illustrate some of the common ways of working with queues, for example, handling customers in a queue for coffee, as shown in the following code:

```
Queue<string> coffee = new();

coffee.Enqueue("Damir"); // Front of the queue.
coffee.Enqueue("Andrea");
coffee.Enqueue("Ronald");
coffee.Enqueue("Amin");
coffee.Enqueue("Irina"); // Back of the queue.

OutputCollection("Initial queue from front to back", coffee);

// Server handles next person in queue.
string served = coffee.Dequeue();
WriteLine($"Served: {served}.");

// Server handles next person in queue.
served = coffee.Dequeue();
WriteLine($"Served: {served}.");
OutputCollection("Current queue from front to back", coffee);

WriteLine($"{coffee.Peek()} is next in line.");
OutputCollection("Current queue from front to back", coffee);
```

2. Run the code and view the result, as shown in the following output:

```
Initial queue from front to back:
  Damir
  Andrea
  Ronald
  Amin
  Irina
Served: Damir.
Served: Andrea.
Current queue from front to back:
  Ronald
  Amin
  Irina
Ronald is next in line.
```

```
Current queue from front to back:
  Ronald
  Amin
  Irina
```

3. In `Program.Helpers.cs`, in the partial `Program` class, add a static method named `OutputPQ`, as shown in the following code:

```
private static void OutputPQ<TElement, TPriority>(string title,
  IEnumerable<(TElement Element, TPriority Priority)> collection)
{
  WriteLine($"{title}:");
  foreach ((TElement, TPriority) item in collection)
  {
    WriteLine($"  {item.Item1}: {item.Item2}");
  }
}
```

Note that the `OutputPQ` method is generic. You can specify the two types used in the tuples that are passed in as `collection`.

4. In `Program.cs`, add some statements to illustrate some of the common ways of working with priority queues, as shown in the following code:

```
PriorityQueue<string, int> vaccine = new();

// Add some people.
// 1 = High priority people in their 70s or poor health.
// 2 = Medium priority e.g. middle-aged.
// 3 = Low priority e.g. teens and twenties.

vaccine.Enqueue("Pamela", 1);
vaccine.Enqueue("Rebecca", 3);
vaccine.Enqueue("Juliet", 2);
vaccine.Enqueue("Ian", 1);

OutputPQ("Current queue for vaccination", vaccine.UnorderedItems);

WriteLine($"{vaccine.Dequeue()} has been vaccinated.");
WriteLine($"{vaccine.Dequeue()} has been vaccinated.");
OutputPQ("Current queue for vaccination", vaccine.UnorderedItems);
```

```
WriteLine($"{vaccine.Dequeue()} has been vaccinated.");

WriteLine("Adding Mark to queue with priority 2.");
vaccine.Enqueue("Mark", 2);

WriteLine($"{vaccine.Peek()} will be next to be vaccinated.");
OutputPQ("Current queue for vaccination", vaccine.UnorderedItems);
```

5. Run the code and view the result, as shown in the following output:

```
Current queue for vaccination:
  Pamela: 1
  Rebecca: 3
  Juliet: 2
  Ian: 1
Pamela has been vaccinated.
Ian has been vaccinated.
Current queue for vaccination:
  Juliet: 2
  Rebecca: 3
Juliet has been vaccinated.
Adding Mark to queue with priority 2
Mark will be next to be vaccinated.
Current queue for vaccination:
  Mark: 2
  Rebecca: 3
```

.NET 9 introduced a `Remove` method to `PriorityQueue`. The method performs a linear-time scan through each element in the heap, removing the first occurrence that matches the specified element parameter. If there are duplicate entries, the method removes one of them in a non-deterministic manner, without considering any priority.

Collection add and remove methods

Each collection has a different set of methods to "add" and "remove" items, as shown in *Table 8.14*:

Collection	"Add" methods	"Remove" methods	Description
`List`	`Add`, `Insert`	`Remove`, `RemoveAt`	Lists are ordered so items have an integer index position. `Add` will add a new item at the end of the list. `Insert` will add a new item at the index position specified.

Dictionary	Add	Remove	Dictionaries are not ordered so items do not have integer index positions. You can check if a key has been used by calling the ContainsKey method.
Stack	Push	Pop	Stacks always add a new item at the top of the stack using the Push method. The first item is at the bottom. Items are always removed from the top of the stack using the Pop method. Call the Peek method to see this value without removing it. Stacks are LIFO.
Queue	Enqueue	Dequeue	Queues always add a new item at the end of the queue using the Enqueue method. The first item is at the front of the queue. Items are always removed from the front of the queue using the Dequeue method. Call the Peek method to see this value without removing it. Queues are FIFO.

Table 8.14: Collection "add" and "remove" methods

Sorting collections

A `List<T>` class can be sorted by manually calling its `Sort` method (but remember that the indexes of each item will change). Manually sorting a list of `string` values or other built-in types will work without extra effort on your part, but if you create a collection of your own type, then that type must implement an interface named `IComparable`. You learned how to do this in *Chapter 6, Implementing Interfaces and Inheriting Classes.*

A `Stack<T>` or `Queue<T>` collection cannot be sorted because you wouldn't usually want that functionality; for example, you would probably never sort a queue of guests checking into a hotel. But sometimes, you might want to sort a dictionary or a set.

Sometimes, it would be useful to have an automatically sorted collection, that is, one that maintains the items in a sorted order as you add and remove them.

There are multiple auto-sorting collections to choose from. The differences between these sorted collections are often subtle but can have an impact on the memory requirements and performance of your application, so it is worth putting effort into picking the most appropriate option for your requirements.

Since the early days of .NET, the `OrderedDictionary` type has supported this, but it was non-generic, with keys and values typed as `System.Object`. .NET 9 introduces the long-awaited `OrderedDictionary<TKey, TValue>` collection, offering an efficient and generic solution for these scenarios.

Some other common auto-sorting collections are shown in *Table 8.15*:

Collection	Description
`SortedDictionary<TKey, TValue>`	This represents a collection of key-value pairs that are sorted by key. Internally, it maintains a binary tree for items.
`SortedList<TKey, TValue>`	This represents a collection of key-value pairs that are sorted by key. The name is misleading because this is not a list. Compared to `SortedDictionary<TKey, TValue>`, retrieval performance is similar, it uses less memory, and insert and remove operations are slower for unsorted data. If it is populated from sorted data, then it is faster. Internally, it maintains a sorted array with a binary search to find elements.
`SortedSet<T>`	This represents a collection of unique objects that are maintained in a sorted order.

Table 8.15: Common auto-sorting collections

Specialized collections

There are a few other collections for special situations.

The `System.Collections.BitArray` collection manages a compact array of bit values, which are represented as Booleans, where `true` indicates that the bit is on (value is 1) and `false` indicates that the bit is off (value is 0).

The `System.Collections.Generics.LinkedList<T>` collection represents a doubly linked list where every item has a reference to its previous and next items. They provide better performance compared to `List<T>` for scenarios where you will frequently insert and remove items from the middle of the list. In a `LinkedList<T>`, the items do not have to be rearranged in memory.

Read-only, immutable, and frozen collections

When we looked at the generic collection interface, we saw that it has a property named `IsReadOnly`. This is useful when we want to pass a collection to a method but not allow it to make changes.

For example, we might define a method as shown in the following code:

```
void ReadCollection<T>(ICollection<T> collection)
{
  // We can check if the collection is read-only.
  if (collection.IsReadOnly)
  {
    // Read the collection.
  }
  else
  {
```

```
        WriteLine("You have given me a collection that I could change!");
      }
    }
```

Generic collections, like `List<T>` and `Dictionary<TKey, TValue>`, have an `AsReadOnly` method to create a `ReadOnlyCollection<T>` that references the original collection. Although the `ReadOnlyCollection<T>` has to have an `Add` and a `Remove` method because it implements `ICollection<T>`, it throws a `NotImplementedException` to prevent changes.

If the original collection has items added or removed, the `ReadOnlyCollection<T>` will see those changes. You can think of a `ReadOnlyCollection<T>` as a protected view of a collection.

Let's see how we can make sure a collection is read-only:

1. In the `WorkingWithCollections` project, in `Program.Helpers.cs`, add a method that should only be given a read-only dictionary with `string` for the type of key and value, but the naughty method tries to call `Add`, as shown in the following code:

```
private static void UseDictionary(
  IDictionary<string, string> dictionary)
{
  WriteLine($"Count before is {dictionary.Count}.");
  try
  {
    WriteLine("Adding new item with GUID values.");
    // Add method with return type of void.
    dictionary.Add(
      key: Guid.NewGuid().ToString(),
      value: Guid.NewGuid().ToString());
  }
  catch (NotSupportedException)
  {
    WriteLine("This dictionary does not support the Add method.");
  }
  WriteLine($"Count after is {dictionary.Count}.");
}
```

Note the type of parameter is `IDictionary<TKey, TValue>`. Using an interface provides more flexibility because we can pass either a `Dictionary<TKey, TValue>`, a `ReadOnlyDictionary<TKey, TValue>`, or anything else that implements that interface.

2. In `Program.cs`, add statements to pass the `keywords` dictionary to this naughty method, as shown in the following code:

```
UseDictionary(keywords);
```

3. Run the code, view the result, and note that the naughty method was able to add a new key-value pair, so the count has incremented, as shown in the following output:

```
Count before is 3.
Adding new item with GUID values.
Count after is 4.
```

4. In `Program.cs`, comment out the `UseDictionary` statement, and then add a statement to pass the dictionary converted into a read-only collection, as shown in the following code:

```
//UseDictionary(keywords);
UseDictionary(keywords.AsReadOnly());
```

5. Run the code, view the result, and note that this time, the method was not able to add an item, so the count is the same, as shown in the following output:

```
Count before is 3.
Adding new item with GUID values.
This dictionary does not support the Add method.
Count after is 3.
```

6. At the top of `Program.cs`, import the `System.Collections.Immutable` namespace, as shown in the following code:

```
using System.Collections.Immutable; // To use ImmutableDictionary<T, T>.
```

7. In `Program.cs`, comment out the `AsReadOnly` statement and then add a statement to pass the keywords converted into an immutable dictionary, as shown highlighted in the following code:

```
//UseDictionary(keywords.AsReadOnly());
UseDictionary(keywords.ToImmutableDictionary());
```

8. Run the code, view the result, and note that this time, the method was also not able to add a default value, so the count is the same – it is the same behavior as using a read-only collection, so what's the point of immutable collections?

If you import the `System.Collections.Immutable` namespace, then any collection that implements `IEnumerable<T>` is given six extension methods to convert it into an immutable collection, like a list, dictionary, set, and so on.

Although the immutable collection will have a method named `Add`, it does not add an item to the original immutable collection! Instead, it returns a new immutable collection with the new item in it. The original immutable collection still only has the original items in it.

Let's see an example:

1. In `Program.cs`, add statements to convert the keywords dictionary into an immutable dictionary, and then add a new keyword definition to it by randomly generating GUID values, as shown in the following code:

```
ImmutableDictionary<string, string> immutableKeywords =
  keywords.ToImmutableDictionary();

// Call the Add method with a return value.
ImmutableDictionary<string, string> newDictionary =
  immutableKeywords.Add(
    key: Guid.NewGuid().ToString(),
    value: Guid.NewGuid().ToString());

OutputCollection("Immutable keywords dictionary", immutableKeywords);
OutputCollection("New keywords dictionary", newDictionary);
```

2. Run the code, view the result, and note that the immutable keywords dictionary does not get modified when you call the `Add` method on it; instead, it returns a new dictionary with all the existing keywords plus the newly added keyword, as shown in the following output:

```
Immutable keywords dictionary:
  [float, Single precision floating point number]
  [long, 64-bit integer data type]
  [int, 32-bit integer data type]
New keywords dictionary:
  [d0e099ff-995f-4463-ae7f-7b59ed3c8d1d, 3f8e4c38-c7a3-4b20-acb3-
01b2e3c86e8c]
  [float, Single precision floating point number]
  [long, 64-bit integer data type]
  [int, 32-bit integer data type]
```

Newly added items will not always appear at the top of the dictionary, as shown in the output above. Internally, the order is defined by the hash of the key. This is why dictionaries are sometimes called hash tables.

Good practice: To improve performance, many applications store a shared copy of commonly accessed objects in a central cache. To safely allow multiple threads to work with those objects knowing they won't change, you should make them immutable or use a concurrent collection type, which you can read about at the following link: `https://learn.microsoft.com/en-us/dotnet/api/system.collections.concurrent`.

The generic collections have some potential performance issues related to how they are designed.

First, being generic, the types of items or types used for keys and values for a dictionary have a big effect on performance, depending on what they are. Since they could be any type, the .NET team cannot optimize the algorithm. `string` and `int` types are the most used in real life. If the .NET team could rely on those always being the types used, then they could greatly improve performance.

Second, collections are dynamic, meaning that new items can be added, and existing items can be removed at any time. Even more optimizations could be made if the .NET team knew that no more changes would be made to the collection.

.NET 8 introduced a new concept: frozen collections. Hmmm, we already have immutable collections, so what is different about frozen collections? Are they delicious like ice cream? The idea is that 95% of the time, a collection is populated and then never changed. So, if we could optimize them at the time of creation, then those optimizations could be made, adding some time and effort upfront, but then after that, performance for reading the collection could be greatly improved.

In .NET 8, there are only two frozen collections: `FrozenDictionary<TKey, TValue>` and `FrozenSet<T>`. More may come in future versions of .NET, but these are the two most common scenarios that would benefit from the frozen concept.

Let's go:

1. At the top of `Program.cs`, import the `System.Collections.Frozen` namespace, as shown in the following code:

```
using System.Collections.Frozen; // To use FrozenDictionary<T, T>.
```

2. At the bottom of `Program.cs`, add statements to convert the keywords dictionary into a frozen dictionary, output its items, and then look up the definition of `long`, as shown in the following code:

```
// Creating a frozen collection has an overhead to perform the
// sometimes complex optimizations.
FrozenDictionary<string, string> frozenKeywords =
  keywords.ToFrozenDictionary();

OutputCollection("Frozen keywords dictionary", frozenKeywords);

// Lookups are faster in a frozen dictionary.
WriteLine($"Define long: {frozenKeywords["long"]}");
```

3. Run the code and view the result, as shown in the following output:

```
Frozen keywords dictionary:
  [int, 32-bit integer data type]
  [long, 64-bit integer data type]
```

```
  [float, Single precision floating point number]
Define long: 64-bit integer data type
```

What the `Add` method does depends on the type, as summarized in the following list:

- `List<T>`: This adds a new item to the end of the existing list.
- `Dictionary<TKey, TValue>`: This adds a new item to the existing dictionary in a position determined by its internal structure.
- `ReadOnlyCollection<T>`: This throws a not-supported exception.
- `ImmutableList<T>`: This returns a new list with the new item in it. This does not affect the original list.
- `ImmutableDictionary<TKey, TValue>`: This returns a new dictionary with the new item in it. This does not affect the original dictionary.
- `FrozenDictionary<TKey, TValue>`: This does not exist.

More information: The documentation for frozen collections is found at the following link: `https://learn.microsoft.com/en-us/dotnet/api/system.collections.frozen`.

You've seen that you can create read-only lists and dictionaries, but what about sets? In .NET 9, there is now a `ReadOnlySet<T>` to address this.

Initializing collections using collection expressions

Introduced with C# 12 is a new consistent syntax for initializing arrays, collections, and span variables.

With C# 11 and earlier, you would have to declare and initialize an array, collection, or span of `int` values using the following code:

```
int[] numbersArray11 = { 1, 3, 5 };
List<int> numbersList11 = new() { 1, 3, 5 };
Span<int> numbersSpan11 = stackalloc int[] { 1, 3, 5 };
```

Starting with C# 12, you can now consistently use square brackets, and the compiler will do the right thing, as shown in the following code:

```
int[] numbersArray12 = [ 1, 3, 5 ];
List<int> numbersList12 = [ 1, 3, 5 ];
Span<int> numbersSpan12 = [ 1, 3, 5 ];
```

More information: You can learn more about collection expressions at the following link: `https://learn.microsoft.com/en-us/dotnet/csharp/language-reference/proposals/csharp-12.0/collection-expressions`.

Collection expressions do not work with all collections. For example, they do not work with dictionaries or multidimensional arrays. The documentation lists the types that a collection expression can be converted to: `https://learn.microsoft.com/en-us/dotnet/csharp/language-reference/proposals/csharp-12.0/collection-expressions#conversions`.

Using the spread element

The **spread element** `..` can be prefixed before any expression that can be enumerated to evaluate it in a collection expression. For example, any type that can be enumerated using `foreach`, like an array or collection, can be evaluated using the spread element.

The use of the spread element `..` in a collection expression replaces its argument with the elements from that collection. You can combine spread elements with individual elements in a collection expression.

Microsoft official documentation uses both the **spread element** and **spread operator** to refer to the same language feature. I prefer the element because it is used in collection expressions to represent an element within the defined collection.

Let's see an example of using the spread element, as shown in the following code:

```
int[] row0 = [1, 2, 3];
int[] row1 = [4, 5];
int[] row2 = [6, 7, 8, 9];

// Use the spread element to combine the three arrays and an integer into a
single array.
int[] combinedRows = [..row0, ..row1, ..row2, 10];

foreach (int number in combinedRows)
{
  Console.Write($"{number}, ");
}
```

The output would be:

```
1, 2, 3, 4, 5, 6, 7, 8, 9, 10,
```

More information: You can learn more about the spread element at the following link: `https://learn.microsoft.com/en-us/dotnet/csharp/language-reference/operators/collection-expressions#spread-element`.

Warning! Be careful not to confuse the spread element `..` that must be applied before an enumerable expression, with the range operator `..` that is used to define a `Range`. You will learn about ranges in the next major section in this chapter, *Working with spans, indexes, and ranges*. There is a discussion about the design decision around the spread element at the following link: `https://learn.microsoft.com/en-us/dotnet/csharp/language-reference/proposals/csharp-12.0/collection-expressions#drawbacks`.

Good practice with collections

Let's end this section about collections by reviewing some good practices that you should follow when working with collections.

Presizing collections

Since .NET 1.1, types like `StringBuilder` have had a method named `EnsureCapacity` that can presize its internal storage array to the expected final size of the `string`. This improves performance because it does not have to repeatedly increment the size of the array as more characters are appended.

Since .NET Core 2.1, types like `Dictionary<T>` and `HashSet<T>` have also had `EnsureCapacity`.

In .NET 6 and later, collections like `List<T>`, `Queue<T>`, and `Stack<T>` now have an `EnsureCapacity` method too, as shown in the following code:

```
List<string> names = new();
names.EnsureCapacity(10_000);
// Load ten thousand names into the list.
```

Passing collections to methods

Let's say you need to create a method to process a collection. For maximum flexibility, you could declare the input parameter to be `IEnumerable<T>` and make the method generic, as shown in the following code:

```
void ProcessCollection<T>(IEnumerable<T> collection)
{
  // Process the items in the collection,
  // perhaps using a foreach statement.
}
```

I could pass an array, a list, a queue, or a stack, containing any type, like `int`, `string`, `Person`, or anything else that implements `IEnumerable<T>`, into this method and it will process the items. However, the flexibility to pass any collection to this method comes at a performance cost.

One of the performance problems with `IEnumerable<T>` is also one of its benefits: deferred execution, also known as lazy loading. Types that implement this interface do not have to implement deferred execution, but many do.

But the worst performance problem with `IEnumerable<T>` is that the iteration must allocate an object on the heap. To avoid this memory allocation, you should define your method using a concrete type, as shown highlighted in the following code:

```
void ProcessCollection<T>(List<T> collection)
{
  // Process the items in the collection,
  // perhaps using a foreach statement.
}
```

This will use the `List<T>.Enumerator GetEnumerator()` method, which returns a `struct`, instead of the `IEnumerator<T> GetEnumerator()` method, which returns a reference type. Your code will be two to three times faster and require less memory. As with all recommendations related to performance, you should confirm the benefit by running performance tests on your actual code in a product environment.

Returning collections from members

Collections are reference types, which means they can be `null`. You might define methods or properties that return `null`, as shown in the following code:

```
public class Vehicle
{
  public List<Person>? GetPassengers()
  {
    ICollection<Person> passengers = GetFromDatabase();
    if (passengers.Count > 0)
    {
      return passengers;
    }
    else
    {
      return null;
    }
  }

  public List<Person>? Passengers
  {
    get
    {
      ICollection<Person> passengers = GetFromDatabase();
      if (passengers.Count > 0)
      {
        return passengers;
      }
```

```
          else
          {
            return null;
          }
        }
      }
    }
```

This can cause issues if a developer calls your methods and properties that return a collection without checking for `null`, as shown in the following code:

```
var people = car.GetPassengers();

// Accessing people could throw a NullReferenceException!
WriteLine($"There are {people.Count} people.");

foreach (Person p in car.Passengers)
{
  // Process each person.
}
```

In your implementations of methods and properties that return collections, return an empty collection or array instead of `null`, as shown in the following code:

```
// Return an empty sequence instead.
return Enumerable.Empty<Person>();

// Or an empty array.
return Array,Empty<Person>();

// Or an empty collection expression.
return [];
```

Working with spans, indexes, and ranges

One of Microsoft's goals with .NET Core 2.1 was to improve performance and resource usage. A key .NET feature that enables this is the `Span<T>` type.

Using memory efficiently using spans

When manipulating arrays, you will often create new copies or subsets of existing ones so that you can process just the subset. This is not efficient because duplicate objects must be created in memory.

If you need to work with a subset of an array, use a **span** because it is like a window into the original array. This is more efficient in terms of memory usage and improves performance. Spans only work with arrays, not collections, because the memory must be contiguous.

Before we look at spans in more detail, we need to understand some related objects: indexes and ranges.

Identifying positions with the Index type

C# 8 introduced two features for identifying an item's index position within an array and a range of items using two indexes.

You learned in the previous section that objects in a list can be accessed by passing an integer into their indexer, as shown in the following code:

```
int index = 3;
Person p = people[index]; // Fourth person in array.
char letter = name[index]; // Fourth letter in name.
```

The `Index` value type is a more formal way of identifying a position, and supports counting from the end, as shown in the following code:

```
// Two ways to define the same index, 3 in from the start.
Index i1 = new(value: 3); // Counts from the start
Index i2 = 3; // Using implicit int conversion operator.

// Two ways to define the same index, 5 in from the end.
Index i3 = new(value: 5, fromEnd: true);
Index i4 = ^5; // Using the caret ^ operator.
```

Identifying ranges with the Range type

The `Range` value type uses `Index` values to indicate the start and end of its range, using its constructor, C# syntax, or its static methods, as shown in the following code:

```
Range r1 = new(start: new Index(3), end: new Index(7));
Range r2 = new(start: 3, end: 7); // Using implicit int conversion.
Range r3 = 3..7; // Using C# 8.0 or later syntax.
Range r4 = Range.StartAt(3); // From index 3 to last index.
Range r5 = 3..; // From index 3 to last index.
Range r6 = Range.EndAt(3); // From index 0 to index 3.
Range r7 = ..3; // From index 0 to index 3.
```

Extension methods have been added to `string` values (which internally use an array of `char`), `int` arrays, and spans to make ranges easier to work with. These extension methods accept a range as a parameter and return a `Span<T>`. This makes them very memory-efficient.

Warning! The range operator `..` (introduced with C# 8) and the spread element `..` (introduced with C# 12) look identical. Be careful to differentiate them when reading code. The range operator should always appear next to an integer, either before, after, or both. The spread element should always appear before a named variable that is enumerable.

Using indexes, ranges, and spans

Let's explore using indexes and ranges to return spans:

1. Use your preferred code editor to add a new **Console App** / `console` project named `WorkingWithRanges` to the `Chapter08` solution.
2. In `Program.cs`, delete the existing statements and then add statements to compare using the `string` type's `Substring` method with ranges to extract parts of someone's name, as shown in the following code:

```
string name = "Samantha Jones";

// Getting the lengths of the first and last names.
int lengthOfFirst = name.IndexOf(' ');
int lengthOfLast = name.Length - lengthOfFirst - 1;

// Using Substring.
string firstName = name.Substring(
  startIndex: 0,
  length: lengthOfFirst);

string lastName = name.Substring(
  startIndex: name.Length - lengthOfLast,
  length: lengthOfLast);

WriteLine($"First: {firstName}, Last: {lastName}");

// Using spans.
ReadOnlySpan<char> nameAsSpan = name.AsSpan();
ReadOnlySpan<char> firstNameSpan = nameAsSpan[0..lengthOfFirst];
ReadOnlySpan<char> lastNameSpan = nameAsSpan[^lengthOfLast..];

WriteLine($"First: {firstNameSpan}, Last: {lastNameSpan}");
```

3. Run the code and view the result, as shown in the following output:

```
First: Samantha, Last: Jones
First: Samantha, Last: Jones
```

Using spans for efficient text handling

One of the common problems with working with text is that copies of `string` values are often allocated in memory during processing. It would be more efficient to reuse the original values when possible.

Spans enable this by creating data structures that point to the original parts of a string.

For example, consider a `string` value containing integers separated by addition symbols that need to be summed up:

With .NET 9 and later, we can now create a span over the `string` value, and then use ranges to efficiently divide the integer values and pass them to the `int.Parse` method, as shown in the following code:

```
ReadOnlySpan<char> text = "12+23+456".AsSpan();

int sum = 0;
foreach (Range r in text.Split('+'))
{
  sum += int.Parse(text[r]);
}
WriteLine($"Sum using Split: {sum}");
```

If you set a breakpoint in the loop, then note the range object named `r` is first `[0..2]`, as shown in *Figure 8.6*, then `{3..5}`, and finally `{6..9}`:

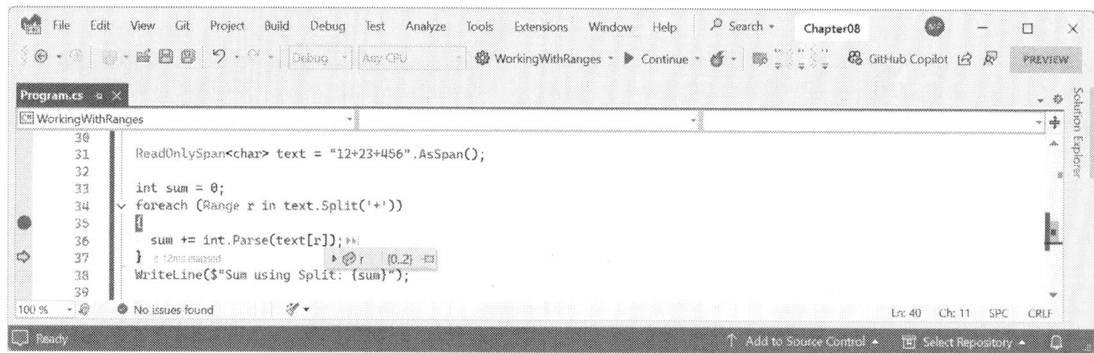

Figure 8.6: A span and range used to split a string efficiently

The range values efficiently define windows into the original string without allocating new memory resources.

Practicing and exploring

Test your knowledge and understanding by answering some questions, getting some hands-on practice, and exploring, with deeper research into the topics in this chapter.

Exercise 8.1 – online material

Online material could be created by Microsoft or third parties, or extra content for this book.

Working with network resources

If you are interested in some low-level types for working with network resources, then you can read an online-only section found at the following link:

`https://github.com/markjprice/cs13net9/blob/main/docs/ch08-network-resources.md`

There is a useful online book about networking with .NET by Chris Woodruff called *Beyond Boundaries – Networking Programming with C# 12 and .NET 8*, which you can find at the following link: `https://csharp-networking.com`.

Collection expressions

You can learn more about how to refactor existing code to make the most of collection expressions at the following link:

`https://devblogs.microsoft.com/dotnet/refactor-your-code-with-collection-expressions/`

Exercise 8.2 – practice exercises

Practice exercises go deeper into the topics for this chapter.

Regular expressions

In the `Chapter08` solution, create a console app named `Exercise_RegularExpressions` that prompts the user to enter a regular expression and then prompts the user to enter some input, and compare the two for a match until the user presses *Esc*, as shown in the following output:

```
The default regular expression checks for at least one digit.
Enter a regular expression (or press ENTER to use the default): ^[a-z]+$
Enter some input: apples
apples matches ^[a-z]+$? True
Press ESC to end or any key to try again.
Enter a regular expression (or press ENTER to use the default): ^[a-z]+$
Enter some input: abc123xyz
abc123xyz matches ^[a-z]+$? False
Press ESC to end or any key to try again.
```

Extension methods

In the `Chapter08` solution, create a class library named `Exercise_NumbersAsWordsLib` and projects to test it. It should define extension methods that extend number types such as `BigInteger` and `int` with a method named `ToWords` that returns a `string` describing the number.

For example, `18,000,000` would be eighteen million, and `18,456,002,032,011,000,007` would be eighteen quintillion, four hundred and fifty-six quadrillion, two trillion, thirty-two billion, eleven million, and seven.

You can read more about names for large numbers at the following link: `https://en.wikipedia.org/wiki/Names_of_large_numbers`.

Create projects to test your class library using xUnit and as an interactive console app.

Exercise 8.3 – test your knowledge

Use the web to answer the following questions:

1. What is the maximum number of characters that can be stored in a `string` variable?
2. When and why should you use a `SecureString` type?
3. When is it appropriate to use a `StringBuilder` class?
4. When should you use a `LinkedList<T>` class?
5. When should you use a `SortedDictionary<T>` class rather than a `SortedList<T>` class?
6. In a regular expression, what does `$` mean?
7. In a regular expression, how can you represent digits?
8. Why should you *not* use the official standard for email addresses to create a regular expression to validate a user's email address?
9. What characters are output when the following code runs?

```
string city = "Aberdeen";
ReadOnlySpan<char> citySpan = city.AsSpan()[^5..^0];
WriteLine(citySpan.ToString());
```

10. How could you check that a web service is available before calling it?

Exercise 8.4 – explore topics

Use the following link to learn more details about the topics covered in this chapter:

https://github.com/markjprice/cs13net9/blob/main/docs/book-links.md#chapter-8---working-with-common-net-types

Summary

In this chapter, you explored:

- Choices for types to store and manipulate numbers.
- Handling text, including using regular expressions for validating input.
- Collections to use for storing multiple items.
- Working with indexes, ranges, and spans.

In the next chapter, we will manage files and streams, encode and decode text, and perform serialization.

Learn more on Discord

To join the Discord community for this book – where you can share feedback, ask questions to the author, and learn about new releases – follow the QR code below:

https://packt.link/csharp13dotnet9

Leave a Review!

Thank you for purchasing this book from Packt Publishing—we hope you enjoy it! Your feedback is invaluable and helps us improve and grow. Once you've completed reading it, please take a moment to leave an Amazon review; it will only take a minute, but it makes a big difference for readers like you.

Scan the QR or visit the link to receive a free ebook of your choice.

https://packt.link/NzOWQ

Working with Files, Streams, and Serialization

This chapter is about reading and writing to files and streams, text encoding, and serialization. Applications that do not interact with a filesystem are extraordinarily rare. As a .NET developer, almost every application that you build will need to manage the filesystem and create, open, read, and write to and from files. Most of those files will contain text, so it is important to understand how text is encoded. And finally, after working with objects in memory, you will need to store them somewhere permanently for later reuse. You do that using a technique called serialization.

In this chapter, we will cover the following topics:

- Managing a filesystem
- Reading and writing with streams
- Encoding and decoding text
- Serializing object graphs

Managing a filesystem

Your applications will often need to perform input and output operations with files and directories in different environments. The `System` and `System.IO` namespaces contain classes for this purpose.

Handling cross-platform environments and filesystems

Let's explore how to handle cross-platform environments and the differences between Windows, Linux, and macOS. Paths are different for Windows, macOS, and Linux, so we will start by exploring how .NET handles this:

1. Use your preferred code editor to create a new project, as defined in the following list:
    - Project template: **Console App** / `console`
    - Project file and folder: `WorkingWithFileSystems`
    - Solution file and folder: `Chapter09`

2. In the project file, add a package reference for `Spectre.Console`, and then add elements to import the following classes statically and globally, `System.Console`, `System.IO.Directory`, `System.IO.Path`, and `System.Environment`, as shown in the following markup:

```
<ItemGroup>
  <PackageReference Include="Spectre.Console" Version="0.47.0" />
</ItemGroup>

<ItemGroup>
  <Using Include="System.Console" Static="true" />
  <Using Include="System.IO.Directory" Static="true" />
  <Using Include="System.IO.Path" Static="true" />
  <Using Include="System.Environment" Static="true" />
</ItemGroup>
```

3. Build the `WorkingWithFileSystems` project to restore packages.
4. Add a new class file named `Program.Helpers.cs`.
5. In `Program.Helpers.cs`, add a partial `Program` class with a `SectionTitle` method, as shown in the following code:

```
// null namespace to merge with auto-generated Program.

partial class Program
{
  private static void SectionTitle(string title)
  {
    WriteLine();
    ConsoleColor previousColor = ForegroundColor;
    // Use a color that stands out on your system.
    ForegroundColor = ConsoleColor.DarkYellow;
    WriteLine($"*** {title} ***");
    ForegroundColor = previousColor;
  }
}
```

6. In `Program.cs`, add statements to use a `Spectre.Console` table to do the following:
   - Output the path and directory separation characters.
   - Output the path of the current directory.
   - Output some special paths for system files, temporary files, and documents:

```
using Spectre.Console; // To use Table.

#region Handling cross-platform environments and filesystems
```

```
SectionTitle("Handling cross-platform environments and filesystems");

// Create a Spectre Console table.
Table table = new();

// Add two columns with markup for colors.
table.AddColumn("[blue]MEMBER[/]");
table.AddColumn("[blue]VALUE[/]");

// Add rows.
table.AddRow("Path.PathSeparator", PathSeparator.ToString());
table.AddRow("Path.DirectorySeparatorChar",
  DirectorySeparatorChar.ToString());
table.AddRow("Directory.GetCurrentDirectory()",
  GetCurrentDirectory());
table.AddRow("Environment.CurrentDirectory", CurrentDirectory);
table.AddRow("Environment.SystemDirectory", SystemDirectory);
table.AddRow("Path.GetTempPath()", GetTempPath());
table.AddRow("");
table.AddRow("GetFolderPath(SpecialFolder", "");
table.AddRow("  .System)", GetFolderPath(SpecialFolder.System));
table.AddRow("  .ApplicationData)",
  GetFolderPath(SpecialFolder.ApplicationData));
table.AddRow("  .MyDocuments)",
  GetFolderPath(SpecialFolder.MyDocuments));
table.AddRow("  .Personal)",
  GetFolderPath(SpecialFolder.Personal));

// Render the table to the console
AnsiConsole.Write(table);

#endregion
```

The `Environment` type has many other useful members that we did not use in this code, including the `OSVersion` and `ProcessorCount` properties.

7. Run the code and view the result, as shown using Visual Studio on Windows in *Figure 9.1*:

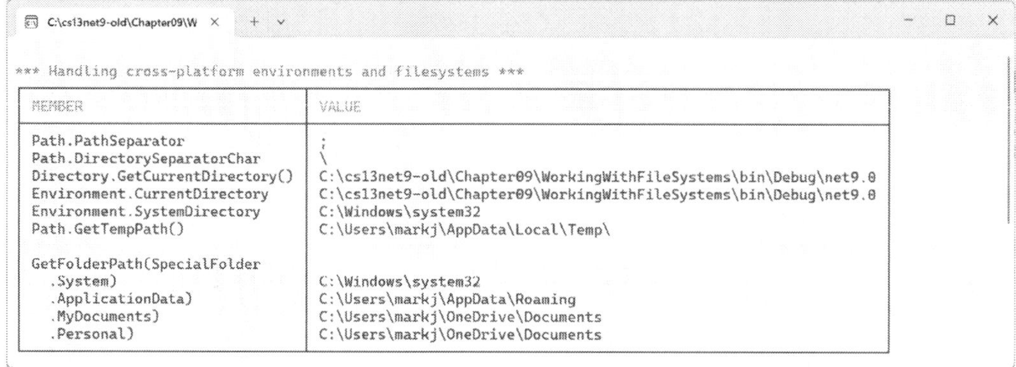

Figure 9.1: Showing filesystem information with Visual Studio on Windows

More Information: You can learn more about using Spectre Console tables at the following link: `https://spectreconsole.net/widgets/table`.

When running the console app using `dotnet run` on a Mac, the path and directory separator characters are different, and the `CurrentDirectory` will be the project folder, not a folder inside `bin`, as shown in *Figure 9.2*:

Figure 9.2: Showing filesystem information with the CLI on macOS

Good Practice: Windows uses a backslash (\) for the directory separator character. macOS and Linux use a forward slash (/) for the directory separator character. Do not assume which character is used in your code when combining paths; use `Path.DirectorySeparatorChar`.

In future sections of this chapter, we will create directories and files in the `Personal` special folder, so make a note of where that is for your operating system. For example, if you're using Linux, it should be `$USER/Documents`.

Managing drives

To manage drives, use the `DriveInfo` type, which has a static method that returns information about all the drives connected to your computer. Each drive has a drive type.

Let's explore drives:

1. In `Program.cs`, write statements to get all the drives and output their name, type, size, available free space, and format, but only if the drive is ready, as shown in the following code:

```
SectionTitle("Managing drives");

Table drives = new();

drives.AddColumn("[blue]NAME[/]");
drives.AddColumn("[blue]TYPE[/]");
drives.AddColumn("[blue]FORMAT[/]");
drives.AddColumn(new TableColumn(
  "[blue]SIZE (BYTES)[/]").RightAligned());
drives.AddColumn(new TableColumn(
  "[blue]FREE SPACE[/]").RightAligned());

foreach (DriveInfo drive in DriveInfo.GetDrives())
{
  if (drive.IsReady)
  {
    drives.AddRow(drive.Name, drive.DriveType.ToString(),
      drive.DriveFormat, drive.TotalSize.ToString("N0"),
      drive.AvailableFreeSpace.ToString("N0"));
  }
  else
  {
    drives.AddRow(drive.Name, drive.DriveType.ToString(),
      string.Empty, string.Empty, string.Empty);
  }
}

AnsiConsole.Write(drives);
```

Good Practice: Check that a drive is ready before reading properties such as `TotalSize`, or you will see an exception thrown with removable drives.

On Linux, by default, your console app will only have permission to read the `Name` and `DriveType` properties when run as a normal user. An `UnauthorizedAccessException` is thrown for `DriveFormat`, `TotalSize`, and `AvailableFreeSpace`. Run the console app as a superuser to avoid this issue, as shown in the following command: `sudo dotnet run`. Using `sudo` is fine in a development environment, but in a production environment, it's recommended to edit your permissions to avoid running with elevated permissions. On Linux, the name and drive format columns might also need to be wider, for example, 55 and 12 characters wide, respectively.

2. Run the code and view the result, as shown in *Figure 9.3*:

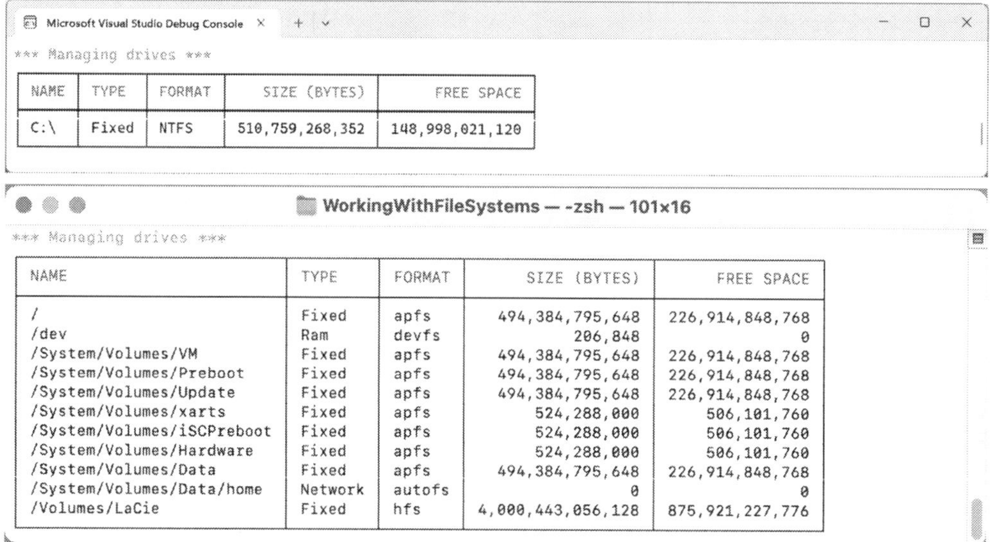

NAME	TYPE	FORMAT	SIZE (BYTES)	FREE SPACE
C:\	Fixed	NTFS	510,759,268,352	148,998,021,120

NAME	TYPE	FORMAT	SIZE (BYTES)	FREE SPACE
/	Fixed	apfs	494,384,795,648	226,914,848,768
/dev	Ram	devfs	206,848	0
/System/Volumes/VM	Fixed	apfs	494,384,795,648	226,914,848,768
/System/Volumes/Preboot	Fixed	apfs	494,384,795,648	226,914,848,768
/System/Volumes/Update	Fixed	apfs	494,384,795,648	226,914,848,768
/System/Volumes/xarts	Fixed	apfs	524,288,000	506,101,760
/System/Volumes/iSCPreboot	Fixed	apfs	524,288,000	506,101,760
/System/Volumes/Hardware	Fixed	apfs	524,288,000	506,101,760
/System/Volumes/Data	Fixed	apfs	494,384,795,648	226,914,848,768
/System/Volumes/Data/home	Network	autofs	0	0
/Volumes/LaCie	Fixed	hfs	4,000,443,056,128	875,921,227,776

Figure 9.3: Showing drive information on Windows and macOS

Managing directories

To manage directories, use the `Directory`, `Path`, and `Environment` static classes. These types include many members to work with the filesystem.

When constructing custom paths, you must be careful to write your code so that it makes no assumptions about the platform, for example, what to use for the directory separator character:

1. In `Program.cs`, write statements to do the following:
    - Define a custom path under the user's home directory by creating an array of strings for the directory names, and then properly combine them with the `Path` type's `Combine` method.
    - Check for the existence of the custom directory path using the `Exists` method of the `Directory` class.
    - Create and then delete the directory, including the files and subdirectories within it, using the `CreateDirectory` and `Delete` methods of the `Directory` class:

    ```
    SectionTitle("Managing directories");

    string newFolder = Combine(
      GetFolderPath(SpecialFolder.Personal), "NewFolder");

    WriteLine($"Working with: {newFolder}");

    // We must explicitly say which Exists method to use
    // because we statically imported both Path and Directory.
    WriteLine($"Does it exist? {Path.Exists(newFolder)}");

    WriteLine("Creating it...");
    CreateDirectory(newFolder);

    // Let's use the Directory.Exists method this time.
    WriteLine($"Does it exist? {Directory.Exists(newFolder)}");
    Write("Confirm the directory exists, and then press any key.");
    ReadKey(intercept: true);

    WriteLine("Deleting it...");
    Delete(newFolder, recursive: true);
    WriteLine($"Does it exist? {Path.Exists(newFolder)}");
    ```

In .NET 6 and earlier, only the `Directory` class had an `Exists` method. In .NET 7 or later, the `Path` class also has an `Exists` method. Both can be used to check for the existence of a path.

2. Run the code, view the result, and use your favorite file management tool to confirm that the directory has been created before pressing *Enter* to delete it, as shown in the following output:

```
Working with: C:\Users\markj\OneDrive\Documents\NewFolder
Does it exist? False
Creating it...
Does it exist? True
Confirm the directory exists, and then press any key.
Deleting it...
Does it exist? False
```

Managing files

When working with files, you can statically import the file type, just as we did for the directory type. However, for the next example, we will not do so because it has some of the same methods as the directory type, and they would conflict. The file type has a short enough name not to matter in this case. The steps are as follows:

1. In `Program.cs`, write statements to do the following:
    a. Check for the existence of a file.
    b. Create a text file.
    c. Write a line of text to the file.
    d. Close the file to release system resources and file locks (this would normally be done inside a `try-finally` statement block to ensure that the file is closed, even if an exception occurs when writing to it).
    e. Copy the file to a backup.
    f. Delete the original file.
    g. Read the backup file's contents and then close it:

```
SectionTitle("Managing files");

// Define a directory path to output files starting
// in the user's folder.
string dir = Combine(
  GetFolderPath(SpecialFolder.Personal), "OutputFiles");

CreateDirectory(dir);

// Define file paths.
string textFile = Combine(dir, "Dummy.txt");
string backupFile = Combine(dir, "Dummy.bak");
WriteLine($"Working with: {textFile}");
```

```
WriteLine($"Does it exist? {File.Exists(textFile)}");

// Create a new text file and write a line to it.
StreamWriter textWriter = File.CreateText(textFile);
textWriter.WriteLine("Hello, C#!");
textWriter.Close(); // Close file and release resources.
WriteLine($"Does it exist? {File.Exists(textFile)}");

// Copy the file, and overwrite if it already exists.
File.Copy(sourceFileName: textFile,
  destFileName: backupFile, overwrite: true);

WriteLine(
  $"Does {backupFile} exist? {File.Exists(backupFile)}");

Write("Confirm the files exist, and then press any key.");
ReadKey(intercept: true);

// Delete the file.
File.Delete(textFile);
WriteLine($"Does it exist? {File.Exists(textFile)}");

// Read from the text file backup.
WriteLine($"Reading contents of {backupFile}:");
StreamReader textReader = File.OpenText(backupFile);
WriteLine(textReader.ReadToEnd());
textReader.Close();
```

2. Run the code and view the result, as shown in the following output:

```
Working with: C:\Users\markj\OneDrive\Documents\OutputFiles\Dummy.txt
Does it exist? False
Does it exist? True
Does C:\Users\markj\OneDrive\Documents\OutputFiles\Dummy.bak exist? True
Confirm the files exist, and then press any key.
Does it exist? False
Reading contents of C:\Users\markj\OneDrive\Documents\OutputFiles\Dummy.
bak:
Hello, C#!
```

Managing paths

Sometimes, you need to work with parts of a path; for example, you might want to extract just the folder name, the filename, or the extension. Sometimes, you need to generate temporary folders and filenames. You can do this with static methods of the `Path` class:

1. In `Program.cs`, add the following statements:

```
SectionTitle("Managing paths");

WriteLine($"Folder Name: {GetDirectoryName(textFile)}");
WriteLine($"File Name: {GetFileName(textFile)}");
WriteLine("File Name without Extension: {0}",
  GetFileNameWithoutExtension(textFile));
WriteLine($"File Extension: {GetExtension(textFile)}");
WriteLine($"Random File Name: {GetRandomFileName()}");
WriteLine($"Temporary File Name: {GetTempFileName()}");
```

2. Run the code and view the result, as shown in the following output:

```
Folder Name: C:\Users\markj\OneDrive\Documents\OutputFiles
File Name: Dummy.txt
File Name without Extension: Dummy
File Extension: .txt
Random File Name: u45w1zki.co3
Temporary File Name:
C:\Users\markj\AppData\Local\Temp\tmphdmipz.tmp
```

`GetTempFileName` creates a zero-byte file and returns its name, ready for you to use. `GetRandomFileName` just returns a filename; it doesn't create the file.

Getting file information

To get more information about a file or directory, for example, its size or when it was last accessed, you can create an instance of the `FileInfo` or `DirectoryInfo` class.

`FileInfo` and `DirectoryInfo` both inherit from `FileSystemInfo`, so they both have members such as `LastAccessTime` and `Delete`, as well as extra members specific to themselves, as shown in *Table 9.1*:

Class	Members
FileSystemInfo	Fields: FullPath and OriginalPath Properties: Attributes, CreationTime, CreationTimeUtc, Exists, Extension, FullName, LastAccessTime, LastAccessTimeUtc, LastWriteTime, LastWriteTimeUtc, and Name Methods: Delete, GetObjectData, and Refresh
DirectoryInfo	Properties: Parent and Root Methods: Create, CreateSubdirectory, EnumerateDirectories, EnumerateFiles, EnumerateFileSystemInfos, GetAccessControl, GetDirectories, GetFiles, GetFileSystemInfos, MoveTo, and SetAccessControl
FileInfo	Properties: Directory, DirectoryName, IsReadOnly, and Length Methods: AppendText, CopyTo, Create, CreateText, Decrypt, Encrypt, GetAccessControl, MoveTo, Open, OpenRead, OpenText, OpenWrite, Replace, and SetAccessControl

Table 9.1: Classes to get information about files and directories

Let's write some code that uses a `FileInfo` instance to efficiently perform multiple actions on a file:

1. In `Program.cs`, add statements to create an instance of `FileInfo` for the backup file, and write information about it to the console, as shown in the following code:

```
SectionTitle("Getting file information");

FileInfo info = new(backupFile);
WriteLine($"{backupFile}:");
WriteLine($"  Contains {info.Length} bytes.");
WriteLine($"  Last accessed: {info.LastAccessTime}");
WriteLine($"  Has readonly set to {info.IsReadOnly}.");
```

2. Run the code and view the result, as shown in the following output:

```
C:\Users\markj\OneDrive\Documents\OutputFiles\Dummy.bak:
  Contains 12 bytes.
  Last accessed: 13/07/2023 12:11:12
  Has readonly set to False.
```

The number of bytes might be different on your operating system because operating systems can use different line endings.

Controlling how you work with files

When working with files, you often need to control how they are opened. The `File.Open` method has overloads to specify additional options using `enum` values.

The `enum` types are as follows:

- `FileMode`: This controls what you want to do with a file, like `CreateNew`, `OpenOrCreate`, or `Truncate`.
- `FileAccess`: This controls what level of access you need, like `ReadWrite`.
- `FileShare`: This controls locks on a file to allow other processes the specified level of access, like `Read`.

You might want to open a file and read from it, allowing other processes to read it too, as shown in the following code:

```
FileStream file = File.Open(pathToFile,
  FileMode.Open, FileAccess.Read, FileShare.Read);
```

There is also an `enum` for attributes of a file, `FileAttributes`, that checks a `FileSystemInfo`-derived type's `Attributes` property for values like `Archive` and `Encrypted`. For example, you could check a file or directory's attributes, as shown in the following code:

```
FileInfo info = new(backupFile);
WriteLine("Is the backup file compressed? {0}",
  info.Attributes.HasFlag(FileAttributes.Compressed));
```

All the methods of the `File` class that manage a file have a parameter to specify the path to the file as a `string` value. Additional parameters of the `File` class, its methods, and what the methods return are shown in *Table 9.2*:

Method	Special parameters	Returns	Notes
`Open`	`FileMode`, `FileAccess`, `FileShare`	`FileStream`	Byte-level access to a file.
`OpenWrite`		`FileStream`	Overwrites from the start but does not truncate.
`Create`	`FileOptions`	`FileStream`	Overwrites and truncates.
`OpenText`		`StreamReader`	For reading a text file.
`CreateText`		`StreamWriter`	Overwrites and truncates.
`AppendText`		`StreamWriter`	Creates the file if it does not exist.
`ReadAllLines`	`Encoding`	`string[]`	Warning! This uses a lot of memory for large files.

ReadAllText	Encoding	string	Warning! This uses a lot of memory for large files.
WriteAllText	string, Encoding	void	
AppendAllText	string, Encoding	void	
WriteAllLines	string[], IEnumerable<string>, Encoding	void	
AppendAllLines	string[], IEnumerable<string>, Encoding	void	

Table 9.2: File class and its methods

Now that you've learned some common ways to work with the directories and files in a filesystem, we next need to learn how to read and write data stored in a file, that is, how to work with streams.

Reading and writing with streams

In *Chapter 10, Working with Data Using Entity Framework Core*, you will use a file named `Northwind.db`, but you will not work with the file directly. Instead, you will interact with the SQLite database engine, which, in turn, will read and write to the file. In scenarios where there is no other system that "owns" the file and does the reading and writing for you, you will use a file stream to work directly with the file.

A **stream** is a sequence of bytes that can be read from and written to. Although files can be processed rather like arrays, with random access provided by knowing the position of a byte within the file, it is more efficient to process a file as a stream in which the bytes can be accessed in sequential order. When a human does the processing, they tend to need random access so that they can jump around the data, make changes, and then return to the data they worked on earlier. When an automated system does the processing, it tends to be able to work sequentially and only needs to "touch" the data once.

Streams can also be used to process terminal input and output and networking resources, such as sockets and ports, that do not provide random access and cannot seek (that is, move) to a position. You can write code to process some arbitrary bytes without knowing or caring where they come from. Your code simply reads or writes to a stream, and another piece of code handles where the bytes are stored.

Understanding abstract and concrete streams

There is an `abstract` class named `Stream` that represents any type of stream. Remember that an `abstract` class cannot be instantiated using `new`; it can only be inherited. This is because it is only partially implemented.

There are many concrete classes that inherit from this base class, including `FileStream`, `MemoryStream`, `BufferedStream`, `GZipStream`, and `SslStream`. They all work the same way. All streams implement `IDisposable`, so they have a `Dispose` method to release unmanaged resources.

Some of the common members of the `Stream` class are described in *Table 9.3*:

Member	Description
`CanRead` and `CanWrite`	These properties determine if you can read from and write to the stream.
`Length` and `Position`	These properties determine the total number of bytes and the current position within the stream. These properties may throw a `NotSupportedException` for some types of streams, for example, if `CanSeek` returns `false`.
`Close` and `Dispose`	This method closes the stream and releases its resources. You can call either method, since the implementation of `Dispose` calls `Close`!
`Flush`	If the stream has a buffer, then this method writes the bytes in the buffer to the stream, and the buffer is cleared.
`CanSeek`	This property determines if the `Seek` method can be used.
`Seek`	This method moves the current position to the one specified in its parameter.
`Read` and `ReadAsync`	These methods read a specified number of bytes from the stream into a byte array and advance the position.
`ReadByte`	This method reads the next byte from the stream and advances the position.
`Write` and `WriteAsync`	These methods write the contents of a byte array into the stream.
`WriteByte`	This method writes a byte to the stream.

Table 9.3: Common members of the Stream class

Understanding storage streams

Some storage streams that represent a location where bytes will be stored are described in *Table 9.4*:

Namespace	Class	Description
`System.IO`	`FileStream`	Bytes stored in the filesystem
`System.IO`	`MemoryStream`	Bytes stored in memory in the current process
`System.Net.Sockets`	`NetworkStream`	Bytes stored at a network location

Table 9.4: Storage stream classes

`FileStream` has been rewritten in .NET 6 to have much higher performance and reliability on Windows. You can read more about this at the following link: `https://devblogs.microsoft.com/dotnet/file-io-improvements-in-dotnet-6/`.

Understanding function streams

Function streams cannot exist on their own and can only be "plugged into" other streams to add functionality. Some are described in *Table 9.5*:

Namespace	Class	Description
`System.Security.Cryptography`	`CryptoStream`	This encrypts and decrypts the stream.
`System.IO.Compression`	`GZipStream`, `DeflateStream`	These compress and decompress the stream.
`System.Net.Security`	`AuthenticatedStream`	This sends credentials across the stream.

Table 9.5: Function stream classes

Understanding stream helpers

Although there will be occasions where you need to work with streams at a low level, most often, you can plug helper classes into the chain to make things easier. All the helper types for streams implement `IDisposable`, so they have a `Dispose` method to release unmanaged resources.

Some helper classes to handle common scenarios are described in *Table 9.6*:

Namespace	Class	Description
`System.IO`	`StreamReader`	This reads from the underlying stream as plain text.
`System.IO`	`StreamWriter`	This writes to the underlying stream as plain text.
`System.IO`	`BinaryReader`	This reads from streams as .NET types. For example, the `ReadDecimal` method reads the next 16 bytes from the underlying stream as a `decimal` value, and the `ReadInt32` method reads the next 4 bytes as an `int` value.
`System.IO`	`BinaryWriter`	This writes to streams as .NET types. For example, the `Write` method with a `decimal` parameter writes 16 bytes to the underlying stream, and the `Write` method with an `int` parameter writes 4 bytes.
`System.Xml`	`XmlReader`	This reads from the underlying stream using the XML format.
`System.Xml`	`XmlWriter`	This writes to the underlying stream using the XML format.

Table 9.6: Stream helper classes

Building a stream pipeline

It is very common to combine a helper, like `StreamWriter`, and multiple function streams, like `GZipStream` and `CryptoStream`, with a storage stream, like `FileStream`, into a pipeline, as shown in *Figure 9.4*:

Figure 9.4: Writing plain text, and then encrypting and compressing it into a file stream

Your code would just call a simple helper method, like `WriteLine`, to send a `string` value, like `"Hello"`, through the pipeline until it arrives at its final destination, having been compressed and then encrypted so that it gets written to the file as `"G7x"` (or whatever it would be).

Good Practice: *"Good encryption will generate relatively uncompressible data. If you switch the order of the operations so that you first compress and then encrypt, not only should you end up with a smaller file, but the encryption will most likely take less time as it'll be operating on less data."* From an article by Stephen Toub: `https://learn.microsoft.com/en-us/archive/msdn-magazine/2008/february/net-matters-stream-pipeline`.

Writing to text streams

When you open a file to read or write to it, you use resources outside of .NET. These are called **unmanaged resources** and must be disposed of when you are done working with them.

To deterministically control when these resources are disposed of, we can call the `Dispose` method. When the `Stream` class was first designed, all cleanup code was expected to go in the `Close` method. But later, the concept of `IDisposable` was added to .NET, and `Stream` had to implement a `Dispose` method. Later, the `using` statement was added to .NET, which can automatically call `Dispose`. So today, you can call either `Close` or `Dispose`, which actually do the same thing.

Let's type some code to write text to a stream:

1. Use your preferred code editor to add a new **Console App** / `console` project, named `WorkingWithStreams`, to the `Chapter09` solution:
2. In the project file, add an element to import the `System.Console`, `System.Environment`, and `System.IO.Path` classes, statically and globally.
3. Add a new class file named `Program.Helpers.cs`.
4. In `Program.Helpers.cs`, add a partial `Program` class with a `SectionTitle` and an `OutputFileInfo` method, as shown in the following code:

```
// null namespace to merge with auto-generated Program.

partial class Program
{
```

```
    private static void SectionTitle(string title)
    {
      ConsoleColor previousColor = ForegroundColor;
      ForegroundColor = ConsoleColor.DarkYellow;
      WriteLine($"*** {title} ***");
      ForegroundColor = previousColor;
    }

    private static void OutputFileInfo(string path)
    {
      WriteLine("**** File Info ****");
      WriteLine($"File: {GetFileName(path)}");
      WriteLine($"Path: {GetDirectoryName(path)}");
      WriteLine($"Size: {new FileInfo(path).Length:N0} bytes.");
      WriteLine("/-----------------");
      WriteLine(File.ReadAllText(path));
      WriteLine("-----------------/");
    }
  }
```

5. Add a new class file named `Viper.cs`.
6. In `Viper.cs`, define a static class named `Viper` with a static array of `string` values, named `Callsigns`, as shown in the following code:

```
namespace Packt.Shared;

public static class Viper
{
  // Define an array of Viper pilot call signs.
  public static string[] Callsigns = new[]
  {
    "Husker", "Starbuck", "Apollo", "Boomer",
    "Bulldog", "Athena", "Helo", "Racetrack"
  };
}
```

7. In `Program.cs`, delete the existing statements, and then import the namespace to work with the `Viper` class, as shown in the following code:

```
using Packt.Shared; // To use Viper.
```

8. In `Program.cs`, add statements to enumerate the `Viper` call signs, writing each one on its own line in a single text file, as shown in the following code:

```
SectionTitle("Writing to text streams");

// Define a file to write to.
string textFile = Combine(CurrentDirectory, "streams.txt");

// Create a text file and return a helper writer.
StreamWriter text = File.CreateText(textFile);

// Enumerate the strings, writing each one to the stream
// on a separate line.
foreach (string item in Viper.Callsigns)
{
  text.WriteLine(item);
}
text.Close(); // Release unmanaged file resources.

OutputFileInfo(textFile);
```

Calling `Close` on the stream writer helper will call `Close` on the underlying stream. This in turn calls `Dispose` to release unmanaged file resources.

9. Run the code and view the result, as shown in the following output:

```
**** File Info ****
File: streams.txt
Path: C:\cs13net9\Chapter09\WorkingWithStreams\bin\Debug\net9.0
Size: 68 bytes.
/------------------
Husker
Starbuck
Apollo
Boomer
Bulldog
Athena
Helo
Racetrack

------------------/
```

10. Open the file that was created, and confirm that it contains the list of call signs as well as a blank line, as we are effectively calling `WriteLine` twice: once when we write the last call sign to the file, and again when we read the whole file and write it out to the console.

Remember that if you run the project at the command prompt using `dotnet run`, then the path will be the project folder. It will not include `bin\Debug\net9.0`.

Writing to XML streams

There are two ways to write an XML element, as follows:

- `WriteStartElement` and `WriteEndElement`: Use this pair when an element might have child elements.
- `WriteElementString`: Use this when an element does not have children.

Now, let's try storing the Viper pilot call signs array of `string` values in an XML file:

1. At the top of `Program.cs`, import the `System.Xml` namespace, as shown in the following code:

```
using System.Xml; // To use XmlWriter and so on.
```

2. At the bottom of `Program.cs`, add statements that enumerate the call signs, writing each one as an element in a single XML file, as shown in the following code:

```
SectionTitle("Writing to XML streams");

// Define a file path to write to.
string xmlFile = Combine(CurrentDirectory, "streams.xml");

// Declare variables for the filestream and XML writer.
FileStream? xmlFileStream = null;
XmlWriter? xml = null;

try
{
  xmlFileStream = File.Create(xmlFile);

  // Wrap the file stream in an XML writer helper and tell it
  // to automatically indent nested elements.
  xml = XmlWriter.Create(xmlFileStream,
    new XmlWriterSettings { Indent = true });

  // Write the XML declaration.
```

```
    xml.WriteStartDocument();

    // Write a root element.
    xml.WriteStartElement("callsigns");

    // Enumerate the strings, writing each one to the stream.
    foreach (string item in Viper.Callsigns)
    {
      xml.WriteElementString("callsign", item);
    }

    // Write the close root element.
    xml.WriteEndElement();
  }
  catch (Exception ex)
  {
    // If the path doesn't exist the exception will be caught.
    WriteLine($"{ex.GetType()} says {ex.Message}");
  }
  finally
  {
    if (xml is not null)
    {
      xml.Close();
      WriteLine("The XML writer's unmanaged resources have been
  disposed.");
    }

    if (xmlFileStream is not null)
    {
      xmlFileStream.Close();
      WriteLine("The file stream's unmanaged resources have been
  disposed.");
    }
  }

  OutputFileInfo(xmlFile);
```

3. Optionally, right-click in the `Close` method of `xmlFileStream`, select **Go To Implementation**, and note the implementations of the `Dispose`, `Close`, and `Dispose(bool)` methods, as shown in the following code:

```
public void Dispose() => Close();

public virtual void Close()
{
  // When initially designed, Stream required that all cleanup logic
  // went into Close(), but this was thought up before IDisposable
  // was added and never revisited. All subclasses
  // should put their cleanup now in Dispose(bool).
  Dispose(true);
  GC.SuppressFinalize(this);
}

protected virtual void Dispose(bool disposing)
{
  // Note: Never change this to call other virtual methods on Stream
  // like Write, since the state on subclasses has already been
  // torn down.  This is the last code to run on cleanup for a stream.
}
```

The `Close` and `Dispose(bool)` methods are `virtual` in the `Stream` class because they are designed to be overridden in a derived class, like `FileStream`, to do the work of releasing unmanaged resources.

4. Run the code and view the result, as shown in the following output:

```
**** File Info ****
The XML writer's unmanaged resources have been disposed.
The file stream's unmanaged resources have been disposed.
File: streams.xml
Path: C:\cs13net9\Chapter09\WorkingWithStreams\bin\Debug\net9.0
Size: 320 bytes.
/------------------
<?xml version="1.0" encoding="utf-8"?>
<callsigns>
  <callsign>Husker</callsign>
  <callsign>Starbuck</callsign>
  <callsign>Apollo</callsign>
  <callsign>Boomer</callsign>
  <callsign>Bulldog</callsign>
  <callsign>Athena</callsign>
  <callsign>Helo</callsign>
```

```
  <callsign>Racetrack</callsign>
</callsigns>
-------------------/
```

Good Practice: Before calling the `Dispose` method, check that the object is not `null`.

Simplifying disposal by using the using statement

You can simplify the code that needs to check for a `null` object and then call its `Dispose` method by using the `using` statement. Unless you need a greater level of control, I would generally recommend using `using` rather than manually calling `Dispose`, as it's less code to write.

Confusingly, there are two uses for the `using` keyword: importing a namespace and generating a `finally` statement that calls `Dispose` on an object implementing `IDisposable`.

The compiler changes a `using` statement block into a `try`-`finally` statement without a `catch` statement. You can use nested `try` statements, so if you do want to catch any exceptions, you can, as shown in the following code example:

```
using (FileStream file2 = File.OpenWrite(
  Path.Combine(path, "file2.txt")))
{
  using (StreamWriter writer2 = new StreamWriter(file2))
  {
    try
    {
      writer2.WriteLine("Welcome, .NET!");
    }
    catch(Exception ex)
    {
      WriteLine($"{ex.GetType()} says {ex.Message}");
    }
  } // Automatically calls Dispose if the object is not null.
} // Automatically calls Dispose if the object is not null.
```

You can even simplify the code further by not explicitly specifying the braces and indentation for the `using` statements, as shown in the following code:

```
using FileStream file2 = File.OpenWrite(
  Path.Combine(path, "file2.txt"));

using StreamWriter writer2 = new(file2);
```

```
try
{
  writer2.WriteLine("Welcome, .NET!");
}
catch(Exception ex)
{
  WriteLine($"{ex.GetType()} says {ex.Message}");
}
```

To make this clearer, let's review a simpler example. You can use a `using` block to ensure that the `Dispose` method is called at the end of the scope, as shown in the following code:

```
using (ObjectWithUnmanagedResources thing = new())
{
  // Statements that use thing.
}
```

More Information: You can learn how the preceding code is lowered into a `try-catch` block at the following link: `https://github.com/markjprice/cs13net9/blob/main/docs/ch06-memory.md#ensuring-that-dispose-is-called`. If you haven't already, you should also read the following online section: `https://github.com/markjprice/cs13net9/blob/main/docs/ch06-memory.md#releasing-unmanaged-resources`.

You can also use a simplified syntax without braces, as shown in the following code:

```
using ObjectWithUnmanagedResources thing = new();

// Statements that use thing.

// Dispose called at the end of the container scope e.g. method.
```

In the preceding code example, there is no explicit block defined by braces, so an implicit block is defined that ends at the end of the containing scope.

More Information: You can learn more about this at the following links: `https://learn.microsoft.com/en-us/dotnet/csharp/language-reference/proposals/csharp-8.0/using` and `https://learn.microsoft.com/en-us/dotnet/csharp/language-reference/language-specification/statements#1314-the-using-statement`.

Compressing streams

XML is relatively verbose, so it takes up more space in bytes than plain text. Let's see how we can squeeze the XML using a common compression algorithm, known as GZIP.

In .NET Core 2.1, Microsoft introduced an implementation of the Brotli compression algorithm. In performance, Brotli is like the algorithm used in DEFLATE and GZIP, but the output is about 20% denser.

Let's compare the two compression algorithms:

1. Add a new class file named `Program.Compress.cs`.
2. In `Program.Compress.cs`, write statements to use instances of `GZipStream` or `BrotliStream` to create a compressed file that contains the same XML elements as before, and then decompress it while reading it and outputting to the console, as shown in the following code:

```
using Packt.Shared; // To use Viper.
using System.IO.Compression; // To use BrotliStream, GZipStream.
using System.Xml; // To use XmlWriter, XmlReader.

partial class Program
{
  private static void Compress(string algorithm = "gzip")
  {
    // Define a file path using the algorithm as file extension.
    string filePath = Combine(
      CurrentDirectory, $"streams.{algorithm}");

    FileStream file = File.Create(filePath);
    Stream compressor;
    if (algorithm == "gzip")
    {
      compressor = new GZipStream(file, CompressionMode.Compress);
    }
    else
    {
      compressor = new BrotliStream(file, CompressionMode.Compress);
    }

    using (compressor)
    {
      using (XmlWriter xml = XmlWriter.Create(compressor))
      {
        xml.WriteStartDocument();
        xml.WriteStartElement("callsigns");
```

```
    foreach (string item in Viper.Callsigns)
    {
      xml.WriteElementString("callsign", item);
    }
  }
} // Also closes the underlying stream.

OutputFileInfo(filePath);

// Read the compressed file.
WriteLine("Reading the compressed XML file:");
file = File.Open(filePath, FileMode.Open);
Stream decompressor;
if (algorithm == "gzip")
{
  decompressor = new GZipStream(
    file, CompressionMode.Decompress);
}
else
{
  decompressor = new BrotliStream(
    file, CompressionMode.Decompress);
}

using (decompressor)

using (XmlReader reader = XmlReader.Create(decompressor))

while (reader.Read())
{
  // Check if we are on an element node named callsign.
  if ((reader.NodeType == XmlNodeType.Element)
    && (reader.Name == "callsign"))
  {
    reader.Read(); // Move to the text inside element.
    WriteLine($"{reader.Value}"); // Read its value.
  }

  // Alternative syntax with property pattern matching:
  // if (reader is { NodeType: XmlNodeType.Element,
  //   Name: "callsign" })
```

```
    }
  }
}
```

The code that uses the `decompressor` object does not use the simplified `using` syntax. Instead, it uses the fact that `using` blocks can omit their braces for a single "statement," just like `if` statements. Remember that `if` statements can have explicit braces even if only one statement is executed within a block, as shown in the following code:

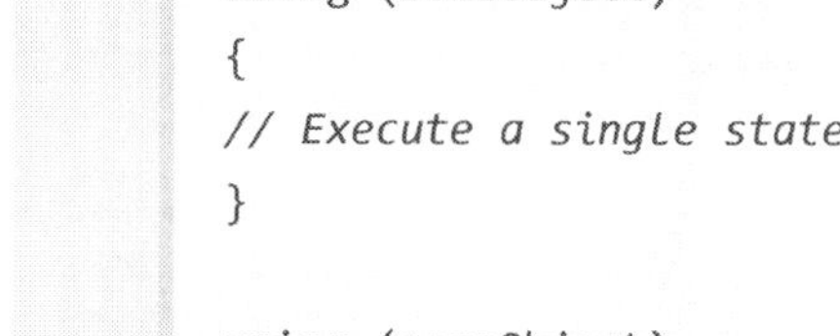

```
if (c = 1)
{
// Execute a single statement.
}

if (c = 1)
// Execute a single statement.

using (someObject)
{
// Execute a single statement.
}

using (someObject)
// Execute a single statement
```

In the preceding code, `using (XmlReader reader = XmlReader.Create(decompressor))` and the entire `while (reader.Read()) { ... }` block are equivalent to single statements, so we can remove the braces and the code works as expected.

3. In `Program.cs`, add calls to `Compress` with parameters to use the `gzip` and `brotli` algorithms, as shown in the following code:

```
SectionTitle("Compressing streams");
Compress(algorithm: "gzip");
Compress(algorithm: "brotli");
```

4. Run the code, and compare the sizes of the XML file and the compressed XML file using the `gzip` and `brotli` algorithms, as shown in the following output:

```
**** File Info ****
File: streams.gzip
Path: C:\cs13net9\Chapter09\WorkingWithStreams\bin\Debug\net9.0
```

```
Size: 151 bytes.
/------------------
?
z?{??}
En?BYjQqf~???????Bj^r~Jf^??RiI??????MrbNNqfz^1?i?QZ??Zd?@H?$%?&gc?t,
?????*????H?????t?&?d??%b??H?aUPbrjIQ"??b;????9
------------------/
Reading the compressed XML file:
Husker
Starbuck
Apollo
Boomer
Bulldog
Athena
Helo
Racetrack
**** File Info ****
File: streams.brotli
Path: C:\cs13net9\Chapter09\WorkingWithStreams\bin\Debug\net9.0
Size: 117 bytes.
/-------------------
 ??d?&?_????\@?Gm????/?h>?6????? ??^?__???wE?'?t<J??]??
???b?\fA?>?+??F??]
?T?\?~??A?J?Q?q6 ?-??
???
--------------------/
Reading the compressed XML file:
Husker
Starbuck
Apollo
Boomer
Bulldog
Athena
Helo
Racetrack
```

To summarize the file sizes:

- Uncompressed: 320 bytes
- GZIP-compressed: 151 bytes
- Brotli-compressed: 117 bytes

As well as choosing a compression mode, you can also choose a compression level. You can learn more about this at the following link: `https://learn.microsoft.com/en-us/dotnet/api/system.io.compression.compressionlevel`.

Reading and writing with random access handles

For the first 20 years of .NET's life, the only API to work directly with files was the one for stream classes. These work great for automated tasks that only need to process data sequentially. But when a human interacts with the data, they often want to jump around and return multiple times to the same location.

With .NET 6 and later, there is a new API for working with files without needing a file stream and in a random access way. Let's see a simple example:

1. Use your preferred code editor to add a new **Console App** / `console` project named `WorkingWithRandomAccess` to the `Chapter09` solution:
2. In the project file, add an element to import the `System.Console` class statically and globally.
3. In `Program.cs`, delete the existing statements, and then get a handle to a file named `coffee.txt`, as shown in the following code:

```
using Microsoft.Win32.SafeHandles; // To use SafeFileHandle.
using System.Text; // To use Encoding.

using SafeFileHandle handle =
  File.OpenHandle(path: "coffee.txt",
    mode: FileMode.OpenOrCreate,
    access: FileAccess.ReadWrite);
```

4. Write some text encoded as a byte array, and then store it in a read-only memory buffer to the file, as shown in the following code:

```
string message = "Café £4.39";
ReadOnlyMemory<byte> buffer = new(Encoding.UTF8.GetBytes(message));
await RandomAccess.WriteAsync(handle, buffer, fileOffset: 0);
```

5. To read from the file, get the length of the file, allocate a memory buffer for the contents using that length, and then read the file, as shown in the following code:

```
long length = RandomAccess.GetLength(handle);
Memory<byte> contentBytes = new(new byte[length]);
await RandomAccess.ReadAsync(handle, contentBytes, fileOffset: 0);
string content = Encoding.UTF8.GetString(contentBytes.ToArray());
WriteLine($"Content of file: {content}");
```

6. Run the code, and note the content of the file, as shown in the following output:

```
Content of file: Café £4.39
```

Encoding and decoding text

Text characters can be represented in different ways. For example, the alphabet can be encoded using Morse code into a series of dots and dashes for transmission over a telegraph line.

In a similar way, text inside a computer is stored as bits (ones and zeros) representing a code point within a code space. Most code points represent a single character, but they can also have other meanings, like formatting.

For example, ASCII has a code space with 128 code points. .NET uses a standard called **Unicode** to encode text internally. Unicode has more than 1 million code points.

Sometimes, you will need to move text outside .NET for use by systems that do not use Unicode or a variation of it, so it is important to learn how to convert between encodings.

Some common text encodings used by computers are shown in *Table 9.7*:

Encoding	Description
ASCII	This encodes a limited range of characters using the lower 7 bits of a byte.
UTF-8	This represents each Unicode code point as a sequence of 1 to 4 bytes.
UTF-7	This is designed to be more efficient over 7-bit channels than UTF-8, but it has security and robustness issues, so UTF-8 is recommended over UTF-7.
UTF-16	This represents each Unicode code point as a sequence of one or two 16-bit integers.
UTF-32	This represents each Unicode code point as a 32-bit integer and is, therefore, a fixed-length encoding, unlike the other Unicode encodings, which are all variable-length encodings.
ANSI/ISO encodings	This provides support for a variety of code pages that are used to support a specific language or group of languages.

Table 9.7: Common text encodings

Good Practice: In most cases today, UTF-8 is a good default, which is why it is literally the default encoding, that is, `Encoding.Default`. You should avoid using `Encoding.UTF7` because it is not secure. Due to this, the C# compiler will warn you when you try to use UTF-7. Of course, you might need to generate text using that encoding for compatibility with another system, so it needs to remain an option in .NET.

Encoding strings as byte arrays

Let's explore text encodings:

1. Use your preferred code editor to add a new **Console App** / `console` project, named `WorkingWithEncodings`, to the `Chapter09` solution.
2. In the project file, add an element to statically and globally import the `System.Console` class.
3. In `Program.cs`, delete the existing statements, import the `System.Text` namespace, add statements to encode a `string` using an encoding chosen by the user, loop through each byte, and then decode the encoded value back into a `string` and output it, as shown in the following code:

```
using System.Text; // To use Encoding.

WriteLine("Encodings");
WriteLine("[1] ASCII");
WriteLine("[2] UTF-7");
WriteLine("[3] UTF-8");
WriteLine("[4] UTF-16 (Unicode)");
WriteLine("[5] UTF-32");
WriteLine("[6] Latin1");
WriteLine("[any other key] Default encoding");
WriteLine();

Write("Press a number to choose an encoding.");
ConsoleKey number = ReadKey(intercept: true).Key;
WriteLine(); WriteLine();

Encoding encoder = number switch
{
  ConsoleKey.D1 or ConsoleKey.NumPad1 => Encoding.ASCII,
  ConsoleKey.D2 or ConsoleKey.NumPad2 => Encoding.UTF7,
  ConsoleKey.D3 or ConsoleKey.NumPad3 => Encoding.UTF8,
  ConsoleKey.D4 or ConsoleKey.NumPad4 => Encoding.Unicode,
  ConsoleKey.D5 or ConsoleKey.NumPad5 => Encoding.UTF32,
  ConsoleKey.D6 or ConsoleKey.NumPad6 => Encoding.Latin1,
  _ => Encoding.Default
};

// Define a string to encode
```

```
string message = "Café £4.39";
WriteLine($"Text to encode: {message}  Characters: {message.Length}.");

// Encode the string into a byte array.
byte[] encoded = encoder.GetBytes(message);

// Check how many bytes the encoding needed.
WriteLine("{0} used {1:N0} bytes.",
  encoder.GetType().Name, encoded.Length);
WriteLine();

// Enumerate each byte.
WriteLine("BYTE | HEX | CHAR");
foreach (byte b in encoded)
{
  WriteLine($"{b,4} | {b,3:X} | {(char)b,4}");
}

// Decode the byte array back into a string and display it.
string decoded = encoder.GetString(encoded);
WriteLine($"Decoded: {decoded}");
```

4. Run the code, press *1* to choose ASCII, and note that when outputting the bytes, the pound sign (£) and accented e (é) cannot be represented in ASCII, so it uses a question mark instead:

```
Text to encode: Café £4.39  Characters: 10
ASCIIEncodingSealed used 10 bytes.

BYTE | HEX | CHAR
  67 |  43 |    C
  97 |  61 |    a
 102 |  66 |    f
  63 |  3F |    ?
  32 |  20 |
  63 |  3F |    ?
  52 |  34 |    4
  46 |  2E |    .
  51 |  33 |    3
  57 |  39 |    9
Decoded: Caf? ?4.39
```

5. Rerun the code and press *3* to choose UTF-8. Note that UTF-8 requires 2 extra bytes for the two characters that need 2 bytes each (12 bytes instead of 10 bytes in total), but it can encode and decode the é and £ characters:

```
Text to encode: Café £4.39  Characters: 10
UTF8EncodingSealed used 12 bytes.

BYTE | HEX | CHAR
  67 |  43 |    C
  97 |  61 |    a
 102 |  66 |    f
 195 |  C3 |    Ã
 169 |  A9 |    ©
  32 |  20 |
 194 |  C2 |    Â
 163 |  A3 |    £
  52 |  34 |    4
  46 |  2E |    .
  51 |  33 |    3
  57 |  39 |    9
Decoded: Café £4.39
```

6. Rerun the code and press *4* to choose Unicode (UTF-16). Note that UTF-16 requires 2 bytes for every character, so 20 bytes in total, and it can encode and decode the é and £ characters. This encoding is used internally by .NET to store `char` and `string` values.

Encoding and decoding text in files

When using stream helper classes, such as `StreamReader` and `StreamWriter`, you can specify the encoding you want to use. As you write to the helper, the text will be automatically encoded, and as you read from the helper, the bytes will be automatically decoded.

To specify an encoding, pass the encoding as a second parameter to the helper type's constructor, as shown in the following code:

```
StreamReader reader = new(stream, Encoding.UTF8);
StreamWriter writer = new(stream, Encoding.UTF8);
```

Good Practice: Often, you won't have the choice of which encoding to use because you will generate a file for use by another system. However, if you do, pick one that uses the least number of bytes but can store every character you need.

Serializing object graphs

An **object graph** is a structure of multiple objects that are related to each other, either through a direct reference or indirectly through a chain of references.

Serialization is the process of converting a live object graph into a sequence of bytes using a specified format. **Deserialization** is the reverse process.

You would use serialization to save the current state of a live object so that you can recreate it in the future, for example, saving the current state of a game so that you can continue at the same place tomorrow. The stream produced from a serialized object is usually stored in a file or database.

There are dozens of formats you can choose for serialization, but the two most common text-based human-readable formats are **eXtensible Markup Language** (**XML**) and **JavaScript Object Notation** (**JSON**). There are also more efficient binary formats like Protobuf, used by gRPC.

Good Practice: JSON is more compact and is best for web and mobile applications. XML is more verbose but is better supported in more legacy systems. Use JSON to minimize the size of serialized object graphs. JSON is also a good choice when sending object graphs to web applications and mobile applications, as it is the native serialization format for JavaScript, and mobile apps often make calls over limited bandwidth, so the number of bytes is important.

.NET has multiple classes that will serialize to and from XML and JSON. We will start by looking at `XmlSerializer` and `JsonSerializer`.

Serializing as XML

Let's start by looking at XML, probably the world's most used serialization format (for now). To show a typical example, we will define a custom class to store information about a person and then create an object graph, using a list of `Person` instances with nesting:

1. Use your preferred code editor to add a new **Console App** / `console` project named `WorkingWithSerialization` to the `Chapter09` solution.
2. In the project file, add elements to statically and globally import the following classes:
    - System.Console to use ForegroundColor and WriteLine
    - System.Environment to use CurrentDirectory
    - System.IO.Path to use Combine, GetFileName, and GetDirectoryName
3. Add a new class file named `Program.Helpers.cs`.
4. In `Program.Helpers.cs`, add a partial `Program` class with a `SectionTitle` and an `OutputFileInfo` method, as shown in the following code:

```
// null namespace to merge with auto-generated Program.

partial class Program
```

```
{
  private static void SectionTitle(string title)
  {
    ConsoleColor previousColor = ForegroundColor;
    ForegroundColor = ConsoleColor.DarkYellow;
    WriteLine($"*** {title} ***");
    ForegroundColor = previousColor;
  }

  private static void OutputFileInfo(string path)
  {
    WriteLine("**** File Info ****");
    WriteLine($"File: {GetFileName(path)}");
    WriteLine($"Path: {GetDirectoryName(path)}");
    WriteLine($"Size: {new FileInfo(path).Length:N0} bytes.");
    WriteLine("/------------------");
    WriteLine(File.ReadAllText(path));
    WriteLine("------------------/");
  }
}
```

5. Add a new class file, named `Person.cs`, to define a `Person` class with a `Salary` property that is `protected`, meaning it is only accessible to itself and derived classes. To populate the salary, the class has a constructor with a single parameter to set the initial salary, as shown in the following code:

```
namespace Packt.Shared;

public class Person
{
  public Person(decimal initialSalary)
  {
    Salary = initialSalary;
  }

  public string? FirstName { get; set; }
  public string? LastName { get; set; }
  public DateTime DateOfBirth { get; set; }
  public HashSet<Person>? Children { get; set; }
  protected decimal Salary { get; set; }
}
```

6. In `Program.cs`, delete the existing statements, and then import namespaces to work with XML serialization and the `Person` class, as shown in the following code:

```
using System.Xml.Serialization; // To use XmlSerializer.
using Packt.Shared; // To use Person.
```

7. In `Program.cs`, add statements to create an object graph of `Person` instances, as shown in the following code:

```
List<Person> people = new()
{
  new(initialSalary: 30_000M)
  {
    FirstName = "Alice",
    LastName = "Smith",
    DateOfBirth = new(year: 1974, month: 3, day: 14)
  },
  new(initialSalary: 40_000M)
  {
    FirstName = "Bob",
    LastName = "Jones",
    DateOfBirth = new(year: 1969, month: 11, day: 23)
  },
  new(initialSalary: 20_000M)
  {
    FirstName = "Charlie",
    LastName = "Cox",
    DateOfBirth = new(year: 1984, month: 5, day: 4),
    Children = new()
    {
      new(initialSalary: 0M)
      {
        FirstName = "Sally",
        LastName = "Cox",
        DateOfBirth = new(year: 2012, month: 7, day: 12)
      }
    }
  }
};

SectionTitle("Serializing as XML");

// Create serializer to format a "List of Person" as XML.
```

```
XmlSerializer xs = new(type: people.GetType());

// Create a file to write to.
string path = Combine(CurrentDirectory, "people.xml");

using (FileStream stream = File.Create(path))
{
  // Serialize the object graph to the stream.
  xs.Serialize(stream, people);
} // Closes the stream.

OutputFileInfo(path);
```

8. Run the code, view the result, and note that an exception is thrown, as shown in the following output:

```
Unhandled Exception: System.InvalidOperationException: Packt.Shared.
Person cannot be serialized because it does not have a parameterless
constructor.
```

9. In `Person.cs`, add a statement to define a parameterless constructor, as shown in the following code:

```
// A parameterless constructor is required for XML serialization.
public Person() { }
```

The constructor does not need to do anything, but it must exist so that the `XmlSerializer` can call it to instantiate new `Person` instances during the deserialization process.

10. Run the code and view the result, and note that the object graph is serialized as XML elements, like `<FirstName>Bob</FirstName>`, and that the `Salary` property is not included because it is not a `public` property, as shown in the following output:

```
**** File Info ****
File: people.xml
Path: C:\cs13net9\Chapter09\WorkingWithSerialization\bin\Debug\net9.0
Size: 793 bytes.
/------------------
<?xml version="1.0" encoding="utf-8"?>
<ArrayOfPerson xmlns:xsi="http://www.w3.org/2001/XMLSchema-instance"
xmlns:xsd="http://www.w3.org/2001/XMLSchema">
  <Person>
    <FirstName>Alice</FirstName>
```

```
    <LastName>Smith</LastName>
    <DateOfBirth>1974-03-14T00:00:00</DateOfBirth>
  </Person>
  <Person>
    <FirstName>Bob</FirstName>
    <LastName>Jones</LastName>
    <DateOfBirth>1969-11-23T00:00:00</DateOfBirth>
  </Person>
  <Person>
    <FirstName>Charlie</FirstName>
    <LastName>Cox</LastName>
    <DateOfBirth>1984-05-04T00:00:00</DateOfBirth>
    <Children>
      <Person>
        <FirstName>Sally</FirstName>
        <LastName>Cox</LastName>
        <DateOfBirth>2012-07-12T00:00:00</DateOfBirth>
      </Person>
    </Children>
  </Person>
</ArrayOfPerson>
------------------/
```

Generating compact XML

We could make the XML more compact using attributes instead of elements for some fields:

1. At the top of `Person.cs`, import the `System.Xml.Serialization` namespace so that you can decorate some properties with the `[XmlAttribute]` attribute, as shown in the following code:

   ```
   using System.Xml.Serialization; // To use [XmlAttribute].
   ```

2. In `Person.cs`, decorate the first name, last name, and date of birth properties with the `[XmlAttribute]` attribute, and set a short name for each property, as highlighted in the following code:

   ```
   [XmlAttribute("fname")]
   public string? FirstName { get; set; }

   [XmlAttribute("lname")]
   public string? LastName { get; set; }

   [XmlAttribute("dob")]
   public DateTime DateOfBirth { get; set; }
   ```

3. Run the code, and note that the size of the file has reduced from 793 to 488 bytes, a space-saving of more than a third. This reduction was achieved by outputting property values as XML attributes, as shown in the following output:

```
**** File Info ****
File: people.xml
Path: C:\cs13net9\Chapter09\WorkingWithSerialization\bin\Debug\net9.0
Size: 488 bytes.
/------------------
<?xml version="1.0" encoding="utf-8"?>
<ArrayOfPerson xmlns:xsi="http://www.w3.org/2001/XMLSchema-instance"
xmlns:xsd="http://www.w3.org/2001/XMLSchema">
  <Person fname="Alice" lname="Smith" dob="1974-03-14T00:00:00" />
  <Person fname="Bob" lname="Jones" dob="1969-11-23T00:00:00" />
  <Person fname="Charlie" lname="Cox" dob="1984-05-04T00:00:00">
    <Children>
      <Person fname="Sally" lname="Cox" dob="2012-07-12T00:00:00" />
    </Children>
  </Person>
</ArrayOfPerson>
------------------/
```

Deserializing XML files

Now, let's try deserializing the XML file back into live objects in memory:

1. In `Program.cs`, add statements to open the XML file, and then deserialize it, as shown in the following code:

```
SectionTitle("Deserializing XML files");

using (FileStream xmlLoad = File.Open(path, FileMode.Open))
{
  // Deserialize and cast the object graph into a "List of Person".
  List<Person>? loadedPeople =
    xs.Deserialize(xmlLoad) as List<Person>;

  if (loadedPeople is not null)
  {
```

```
        foreach (Person p in loadedPeople)
        {
          WriteLine("{0} has {1} children.",
            p.LastName, p.Children?.Count ?? 0);
        }
      }
    }
```

2. Run the code, and note that the people are loaded successfully from the XML file and then enumerated, as shown in the following output:

```
Smith has 0 children.
Jones has 0 children.
Cox has 1 children.
```

More Information: There are many other attributes defined in the `System.Xml.Serialization` namespace that can be used to control the XML generated. A good place to start is the official documentation for the `XmlAttributeAttribute` class found here: `https://learn.microsoft.com/en-us/dotnet/api/system.xml.serialization.xmlattributeattribute`. Do not get this class confused with the `XmlAttribute` class in the `System.Xml` namespace. That is used to represent an XML attribute when reading and writing XML, using `XmlReader` and `XmlWriter`.

If you don't use any annotations, `XmlSerializer` performs a case-insensitive match using the property name when deserializing.

Good Practice: When using `XmlSerializer`, remember that only the public fields and properties are included, and the type must have a parameterless constructor. You can customize the output with attributes.

Serializing with JSON

One of the most popular .NET libraries to work with the JSON serialization format is **Newtonsoft.Json**, known as **Json.NET**. It is mature and powerful.

`Newtonsoft.Json` is so popular that it overflowed the bounds of the 32-bit integer used for the download count in the NuGet package manager, as shown in the following post in *Figure 9.5*:

Figure 9.5: Negative 2 billion downloads for Newtonsoft.Json in August 2022

Let's see it in action:

1. In the `WorkingWithSerialization` project, add a package reference for the latest version of `Newtonsoft.Json`, as shown in the following markup:

```
<ItemGroup>
  <PackageReference Include="Newtonsoft.Json" Version="13.0.3" />
</ItemGroup>
```

2. Build the `WorkingWithSerialization` project to restore packages.
3. In `Program.cs`, add statements to create a text file, and then serialize the people into the file as JSON, as shown in the following code:

```
SectionTitle("Serializing with JSON");

// Create a file to write to.
string jsonPath = Combine(CurrentDirectory, "people.json");

using (StreamWriter jsonStream = File.CreateText(jsonPath))
{
  Newtonsoft.Json.JsonSerializer jss = new();

  // Serialize the object graph into a string.
  jss.Serialize(jsonStream, people);
} // Closes the file stream and release resources.

OutputFileInfo(jsonPath);
```

4. Run the code, and note that JSON requires fewer than half the number of bytes compared to XML with elements. It's even smaller than the XML file, which uses attributes (366 compared to 488), as shown in the following output:

```
**** File Info ****
File: people.json
Path: C:\cs13net9\Chapter09\WorkingWithSerialization\bin\Debug\net9.0
Size: 366 bytes.
/------------------
[{"FirstName":"Alice","LastName":"Smith","DateOfBirth":"1974-03-14T00:00:00","Children":null},{"FirstName":"Bob","LastName":"Jones","DateOfBirth":"1969-11-23T00:00:00","Children":null},{"FirstName":"Charlie","LastName":"Cox","DateOfBirth":"1984-05-04T00:00:00","Children":[{"FirstName":"Sally","LastName":"Cox","DateOfBirth":"2012-07-12T00:00:00","Children":null}]}]
------------------/
```

High-performance JSON processing

.NET Core 3 introduced a new namespace to work with JSON, `System.Text.Json`, which is optimized for performance by leveraging APIs like `Span<T>`.

Also, older libraries like `Json.NET` are implemented by reading UTF-16. It would be more performant to read and write JSON documents using UTF-8 because most network protocols, including HTTP, use UTF-8, and you can avoid transcoding UTF-8 to and from Json.NET's Unicode `string` values.

With the new API, Microsoft achieved between 1.3x and 5x improvement, depending on the scenario.

The original author of Json.NET, James Newton-King, joined Microsoft and is working with them to develop their new JSON types. As he says in a comment discussing the new JSON APIs, *"Json.NET isn't going away,"* as shown in *Figure 9.6*:

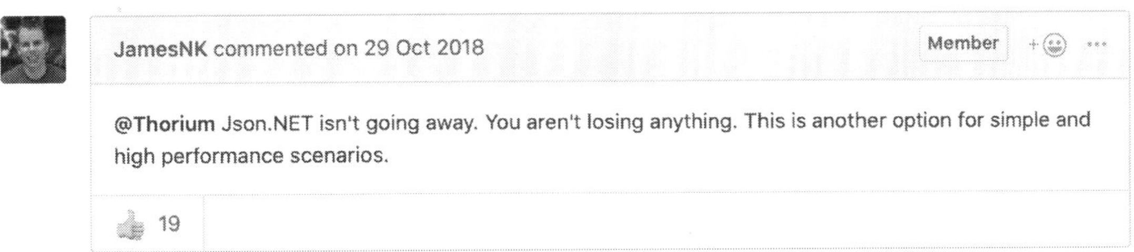

Figure 9.6: A comment by the original author of Json.NET

Deserializing JSON files

Let's see how to use the modern JSON APIs to deserialize a JSON file:

1. In the `WorkingWithSerialization` project, at the top of `Program.cs`, import the new JSON class to perform serialization, using an alias to avoid conflicting names with the Json.NET one we used before, as shown in the following code:

```
using FastJson = System.Text.Json.JsonSerializer;
```

2. In `Program.cs`, add statements to open the JSON file, deserialize it, and output the names and counts of the children of the people, as shown in the following code:

```
SectionTitle("Deserializing JSON files");

await using (FileStream jsonLoad = File.Open(jsonPath, FileMode.Open))
{
  // Deserialize object graph into a "List of Person".
  List<Person>? loadedPeople =
    await FastJson.DeserializeAsync(utf8Json: jsonLoad,
      returnType: typeof(List<Person>)) as List<Person>;

  if (loadedPeople is not null)
  {
    foreach (Person p in loadedPeople)
    {
      WriteLine("{0} has {1} children.",
        p.LastName, p.Children?.Count ?? 0);
    }
  }
}
```

3. Run the code and view the result, as shown in the following output:

```
Smith has 0 children.
Jones has 0 children.
Cox has 1 children.
```

Good Practice: Choose Json.NET for developer productivity and a large feature set, or `System.Text.Json` for performance. You can review a list of the differences at the following link: `https://learn.microsoft.com/en-us/dotnet/standard/serialization/system-text-json-migrate-from-newtonsoft-how-to#table-of-differences-between-newtonsoftjson-and-systemtextjson`.

JSON schema exporter

.NET 9 introduced a class to make it easy to generate JSON schemas for .NET types. As a result, serialization is simplified and data conforms to expected formats.

Let's see a working example:

1. In the `WorkingWithSerialization` project, at the top of `Program.cs`, import the namespaces to perform JSON serialization and work with JSON schemas, as shown in the following code:

```
using System.Text.Json; // To use JsonSerializerOptions.
using System.Text.Json.Schema; // To use JsonSchemaExporter.
```

2. In `Program.cs`, add statements to export the JSON schema for the `Person` class, as shown in the following code:

```
SectionTitle("JSON schema exporter");

WriteLine(JsonSchemaExporter.GetJsonSchemaAsNode(
  JsonSerializerOptions.Default, typeof(Person)));
```

3. Run the code and view the result, as shown in the following output:

```
{
  "type": [
    "object",
    "null"
  ],
  "properties": {
    "FirstName": {
      "type": [
        "string",
        "null"
      ]
    },
    "LastName": {
      "type": [
        "string",
        "null"
      ]
    },
    "DateOfBirth": {
      "type": "string",
      "format": "date-time"
    },
```

```
    "Children": {
      "type": [
        "array",
        "null"
      ],
      "items": {
        "$ref": "#"
      }
    }
  }
}
```

Controlling JSON processing

There are many options to take control of how JSON is processed, as shown in the following list:

- Including and excluding fields
- Setting a casing policy
- Selecting a case-sensitivity policy
- Choosing between compact and prettified whitespace

Let's see some in action:

1. Use your preferred code editor to add a new **Console App** / `console` project, named `ControllingJson`, to the `Chapter09` solution.
2. In the project file, add elements to statically and globally import the `System.Console`, `System.Environment`, and `System.IO.Path` classes.
3. In the `ControllingJson` project, add a new class file named `Book.cs`.
4. In `Book.cs`, define a class named `Book`, as shown in the following code:

```
using System.Text.Json.Serialization; // To use [JsonInclude].

namespace Packt.Shared;

public class Book
{
  // Constructor to set non-nullable property.
  public Book(string title)
  {
    Title = title;
  }

  // Properties.
  public string Title { get; set; }
```

```
  public string? Author { get; set; }

  // Fields.
  [JsonInclude] // Include this field.
  public DateTime PublishDate;

  [JsonInclude] // Include this field.
  public DateTimeOffset Created;

  public ushort Pages;
}
```

5. In `Program.cs`, delete the existing statements, and then import the namespaces to work with high-performance JSON and `Book`, as shown in the following code:

```
using Packt.Shared; // To use Book.
using System.Text.Json; // To use JsonSerializer.
```

6. In `Program.cs`, add statements to create an instance of the `Book` class and serialize it to JSON, as shown in the following code:

```
Book csharpBook = new(title:
  "C# 13 and .NET 9 - Modern Cross-Platform Development Fundamentals")
{
  Author = "Mark J Price",
  PublishDate = new(year: 2024, month: 11, day: 12),
  Pages = 823,
  Created = DateTimeOffset.UtcNow,
};

JsonSerializerOptions options = new()
{
  IncludeFields = true, // Includes all fields.
  PropertyNameCaseInsensitive = true,
  WriteIndented = true,
  PropertyNamingPolicy = JsonNamingPolicy.CamelCase,
};

string path = Combine(CurrentDirectory, "book.json");

using (Stream fileStream = File.Create(path))
{
  JsonSerializer.Serialize(
```

```
    utf8Json: fileStream, value: csharpBook, options);
}

WriteLine("**** File Info ****");
WriteLine($"File: {GetFileName(path)}");
WriteLine($"Path: {GetDirectoryName(path)}");
WriteLine($"Size: {new FileInfo(path).Length:N0} bytes.");
WriteLine("/------------------");
WriteLine(File.ReadAllText(path));
WriteLine("------------------/");
```

7. Run the code and view the result, as shown in the following output:

```
**** File Info ****
File: book.json
Path: C:\cs13net9\Chapter09\ControllingJson\bin\Debug\net9.0
Size: 221 bytes.
/------------------
{
  "title": "C# 13 and .NET 9 - Modern Cross-Platform Development
Fundamentals",
  "author": "Mark J Price",
  "publishDate": "2024-11-12T00:00:00",
  "created": "2024-07-13T14:29:07.119631+00:00",
  "pages": 823
}
------------------/
```

Note the following:

- The JSON file is 221 bytes.
- The member names use camelCasing, for example, `publishDate`. This is best for subsequent processing in a browser with JavaScript.
- All fields are included due to the options set, including `pages`.
- JSON is prettified for easier human legibility.
- `DateTime` and `DateTimeOffset` values are stored as a single standard `string` format.

8. In `Program.cs`, when setting the `JsonSerializerOptions`, comment out the setting of a casing policy, write with an indent, and include fields.
9. Run the code and view the result, as shown in the following output:

```
**** File Info ****
File: book.json
Path: C:\cs13net9\Chapter09\ControllingJson\bin\Debug\net9.0
```

```
Size: 184 bytes.
/-----------------
{"Title":"C# 13 and .NET 9 - Modern Cross-Platform Development
Fundamentals","Author":"Mark J Price","PublishDate":"2024-11-
12T00:00:00","Created":"2024-07-13T14:30:29.2205861+00:00"}
-----------------/
```

Note the following:

- The JSON file has about a 20% reduction.
- The member names use normal casing, for example, `PublishDate`.
- The `Pages` field is missing. The other fields are included due to the `[JsonInclude]` attribute on the `PublishDate` and `Created` fields.

A warning about binary serialization using BinaryFormatter

The `BinaryFormatter` class is a serialization class that can convert a live object, or a graph of connected objects, into a binary format that can be stored or transmitted and later reconstructed. This class has been part of .NET since its early versions and is in the `System.Runtime.Serialization.Formatters.Binary` namespace. But `BinaryFormatter` is problematic for several reasons, primarily related to security and robustness.

For example, `BinaryFormatter` is vulnerable to deserialization attacks, where malicious input can execute arbitrary code during the deserialization process. This makes any application using `BinaryFormatter` susceptible to **remote code execution** (**RCE**) if an attacker can control the serialized data. `BinaryFormatter` does not perform any validation or checks on the data it deserializes, making it inherently insecure when handling untrusted data.

`BinaryFormatter` does not support many modern serialization features, like JSON or XML serialization capabilities, which are more secure and flexible.

Good Practice: Microsoft recommends using serialization libraries like `System.Text.Json`, `System.Xml.Serialization`, or `Google.Protobuf`.

In the first version of .NET Core, the Microsoft team removed `BinaryFormatter` entirely due to its known risks, but without a clear path to using something safer, customers demanded that it be brought back, which the team did with .NET Core 1.1. Since then, the team has disabled it by default but allowed developers to explicitly re-enable it, by setting flags if they accept the risks.

With .NET 9, the flags to enable `BinaryFormatter` have been removed, and any use of it will throw an exception. But there is still a way to re-enable it, by referencing a NuGet package marked as being permanently vulnerable and setting some configuration.

More Information: You can learn more about why and how the `BinaryFormatter` class is being removed in .NET 9 at the following link: https://devblogs.microsoft.com/dotnet/binaryformatter-removed-from-dotnet-9/.

Practicing and exploring

Test your knowledge and understanding by answering some questions, getting some hands-on practice, and exploring this chapter's topics with more in-depth research.

Exercise 9.1 – Online material

Online material can be extra content written by me for this book, or it can be references to content created by Microsoft or third parties.

Working with tar archives

If you use Linux, then you will be interested in how to programmatically work with tar archives. I have written an online-only section to introduce you to them that can be found at the following link:

https://github.com/markjprice/cs13net9/blob/main/docs/ch09-tar-archives.md

Migrating from Newtonsoft to new JSON

If you have existing code that uses the Newtonsoft `Json.NET` library and you want to migrate to the new `System.Text.Json` namespace, then Microsoft has specific documentation for that, which you can find at the following link:

https://learn.microsoft.com/en-us/dotnet/standard/serialization/system-text-json-migrate-from-newtonsoft-how-to

Exercise 9.2 – Practice exercises

Practice exercises go deeper into the topics for this chapter.

Serializing as XML

In the `Chapter09` solution, create a console app named `Exercise_SerializingShapes` that creates a list of shapes, uses serialization to save it to the filesystem with XML, and then deserializes it back:

```
// Create a list of Shapes to serialize.
List<Shape> listOfShapes = new()
{
  new Circle { Colour = "Red", Radius = 2.5 },
  new Rectangle { Colour = "Blue", Height = 20.0, Width = 10.0 },
  new Circle { Colour = "Green", Radius = 8.0 },
  new Circle { Colour = "Purple", Radius = 12.3 },
  new Rectangle { Colour = "Blue", Height = 45.0, Width = 18.0 }
};
```

Shapes should have a read-only property named `Area` so that when you deserialize, you can output a list of shapes, including their areas, as shown here:

```
List<Shape> loadedShapesXml =
  serializerXml.Deserialize(fileXml) as List<Shape>;

foreach (Shape item in loadedShapesXml)
{
  WriteLine("{0} is {1} and has an area of {2:N2}",
    item.GetType().Name, item.Colour, item.Area);
}
```

This is what your output should look like when you run your console application:

```
Loading shapes from XML:
Circle is Red and has an area of 19.63
Rectangle is Blue and has an area of 200.00
Circle is Green and has an area of 201.06
Circle is Purple and has an area of 475.29
Rectangle is Blue and has an area of 810.00
```

Exercise 9.3 – Test your knowledge

Answer the following questions:

1. What is the difference between using the `File` class and the `FileInfo` class?
2. What is the difference between the `ReadByte` method and the `Read` method of a stream?
3. When would you use the `StringReader`, `TextReader`, and `StreamReader` classes?
4. What does the `DeflateStream` type do?
5. How many bytes per character does UTF-8 encoding use?
6. What is an object graph?
7. What is the best serialization format to choose to minimize space requirements?
8. What is the best serialization format to choose for cross-platform compatibility?
9. Why is it bad to use a `string` value like `"\Code\Chapter01"` to represent a path, and what should you do instead?
10. Where can you find information about NuGet packages and their dependencies?

Exercise 9.4 – Explore topics

Use the links on the following page to learn more details about the topics covered in this chapter:

https://github.com/markjprice/cs13net9/blob/main/docs/book-links.md#chapter-9---working-with-files-streams-and-serialization

Summary

In this chapter, you learned how to:

- Read from and write to text files.
- Read from and write to XML files.
- Compress and decompress files.
- Encode and decode text.
- Serialize an object graph into JSON and XML.
- Deserialize an object graph from JSON and XML.

In the next chapter, you will learn how to work with databases using Entity Framework Core.

10

Working with Data Using Entity Framework Core

This chapter is about reading from and writing to relational data stores, such as SQLite and SQL Server, by using the object-to-data store mapping technology named **Entity Framework Core** (**EF Core**).

This chapter will cover the following topics:

- Understanding modern databases
- Setting up EF Core in a .NET project
- Defining EF Core models
- Querying EF Core models

Understanding modern databases

Two of the most common places to store data are in a **relational database management system** (**RDBMS**), such as SQL Server, PostgreSQL, MySQL, and SQLite, or a **NoSQL** database, such as Azure Cosmos DB, Redis, MongoDB, and Apache Cassandra.

Relational databases were invented in the 1970s. They are queried with **Structured Query Language** (**SQL**). At the time, data storage costs were high, so they reduced data duplication as much as possible. Data is stored in tabular structures with rows and columns that are tricky to refactor once in production. They can be difficult and expensive to scale.

NoSQL databases do not just mean "no SQL"; they can also mean "not only SQL." They were invented in the 2000s, after the internet and the web had become popular, and adopted many of the learnings from that era of software. They are designed for massive scalability, high performance, and making programming easier by providing maximum flexibility and allowing schema changes at any time because they do not enforce a structure.

If you know nothing about relational databases, then you should read the database primer that I wrote at the following link:

`https://github.com/markjprice/cs13net9/blob/main/docs/ch10-database-primer.md`

Understanding legacy Entity Framework

Entity Framework (**EF**) was first released as part of .NET Framework 3.5 with Service Pack 1 back in late 2008. Since then, EF has evolved, as Microsoft has observed how programmers use **object-relational mapping** (**ORM**) tools in the real world.

ORMs use a mapping definition to associate columns in tables with properties in classes. Then, a programmer can interact with objects of different types in a way that they are familiar with, instead of having to deal with knowing how to store the values in a relational table or another structure provided by a NoSQL data store.

The version of EF included with .NET Framework is **Entity Framework 6** (**EF6**). It is mature and stable and supports an EDMX (XML file) way of defining the model, as well as complex inheritance models and a few other advanced features.

EF 6.3 and later versions have been extracted from .NET Framework as a separate package, so they can be supported on .NET Core 3 and later. This enables existing projects like web applications and services to be ported and run cross-platform. However, EF6 should be considered legacy technology because it has some limitations when running cross-platform and no new features will be added to it.

Using the legacy Entity Framework 6.3 or later

To use the legacy Entity Framework in a .NET Core 3 or later project, you must add a package reference to it in your project file, as shown in the following markup:

```
<PackageReference Include="EntityFramework" Version="6.5.1" />
```

Good Practice: Only use legacy EF6 if you must; for example, you might use it to migrate a **Windows Presentation Foundation** (**WPF**) app that uses EF6 on .NET Framework to modern .NET. This book is about modern cross-platform development, so in the rest of this chapter, I will only cover the modern EF Core. You will not need to reference the legacy EF6 package as shown above in the projects for this chapter.

Understanding Entity Framework Core

The truly cross-platform version, **EF Core**, is different from the legacy Entity Framework. Although EF Core has a similar name, you should be aware of how it varies from EF6. The latest EF Core is version 9, to match .NET 9.

EF Core 9 targets .NET 8 or later because the EF Core team wants as many developers as possible to benefit from new features in future releases even if you must target only long-term support releases of .NET. This means that you can use all the new features of EF Core 9 with either .NET 8 or .NET 9. But when EF Core 10 is released in November 2025, your projects will need to target .NET 10 to use it.

EF Core 3 and later only work with platforms that support .NET Standard 2.1, meaning .NET Core 3 and later. EF Core 3 and later do not support .NET Standard 2.0 platforms like .NET Framework 4.8.

As well as traditional RDBMSs, EF Core supports modern cloud-based, non-relational, schema-less data stores, such as Azure Cosmos DB and MongoDB, sometimes with third-party providers.

EF Core has so many improvements in each release that this chapter cannot cover them all. In this chapter, I will focus on the fundamentals that all .NET developers should know and some of the most useful new features. You can learn more about EF Core and how to use it with SQL Server in my companion book, *Apps and Services with .NET 8*, or by reading the official documentation, found at the following link:

`https://learn.microsoft.com/en-us/ef/core/`

You can keep up with the latest EF Core news at the following link:

`https://aka.ms/efnews`

Understanding Database First and Code First

There are two approaches to working with EF Core:

- **Database First**: A database already exists, so you build a model that matches its structure and features. This is the most common scenario in real life. You will see an example of this throughout this chapter.
- **Code First**: No database exists, so you build a model and then use EF Core to create a database that matches its structure and features. You will see an example of this if you complete the online-only section linked to in one of the exercises at the end of this chapter.

Performance improvements in EF Core

The EF Core team continues to work hard on improving the performance of EF Core. For example, if EF Core identifies that only a single statement will be executed against the database when `SaveChanges` is called, then it does not create an explicit transaction as earlier versions do. That gives a 25% performance improvement to a common scenario.

There is too much information about all the recent performance improvements to cover in this chapter, and you get all the benefits without needing to know how they work anyway. If you are interested (and it is fascinating what they looked at and how they took advantage of some cool SQL Server features in particular), then I recommend that you read the following posts from the EF Core team:

- Announcing Entity Framework Core 7 Preview 6: Performance Edition: `https://devblogs.microsoft.com/dotnet/announcing-ef-core-7-preview6-performance-optimizations/`
- Announcing Entity Framework Core 6.0 Preview 4: Performance Edition: `https://devblogs.microsoft.com/dotnet/announcing-entity-framework-core-6-0-preview-4-performance-edition/`

Using a sample relational database

To learn how to manage an RDBMS using .NET, it would be useful to have a sample one so that you can practice on one that has a medium complexity and a decent number of sample records. Microsoft offers several sample databases, most of which are too complex for our needs, so instead, we will use a database that was first created in the early 1990s known as `Northwind`.

Let's take a minute to look at a diagram of the `Northwind` database. You can use the diagram in *Figure 10.1* to refer to as we write code and queries throughout this book:

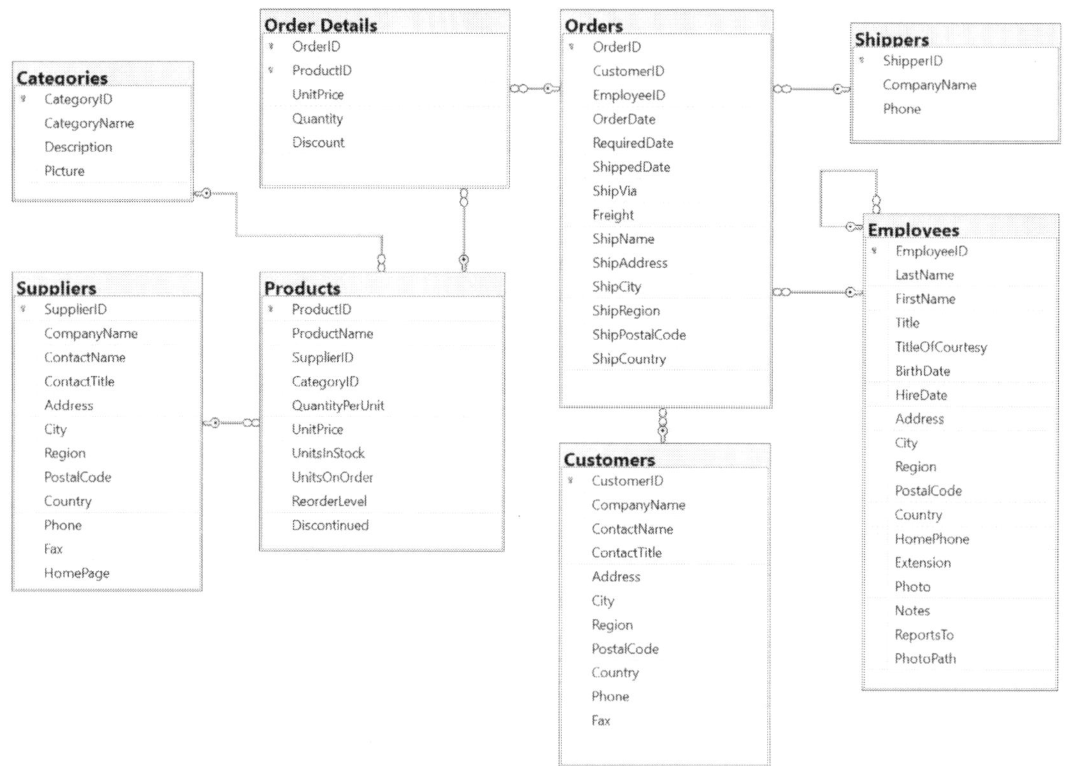

Figure 10.1: The Northwind database tables and relationships

You will write code to work with the `Categories` and `Products` tables later in this chapter, and other tables in later chapters. But before we do, note the following:

- Each category has a unique identifier, name, description, and picture.
- Each product has a unique identifier, name, unit price, units in stock, and other fields.
- Each product is associated with a category by storing the category's unique identifier.
- The relationship between `Categories` and `Products` is one-to-many, meaning each category can have zero or more products. This is indicated in *Figure 10.1* by an infinity symbol at one end (meaning many) and a yellow key at the other end (meaning one).

Using SQLite

SQLite is a small, fast, cross-platform, self-contained RDBMS that is available in the public domain. It's the most common RDBMS for mobile platforms such as iOS (iPhone and iPad) and Android. SQLite is the most used database engine in the world and there are more than one trillion SQLite databases in active use. You can read more about this at `https://www.sqlite.org/mostdeployed.html`.

I decided to demonstrate databases using SQLite in this book since its important themes are cross-platform development and fundamental skills that only need basic database capabilities. I recommend that you initially complete the book code tasks using SQLite. If you also want to try the code tasks using SQL Server, then I provide documentation to do so in the online-only sections in the GitHub repository for this book.

Using SQL Server or other SQL systems

Enterprises that standardize on Windows tend to also use SQL Server as their database. If you would prefer to use SQL Server, please see the online instructions at the following link:

`https://github.com/markjprice/cs13net9/blob/main/docs/sql-server/README.md`

If you would prefer to use a different SQL system, then the SQL scripts that I provide should work with most SQL systems, for example, PostgreSQL or MySQL. However, I have not written step-by-step instructions for them and I make no guarantees they will work.

My recommendation is to complete this book using SQLite, so you focus on learning what's taught in the book about EF Core rather than adding the complication of trying to use a different database system. Learning is hard enough; don't bite off more than you can chew, and don't make it harder on yourself. Once you've learned what's in the book, you can always repeat it with a different database system.

Setting up SQLite CLI tools for Windows

On Windows, we need to download SQLite CLI tools and add the folder for SQLite to the system path so it will be found when we enter commands at a command prompt or terminal:

1. Start your favorite browser and navigate to `https://www.sqlite.org/download.html`.
2. Scroll down the page to the **Precompiled Binaries for Windows** section.

3. Click **sqlite-tools-win32-x86-3460100.zip** (the file might have a higher version number), as shown in the following screenshot:

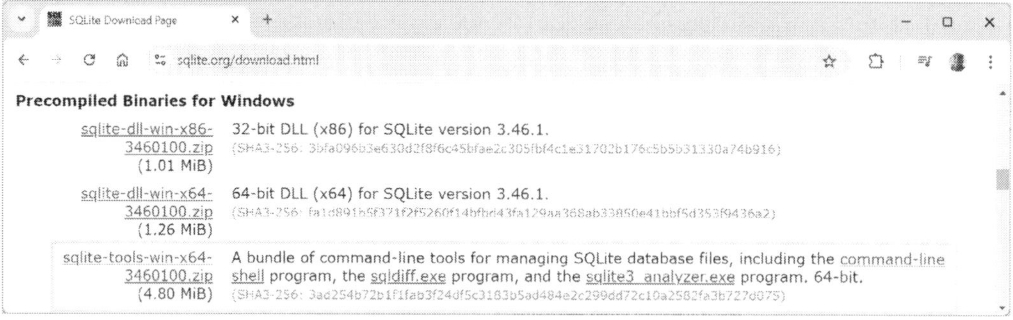

Figure 10.2: Downloading SQLite for Windows

4. Extract the ZIP file into a folder named `C:\Sqlite\`. Make sure that the three extracted files, including `sqlite3.exe`, are directly inside the `C:\SQLite` folder or the executable will not be found later when you try to use it.
5. In the Windows **Start** menu, navigate to **Settings**.
6. Search for `environment` and choose **Edit the system environment variables**. On non-English versions of Windows, please search for the equivalent word in your local language to find the setting.
7. Click the **Environment Variables** button.
8. In **System variables**, select **Path** in the list, and then click **Edit...**.
9. If `C:\SQLite` is not already in the path, then click **New**, enter `C:\Sqlite`, and press *Enter*.
10. Click **OK**, then **OK**, then **OK** again, and then close **Settings**.
11. To confirm that the path to SQLite has been configured correctly, at any command prompt or terminal, enter the following command to start SQLite:

```
sqlite3
```

12. Note the result, as shown in the following output:

```
SQLite version 3.42.0 2023-05-16 12:36:15
Enter ".help" for usage hints.
Connected to a transient in-memory database.
Use ".open FILENAME" to reopen on a persistent database.
sqlite>
```

13. To exit the SQLite command prompt, do the following:

    - On Windows, press *Ctrl* + *C* twice.
    - On macOS, press *Ctrl* + *D*.

Setting up SQLite for macOS and Linux

On macOS, SQLite is included in the `/usr/bin/` directory as a command-line application named `sqlite3`.

On Linux, you can get set up with SQLite using the following command:

```
sudo apt-get install sqlite3
```

SQLite can be downloaded and installed for other OSs from the following link:

`https://www.sqlite.org/download.html`

Setting up EF Core in a .NET project

Now that we have a database system set up, we can create a database and .NET project that uses it.

Creating a console app for working with EF Core

First, we will create a console app project for this chapter.

Use your preferred code editor to create a new project, as defined in the following list:

- Project template: **Console App** / `console`
- Project file and folder: `WorkingWithEFCore`
- Solution file and folder: `Chapter10`

Creating the Northwind sample database for SQLite

Now we can create the `Northwind` sample database for SQLite using an SQL script:

1. If you have not previously cloned or downloaded the ZIP for the GitHub repository for this book, then do so now using the following link: `https://github.com/markjprice/cs13net9`.
2. Copy the script to create the `Northwind` database for SQLite from the following path in your local Git repository or where you extracted the ZIP: `/scripts/sql-scripts/Northwind4SQLite.sql` into the `WorkingWithEFCore` folder.
3. Start a command prompt or terminal in the `WorkingWithEFCore` project folder:
    - On Windows, start **File Explorer**, right-click the `WorkingWithEFCore` folder, and select **New Command Prompt at Folder** or **Open in Windows Terminal**.
    - On macOS, start **Finder**, right-click the `WorkingWithEFCore` folder, and select **New Terminal at Folder**.
4. Enter the command to execute the SQL script using SQLite to create the `Northwind.db` database, as shown here:

    ```
    sqlite3 Northwind.db -init Northwind4SQLite.sql
    ```

5. Be patient because this command might take a while to create the database structure. Eventually, you will see the SQLite command prompt, as shown in the following output:

```
-- Loading resources from Northwind4SQLite.sql
SQLite version 3.42.0 2023-05-16 12:36:15
Enter ".help" for usage hints.
sqlite>
```

6. To exit the SQLite command prompt, do the following:
    - On Windows, press *Ctrl* + *C* twice.
    - On macOS or Linux, press *Ctrl* + *D*.
7. You can leave the command prompt or terminal window open because you will use it again soon.

If you are using Visual Studio

If you are using VS Code and the `dotnet run` command, the compiled application executes in the `WorkingWithEFCore` folder, allowing it to locate the database file stored therein. But if you are using Visual Studio or Rider, then the compiled application executes in the `WorkingWithEFCore\bin\Debug\net9.0` folder, so it will not find the database file because it is not in that directory.

Let's tell Visual Studio to copy the database file to the directory that it runs the code in so that it can find the file, but only if the database file is newer or is missing so it will not overwrite any database changes we make during runtime:

1. In **Solution Explorer**, right-click the `Northwind.db` file and select **Properties**.
2. In **Properties**, set **Copy to Output Directory** to **Copy if newer**.
3. In `WorkingWithEFCore.csproj`, note the new elements, as shown in the following markup:

```
<ItemGroup>
  <None Update="Northwind.db">
    <CopyToOutputDirectory>PreserveNewest</CopyToOutputDirectory>
  </None>
</ItemGroup>
```

If you prefer to overwrite the data changes every time you start the project, then set **CopyToOutputDirectory** to **Always**.

You could make the preceding changes to the project file manually instead of using the **Properties** window. In fact, any change made to the project file by any tool can also be made manually just by editing the XML. Tools like the **Properties** window just read the files in a project and show an alternative view. This is why in *Chapter 1*, I stressed using multiple code editors when learning .NET. The danger is that if you only use Visual Studio, then you may start to think that Visual Studio *is* .NET development. It is not.

The real .NET development is the contents of the source code files, like `.cs`, and the project file `.csproj`, which are then compiled by the command-line interface `dotnet`. Any other tool you use is just an additional layer on top of that.

Interestingly, a future feature in .NET that Microsoft is actively looking at is "implicit project files." This would mean having a folder that contains only one or more `.cs` files and the `.csproj` file would not need to exist because its content could be implied using defaults. For example, it might default to the current SDK and its target .NET version, and so on. But it gets complicated, so we will have to wait and see if this ever happens.

Managing the Northwind sample database with SQLiteStudio

You can use a cross-platform graphical database manager named **SQLiteStudio** to easily manage SQLite databases:

1. Navigate to `https://sqlitestudio.pl`, and then download and install the application.
2. Start **SQLiteStudio**.
3. Navigate to **Database** | **Add a database**.
4. In the **Database** dialog, in the **File** section, click on the yellow folder button to browse for an existing database file on the local computer, select the `Northwind.db` file in the `WorkingWithEFCore` project folder, and then click **OK**, as shown in *Figure 10.3*:

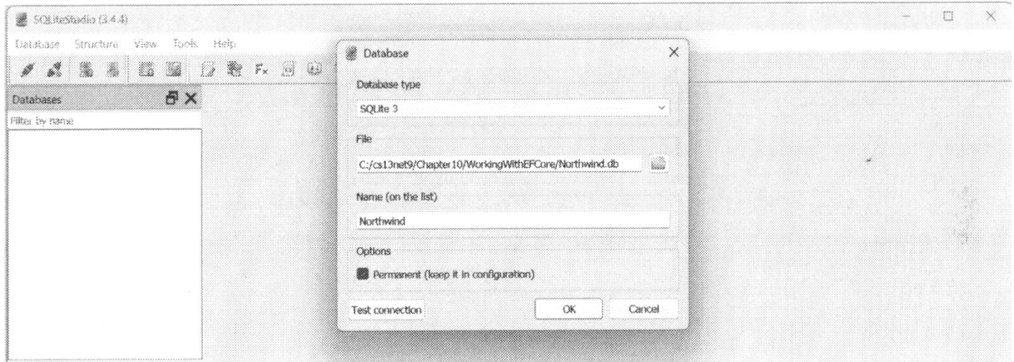

Figure 10.3: Adding the Northwind.db database file to SQLiteStudio

5. If you cannot see the database, then navigate to **View** | **Databases**.
6. In the **Databases** window, right-click on the `Northwind` database and choose **Connect to the database** (or just double-click `Northwind`). You will see the 10 tables that were created by the script. (The script for SQLite is simpler than the one for SQL Server; it does not create as many tables or other database objects.)
7. Right-click on the **Products** table and choose **Edit the table**, or just double-click the table.

8. In the table editor window, note the structure of the `Products` table, including column names, data types, keys, and constraints, as shown in *Figure 10.4*:

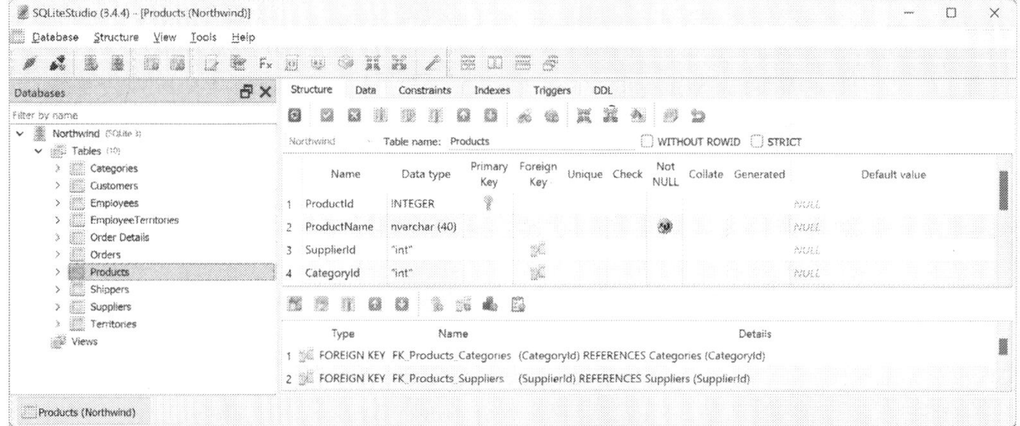

Figure 10.4: The table editor in SQLiteStudio showing the structure of the Products table

9. In the table editor window, click the **Data** tab, and you will see 77 products, as shown in *Figure 10.5*:

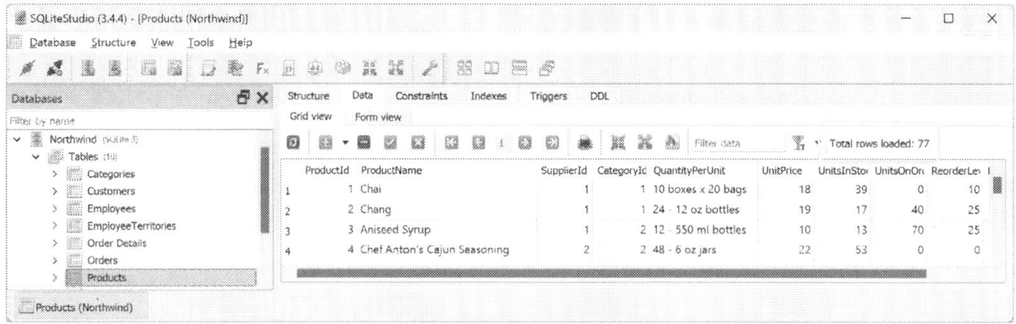

Figure 10.5: The Data tab showing the 77 rows in the Products table

10. In the **Database** window, right-click `Northwind` and select **Disconnect from the database.**
11. Quit SQLiteStudio.

Using the lightweight ADO.NET database providers

Before Entity Framework, there was **ADO.NET**. Compared to EF, this is a simpler and more efficient API for working with databases. It provides abstract classes like `DbConnection`, `DbCommand`, and `DbReader`, and provider-specific implementations of them like `SqlConnection` and `SqlCommand`.

In this chapter, if you choose to use SQL Server, then you should use the `SqlConnectionStringBuilder` class to help write a valid connection string. This is because it has properties for all possible parts of a database connection string that you set individually and then it returns the complete string. You should also get sensitive secrets like passwords from an environment variable or a secret management system instead of writing them in your source code.

For SQLite, the connection string is so simple that you do not need to use the `SqliteConnectionStringBuilder` class.

The EF Core database providers for SQLite and SQL Server are built on top of the ADO.NET libraries, so EF Core is always inherently slower than ADO.NET.

If you want to use native **ahead-of-time** (**AOT**) publishing, be aware that EF Core does not yet support it. This means you can only use the ADO.NET libraries if you plan to compile to native code. The EF Core team is investigating how they can support native AOT, but it is challenging so it has not happened with EF Core 9 this year. Hopefully, it will happen for EF Core 10 in 2025.

Apart from `SqlConnectionStringBuilder`, this book does not cover using the ADO.NET library, but I do cover examples of how to publish native AOT minimal API web services using the ADO.NET for SQL Server library in the companion book, *Apps and Services with .NET 8*.

You can learn more about the ADO.NET for SQLite library at the following link:

`https://learn.microsoft.com/en-us/dotnet/standard/data/sqlite/`

You can learn more about the ADO.NET for SQL Server library at the following link:

`https://learn.microsoft.com/en-us/sql/connect/ado-net/microsoft-ado-net-sql-server`

Warning! All `System.Data.SqlClient` users are encouraged to transition to `Microsoft.Data.SqlClient`. You can read the announcement about how `System.Data.SqlClient` package is now deprecated at the following link:

`https://techcommunity.microsoft.com/t5/sql-server-blog/announcement-system-data-sqlclient-package-is-now-deprecated/ba-p/4227205`

Choosing an EF Core database provider

Before we dive into the practicalities of managing data using EF Core, let's briefly talk about choosing between EF Core database providers. To manage data in a specific database, we need classes that know how to efficiently talk to that database.

EF Core database providers are sets of classes that are optimized for a specific data store. There is even a provider for storing the data in the memory of the current process, which can be useful for high-performance unit testing since it avoids hitting an external system.

They are distributed as NuGet packages, as shown in *Table 10.1*:

To manage this data store	Reference this NuGet package
SQL Server 2012 or later	`Microsoft.EntityFrameworkCore.SqlServer`
SQLite 3.7 or later	`Microsoft.EntityFrameworkCore.SQLite`
In-memory	`Microsoft.EntityFrameworkCore.InMemory`
Azure Cosmos DB SQL API	`Microsoft.EntityFrameworkCore.Cosmos`
MySQL	`MySQL.EntityFrameworkCore`

Oracle DB 11.2	Oracle.EntityFrameworkCore
PostgreSQL	Npgsql.EntityFrameworkCore.PostgreSQL

Table 10.1: NuGet packages for common EF Core database providers

You can reference as many EF Core database providers in the same project as you need. Each package includes the common shared types, as well as provider-specific types.

Connecting to a named SQLite database

To connect to an SQLite database, we just need to know the database path and filename, set using the legacy parameter `Filename` or the modern equivalent, `Data Source`. The path can be relative to the current directory or an absolute path. We specify this information in a **connection string**.

Defining the Northwind database context class

A class named `Northwind` will be used to represent the database. To use EF Core, the class must inherit from `DbContext`. The `DbContext` class understands how to communicate with databases and dynamically generate SQL statements to query and manipulate data.

Your `DbContext`-derived class should have an overridden method named `OnConfiguring`, which will set the database connection string.

We will create a project that uses SQLite, but feel free to use SQL Server or some other database system if you feel comfortable doing so instead:

1. In the `WorkingWithEFCore` project, add a package reference to the EF Core provider for SQLite and globally and statically import the `System.Console` class for all C# files, as shown in the following markup:

```
<ItemGroup>
  <Using Include="System.Console" Static="true" />
</ItemGroup>

<ItemGroup>
  <PackageReference Version="9.0.0"
    Include="Microsoft.EntityFrameworkCore.Sqlite" />
</ItemGroup>
```

2. Build the `WorkingWithEFCore` project to restore packages.

After February 2025, you will be able to try out previews of EF Core 10 by specifying version `10.0-*`, so you must also install a preview of .NET 10 SDK. The target framework for your project must be `net10.0`. By using a wildcard, you will automatically download the latest monthly preview when you restore the packages for the project. Once the EF Core 10 GA version is released in November 2025, change the package version to `10.0.0`. After February 2026, you will be able to do the same with EF Core 11 and that will continue to require a project targeting `net10.0`.

3. In the project folder, add a new class file named `NorthwindDb.cs`.
4. In `NorthwindDb.cs`, import the main namespace for EF Core, define a class named `NorthwindDb`, and make the class inherit from `DbContext`. Then, in an `OnConfiguring` method, configure the options builder to use SQLite with an appropriate database connection string, as shown in the following code:

```
using Microsoft.EntityFrameworkCore; // To use DbContext and so on.

namespace Northwind.EntityModels;

// This manages interactions with the Northwind database.
public class NorthwindDb : DbContext
{
  protected override void OnConfiguring(
    DbContextOptionsBuilder optionsBuilder)
  {
    string databaseFile = "Northwind.db";
    string path = Path.Combine(
      Environment.CurrentDirectory, databaseFile);

    string connectionString = $"Data Source={path}";
    WriteLine($"Connection: {connectionString}");
    optionsBuilder.UseSqlite(connectionString);
  }
}
```

Good Practice: Some code editors like Visual Studio will automatically add a call to the base class member when you override a member. Although this is generally good practice, it is unnecessary in this case because the base implementation does nothing. To decide if you need to keep the call or not, view the tooltip or definition of the member. If you hover over the `OnConfiguring` method name, its tooltip tells you "The base implementation does nothing." If you **Go To Definition** (*F12*) of the method, you will see that it does nothing, as shown in the following code:

```
protected internal virtual void OnConfiguring(
DbContextOptionsBuilder optionsBuilder)
{
}
```

The method only exists in the base class so that subclasses can override it and then EF Core can call your code when it needs to configure the data context. It would be a waste for your overridden method implementation to call the base implementation. If your code editor adds the call to the base class automatically, then you should delete the statement to make your code match the code in the book. The same applies later in the book when you override the `OnModelCreating` method.

5. In `Program.cs`, delete the existing statements. Then, import the `Northwind.EntityModels` namespace and output the database provider, as shown in the following code:

```
using Northwind.EntityModels; // To use Northwind.

using NorthwindDb db = new();
WriteLine($"Provider: {db.Database.ProviderName}");
// Disposes the database context.
```

6. Run the console app and note the output showing the database connection string and which database provider you are using, as shown in the following output:

```
Connection: Data Source=C:\cs13net9\Chapter10\WorkingWithEFCore\bin\
Debug\net9.0\Northwind.db
Provider: Microsoft.EntityFrameworkCore.Sqlite
```

You now know how to connect to a database by defining the EF Core data context. Next, we need to define a model that represents the tables in the database.

Defining EF Core models

EF Core uses a combination of **conventions, annotation attributes**, and **Fluent API** statements to build an **entity model** at runtime, which enables any actions performed on the classes to later be automatically translated into actions performed on the actual database. An **entity class** represents the structure of a table, and an instance of the class represents a row in that table.

First, we will review the three ways to define a model, with code examples, and then we will create some classes that implement those techniques.

Using EF Core conventions to define the model

The code we will write will use the following conventions:

- The name of a table is assumed to match the name of a `DbSet<T>` property in the `DbContext` class, for example, `Products`.
- The names of the columns are assumed to match the names of properties in the entity model class, for example, `ProductId`.
- The `string` .NET type is assumed to be a `nvarchar` type in the database.
- The `int` .NET type is assumed to be an `int` type in the database.
- The primary key is assumed to be a property that is named `Id` or `ID`, or when the entity model class is named `Product`, then the property can be named `ProductId` or `ProductID`. If this property is an integer type or the `Guid` type, then it is also assumed to be an `IDENTITY` column (a column type that automatically assigns a value when inserting).

Good Practice: There are many other conventions that you should know, and you can even define your own, but that is beyond the scope of this book. You can read about them at the following link:

`https://learn.microsoft.com/en-us/ef/core/modeling/`

Using EF Core annotation attributes to define the model

Conventions often aren't enough to completely map the classes to the database objects. A simple way of making your model smarter is to apply annotation attributes. Some common attributes recognized by EF Core are shown in *Table 10.2*:

Attribute	Description
`[Required]`	This ensures the value is not null. In .NET 8, it has a `DisallowAllDefaultValues` parameter to prevent value types from having their default value. For example, an `int` cannot be `0`.
`[StringLength(50)]`	This ensures the value is up to 50 characters in length.
`[Column(TypeName = "money", Name = "UnitPrice")]`	This specifies the column type and column name used in the table.

Table 10.2: Common EF Core annotation attributes

Warning! `[StringLength]` is not honored by all EF Core data providers. For example, although SQL Server honors it, SQLite does not. For SQLite, use `[Column(TypeName = "text(50)")]` instead.

Some additional attributes that can be used to validate entities and are recognized by platforms like ASP.NET Core and Blazor for validation are shown in *Table 10.3*:

Attribute	Description
`[RegularExpression(expression)]`	This ensures the value matches the specified regular expression.
`[EmailAddress]`	This ensures the value contains one @ symbol, but not as the first or last character. It does not use a regular expression.
`[Range(1, 10)]`	This ensures a `double`, `int`, or `string` value within a specified range. New in .NET 8 are the parameters `MinimumIsExclusive` and `MaximumIsExclusive`.
`[Length(10, 20)]`	This ensures a string or collection is within a specified length range, for example, a minimum of 10 characters or items, and a a maximum of 20 characters or items.

[Base64String]	This ensures the value is a well-formed Base64 string.
[AllowedValues]	This ensures the value is one of the items in the `params` array of objects. For example, `"alpha"`, `"beta"`, `"gamma"`, or `1, 2, 3`.
[DeniedValues]	Ensures value is not one of the items in the `params` array of objects. For example, `"alpha"`, `"beta"`, `"gamma"`, or `1, 2, 3`.

Table 10.3: Validation annotation attributes

Why does the `EmailAddress` attribute seem so basic? According to a comment by ajcvickers, who also closed the GitHub issue, *"The check is intentionally naive because doing something infallible is very hard. The email really should be validated in some other way, such as through an email confirmation flow where an email is actually sent. The validation attribute is designed only to catch egregiously wrong values such as for a U.I."* You can read the debate at the following link:

`https://github.com/dotnet/runtime/issues/27592`

For example, in the database, the maximum length of a product name is 40, and the value cannot be null, as shown highlighted in the following **data definition language** (**DDL**) code from the `Northwind4SQLite.sql` script file, which defines how to create a table named `Products` with its columns, data types, keys, and other constraints:

```
CREATE TABLE Products (
    ProductId       INTEGER       PRIMARY KEY,
    ProductName     NVARCHAR (40) NOT NULL,
    SupplierId      "INT",
    CategoryId      "INT",
    QuantityPerUnit NVARCHAR (20),
    UnitPrice       "MONEY"       CONSTRAINT DF_Products_UnitPrice DEFAULT (0),
    UnitsInStock    "SMALLINT"    CONSTRAINT DF_Products_UnitsInStock DEFAULT
(0),
    UnitsOnOrder    "SMALLINT"    CONSTRAINT DF_Products_UnitsOnOrder DEFAULT
(0),
    ReorderLevel    "SMALLINT"    CONSTRAINT DF_Products_ReorderLevel DEFAULT
(0),
    Discontinued    "BIT"         NOT NULL
                                  CONSTRAINT DF_Products_Discontinued DEFAULT
(0),
    CONSTRAINT FK_Products_Categories FOREIGN KEY (
        CategoryId
    )
    REFERENCES Categories (CategoryId),
```

```
    CONSTRAINT FK_Products_Suppliers FOREIGN KEY (
        SupplierId
    )
    REFERENCES Suppliers (SupplierId),
    CONSTRAINT CK_Products_UnitPrice CHECK (UnitPrice >= 0),
    CONSTRAINT CK_ReorderLevel CHECK (ReorderLevel >= 0),
    CONSTRAINT CK_UnitsInStock CHECK (UnitsInStock >= 0),
    CONSTRAINT CK_UnitsOnOrder CHECK (UnitsOnOrder >= 0)
);
```

In a `Product` class, we could apply attributes to specify this, as shown in the following code:

```
[Required]
[StringLength(40)]
public string ProductName { get; set; }
```

When there isn't an obvious map between .NET types and database types, an attribute can be used.

For example, in the database, the column type of `UnitPrice` for the `Products` table is `money`. .NET does not have a `money` type, so it should use `decimal` instead, as shown in the following code:

```
[Column(TypeName = "money")]
public decimal? UnitPrice { get; set; }
```

Using the EF Core Fluent API to define the model

The last way that the model can be defined is by using the Fluent API. This API can be used instead of attributes, as well as in addition to them. For example, to define the `ProductName` property, instead of decorating the property with two attributes, an equivalent Fluent API statement could be written in the `OnModelCreating` method of the database context class, as shown in the following code:

```
modelBuilder.Entity<Product>()
  .Property(product => product.ProductName)
  .IsRequired()
  .HasMaxLength(40);
```

This keeps the entity model class simpler.

Understanding data seeding with the Fluent API

Another benefit of the Fluent API is to provide initial data to populate a database. EF Core automatically works out which insert, update, or delete operations must be executed.

For example, if we wanted to make sure that a new database had at least one row in the `Products` table, then we would call the `HasData` method, as shown in the following code:

```
modelBuilder.Entity<Product>()
  .HasData(new Product
```

```
{
  ProductId = 1,
  ProductName = "Chai",
  UnitPrice = 8.99M
});
```

Calls to HasData take effect either during a data migration executed by the command `dotnet ef database update` or when you call the `Database.EnsureCreated` method.

Our model will map to an existing database that is already populated with data, so we will not need to use this technique in our code.

Building EF Core models for the Northwind tables

Now that you've learned about the ways to define EF Core models, let's build models to represent two of the tables in the `Northwind` database. For reuse, we will do this in a separate class library project.

The two entity classes will refer to each other, so to avoid compiler errors, we will create the classes without any members first:

1. Use your preferred code editor to create a new project, as defined in the following list:
   - **Project template: Class Library** / `classlib`
   - **Project file and folder:** `Northwind.EntityModels`
   - **Solution file and folder:** `Chapter10`
2. In the `Northwind.EntityModels` project, delete the file named `Class1.cs` and then add two class files named `Category.cs` and `Product.cs`.
3. In `Category.cs`, define a class named `Category`, as shown in the following code:

   ```
   namespace Northwind.EntityModels;

   public class Category
   {
   }
   ```

4. In `Product.cs`, define a class named `Product`, as shown in the following code:

   ```
   namespace Northwind.EntityModels;

   public class Product
   {
   }
   ```

5. In the `WorkingWithEFCore` project, add a project reference to the `Northwind.EntityModels` project, as shown in the following markup:

```
<ItemGroup>
  <ProjectReference Include="..\Northwind.EntityModels\
Northwind.EntityModels.csproj" />
</ItemGroup>
```

The project reference path and filename must all go on one line.

6. Build the `WorkingWithEFCore` project.

Defining the Category and Product entity classes

The `Category` class, also known as an entity model, will be used to represent a row in the `Categories` table. This table has four columns, as shown in the following DDL taken from the `Northwind4SQLite.sql` script file:

```
CREATE TABLE Categories (
  CategoryId   INTEGER       PRIMARY KEY,
  CategoryName NVARCHAR (15) NOT NULL,
  Description  "NTEXT",
  Picture      "IMAGE"
);
```

We will use conventions to define the following:

- Three of the four properties (we will not map the `Picture` column)
- The primary key
- The one-to-many relationship to the `Products` table

To map the `Description` column to the correct database type, we will need to decorate the `string` property with the `Column` attribute.

Later in this chapter, we will use the Fluent API to define that `CategoryName` cannot be null and is limited to a maximum of 15 characters.

Let's go:

1. In the `Northwind.EntityModels` project, modify the `Category` entity model class, as shown highlighted in the following code:

```
using System.ComponentModel.DataAnnotations.Schema; // To use [Column].

namespace Northwind.EntityModels;
```

```
public class Category
{
  // These properties map to columns in the database.
  public int CategoryId { get; set; } // The primary key.

  public string CategoryName { get; set; } = null!;

  [Column(TypeName = "ntext")]
  public string? Description { get; set; }

  // Defines a navigation property for related rows.
  public virtual ICollection<Product> Products { get; set; }
    // To enable developers to add products to a Category, we must
    // initialize the navigation property to an empty collection.
    // This also avoids an exception if we get a member like Count.
    = new HashSet<Product>();
}
```

Note the following:

- The `Category` class will be in the `Northwind.EntityModels` namespace.
- The `CategoryId` property follows the primary key naming convention, so it will be mapped to a column marked as the primary key with an index.
- The `CategoryName` property maps to a column that does not allow database NULL values, so it is a non-nullable string. To disable nullability warnings, we have assigned the null-forgiving operator.
- The `Description` property maps to a column with the `ntext` data type instead of the default mapping for `string` values to `nvarchar`.
- We initialize the collection of `Product` objects to a new, empty `HashSet`. A hash set is more efficient than a list because it is unordered. If you do not initialize `Products`, then it will be null and if you try to get its `Count`, then you will get an exception.

2. Modify the `Product` class, as shown highlighted in the following code:

```
using System.ComponentModel.DataAnnotations; // To use [Required].
using System.ComponentModel.DataAnnotations.Schema; // To use [Column].

namespace Northwind.EntityModels;

public class Product
{
  public int ProductId { get; set; } // The primary key.
```

```
  [Required]
  [StringLength(40)]
  public string ProductName { get; set; } = null!;

  // Property name is different from the column name.
  [Column("UnitPrice", TypeName = "money")]
  public decimal? Cost { get; set; }

  [Column("UnitsInStock")]
  public short? Stock { get; set; }

  public bool Discontinued { get; set; }

  // These two properties define the foreign key relationship
  // to the Categories table.
  public int CategoryId { get; set; }
  public virtual Category Category { get; set; } = null!;
}
```

Note the following:

- The `Product` class will be used to represent a row in the `Products` table, which has 10 columns.
- You do not need to include all columns from a table as properties of a class. We will only map six properties: `ProductId`, `ProductName`, `UnitPrice`, `UnitsInStock`, `Discontinued`, and `CategoryId`.
- Columns that are not mapped to properties cannot be read or set using the class instances. If you use the class to create a new object, then the new row in the table will have `NULL` or some other default value for the unmapped column values in that row. You must make sure that those missing columns are optional or have default values set by the database, or else an exception will be thrown at runtime. In this scenario, the rows already have data values and I have decided that I do not need to read those values in this application.
- We can rename a column by defining a property with a different name, like `Cost`, and then decorating the property with the `[Column]` attribute and specifying its column name, like `UnitPrice`.
- The final property, `CategoryId`, is associated with a `Category` property that will be used to map each product to its parent category.

The two properties that relate to the two entities, `Category.Products` and `Product.Category`, are both marked as `virtual`. This allows EF Core to inherit and override the properties to provide extra features, such as lazy loading.

Adding tables to the Northwind database context class

Inside your DbContext-derived class, you must define at least one property of the DbSet<T> type. These properties represent the tables. To tell EF Core what columns each table has, the DbSet<T> properties use generics to specify a class that represents a row in the table. That entity model class has properties that represent its columns.

The DbContext-derived class can optionally have an overridden method named OnModelCreating. This is where you can write Fluent API statements as an alternative to decorating your entity classes with attributes.

Let's write the code:

1. In the WorkingWithEFCore project, modify the NorthwindDb class to add statements to define two properties for the two tables and an OnModelCreating method, as shown highlighted in the following code:

```
public class NorthwindDb : DbContext
{
  // These two properties map to tables in the database.
  public DbSet<Category>? Categories { get; set; }
  public DbSet<Product>? Products { get; set; }

  protected override void OnConfiguring(
    DbContextOptionsBuilder optionsBuilder)
  {
    ...
  }

  protected override void OnModelCreating(
    ModelBuilder modelBuilder)
  {
    // Example of using Fluent API instead of attributes
    // to limit the length of a category name to 15.
    modelBuilder.Entity<Category>()
      .Property(category => category.CategoryName)
      .IsRequired() // NOT NULL
      .HasMaxLength(15);

    // Some SQLite-specific configuration.
    if (Database.ProviderName?.Contains("Sqlite") ?? false)
    {
```

```
        // To "fix" the lack of decimal support in SQLite.
        modelBuilder.Entity<Product>()
          .Property(product => product.Cost)
          .HasConversion<double>();
      }
    }
}
```

The `decimal` type is not supported by the SQLite database provider for sorting and other operations. We can fix this by telling the model that `decimal` values can be treated as `double` values when using the SQLite database provider. This does not actually perform any conversion at runtime.

Now that you have seen some examples of defining an entity model manually, let's look at a tool that can do some of the work for you.

Setting up the dotnet-ef tool

The .NET CLI tool named `dotnet` can be extended with capabilities useful for working with EF Core. It can perform design-time tasks like creating and applying migrations from an older model to a newer model and generating code for a model from an existing database.

The `dotnet-ef` command-line tool is not automatically installed. You must install this package as either a **global** or **local tool**. If you have already installed an older version of the tool, then you should update it to the latest version:

1. At a command prompt or terminal, check if you have already installed `dotnet-ef` as a global tool, as shown in the following command:

   ```
   dotnet tool list --global
   ```

2. Check in the list if an older version of the tool has been installed, like the one for .NET 7, as shown in the following output:

   ```
   Package Id      Version     Commands
   -------------------------------------
   dotnet-ef       7.0.0       dotnet-ef
   ```

3. If an old version is already installed, then update the tool, as shown in the following command:

   ```
   dotnet tool update --global dotnet-ef
   ```

4. If it is not already installed, then install the latest version, as shown in the following command:

   ```
   dotnet tool install --global dotnet-ef
   ```

If necessary, follow any OS-specific instructions to add the `dotnet tools` directory to your `PATH` environment variable, as described in the output of installing the `dotnet-ef` tool.

By default, the latest GA release of .NET will be used to install the tool. To explicitly set a version, for example, to use a preview, add the `--version` switch. As another example, to update to the latest .NET 9 preview version available from February 2024 to October 2024, use the following command with a version wildcard:

```
dotnet tool update --global dotnet-ef --version 9.0-*
```

Once the .NET 9 GA release happens in November 2024, you can just use the command without the `--version` switch to upgrade.

You can also remove the tool, as shown in the following command:

```
dotnet tool uninstall --global dotnet-ef
```

Scaffolding models using an existing database

Scaffolding is the process of using a tool to create classes that represent the model of an existing database using reverse engineering. A good scaffolding tool allows you to extend the automatically generated classes because they are `partial` and then regenerate those classes without losing your `partial` classes.

If you know that you will never regenerate the classes using the tool, then feel free to change the code for the automatically generated classes as much as you want. The code generated by the tool is just the best approximation.

Good Practice: Do not be afraid to overrule a tool when you know better.

The instructions in this section apply to any data provider, not just SQLite. Whatever data provider you use, you will need to add a reference to the `Microsoft.EntityFrameworkCore.Design` package.

Let's see if the tool generates the same model as we did manually:

1. Add the latest version of the `Microsoft.EntityFrameworkCore.Design` package to the `WorkingWithEFCore` project, as shown highlighted in the following markup:

```
<ItemGroup>
  <PackageReference Version="9.0.0"
                    Include="Microsoft.EntityFrameworkCore.Design">
    <PrivateAssets>all</PrivateAssets>
    <IncludeAssets>runtime; build; native; contentfiles; analyzers; 
buildtransitive</IncludeAssets>
  </PackageReference>
  <PackageReference Version="9.0.0"
                    Include="Microsoft.EntityFrameworkCore.Sqlite" />
</ItemGroup>
```

More Information: If you are unfamiliar with how packages like `Microsoft.EntityFrameworkCore.Design` can manage their assets, then you can learn more at the following link:

`https://learn.microsoft.com/en-us/nuget/consume-packages/package-references-in-project-files#controlling-dependency-assets`

2. Build the `WorkingWithEFCore` project to restore packages.
3. Start a command prompt or terminal in the `WorkingWithEFCore` project folder. Here is an example:
    - If you are using Visual Studio, in **Solution Explorer**, right-click the `WorkingWithEFCore` project and select **Open in Terminal**.
    - On Windows, start **File Explorer**, right-click the `WorkingWithEFCore` folder, and select **New Command Prompt at Folder** or **Open in Windows Terminal**.
    - On macOS, start **Finder**, right-click the `WorkingWithEFCore` folder, and select **New Terminal at Folder**.
    - If you are using Rider, in **Solution Explorer**, right-click the `WorkingWithEFCore` project and select **Open In | Terminal**.

Warning! When I say the `WorkingWithEFCore` project folder, I mean the folder that contains the `WorkingWithEFCore.csproj` project file. If you enter the command in a folder that does not contain a project file, then you will see the following error: `No project was found. Change the current working directory or use the --project option.`

Good Practice: You are about to enter a long command. I recommend that you type from the print book or copy and paste long commands like this from the eBook into a plain text editor like Notepad. Then make sure that the whole command is properly formatted as a single line with correct spacing. Only then should you copy and paste it into the command prompt or terminal. Copying and pasting directly from the eBook is likely to include newline characters and missing spaces that break the command if you aren't careful. Also, remember that all commands are available to copy at `https://github.com/markjprice/cs13net9/blob/main/docs/command-lines.md`. **Warning!** If you copy from a PDF, double-dashes in a command link are sometimes lost, so be careful!

4. At a command prompt or terminal, use the `dotnet-ef` tool to generate a model for the `Categories` and `Products` tables in a new folder named `AutoGenModels`, as shown in the following command:

```
dotnet ef dbcontext scaffold "Data Source=Northwind.db" Microsoft.
EntityFrameworkCore.Sqlite --table Categories --table Products --output-
dir AutoGenModels --namespace WorkingWithEFCore.AutoGen --data-
annotations --context NorthwindDb
```

   Note the following:

   - The command action: `dbcontext scaffold`
   - The connection string: `"Data Source=Northwind.db"`
   - The database provider: `Microsoft.EntityFrameworkCore.Sqlite`
   - The tables to generate models for: `--table Categories --table Products`
   - The output folder: `--output-dir AutoGenModels`
   - The namespace: `--namespace WorkingWithEFCore.AutoGen`
   - To use data annotations as well as the Fluent API: `--data-annotations`
   - To rename the context from `[database_name]Context`: `--context NorthwindDb`

If you prefer to use SQL Server, then the equivalent command is found at the following link:

`https://github.com/markjprice/cs13net9/blob/main/docs/sql-server/README.md#scaffolding-models-using-an-existing-database`

5. Note the build messages and warnings, as shown in the following output:

```
Build started...
Build succeeded.
To protect potentially sensitive information in your connection string,
you should move it out of source code. You can avoid scaffolding the
connection string by using the Name= syntax to read it from configuration
- see https://go.microsoft.com/fwlink/?linkid=2131148. For more
guidance on storing connection strings, see http://go.microsoft.com/
fwlink/?LinkId=723263.
Skipping foreign key with identity '0' on table 'Products' since
principal table 'Suppliers' was not found in the model. This usually
happens when the principal table was not included in the selection set.
```

Reviewing the scaffolded code

Now let's review the scaffolded code:

1. In the `AutoGenModels` folder, note the three class files that were automatically generated: `Category.cs`, `NorthwindDb.cs`, and `Product.cs`.
2. In the `AutoGenModels` folder, in `Category.cs`, note the differences compared to the one you created manually. I have not included namespace imports to save space, as shown in the following code:

```
namespace WorkingWithEFCore.AutoGen;

[Index("CategoryName", Name = "CategoryName")]
public partial class Category
{
  [Key]
  public int CategoryId { get; set; }

  [Column(TypeName = "nvarchar (15)")]
  public string CategoryName { get; set; } = null!;

  [Column(TypeName = "ntext")]
  public string? Description { get; set; }

  [Column(TypeName = "image")]
  public byte[]? Picture { get; set; }

  [InverseProperty("Category")]
  public virtual ICollection<Product> Products { get; set; }
    = new List<Product>();
}
```

Note the following:

- It decorates the entity class with the `[Index]` attribute, which was introduced in EF Core 5. This indicates properties that should have an index when using the Code First approach to generate a database at runtime. Since we are using Database First with an existing database, this is not needed. But if we wanted to recreate a new, empty database from our code, then this information would be needed.
- The table name in the database is `Categories` but the `dotnet-ef` tool uses the **Humanizer** third-party library to automatically singularize the class name to `Category`, which is a more natural name when creating a single entity that represents a row in the table.
- The entity class is declared using the `partial` keyword so that you can create a matching `partial` class for adding additional code. This allows you to rerun the tool and regenerate the entity class without losing that extra code.

- The `CategoryId` property is decorated with the `[Key]` attribute to indicate that it is the primary key for this entity. The data type for this property is `int` for SQLite and `long` for SQL Server. We did not need to decorate the property in our code because we followed the naming primary key convention.
- The `CategoryName` property is decorated with the `[Column(TypeName = "nvarchar (15)")]` attribute, which is only needed if you want to generate a database from the model.
- We chose not to include the `Picture` column as a property because this is a binary object that we will not use in our console app.
- The `Products` property uses the `[InverseProperty]` attribute to define the foreign key relationship to the `Category` property on the `Product` entity class, and it initializes the collection to a new empty list.

3. In the `AutoGenModels` folder, in `Product.cs`, note the differences compared to the one you created manually.
4. In the `AutoGenModels` folder, in `NorthwindDb.cs`, note the differences compared to the one you created manually, as shown in the following edited-for-space code:

```
using Microsoft.EntityFrameworkCore;

namespace WorkingWithEFCore.AutoGen;

public partial class NorthwindDb : DbContext
{
  public NorthwindDb()
  {
  }

  public NorthwindDb(DbContextOptions<NorthwindDb> options)
      : base(options)
  {
  }

  public virtual DbSet<Category> Categories { get; set; }

  public virtual DbSet<Product> Products { get; set; }

  protected override void OnConfiguring(DbContextOptionsBuilder
optionsBuilder)
```

```
#warning To protect potentially sensitive information in your connection
string, you should move it out of source code. You can avoid scaffolding
the connection string by using the Name= syntax to read it from
configuration - see https://go.microsoft.com/fwlink/?linkid=2131148. For
more guidance on storing connection strings, see http://go.microsoft.com/
fwlink/?LinkId=723263.
      => optionsBuilder.UseSqlite("Data Source=Northwind.db");

  protected override void OnModelCreating(ModelBuilder modelBuilder)
  {
    modelBuilder.Entity<Category>(entity =>
    {
      entity.Property(e => e.CategoryId).ValueGeneratedNever();
    });

    modelBuilder.Entity<Product>(entity =>
    {
      entity.Property(e => e.ProductId).ValueGeneratedNever();
      entity.Property(e => e.Discontinued).HasDefaultValueSql("0");
      entity.Property(e => e.ReorderLevel).HasDefaultValueSql("0");
      entity.Property(e => e.UnitPrice).HasDefaultValueSql("0");
      entity.Property(e => e.UnitsInStock).HasDefaultValueSql("0");
      entity.Property(e => e.UnitsOnOrder).HasDefaultValueSql("0");
    });

    OnModelCreatingPartial(modelBuilder);
  }

  partial void OnModelCreatingPartial(ModelBuilder modelBuilder);
}
```

Note the following:

- The `NorthwindDb` data context class is `partial` to allow you to extend it and regenerate it in the future.
- It has two constructors: a default parameter-less one and one that allows options to be passed in. This is useful in apps where you want to specify the connection string at runtime.
- In the `OnConfiguring` method, if options have not been specified in the constructor, then it defaults to using a connection string that looks for the database file in the current folder. It has a compiler warning to remind you that you should not hardcode security information in this connection string.

- In the `OnModelCreating` method, the Fluent API is used to configure the two entity classes, and then a `partial` method named `OnModelCreatingPartial` is invoked. This allows you to implement that `partial` method in your own `partial` `Northwind` class to add your own Fluent API configuration that will not be lost if you regenerate the model classes.

5. Close the automatically generated class files.

Customizing the reverse engineering templates

One of the features introduced with EF Core 7 was the ability to customize the code that is automatically generated by the `dotnet-ef` scaffolding tool. This is an advanced technique, so I do not cover it in this book. Usually, it is easier to just modify the code that is generated by default anyway.

If you would like to learn how to modify the T4 templates used by the `dotnet-ef` scaffolding tool, then you can find that information at the following link:

`https://learn.microsoft.com/en-us/ef/core/managing-schemas/scaffolding/templates`

Configuring preconvention models

Along with support for the `DateOnly` and `TimeOnly` types for use with the SQLite database provider, one of the features introduced with EF Core 6 was configuring preconvention models.

As models become more complex, relying on conventions to discover entity types and their properties and successfully map them to tables and columns becomes harder. It would be useful if you could configure the conventions themselves before they are used to analyze and build a model.

For example, you might want to define a convention to say that all `string` properties should have a maximum length of 50 characters as a default, or any property types that implement a custom interface should not be mapped, as shown in the following code:

```
protected override void ConfigureConventions(
  ModelConfigurationBuilder configurationBuilder)
{
  configurationBuilder.Properties<string>().HaveMaxLength(50);
  configurationBuilder.IgnoreAny<IDoNotMap>();
}
```

In the rest of this chapter, we will use the classes that you manually created.

Querying EF Core models

Now that we have a model that maps to the `Northwind` database and two of its tables, we can write some simple **Language-Integrated Query** (**LINQ**) queries to fetch data. You will learn much more about writing LINQ queries in *Chapter 11, Querying and Manipulating Data Using LINQ*.

LINQ to Entities (a.k.a. **LINQ to EF Core**) is a LINQ provider that converts a LINQ query into SQL to execute against the database. You can write a LINQ query built up over many C# statements.

You can discover the equivalent SQL statement without executing the query against the database by calling `ToQueryString`. This is known as deferred execution. Only when the query is enumerated using `foreach`, or when you call a method like `ToArray` or `ToList` on the LINQ query, will you trigger the execution of the query against the database and the results are returned to your code. This is known as **materialization**.

For now, just write the code and view the results:

1. In the `WorkingWithEFCore` project, add a new class file named `Program.Helpers.cs`.
2. In `Program.Helpers.cs`, add a partial `Program` class with some methods, as shown in the following code:

```
partial class Program
{
  private static void ConfigureConsole(string culture = "en-US",
    bool useComputerCulture = false)
  {
    // To enable Unicode characters like Euro symbol in the console.
    OutputEncoding = System.Text.Encoding.UTF8;

    if (!useComputerCulture)
    {
      CultureInfo.CurrentCulture = CultureInfo.GetCultureInfo(culture);
    }
    WriteLine($"CurrentCulture: {CultureInfo.CurrentCulture.
DisplayName}");
  }

  private static void WriteLineInColor(string text, ConsoleColor color)
  {
    ConsoleColor previousColor = ForegroundColor;
    ForegroundColor = color;
    WriteLine(text);
    ForegroundColor = previousColor;
  }

  private static void SectionTitle(string title)
  {
    WriteLineInColor($"*** {title} ***", ConsoleColor.DarkYellow);
  }

  private static void Fail(string message)
  {
```

```
    WriteLineInColor($"Fail > {message}", ConsoleColor.Red);
  }

  private static void Info(string message)
  {
    WriteLineInColor($"Info > {message}", ConsoleColor.Cyan);
  }
}
```

3. Add a new class file named `Program.Queries.cs`.
4. In `Program.Queries.cs`, define a partial `Program` class with a `QueryingCategories` method, and add statements to do these tasks, as shown in the following code:
   - Create an instance of the `Northwind` class that will manage the database. Database context instances are designed for short lifetimes in a unit of work. They should be disposed of as soon as possible. So, we will wrap our instance in a `using` statement. In *Chapter 13, Building Websites Using ASP.NET Core*, you will learn how to get database context using dependency injection.
   - Create a query for all categories that include their related products. `Include` is an extension method that requires you to import the `Microsoft.EntityFrameworkCore` namespace.
   - Enumerate through the categories, outputting the name and number of products for each one:

```
using Microsoft.EntityFrameworkCore; // To use Include method.
using Northwind.EntityModels; // To use Northwind, Category,
Product.

partial class Program
{
  private static void QueryingCategories()
  {
    using NorthwindDb db = new();

    SectionTitle("Categories and how many products they have");

    // A query to get all categories and their related products.
    // This is a query definition. Nothing has executed against
the database.
    IQueryable<Category>? categories = db.Categories?
      .Include(c => c.Products);
```

```
    // You could call any of the following LINQ methods and nothing
will be executed against the database:
    // Where, GroupBy, Select, SelectMany, OfType, OrderBy, ThenBy,
Join, GroupJoin, Take, Skip, Reverse.
    // Usually, methods that return IEnumerable or IQueryable
support deferred execution.
    // Usually, methods that return a single value do not support
deferred execution.

    if (categories is null || !categories.Any())
    {
      Fail("No categories found.");
      return;
    }

    // Enumerating the query converts it to SQL and executes it
against the database.
    // Execute query and enumerate results.
    foreach (Category c in categories)
    {
      WriteLine($"{c.CategoryName} has {c.Products.Count}
products.");
    }
  }
}
```

Note that the order of the clauses in the `if` statement is important. We must check that `categories` is `null` first. If this is `true`, then the code will never execute the second clause and, therefore, won't throw a `NullReferenceException` when accessing the `Any()` member.

5. In `Program.cs`, comment out the two statements that create a `Northwind` instance and output the database provider name, and then call the `ConfigureConsole` and `QueryingCategories` methods, as shown in the following code:

```
ConfigureConsole();
QueryingCategories();
```

6. Run the code and view the result, as shown in the following partial output:

```
Beverages has 12 products.
Condiments has 12 products.
Confections has 13 products.
Dairy Products has 10 products.
```

```
Grains/Cereals has 7 products.
Meat/Poultry has 6 products.
Produce has 5 products.
Seafood has 12 products.
```

Warning! If you see the following exception, the most likely problem is that the `Northwind.db` file is not being copied to the output directory: `Unhandled exception. Microsoft.Data.Sqlite.SqliteException (0x80004005): SQLite Error 1: 'no such table: Categories'`. Make sure that **Copy to Output Directory** is set, but even when it is, some code editors do not always copy the file when they should. You might need to manually copy the `Northwind.db` file to the appropriate directory.

Filtering included entities

EF Core 5 introduced **filtered includes**, which means you can specify a lambda expression in the `Include` method call to filter which entities are returned in the results:

1. In `Program.Queries.cs`, define a `FilteredIncludes` method and add statements to do these tasks, as shown in the following code:
    - Create an instance of the `Northwind` class that will manage the database.
    - Prompt the user to enter a minimum value for units in stock.
    - Create a query for categories that have products with that minimum number of units in stock.
    - Enumerate through the categories and products, outputting the name and units in stock for each one:

```
private static void FilteredIncludes()
{
  using NorthwindDb db = new();

  SectionTitle("Products with a minimum number of units in stock");

  string? input;
  int stock;

  do
  {
    Write("Enter a minimum for units in stock: ");
    input = ReadLine();
  } while (!int.TryParse(input, out stock));

  IQueryable<Category>? categories = db.Categories?
```

```
    .Include(c => c.Products.Where(p => p.Stock >= stock));

  if (categories is null || !categories.Any())
  {
    Fail("No categories found.");
    return;
  }

  foreach (Category c in categories)
  {
    WriteLine(
      "{0} has {1} products with a minimum {2} units in stock.",
      arg0: c.CategoryName, arg1: c.Products.Count, arg2: stock);

    foreach(Product p in c.Products)
    {
      WriteLine($"  {p.ProductName} has {p.Stock} units in
stock.");
    }
  }
}
```

2. In `Program.cs`, call the `FilteredIncludes` method, as shown in the following code:

```
FilteredIncludes();
```

3. Run the code, enter a minimum value for units in stock, like `100`, and view the result, as shown in the following partial output:

```
Enter a minimum for units in stock: 100
Beverages has 2 products with a minimum of 100 units in stock.
  Sasquatch Ale has 111 units in stock.
  Rhönbräu Klosterbier has 125 units in stock.
Condiments has 2 products with a minimum of 100 units in stock.
  Grandma's Boysenberry Spread has 120 units in stock.
  Sirop d'érable has 113 units in stock.
Confections has 0 products with a minimum of 100 units in stock.
Dairy Products has 1 products with a minimum of 100 units in stock.
  Geitost has 112 units in stock.
Grains/Cereals has 1 products with a minimum of 100 units in stock.
  Gustaf's Knäckebröd has 104 units in stock.
Meat/Poultry has 1 products with a minimum of 100 units in stock.
  Pâté chinois has 115 units in stock.
```

```
Produce has 0 products with a minimum of 100 units in stock.
Seafood has 3 products with a minimum of 100 units in stock.
  Inlagd Sill has 112 units in stock.
  Boston Crab Meat has 123 units in stock.
  Röd Kaviar has 101 units in stock.
```

Unicode characters in the Windows console: There is a limitation with the console provided by Microsoft on versions of Windows before the Windows 10 Fall Creators Update. By default, the console cannot display Unicode characters, for example, the ones in the name Rhönbräu.

If you have this issue, then you can temporarily change the code page (also known as the character set) in a console to Unicode UTF-8 by entering the following command at the prompt before running the app:

```
chcp 65001
```

Filtering and sorting products

Let's explore a more complex query that will filter and sort data:

1. In `Program.Queries.cs`, define a `QueryingProducts` method, and add statements to do the following, as shown in the following code:

    - Create an instance of the `Northwind` class that will manage the database.
    - Prompt the user for a price for products.
    - Create a query for products that cost more than the price using LINQ.
    - Loop through the results, outputting the ID, name, cost (formatted in US dollars), and number of units in stock:

    ```
    private static void QueryingProducts()
    {
      using NorthwindDb db = new();

      SectionTitle("Products that cost more than a price, highest at
    top");

      string? input;
      decimal price;

      do
      {
        Write("Enter a product price: ");
        input = ReadLine();
      } while (!decimal.TryParse(input, out price));
    ```

```
    IQueryable<Product>? products = db.Products?
      .Where(product => product.Cost > price)
      .OrderByDescending(product => product.Cost);

    if (products is null || !products.Any())
    {
      Fail("No products found.");
      return;
    }

    foreach (Product p in products)
    {
      WriteLine(
        "{0}: {1} costs {2:$#,##0.00} and has {3} in stock.",
        p.ProductId, p.ProductName, p.Cost, p.Stock);
    }
  }
```

2. In `Program.cs`, call the `QueryingProducts` method.
3. Run the code, enter `50` when prompted to enter a product price, view the result, and note the descending order by cost, as shown in the following partial output:

```
Enter a product price: 50
38: Côte de Blaye costs $263.50 and has 17 in stock.
29: Thüringer Rostbratwurst costs $123.79 and has 0 in stock.
9: Mishi Kobe Niku costs $97.00 and has 29 in stock.
20: Sir Rodney's Marmalade costs $81.00 and has 40 in stock.
18: Carnarvon Tigers costs $62.50 and has 42 in stock.
59: Raclette Courdavault costs $55.00 and has 79 in stock.
51: Manjimup Dried Apples costs $53.00 and has 20 in stock.
```

4. Run the code, enter `500` when prompted to enter a product price, and view the result, as shown in the following output:

```
Fail > No products found.
```

Getting the generated SQL

You might be wondering how well written the SQL statements are that are generated from the C# queries we write. EF Core 5 introduced a quick and easy way to see the SQL generated:

1. In the `QueryingProducts` method, before using the `foreach` statement to enumerate the query, add a statement to output the generated SQL, as shown in the following code:

```
// Calling ToQueryString does not execute against the database.
// LINQ to Entities just converts the LINQ query to an SQL statement.
Info($"ToQueryString: {products.ToQueryString()}");
```

Warning! The `ToQueryString` method can only work on objects that implement `IQueryable`. This means that if you write a LINQ query using deferred methods like `Where`, `GroupBy`, `Select`, `OrderBy`, `Join`, `Take`, `Skip`, `Reverse`, and so on, then `ToQueryString` can show you the SQL before you run the query. But methods that return a non-`IQueryable` value and immediately execute the query, like a single scalar result like `Count()` or `First()`, do not support `ToQueryString`.

2. Run the code, enter a minimum value for units in stock, like `95`, and view the result, as shown in the following partial output:

```
Enter a minimum for units in stock: 95
Connection: Data Source=C:\cs13net9\Chapter10\WorkingWithEFCore\bin\
Debug\net9.0\Northwind.db
Info > ToQueryString: .param set @__stock_0 95

SELECT "c"."CategoryId", "c"."CategoryName", "c"."Description",
"t"."ProductId", "t"."CategoryId", "t"."UnitPrice", "t"."Discontinued",
"t"."ProductName", "t"."UnitsInStock"
FROM "Categories" AS "c"
LEFT JOIN (
    SELECT "p"."ProductId", "p"."CategoryId", "p"."UnitPrice",
"p"."Discontinued", "p"."ProductName", "p"."UnitsInStock"
    FROM "Products" AS "p"
    WHERE "p"."UnitsInStock" >= @__stock_0
) AS "t" ON "c"."CategoryId" = "t"."CategoryId"
ORDER BY "c"."CategoryId"
Beverages has 2 products with a minimum of 95 units in stock.
  Sasquatch Ale has 111 units in stock.
  Rhönbräu Klosterbier has 125 units in stock.
...
```

Note that the SQL parameter named `@__stock_0` has been set to a minimum stock value of `95`.

If you used SQL Server, the generated SQL will be slightly different. For example, it uses square brackets instead of double quotes around object names, as shown in the following output:

```
Info > ToQueryString: DECLARE @__stock_0 smallint = CAST(95 AS smallint);

SELECT [c].[CategoryId], [c].[CategoryName], [c].[Description], [t].
[ProductId], [t].[CategoryId], [t].[UnitPrice], [t].[Discontinued], [t].
[ProductName], [t].[UnitsInStock]
FROM [Categories] AS [c]
LEFT JOIN (
    SELECT [p].[ProductId], [p].[CategoryId], [p].[UnitPrice], [p].
[Discontinued], [p].[ProductName], [p].[UnitsInStock]
    FROM [Products] AS [p]
    WHERE [p].[UnitsInStock] >= @__stock_0
) AS [t] ON [c].[CategoryId] = [t].[CategoryId]
ORDER BY [c].[CategoryId]
```

Logging EF Core

To monitor the interaction between EF Core and the database, we can enable logging. Logging could be to the console, to `Debug` or `Trace`, or to a file. Enabling logging for EF Core shows all of the SQL commands that are actually executed against the database. `ToQueryString` does not execute against the database.

Good Practice: By default, EF Core logging will exclude any data that is sensitive. You can include this data by calling the `EnableSensitiveDataLogging` method, especially during development. You should disable it before deploying to production. You can also call `EnableDetailedErrors`.

Let's see an example of this in action:

1. In `NorthwindDb.cs`, at the bottom of the `OnConfiguring` method, add statements to log to the console and to include sensitive data like parameter values for commands being sent to the database if we compile the debug configuration, as shown in the following code:

```
optionsBuilder.LogTo(WriteLine) // This is the Console method.
#if DEBUG
  .EnableSensitiveDataLogging() // Include SQL parameters.
  .EnableDetailedErrors()
#endif
;
```

`LogTo` requires an `Action<string>` delegate. EF Core will call this delegate, passing a `string` value for each log message. Passing the `Console` class `WriteLine` method, therefore, tells the logger to write each method to the console.

2. Note that when the solution configuration is **Debug**, the calls to the `EnableSensitiveDataLogging` and `EnableDetailedErrors` methods are included in the compilation, but if you change the solution configuration to **Release**, the method calls are grayed out to indicate that they are not compiled, as shown in *Figure 10.6*:

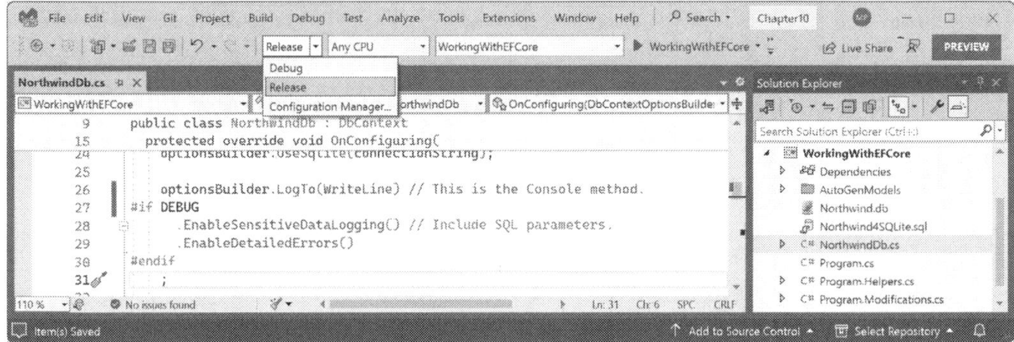

Figure 10.6: Including SQL parameters in logging for debug configuration

3. Run the code and view the log messages, which are shown in the following partial output:

```
warn: 7/16/2023 14:03:40.255 CoreEventId.
SensitiveDataLoggingEnabledWarning[10400] (Microsoft.EntityFrameworkCore.
Infrastructure)
      Sensitive data logging is enabled. Log entries and exception
messages may include sensitive application data; this mode should only be
enabled during development.
...
dbug: 05/03/2023 12:36:11.702 RelationalEventId.ConnectionOpening[20000]
(Microsoft.EntityFrameworkCore.Database.Connection)
      Opening connection to database 'main' on server 'C:\cs13net9\
Chapter10\WorkingWithEFCore\bin\Debug\net9.0\Northwind.db'.
dbug: 05/03/2023 12:36:11.718 RelationalEventId.ConnectionOpened[20001]
(Microsoft.EntityFrameworkCore.Database.Connection)
      Opened connection to database 'main' on server 'C:\cs13net9\
Chapter10\WorkingWithEFCore\bin\Debug\net9.0\Northwind.db'.
dbug: 05/03/2023 12:36:11.721 RelationalEventId.CommandExecuting[20100]
(Microsoft.EntityFrameworkCore.Database.Command)

      Executing DbCommand [Parameters=[], CommandType='Text',
CommandTimeout='30']
      SELECT "c"."CategoryId", "c"."CategoryName", "c"."Description",
"p"."ProductId", "p"."CategoryId", "p"."UnitPrice", "p"."Discontinued",
"p"."ProductName", "p"."UnitsInStock"
      FROM "Categories" AS "c"
      LEFT JOIN "Products" AS "p" ON "c"."CategoryId" = "p"."CategoryId"
```

```
        ORDER BY "c"."CategoryId"
...
```

Your logs might vary from those shown above based on your chosen database provider and code editor, as well as future improvements to EF Core. For now, note that different events, like opening a connection or executing a command, have different event IDs, as shown in the following list:

- `20000 RelationalEventId.ConnectionOpening`: Includes the database file path
- `20001 RelationalEventId.ConnectionOpened`: Includes the database file path
- `20100 RelationalEventId.CommandExecuting`: Includes the SQL statement

Filtering logs by provider-specific values

The event ID values and what they mean will be specific to the EF Core provider. If we want to know how the LINQ query has been translated into SQL statements and is executing, then the event ID to output has an `Id` value of `20100`:

1. At the top of `NorthwindDb.cs`, import the namespace for working for EF Core diagnostics, as shown in the following code:

```
// To use RelationalEventId.
using Microsoft.EntityFrameworkCore.Diagnostics;
```

2. Modify the `LogTo` method call to only output events with an `Id` of `20100`, as shown highlighted in the following code:

```
optionsBuilder.LogTo(WriteLine, // This is the Console method.
  new[] { RelationalEventId.CommandExecuting })
#if DEBUG
  .EnableSensitiveDataLogging()
  .EnableDetailedErrors()
#endif
;
```

3. Run the code and note the following SQL statements that were logged, as shown in the following output, which has been edited for space:

```
dbug: 05/03/2022 12:48:43.153 RelationalEventId.CommandExecuting[20100]
(Microsoft.EntityFrameworkCore.Database.Command)
      Executing DbCommand [Parameters=[], CommandType='Text',
CommandTimeout='30']
      SELECT "c"."CategoryId", "c"."CategoryName", "c"."Description",
"p"."ProductId", "p"."CategoryId", "p"."UnitPrice", "p"."Discontinued",
"p"."ProductName", "p"."UnitsInStock"
      FROM "Categories" AS "c"
      LEFT JOIN "Products" AS "p" ON "c"."CategoryId" = "p"."CategoryId"
      ORDER BY "c"."CategoryId"
```

```
Beverages has 12 products.
Condiments has 12 products.
Confections has 13 products.
Dairy Products has 10 products.
Grains/Cereals has 7 products.
Meat/Poultry has 6 products.
Produce has 5 products.
Seafood has 12 products.
```

Logging with query tags

When logging LINQ queries, it can be tricky to correlate log messages in complex scenarios. EF Core 2.2 introduced the query tags feature to help by allowing you to add SQL comments to the log.

You can annotate a LINQ query using the `TagWith` method, as shown in the following code:

```
IQueryable<Product>? products = db.Products?
  .TagWith("Products filtered by price and sorted.")
  .Where(product => product.Cost > price)
  .OrderByDescending(product => product.Cost);
```

This will add an SQL comment to the log, as shown in the following output:

```
-- Products filtered by price and sorted.
```

Getting a single entity

There are two LINQ methods to get a single entity: `First` and `Single`. It is important to understand the difference between them when using an EF Core database provider. Let's see an example:

1. In `Program.Queries.cs`, define a `GettingOneProduct` method, and add statements to do the following, as shown in the following code:

   - Create an instance of the `Northwind` class that will manage the database.
   - Prompt the user for a product ID.
   - Create a query for products with that product ID using the `First` and `Single` methods.
   - Write an SQL statement for each query to the console:

   ```
   private static void GettingOneProduct()
   {
     using NorthwindDb db = new();

     SectionTitle("Getting a single product");

     string? input;
     int id;
   ```

```
  do
  {
    Write("Enter a product ID: ");
    input = ReadLine();
  } while (!int.TryParse(input, out id));

  // This query is not deferred because the First method does not
return IEnumerable or IQueryable.
  // The LINQ query is immediately converted to SQL and executed
to fetch the first product.
  Product? product = db.Products?
    .First(product => product.ProductId == id);

  Info($"First: {product?.ProductName}");

  if (product is null) Fail("No product found using First.");

  product = db.Products?
    .Single(product => product.ProductId == id);

  Info($"Single: {product?.ProductName}");

  if (product is null) Fail("No product found using Single.");
}
```

LINQ methods that fetch a single entity (`First`, `FirstOrDefault`, `Single`, `SingleOrDefault`, `ElementAt`, and `ElementAtOrDefault`) or return a single scalar value or entity like the aggregate methods (`Count`, `Sum`, `Max`, `Min`, `Average`, `All`, `Any`, and so on) are not deferred. When using the LINQ to Entities provider, any LINQ query that ends with a call to one of these methods is immediately converted to an SQL statement and executed against the database.

2. In `Program.cs`, call the `GettingOneProduct` method.
3. Run the code, enter `1` when prompted to enter a product ID, view the result, and note the SQL statements used by `First` and `Single`, as shown in the following output:

```
Enter a product ID: 1
Connection: Data Source=C:\cs13net9\Chapter10\WorkingWithEFCore\bin\
Debug\net9.0\Northwind.db
dbug: 9/17/2023 18:04:14.210 RelationalEventId.CommandExecuting[20100]
(Microsoft.EntityFrameworkCore.Database.Command)
      Executing DbCommand [Parameters=[@__id_0='1'], CommandType='Text',
CommandTimeout='30']
```

```
      SELECT "p"."ProductId", "p"."CategoryId", "p"."UnitPrice",
"p"."Discontinued", "p"."ProductName", "p"."UnitsInStock"
      FROM "Products" AS "p"
      WHERE "p"."ProductId" > @__id_0
      LIMIT 1
Info > First: Chang
dbug: 9/17/2023 18:04:14.286 RelationalEventId.CommandExecuting[20100]
(Microsoft.EntityFrameworkCore.Database.Command)
      Executing DbCommand [Parameters=[@__id_0='1'], CommandType='Text',
CommandTimeout='30']
      SELECT "p"."ProductId", "p"."CategoryId", "p"."UnitPrice",
"p"."Discontinued", "p"."ProductName", "p"."UnitsInStock"
      FROM "Products" AS "p"
      WHERE "p"."ProductId" > @__id_0
      LIMIT 2
Info > Single: Chang
```

Note that both methods execute the same SQL statement except for the `LIMIT` clauses highlighted in the preceding code. For `First`, it sets `LIMIT 1` but for `Single`, it sets `LIMIT 2`. Why?

For `First`, the query can match one or more entities and only the first will be returned. If there are no matches, an exception is thrown, but you can call `FirstOrDefault` to return `null` if there are no matches.

For `Single`, the query must match only one entity and it will be returned. If there is more than one match, an exception must be thrown. But the only way for EF Core to know if there is more than one match is to request more than one and check. So, it has to set `LIMIT 2` and check if there is a second entity match.

Good Practice: If you do not need to make sure that only one entity matches, use `First` instead of `Single` to avoid retrieving two records.

Pattern matching with Like

EF Core supports common SQL statements, including `Like` for pattern matching:

1. In `Program.Queries.cs`, add a method named `QueryingWithLike`, as shown in the following code, and note the following:
   - We have enabled logging.
   - We prompt the user to enter part of a product name and then use the `EF.Functions.Like` method to search anywhere in the `ProductName` property.

- For each matching product, we output its name, stock, and if it is discontinued:

```
private static void QueryingWithLike()
{
  using NorthwindDb db = new();

  SectionTitle("Pattern matching with LIKE");

  Write("Enter part of a product name: ");
  string? input = ReadLine();

  if (string.IsNullOrWhiteSpace(input))
  {
    Fail("You did not enter part of a product name.");
    return;
  }

  IQueryable<Product>? products = db.Products?
    .Where(p => EF.Functions.Like(p.ProductName, $"%{input}%"));

  if (products is null || !products.Any())
  {
    Fail("No products found.");
    return;
  }

  foreach (Product p in products)
  {
    WriteLine("{0} has {1} units in stock. Discontinued: {2}",
      p.ProductName, p.Stock, p.Discontinued);
  }
}
```

2. In `Program.cs`, comment out the existing methods and call `QueryingWithLike`.
3. Run the code, enter a partial product name, such as `che`, and view the result, as shown in the following edited output:

```
Enter part of a product name: che
dbug: 07/16/2023 13:03:42.793 RelationalEventId.CommandExecuting[20100]
(Microsoft.EntityFrameworkCore.Database.Command)
      Executing DbCommand [Parameters=[@__Format_1='%che%' (Size = 5)],
CommandType='Text', CommandTimeout='30']
```

```
      SELECT "p"."ProductId", "p"."CategoryId", "p"."UnitPrice",
"p"."Discontinued", "p"."ProductName", "p"."UnitsInStock"
      FROM "Products" AS "p"
      WHERE "p"."ProductName" LIKE @__Format_1
Chef Anton's Cajun Seasoning has 53 units in stock. Discontinued: False
Chef Anton's Gumbo Mix has 0 units in stock. Discontinued: True
Queso Manchego La Pastora has 86 units in stock. Discontinued: False
```

More Information: You can learn more about wildcards with `Like` at `https://learn.microsoft.com/en-us/dotnet/framework/data/adonet/ef/language-reference/like-entity-sql`.

Generating a random number in queries

EF Core 6 introduced a useful function, `EF.Functions.Random`, that maps to a database function returning a pseudo-random real number between `0.0` and `1.0`, exclusive.

For example, suppose you're developing a quiz or survey app that needs to display questions in a random order each time a user takes it. You can use `EF.Functions.Random()` to fetch a random selection of questions directly from the database, as shown in the following code:

```
var randomQuestions = await db.Questions
  .OrderBy(q => EF.Functions.Random())
  .Take(10); // Select 10 random questions.
```

During the development and testing phase, generating random data can be useful for simulating various scenarios. For example, creating randomized datasets for performance testing or generating random sets of rows for unit tests by selecting about half of the products randomly, as shown in the following code:

```
var randomDataSample = await db.Products
  .Where(d => EF.Functions.Random() > 0.5);
```

Defining global filters

`Northwind` products can be discontinued, so it might be useful to ensure that discontinued products are never returned in results, even if the programmer does not use `Where` to filter them out in their queries:

1. In `NorthwindDb.cs`, at the bottom of the `OnModelCreating` method, add a global filter to remove discontinued products, as shown in the following code:

```
// A global filter to remove discontinued products.
modelBuilder.Entity<Product>()
  .HasQueryFilter(p => !p.Discontinued);
```

2. In `Program.cs`, uncomment the call to `QueryingWithLike`, and comment out all the other method calls.
3. Run the code, enter the partial product name `che`, view the result, and note that `Chef Anton's Gumbo Mix` is now missing. This is because the SQL statement generated includes a filter for the `Discontinued` column, as shown highlighted in the following output:

```
Enter part of a product name: che
dbug: 05/03/2022 13:34:27.290 RelationalEventId.CommandExecuting[20100]
(Microsoft.EntityFrameworkCore.Database.Command)
      Executing DbCommand [Parameters=[@__Format_1='%che%' (Size = 5)],
CommandType='Text', CommandTimeout='30']
      SELECT "p"."ProductId", "p"."CategoryId", "p"."UnitPrice",
"p"."Discontinued", "p"."ProductName", "p"."UnitsInStock"
      FROM "Products" AS "p"
      WHERE NOT ("p"."Discontinued") AND ("p"."ProductName" LIKE @__
Format_1)
Chef Anton's Cajun Seasoning has 53 units in stock. Discontinued? False
Queso Manchego La Pastora has 86 units in stock. Discontinued? False
Gumbär Gummibärchen has 15 units in stock. Discontinued? False
```

SQL SELECT queries

Usually, you can express all the queries you need to using LINQ. But for times when you cannot, you can use `FromSql` and its related methods.

`FromSql` was introduced with EF Core 7. If you need to execute raw SQL with EF Core 6 or earlier, then you must use the `FromSqlInterpolated` method.

The `FromSql` method allows you to execute raw SQL queries against the database and map the results to your entity classes. `FromSql` can be used to perform `SELECT` queries that return entity types or types that are not part of the EF Core model.

The `FromSql` method is particularly useful in the following situations:

- It allows you to run complex SQL queries that might not be feasible with LINQ.
- Sometimes, raw SQL can be more performant than LINQ for certain types of queries.
- If you are working with a legacy system where specific SQL queries need to be executed.
- You can execute stored procedures that return entities.

The parameter to `FromSql` must be a `FormattableString`, not just a regular `string` value. This is to enforce safe parameterization. Pass parameter values using the interpolated `string` format.

Let's see some examples:

1. In `Program.Queries.cs`, add a method named `GetProductUsingSql`, as shown in the following code:

```
private static void GetProductUsingSql()
{
  using NorthwindDb db = new();

  SectionTitle("Get product using SQL");

  int? rowCount = db.Products?.Count();

  if (rowCount is null)
  {
    Fail("Products table is empty.");
    return;
  }

  int productId = 1;

  Product? p = db.Products?.FromSql(
    $"SELECT * FROM Products WHERE ProductId = {
    productId}").FirstOrDefault();

  if (p is null)
  {
    Fail("Product not found.");
    return;
  }

  WriteLine($"Product: {p.ProductId} - {p.ProductName}");
}
```

2. In `Program.cs`, add a call to `GetProductUsingSql`.
3. Run the code and view the result, as shown in the following output:

```
dbug: 7/27/2024 14:47:07.515 RelationalEventId.CommandExecuting[20100]
(Microsoft.EntityFrameworkCore.Database.Command)
      Executing DbCommand [Parameters=[], CommandType='Text',
CommandTimeout='30']
      SELECT COUNT(*)
      FROM "Products" AS "p"
```

```
      WHERE NOT ("p"."Discontinued")
dbug: 7/27/2024 14:47:07.582 RelationalEventId.CommandExecuting[20100]
(Microsoft.EntityFrameworkCore.Database.Command)
      Executing DbCommand [Parameters=[p0='1'], CommandType='Text',
CommandTimeout='30']
      SELECT "n"."ProductId", "n"."CategoryId", "n"."UnitPrice",
"n"."Discontinued", "n"."ProductName", "n"."UnitsInStock"
      FROM (
          SELECT * FROM Products WHERE ProductId = @p0
      ) AS "n"
      WHERE NOT ("n"."Discontinued")
      LIMIT 1
Product: 1 - Chai
```

`FromSql` can only be called on a `DbSet<T>`, not on a LINQ query.

You can make the SQL even more dynamic for scenarios that need it. For example, if the name of a column might change, as well as the value, then you can use `FromSqlRaw`. But beware! You are responsible for ensuring this `string` value is secure, especially if it originates from an untrusted source. This involves identifying special characters like semicolons, comments, and other SQL constructs, and then properly escaping or rejecting these inputs to prevent potential security risks.

More Information: You can learn more about dynamic SQL and parameters at `https://learn.microsoft.com/en-us/ef/core/querying/sql-queries#dynamic-sql-and-parameters`.

You've now seen many common ways to query data using EF Core. In some online sections, you can look at how data is loaded and tracked and why you might want to control how EF Core does that.

Practicing and exploring

Test your knowledge and understanding by answering some questions, getting some hands-on practice, and exploring this chapter's topics with deeper research.

Exercise 10.1 – online materials

Online materials can be extra content written by me for this book, or they can be references to content created by Microsoft or third parties.

Loading and tracking patterns with EF Core

Learn how data is loaded and tracked with EF Core at the following link:

`https://github.com/markjprice/cs13net9/blob/main/docs/ch10-loading-tracking.md`

Modifying data with EF Core

Learn how data can be modified with EF Core at the following link:

`https://github.com/markjprice/cs13net9/blob/main/docs/ch10-modifying.md`

Working with transactions

Add transactions to the modification code:

`https://github.com/markjprice/cs13net9/blob/main/docs/ch10-transactions.md`

Exploring a Code First EF Core model

Work through an example of a Code First model that generates an empty database, seeds it with sample data, and then queries the data:

`https://github.com/markjprice/cs13net9/blob/main/docs/ch10-code-first.md`

Exploring app secrets

When connecting to a database, you often need to include sensitive secret values like a username or password. These values should never be stored in source code or even in a separate file that might be added to a code repository.

Secrets should be stored locally during development and in secure systems for production. You can use **Secret Manager** during local development and **Azure Key Vault** for cloud production systems. To learn more about app secrets, I have written an online-only section that you can read at the following link:

`https://github.com/markjprice/cs13net9/blob/main/docs/ch10-app-secrets.md`

NoSQL databases

This chapter focused on RDBMSs such as SQL Server and SQLite. If you wish to learn more about NoSQL databases, such as Cosmos DB and MongoDB, and how to use them with EF Core, then I recommend the following links:

- **Welcome to Azure Cosmos DB:** `https://learn.microsoft.com/en-us/azure/cosmos-db/introduction`
- **Use NoSQL databases as a persistence infrastructure:** `https://learn.microsoft.com/en-us/dotnet/standard/microservices-architecture/microservice-ddd-cqrs-patterns/nosql-database-persistence-infrastructure`
- **Document database providers for Entity Framework Core:** `https://github.com/BlueshiftSoftware/EntityFrameworkCore`

Exercise 10.2 – practice exercises

Practice exercises go deeper into the topics for this chapter.

Exporting data using different serialization formats

In the `Chapter10` solution, create a console app named `Exercise_DataSerialization` that queries the `Northwind` database for all the categories and products, and then serializes the data using at least three formats of serialization available to .NET. Which format of serialization uses the least number of bytes?

Exercise 10.3 – test your knowledge

Answer the following questions:

1. What type would you use for the property that represents a table, for example, the `Products` property of a database context?
2. What type would you use for a property that represents a one-to-many relationship, for example, the `Products` property of a `Category` entity?
3. What is the EF Core convention for primary keys?
4. When might you use an annotation attribute in an entity class?
5. Why might you choose the Fluent API in preference to annotation attributes?
6. What does a transaction isolation level of `Serializable` mean?
7. What does the `DbContext.SaveChanges()` method return?
8. What is the difference between eager loading and explicit loading?
9. How should you define an EF Core entity class to match the following table?

```
CREATE TABLE Employees(
  EmpId INT IDENTITY,
  FirstName NVARCHAR(40) NOT NULL,
  Salary MONEY
)
```

10. What benefit do you get from declaring entity navigation properties as `virtual`?

Exercise 10.4 – explore topics

Use this link to learn more about the topics covered in this chapter:

https://github.com/markjprice/cs13net9/blob/main/docs/book-links.md#chapter-10---working-with-data-using-entity-framework-core

Summary

In this chapter, you learned how to do the following:

- Connect to and build entity data models for an existing database.
- Execute a simple LINQ query and process the results.
- Use filtered includes.
- Execute SQL queries directly.

In the next chapter, you will learn how to write more advanced LINQ queries to select, filter, sort, join, and group.

11

Querying and Manipulating Data Using LINQ

This chapter is about **Language INtegrated Query** (**LINQ**) expressions. LINQ is a set of language extensions that enable you to work with sequences of data and then filter, sort, and project them into different outputs.

This chapter will cover the following topics:

- Writing LINQ expressions
- LINQ in practice
- Sorting and more
- Using LINQ with EF Core
- Joining, grouping, and lookups

Writing LINQ expressions

The first question we need to answer is a fundamental one: *Why does LINQ exist?*

Comparing imperative and declarative language features

LINQ was introduced in 2008 with C# 3 and .NET Framework 3. Before that, if a C# and .NET programmer wanted to process a sequence of items, they had to use procedural, aka imperative, code statements. For example, a loop:

1. Set the current position to the first item.
2. Check if the item is one that should be processed by comparing one or more properties against specified values. For example, is the unit price greater than 50, or is the country equal to Belgium?

3. If there's a match, process that item. For example, output one or more of its properties to the user, update one or more properties to new values, delete the item, or perform an aggregate calculation, like counting or summing values.
4. Move on to the next item. Repeat until all items have been processed.

Procedural code tells the compiler *how* to achieve a goal. Do this, then do that. Since the compiler does not know what you are trying to achieve, it cannot help you as much. You are 100% responsible for ensuring that every *how-to* step is correct.

LINQ makes these common tasks much easier, with less opportunity to introduce subtle bugs. Instead of needing to explicitly state each individual action, like move, read, update, and so on, LINQ enables the programmer to use a declarative aka functional style of writing statements.

Declarative, aka functional, code tells the compiler *what* goal to achieve. The compiler works out the best way to achieve it. The statements also tend to be more concise.

Good Practice: If you do not fully understand how LINQ works, then the statements you write can introduce their own subtle bugs! A code teaser doing the rounds in 2022 involved a sequence of tasks and understanding when they are executed (`https://twitter.com/amantinband/status/1559187912218099714`). Most experienced developers got it wrong! To be fair, it is the combination of LINQ behavior with multi-threading behavior that confused most. But by the end of this chapter, you will be better informed to understand why the code was dangerous due to LINQ behavior.

Although we wrote a few LINQ expressions in *Chapter 10, Working with Data Using Entity Framework Core*, they weren't the focus, so I didn't properly explain how LINQ works. Let's now take the time to properly understand them.

LINQ components

LINQ has several parts; some are required, and some are optional:

- **Extension methods (required)**: These include examples such as `Where`, `OrderBy`, and `Select`. These are what provide the functionality of LINQ.
- **LINQ providers (required)**: These include **LINQ to Objects** for processing in-memory objects, **LINQ to Entities** for processing data stored in external databases and modeled with EF Core, and **LINQ to XML** for processing data stored as XML. These providers are the part of LINQ that executes LINQ expressions in a way specific to different types of data.
- **Lambda expressions (optional)**: These can be used instead of named methods to simplify LINQ queries, for example, for the conditional logic of the `Where` method for filtering.
- **LINQ query comprehension syntax (optional)**: These include C# keywords like `from`, `in`, `where`, `orderby`, `descending`, and `select`. These are aliases for some of the LINQ extension methods, and their use can simplify the queries you write, especially if you already have experience with other query languages, such as **Structured Query Language (SQL)**.

When programmers are first introduced to LINQ, they often believe that LINQ query comprehension syntax is LINQ, but ironically, that is one of the parts of LINQ that is optional!

Building LINQ expressions with the Enumerable class

The LINQ extension methods, such as `Where` and `Select`, are appended by the `Enumerable` static class to any type, known as a **sequence**, that implements `IEnumerable<T>`. A sequence contains zero, one, or more items.

For example, an array of any type implements the `IEnumerable<T>` class, where `T` is the type of item in the array. This means that all arrays support LINQ to query and manipulate them.

All generic collections, such as `List<T>`, `Dictionary<TKey, TValue>`, `Stack<T>`, and `Queue<T>`, implement `IEnumerable<T>`, so they can be queried and manipulated with LINQ too.

`Enumerable` defines more than 50 extension methods, as summarized in *Table 11.1* and *Table 11.2*.

These tables will be useful for you for future reference, but for now, you might want to briefly scan them to get a feel for what extension methods exist and come back later to review them properly. An online version of these tables is available at the following link: `https://github.com/markjprice/cs13net9/blob/main/docs/ch11-linq-methods.md`.

First, here are the deferred methods that return a new sequence of `IEnumerable<T>` items:

Method(s)	Description
`Where`	Returns a sequence of items that match a specified filter.
`Index`	Returns a sequence of both the items and their indices. Introduced with .NET 9.
`Select` and `SelectMany`	Project items into a different shape, that is, a different type, and flattens a nested hierarchy of items.
`Skip`	Skips a number of items.
`SkipWhile`	Skips while an expression is `true`.
`SkipLast`	Returns a new enumerable collection that contains elements from the source, with the last count elements of the source collection omitted.
`Take`	Takes a number of items. .NET 6 introduced an overload that can be passed a `Range`, for example, `Take(range: 3..^5)`, meaning take a subset starting three items in from the start and ending five items in from the end, or instead of `Skip(4)`, you could use `Take(4..)`.
`TakeWhile`	Takes items while an expression is `true`.
`TakeLast`	Returns a new enumerable collection that contains the last count elements from the source.

OrderBy, OrderByDescending, ThenBy, and ThenByDescending	Sort items by a specified field or property.
Order and OrderDescending	Sort items by the item itself.
Reverse	Reverses the order of the items.
GroupBy, GroupJoin, and Join	Group and/or join two sequences.
AggregateBy, CountBy, DistinctBy, ExceptBy, IntersectBy, UnionBy, MinBy, and MaxBy	Allow the comparison to be performed on a subset of items rather than all of them. For example, instead of removing duplicates with Distinct by comparing an entire Person object, you could remove duplicates with DistinctBy by comparing just their LastName and DateOfBirth properties. The CountBy and AggregateBy extension methods were introduced with .NET 9.
AsEnumerable	Returns the input sequence typed as IEnumerable<T>. This is useful when the type has its own implementation of any of the LINQ extension methods, like Where, and you want to call the standard LINQ Where method instead.
DefaultIfEmpty	Returns the elements of an IEnumerable<T>, or a default valued singleton collection if the sequence is empty. For example, if the sequence is an empty IEnumerable<int>, it will return an IEnumerable<int> containing a single item, 0.
Cast<T>	Casts items into a specified type. It is useful to convert non-generic objects into a generic type in scenarios where the compiler would otherwise complain.
OfType<T>	Removes items that do not match a specified type.
Distinct	Removes duplicate items.
Except, Intersect, and Union	Perform operations that return sets. Sets cannot have duplicate items. Although the inputs can be any sequence and, thereby, the inputs can have duplicates, the result is always a set.
Chunk	Divides a sequence into sized batches. The size parameter specifies the number of items in each chunk. The last chunk will contain the remaining items and could be smaller than size.
Append, Concat, and Prepend	Perform sequence-combining operations.
Zip	Performs a match operation on two or three sequences based on the position of items; for example, the item at position 1 in the first sequence matches the item at position 1 in the second sequence.

Table 11.1: Deferred LINQ extension methods

Next, here are the non-deferred methods that return a single scalar value, like a single `TSource` item, a number, or a `bool`:

`First`, `FirstOrDefault`, `Last`, and `LastOrDefault`	Get the first or last item in the sequence, throw an exception, or return the default value for the type, for example, `0` for an `int` and `null` for a reference type, if there is not a first or last item.
`Single` and `SingleOrDefault`	Return an item that matches a specific filter, throw an exception, or return the default value for the type if there is not one exact match.
`ElementAt` and `ElementAtOrDefault`	Return an item at a specified index position, throw an exception, or return the default value for the type if there is not an item at that position. .NET 6 introduced overloads that can be passed as an `Index` instead of an `int`, which is more efficient when working with `Span<T>` sequences.
`Aggregate`, `Average`, `Count`, `LongCount`, `Max`, `Min`, and `Sum`	Calculate aggregate values.
`TryGetNonEnumeratedCount`	`Count()` checks if a `Count` property is implemented on the sequence and returns its value, or it enumerates the entire sequence to count its items. Introduced in .NET 6, this method only checks for `Count`; if it is missing, it returns `false` and sets the `out` parameter to `0` to avoid a potentially poor-performing operation.
`SequenceEqual`	Returns `true` or `false`, depending on whether two sequences are equal according to an equality comparer.
`All`, `Any`, and `Contains`	Return `true` if all or any of the items match the filter, or if the sequence contains a specified item. If the sequence is a `List<T>`, they use its native `TrueForAll` method instead of LINQ's `All` method.
`ToArray`, `ToList`, `ToDictionary`, `ToHashSet`, and `ToLookup`	Convert a sequence into an array or collection. These are the only extension methods that force the execution of a LINQ expression immediately, rather than wait for deferred execution, which you will learn about shortly.

Table 11.2: Non-deferred LINQ extension methods

Good Practice: Make sure that you understand and remember the difference between LINQ extension methods that start with `As` and `To`. The `AsEnumerable` method casts the sequence into a different type but does not allocate memory, so that method are fast. Methods that start with `To`, like `ToList`, allocate memory for a new sequence of items, so they can be slow and will always use more memory resources.

The Enumerable class also has some methods that are not extension methods, as shown in *Table 11.3*:

Method	Description
Empty<T>	Returns an empty sequence of the specified type T. It is useful for passing an empty sequence to a method that requires an IEnumerable<T>.
Range	Returns a sequence of integers from the start value with count items. For example, Enumerable.Range(start: 5, count: 3) would contain the integers 5, 6, and 7.
Repeat	Returns a sequence that contains the same element repeated count times. For example, Enumerable.Repeat(element: "5", count: 3) would contain the string values "5", "5", and "5".

Table 11.3: Enumerable non-extension methods

LINQ in practice

Now, we can build a console app to explore practical examples of using LINQ.

Understanding deferred execution

LINQ uses **deferred execution**. It is important to understand that calling most of the above extension methods does not execute a query and get the results. Most of these extension methods return a LINQ expression that represents a *question*, not an *answer*. Let's explore:

1. Use your preferred code editor to create a new project, as defined in the following list:
    - Project template: **Console App** / console
    - Project file and folder: LinqWithObjects
    - Solution file and folder: Chapter11
2. In the project file, globally and statically import the System.Console class.
3. Add a new class file named Program.Helpers.cs.
4. In Program.Helpers.cs, delete any existing statements, and then define a partial Program class with a method to output a section title, as shown in the following code:

```
partial class Program
{
  private static void SectionTitle(string title)
  {
    ConsoleColor previousColor = ForegroundColor;
    ForegroundColor = ConsoleColor.DarkYellow;
    WriteLine($"*** {title} ***");
    ForegroundColor = previousColor;
  }
}
```

5. Add a new class file named Program.Functions.cs.

6. In `Program.Functions.cs`, delete any existing statements, define a partial `Program` class with a method named `DeferredExecution` that is passed an array of `string` values, and then define two queries, as shown in the following code:

```
partial class Program
{
  private static void DeferredExecution(string[] names)
  {
    SectionTitle("Deferred execution");

    // Question: Which names end with an M?
    // (using a LINQ extension method)
    var query1 = names.Where(name => name.EndsWith("m"));

    // Question: Which names end with an M?
    // (using LINQ query comprehension syntax)
    var query2 = from name in names where name.EndsWith("m") select name;
  }
}
```

7. In `Program.cs`, delete the existing statements, add statements to define a sequence of `string` values for people who work in an office, and then pass it as an argument to the `DeferredExecution` method, as shown in the following code:

```
// A string array is a sequence that implements IEnumerable<string>.
string[] names = { "Michael", "Pam", "Jim", "Dwight",
  "Angela", "Kevin", "Toby", "Creed" };

DeferredExecution(names);
```

8. In `Program.Functions.cs`, in the `DeferredExecution` method, to get the answer (in other words, to execute the query), you must **materialize** it by either calling one of the `To` methods, like `ToArray`, `ToDictionary`, or `ToLookup`, or by enumerating the query. Add statements to do this, as shown in the following code:

```
// Answer returned as an array of strings containing Pam and Jim.
string[] result1 = query1.ToArray();

// Answer returned as a list of strings containing Pam and Jim.
List<string> result2 = query2.ToList();

// Answer returned as we enumerate over the results.
foreach (string name in query1)
{
```

```
    WriteLine(name); // outputs Pam
    names[2] = "Jimmy"; // Change Jim to Jimmy.
    // On the second iteration Jimmy does not
    // end with an "m" so it does not get output.
  }
```

9. Run the console app and note the result, as shown in the following output:

```
*** Deferred execution ***
Pam
```

Due to deferred execution, after outputting the first result, `Pam`, if the original array values change, then by the time we loop back around, there are no more matches because `Jim` has become `Jimmy` and does not end with an `m`, so only `Pam` is output.

Before we get too deep into the weeds, let's slow down and look at some common LINQ extension methods and how to use them, one at a time.

Filtering entities using Where

The most common reason for using LINQ is to filter items in a sequence using the `Where` extension method. Let's explore filtering by defining a sequence of names and then applying LINQ operations to it:

1. In the project file, add an element to prevent the `System.Linq` namespace from automatically being imported globally, as shown highlighted in the following markup:

```
<ItemGroup>
  <Using Include="System.Console" Static="true" />
  <Using Remove="System.Linq" />
</ItemGroup>
```

2. In `Program.Functions.cs`, add a new method named `FilteringUsingWhere`, as shown in the following code:

```
private static void FilteringUsingWhere(string[] names)
{
}
```

3. If you are using Visual Studio, navigate to **Tools** | **Options**. In the **Options** dialog box, navigate to **Text Editor** | **C#** | **IntelliSense**, clear the **Show items from unimported namespaces** checkbox, and then click **OK**.

4. In `FilteringUsingWhere`, attempt to call the `Where` extension method on the array of names, as shown in the following code:

```
SectionTitle("Filtering entities using Where");

var query = names.W
```

5. As you type the `W`, note that in older code editors (or code editors with the option to show items from unimported namespaces disabled), the `Where` method is missing from the IntelliSense list of members of a `string` array, as shown in *Figure 11.1*:

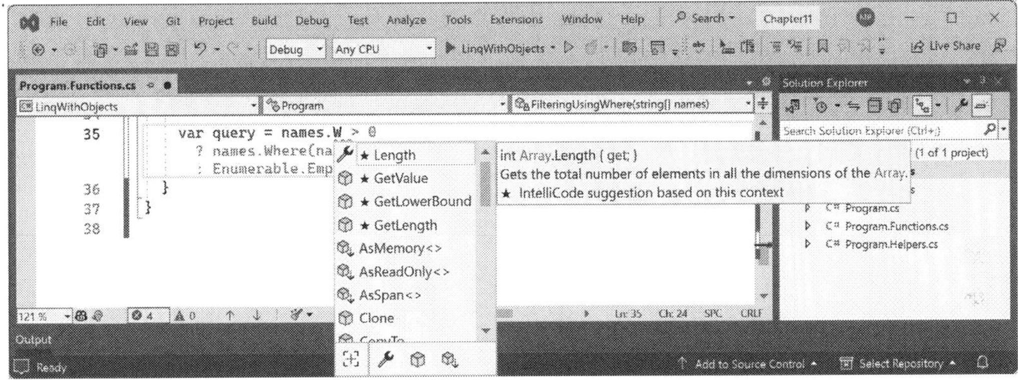

Figure 11.1: IntelliSense with the Where extension method missing

This is because `Where` is an extension method. It does not exist on the array type. To make the `Where` extension method available, we must import the `System.Linq` namespace. This is implicitly imported by default in new .NET 6 and later projects, but we removed it to illustrate the point. Recent versions of code editors are smart enough to suggest using the `Where` method anyway and indicate that they will import the `System.Linq` namespace for you automatically.

6. If you are using Visual Studio, navigate to **Tools** | **Options**. In the **Options** dialog box, navigate to **Text Editor** | **C#** | **IntelliSense**, select the **Show items from unimported namespaces** checkbox, and then click **OK**.
7. In the project file, comment out the element that removed `System.Linq`, as shown in the following code:

```
<!--<Using Remove="System.Linq" />-->
```

8. Save the change and build the project.

9. Retype the W for the Where method, and note that the IntelliSense list now includes the extension methods added by the Enumerable class, as shown in *Figure 11.2*:

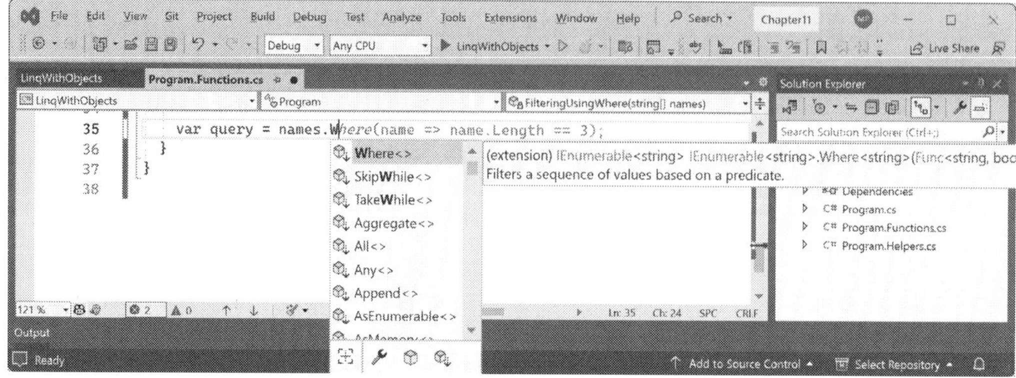

Figure 11.2: IntelliSense showing LINQ extension methods when System.Linq is imported

Interestingly, as you can see in the screenshot of Visual Studio on my computer, GitHub Copilot even suggests autocompleting with a lambda expression, very similar to the one that we will eventually end up writing. But there are some important intermediate steps you need to see before we get to that, so do not insert any GitHub Copilot suggestions if you have that feature enabled.

10. As you type the parentheses for the Where method, IntelliSense tells us that, to call Where, we must pass in an instance of a Func<string, bool> delegate.
11. Enter an expression to create a new instance of a Func<string, bool> delegate, and for now, note that we have not yet supplied a method name because we will define it in the next step, as shown in the following code:

```
var query = names.Where(new Func<string, bool>( ))
```

12. Leave the statement unfinished for now.

The Func<string, bool> delegate tells us that for each string variable passed to the method, the method must return a bool value. If the method returns true, it indicates that we should include the string in the results, and if the method returns false, it indicates that we should exclude it.

Targeting a named method

Let's define a method that only includes names that are longer than four characters:

1. In Program.Functions.cs, add a method that will return true only for names longer than four characters, as shown in the following code:

```
static bool NameLongerThanFour(string name)
{
  // Returns true for a name longer than four characters.
```

```
    return name.Length > 4;
  }
```

2. In the `FilteringUsingWhere` method, pass the method's name into the `Func<string, bool>` delegate, as shown highlighted in the following code:

```
var query = names.Where(
  new Func<string, bool>(NameLongerThanFour));
```

3. In the `FilteringUsingWhere` method, add statements to enumerate the `names` array using `foreach`, as shown in the following code:

```
foreach (string item in query)
{
  WriteLine(item);
}
```

4. In `Program.cs`, comment out the call to `DeferredExecution`, and then pass `names` as an argument to the `FilteringUsingWhere` method, as shown in the following code:

```
// DeferredExecution(names);
FilteringUsingWhere(names);
```

5. Run the code and view the results, noting that only names longer than four letters are listed, as shown in the following output:

```
Michael
Dwight
Angela
Kevin
Creed
```

Simplifying code by removing the explicit delegate instantiation

We can simplify code by deleting the explicit instantiation of the `Func<string, bool>` delegate because the C# compiler can instantiate the delegate for us:

1. To help you learn by seeing progressively improved code, in the `FilteringUsingWhere` method, comment out the query and add a comment about how it works, as shown in the following code:

```
// Explicitly creating the required delegate.
// var query = names.Where(
//   new Func<string, bool>(NameLongerThanFour));
```

2. Enter the query a second time but, this time, without the explicit instantiation of the delegate, as shown in the following code:

```
// The compiler creates the delegate automatically.
var query = names.Where(NameLongerThanFour);
```

3. Run the code, and note that it has the same behavior.

Targeting a lambda expression

We can simplify our code even further using a **lambda expression** in place of a named method.

Although it can look complicated at first, a lambda expression is simply a *nameless function*. It uses the => (read as "goes to") symbol to indicate the return value:

1. Comment out the second query, and then add a third version of the query that uses a lambda expression, as shown in the following code:

   ```
   // Using a lambda expression instead of a named method.
   var query = names.Where(name => name.Length > 4);
   ```

   Note that the syntax for a lambda expression includes all the important parts of the `NameLongerThanFour` method, and nothing more. A lambda expression only needs to define the following:

   - The names of input parameters: `name`
   - A return value expression: `name.Length > 4`

   The type of the `name` input parameter is inferred from the fact that the sequence contains `string` values, and the return type must be a `bool` value, as defined by the delegate, for `Where` to work; therefore, the expression after the => symbol must return a `bool` value. The compiler does most of the work for us, so our code can be as concise as possible.

2. Run the code, and note that it has the same behavior.

Lambda expressions with default parameter values

Introduced with C# 12, you can now provide default values for parameters in lambda expressions, as shown in the following code:

```
var query = names.Where((string name = "Bob") => name.Length > 4);
```

Our purpose for using this lambda expression means that setting a default value is not necessary, but later, you will see more useful examples.

Sorting and more

Other commonly used extension methods are `OrderBy` and `ThenBy`, used to sort a sequence.

Sorting by a single property using OrderBy

Extension methods can be chained if the previous method returns another sequence, that is, a type that implements the `IEnumerable<T>` interface.

Let's continue working with the current project to explore sorting:

1. In the `FilteringUsingWhere` method, append a call to `OrderBy` to the end of the existing query, as shown in the following code:

```
var query = names
  .Where(name => name.Length > 4)
  .OrderBy(name => name.Length);
```

Good Practice: Format the LINQ statement so that each extension method call happens on its own line, making it easier to read.

2. Run the code, and note that the names are now sorted by the shortest first, as shown in the following output:

```
Kevin
Creed
Dwight
Angela
Michael
```

To put the longest name first, you would use `OrderByDescending`.

Sorting by a subsequent property using ThenBy

We might want to sort by more than one property, for example, to sort names of the same length in alphabetical order:

1. In the `FilteringUsingWhere` method, append a call to the `ThenBy` method at the end of the existing query, as highlighted in the following code:

```
var query = names
  .Where(name => name.Length > 4)
  .OrderBy(name => name.Length)
  .ThenBy(name => name);
```

2. Run the code, and note the slight difference in the following sort order. Within a group of names of the same length, the names are sorted alphabetically by the full value of the `string`, so `Creed` comes before `Kevin`, and `Angela` comes before `Dwight`, as shown in the following output:

```
Creed
Kevin
Angela
Dwight
Michael
```

Sorting by the item itself

.NET 7 introduced the Order and OrderDescending extension methods. These simplify ordering by the item itself. For example, if you have a sequence of string values, then before .NET 7, you would have had to call the OrderBy method and pass a lambda that selects the items themselves, as shown in the following code:

```
var query = names.OrderBy(name => name);
```

With .NET 7 or later, we can simplify the statement, as shown in the following code:

```
var query = names.Order();
```

OrderDescending does a similar thing but in descending order.

Remember that the names array contains instances of the string type, which implements the IComparable interface. This is why they can be ordered, aka sorted. If the array contained instances of a complex type, like Person or Product, then those types would have to implement the IComparable interface so that they could be ordered too.

Declaring a query using var or a specified type

While writing a LINQ expression, it is convenient to use var to declare the query object. This is because the return type frequently changes as you work on a LINQ expression. For example, our query started as an IEnumerable<string> and is currently an IOrderedEnumerable<string>:

1. Hover your mouse over the var keyword, and note that its type is IOrderedEnumerable<string>, as shown in *Figure 11.3*:

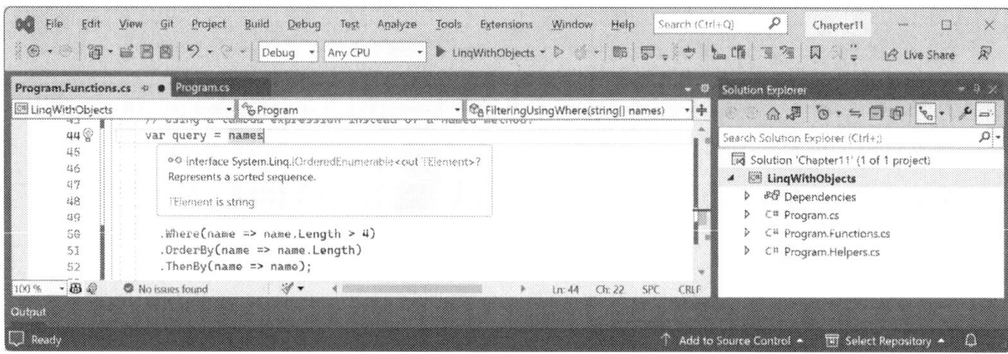

Figure 11.3: Hover over var to see the actual implied type of the query expression

In *Figure 11.3*, I added extra vertical space between names and .Where so that the tooltip did not cover up the query.

2. Replace var with the actual type, as shown highlighted in the following code:

```
IOrderedEnumerable<string> query = names
  .Where(name => name.Length > 4)
  .OrderBy(name => name.Length)
  .ThenBy(name => name);
```

Good practice: Once you have finished working on a query, you could change the declared type from var to the actual type to make it clearer what the type is. This is easy because your code editor can tell you what it is. Doing this is just for clarity. It has no effect on performance because C# changes all var declarations to the actual types at compile time.

3. Run the code, and note that it has the same behavior.

Filtering by type

The Where extension method is great for filtering by values, such as text and numbers. But what if the sequence contains multiple types and you want to filter by a specific type, respecting any inheritance hierarchy?

Imagine that you have a sequence of exceptions. There are hundreds of exception types that form a complex inheritance hierarchy, as partially shown in *Figure 11.4*:

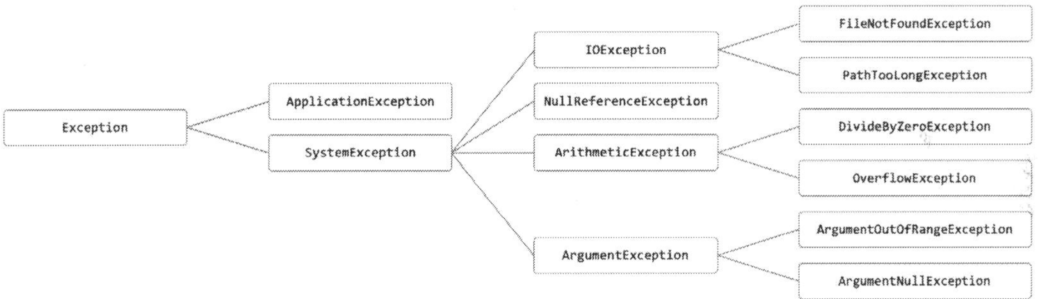

Figure 11.4: A partial exception inheritance hierarchy

Let's explore filtering by type:

1. In Program.Functions.cs, define a new method to list, and then filter exception-derived objects using the OfType<T> extension method to remove exceptions that are not arithmetic exceptions, writing only the arithmetic exceptions to the console, as shown in the following code:

```
static void FilteringByType()
{
  SectionTitle("Filtering by type");

  List<Exception> exceptions = new()
```

```
  {
    new ArgumentException(), new SystemException(),
    new IndexOutOfRangeException(), new InvalidOperationException(),
    new NullReferenceException(), new InvalidCastException(),
    new OverflowException(), new DivideByZeroException(),
    new ApplicationException()
  };

  IEnumerable<ArithmeticException> arithmeticExceptionsQuery =
    exceptions.OfType<ArithmeticException>();

  foreach (ArithmeticException exception in arithmeticExceptionsQuery)
  {
    WriteLine(exception);
  }
}
```

2. In `Program.cs`, comment out the call to `FilteringUsingWhere`, and then add a call to the `FilteringByType` method, as shown in the following code:

```
// FilteringUsingWhere(names);
FilteringByType();
```

3. Run the code, and note that the results only include exceptions of the `ArithmeticException` type or the `ArithmeticException`-derived types, as shown in the following output:

```
System.OverflowException: Arithmetic operation resulted in an overflow.
System.DivideByZeroException: Attempted to divide by zero.
```

Working with sets and bags

Sets are one of the most fundamental concepts in mathematics. A **set** is a collection of one or more unique objects. A **multiset**, aka a **bag**, is a collection of one or more objects that can have duplicates.

You might remember being taught about Venn diagrams in school. Common set operations include the **intersect** or **union** between sets.

Let's write some code that will define three arrays of `string` values for cohorts of apprentices, and then we'll perform some common set and multiset operations on them:

1. In `Program.Functions.cs`, add a method that outputs any sequence of `string` variables as a comma-separated single `string` to the console output, along with an optional description, as shown in the following code:

```
static void Output(IEnumerable<string> cohort,
  string description = "")
{
```

```
    if (!string.IsNullOrEmpty(description))
    {
      WriteLine(description);
    }
    Write(" ");
    WriteLine(string.Join(", ", cohort.ToArray()));
    WriteLine();
  }
```

2. In `Program.Functions.cs`, add a method that defines three arrays of names, outputs them, and then performs various set operations on them, as shown in the following code:

```
static void WorkingWithSets()
{
  string[] cohort1 =
    { "Rachel", "Gareth", "Jonathan", "George" };

  string[] cohort2 =
    { "Jack", "Stephen", "Daniel", "Jack", "Jared" };

  string[] cohort3 =
    { "Declan", "Jack", "Jack", "Jasmine", "Conor" };

  SectionTitle("The cohorts");

  Output(cohort1, "Cohort 1");
  Output(cohort2, "Cohort 2");
  Output(cohort3, "Cohort 3");

  SectionTitle("Set operations");

  Output(cohort2.Distinct(), "cohort2.Distinct()");
  Output(cohort2.DistinctBy(name => name.Substring(0, 2)),
    "cohort2.DistinctBy(name => name.Substring(0, 2)):");
  Output(cohort2.Union(cohort3), "cohort2.Union(cohort3)");
  Output(cohort2.Concat(cohort3), "cohort2.Concat(cohort3)");
  Output(cohort2.Intersect(cohort3), "cohort2.Intersect(cohort3)");
  Output(cohort2.Except(cohort3), "cohort2.Except(cohort3)");
  Output(cohort1.Zip(cohort2,(c1, c2) => $"{c1} matched with {c2}"),
    "cohort1.Zip(cohort2)");
}
```

3. In `Program.cs`, comment out the call to `FilteringByType`, and then add a call to the `WorkingWithSets` method, as shown in the following code:

```
// FilteringByType();
WorkingWithSets();
```

4. Run the code and view the results, as shown in the following output:

```
Cohort 1
  Rachel, Gareth, Jonathan, George
Cohort 2
  Jack, Stephen, Daniel, Jack, Jared
Cohort 3
  Declan, Jack, Jack, Jasmine, Conor

cohort2.Distinct()
  Jack, Stephen, Daniel, Jared
cohort2.DistinctBy(name => name.Substring(0, 2)):
  Jack, Stephen, Daniel
cohort2.Union(cohort3)
  Jack, Stephen, Daniel, Jared, Declan, Jasmine, Conor
cohort2.Concat(cohort3)
  Jack, Stephen, Daniel, Jack, Jared, Declan, Jack, Jack, Jasmine, Conor
cohort2.Intersect(cohort3)
  Jack
cohort2.Except(cohort3)
  Stephen, Daniel, Jared
cohort1.Zip(cohort2)
  Rachel matched with Jack, Gareth matched with Stephen, Jonathan matched
with Daniel, George matched with Jack
```

With `Zip`, if there are unequal numbers of items in the two sequences, then some items will not have a matching partner. Those without a partner, like `Jared`, will not be included in the result.

For the `DistinctBy` example, instead of removing duplicates by comparing the whole name, we define a lambda key selector to remove duplicates by comparing the first two characters, so `Jared` is removed because `Jack` is already a name that starts with `Ja`.

Getting the index as well as items

.NET 9 introduced the `Index` LINQ extension method. In previous versions of .NET, if you wanted to get the index position of each item as well as the items themselves, you could do it using the `Select` method, but it was a bit messy.

Let's see an example of the old way and the new way:

1. In `Program.Functions.cs`, add a method that defines an array of names and outputs them with their index position, using the old way (the `Select` method) and then the new way (the `Index` method), as shown in the following code:

```
static void WorkingWithIndices()
{
  string[] theSeven = { "Homelander",
    "Black Noir", "The Deep", "A-Train",
    "Queen Maeve", "Starlight", "Stormfront" };

  SectionTitle("Working With Indices (old)");

  foreach (var (item, index) in
    theSeven.Select((item, index) => (item, index)))
  {
    WriteLine($"{index}: {item}");
  }

  SectionTitle("Working With Indices (new)");

  foreach (var (index, item) in theSeven.Index())
  {
    WriteLine($"{index}: {item}");
  }
}
```

Warning! Note the order of the two declared variables that will hold the index and the item. When using the `Select` method, you must declare the `item` first and then the `index` second. When using the `Index` method, you must declare the `index` first and then the `item` second.

2. In `Program.cs`, comment out the call to `WorkingWithSets`, and then add a call to the `WorkingWithIndices` method, as shown in the following code:

```
// WorkingWithSets();
WorkingWithIndices();
```

3. Run the code and view the results, as shown in the following output:

```
*** Working With Indices (old) ***
0: Homelander
1: Black Noir
```

```
2: The Deep
3: A-Train
4: Queen Maeve
5: Starlight
6: Stormfront
*** Working With Indices (new) ***
0: Homelander
1: Black Noir
2: The Deep
3: A-Train
4: Queen Maeve
5: Starlight
6: Stormfront
```

So far, we have used the LINQ to Objects provider to work with in-memory objects. Next, we will use the LINQ to Entities provider to work with entities stored in a database.

Using LINQ with EF Core

We have looked at LINQ queries that filter and sort, but none that change the shape of the items in the sequence. This operation is called **projection** because it's about projecting items of one shape into another shape. To learn about projection, it is best to have some more complex types to work with, so in the next project, instead of using `string` sequences, we will use sequences of entities from the Northwind sample database that you were introduced to in *Chapter 10, Working with Data Using Entity Framework Core*.

I will give instructions to use SQLite because it is cross-platform, but if you prefer to use SQL Server, then feel free to do so. I have included some commented code to enable SQL Server if you choose.

Creating a console app for exploring LINQ to Entities

First, we must create a console app and Northwind database to work with:

1. Use your preferred code editor to add a new **Console App** / `console` project, named `LinqWithEFCore`, to the `Chapter11` solution.
2. In the project file, globally and statically import the `System.Console` class.
3. In the `LinqWithEFCore` project, add a package reference to the EF Core provider for SQLite and/or SQL Server, as shown in the following markup:

```
<ItemGroup>
  <!--To use SQLite-->
  <PackageReference Version="9.0.0"
    Include="Microsoft.EntityFrameworkCore.Sqlite" />

  <!--To use SQL Server-->
```

```
  <PackageReference Version="9.0.0"
    Include="Microsoft.EntityFrameworkCore.SqlServer" />
</ItemGroup>
```

4. Build the `LinqWithEFCore` project to restore packages.
5. Copy the `Northwind4Sqlite.sql` file to the `LinqWithEFCore` folder.
6. At a command prompt or terminal in the `LinqWithEFCore` folder, create the Northwind database by executing the following command:

```
sqlite3 Northwind.db -init Northwind4Sqlite.sql
```

7. Be patient because this command might take a while to create the database structure. Eventually, you will see the SQLite command prompt, as shown in the following output:

```
 -- Loading resources from Northwind4Sqlite.sql
SQLite version 3.38.0 2022-02-22 15:20:15
Enter ".help" for usage hints.
sqlite>
```

8. To exit SQLite command mode, press *Ctrl* + *C* twice on Windows or *Cmd* + *D* on macOS or Linux.
9. If you prefer to work with SQL Server, then you should already have the Northwind database created in SQL Server from the previous chapter.

Building an EF Core model

We must define an EF Core model to represent the database and tables that we will work with. We will define the model manually to take complete control, and to prevent a relationship from being automatically defined between the `Categories` and `Products` tables. Later, you will use LINQ to join the two entity sets:

1. In the `LinqWithEFCore` project, add a new folder named `EntityModels`.
2. In the `EntityModels` folder, add three class files to the project, named `NorthwindDb.cs`, `Category.cs`, and `Product.cs`.
3. Modify the class file named `Category.cs`, as shown in the following code:

```
// To use [Required] and [StringLength].
using System.ComponentModel.DataAnnotations;

namespace Northwind.EntityModels;

public class Category
{
  public int CategoryId { get; set; }

  [Required]
  [StringLength(15)]
```

```
  public string CategoryName { get; set; } = null!;

  public string? Description { get; set; }
}
```

4. Modify the class file named `Product.cs`, as shown in the following code:

```
// To use [Required] and [StringLength].
using System.ComponentModel.DataAnnotations;

// To use [Column].
using System.ComponentModel.DataAnnotations.Schema;

namespace Northwind.EntityModels;

public class Product
{
  public int ProductId { get; set; }

  [Required]
  [StringLength(40)]
  public string ProductName { get; set; } = null!;

  public int? SupplierId { get; set; }
  public int? CategoryId { get; set; }

  [StringLength(20)]
  public string? QuantityPerUnit { get; set; }

  // Required for SQL Server provider.
  [Column(TypeName = "money")]
  public decimal? UnitPrice { get; set; }

  public short? UnitsInStock { get; set; }
  public short? UnitsOnOrder { get; set; }
  public short? ReorderLevel { get; set; }
  public bool Discontinued { get; set; }
}
```

We have deliberately not defined any relationships between `Category` and `Product` so that we can see how to manually associate them with each other using LINQ later.

5. Modify the class file named `NorthwindDb.cs`, as shown in the following code:

```
using Microsoft.Data.SqlClient; // To use SqlConnectionStringBuilder.
using Microsoft.EntityFrameworkCore; // To use DbContext, DbSet<T>.

namespace Northwind.EntityModels;

public class NorthwindDb : DbContext
{
  public DbSet<Category> Categories { get; set; } = null!;
  public DbSet<Product> Products { get; set; } = null!;

  protected override void OnConfiguring(
    DbContextOptionsBuilder optionsBuilder)
  {
    #region To use SQLite

    string database = "Northwind.db";
    string dir = Environment.CurrentDirectory;
    string path = string.Empty;

    // The database file will stay in the project folder.
    // We will automatically adjust the relative path to
    // account for running in Visual Studio or CLI.

    if (dir.EndsWith("net9.0"))
    {
      // Running in the <project>\bin\<Debug|Release>\net9.0 directory.
      path = Path.Combine("..", "..", "..", database);
    }
    else
    {
      // Running in the <project> directory.
      path = database;
    }
```

```
    path = Path.GetFullPath(path); // Convert to absolute path.
    WriteLine($"SQLite database path: {path}");

    if (!File.Exists(path))
    {
      throw new FileNotFoundException(
        message: $"{path} not found.", fileName: path);
    }

    // To use SQLite.
    optionsBuilder.UseSqlite($"Data Source={path}");

    #endregion

    #region To use SQL Server

    SqlConnectionStringBuilder builder = new();

    builder.DataSource = ".";
    builder.InitialCatalog = "Northwind";
    builder.IntegratedSecurity = true;
    builder.Encrypt = true;
    builder.TrustServerCertificate = true;
    builder.MultipleActiveResultSets = true;

    string connection = builder.ConnectionString;
    // WriteLine($"SQL Server connection: {connection}");

    // To use SQL Server.
    // optionsBuilder.UseSqlServer(connection);

    #endregion
  }

  protected override void OnModelCreating(
    ModelBuilder modelBuilder)
  {
    if (Database.ProviderName is not null &&
      Database.ProviderName.Contains("Sqlite"))
    {
      // SQLite data provider does not directly support the
```

```
      // decimal type so we can convert to double instead.
      modelBuilder.Entity<Product>()
        .Property(product => product.UnitPrice)
        .HasConversion<double>();
    }
  }
}
```

If you want to use SQL Server, then comment out the statement that calls `UseSqlite` and uncomment the statement that calls `UseSqlServer`.

6. Build the project and fix any compiler errors.

Filtering and sorting sequences

Now, let's write statements to filter and sort sequences of rows from the tables:

1. In the `LinqWithEFCore` project, add a new class file named `Program.Helpers.cs`.
2. In `Program.Helpers.cs`, define a partial `Program` class with a method to configure the console to support special characters, like the Euro currency symbol, and control the current culture, as well as a method to output a section title, as shown in the following code:

```
using System.Globalization; // To use CultureInfo.

partial class Program
{
  private static void ConfigureConsole(string culture = "en-US",
    bool useComputerCulture = false)
  {
    // To enable Unicode characters like Euro symbol in the console.
    OutputEncoding = System.Text.Encoding.UTF8;

    if (!useComputerCulture)
    {
      CultureInfo.CurrentCulture = CultureInfo.GetCultureInfo(culture);
    }
    WriteLine($"CurrentCulture: {CultureInfo.CurrentCulture.
DisplayName}");
  }

  private static void SectionTitle(string title)
  {
```

```
    ConsoleColor previousColor = ForegroundColor;
    ForegroundColor = ConsoleColor.DarkYellow;
    WriteLine($"*** {title} ***");
    ForegroundColor = previousColor;
  }
}
```

3. In the `LinqWithEFCore` project, add a new class file named `Program.Functions.cs`.
4. In `Program.Functions.cs`, define a partial `Program` class, and add a method to filter and sort products, as shown in the following code:

```
using Northwind.EntityModels; // To use NorthwindDb, Category, Product.
using Microsoft.EntityFrameworkCore; // To use DbSet<T>.

partial class Program
{
  private static void FilterAndSort()
  {
    SectionTitle("Filter and sort");

    using NorthwindDb db = new();

    DbSet<Product> allProducts = db.Products;

    IQueryable<Product> filteredProducts =
      allProducts.Where(product => product.UnitPrice < 10M);

    IOrderedQueryable<Product> sortedAndFilteredProducts =
      filteredProducts.OrderByDescending(product => product.UnitPrice);

    WriteLine("Products that cost less than $10:");

    foreach (Product p in sortedAndFilteredProducts)
    {
      WriteLine("{0}: {1} costs {2:$#,##0.00}",
        p.ProductId, p.ProductName, p.UnitPrice);
    }
    WriteLine();
  }
}
```

Note the following about the preceding code:

- `DbSet<T>` implements `IEnumerable<T>`, so LINQ can be used to query and manipulate sequences of entities in models built for EF Core. (Actually, I should say `TEntity` instead of `T`, but the name of this generic type has no functional effect. The only requirement is that the type is a `class`. The name just indicates that the class is expected to be an entity model.)
- The sequences implement `IQueryable<T>` (or `IOrderedQueryable<T>` after a call to an ordering LINQ method) instead of `IEnumerable<T>` or `IOrderedEnumerable<T>`. This is an indication that we are using a LINQ provider that builds the query, using expression trees. They represent code in a tree-like data structure and enable the creation of dynamic queries, which is useful for building LINQ queries for external data providers, like SQLite.
- The LINQ expression will be converted into another query language, such as SQL. Enumerating the query with `foreach` or calling a method such as `ToArray` will force the execution of the query and materialize the results.

5. In `Program.cs`, delete any existing statements, and then call the `ConfigureConsole` and `FilterAndSort` methods, as shown in the following code:

```
ConfigureConsole(); // Sets US English by default.
FilterAndSort();
```

6. Run the project and view the result, as shown in the following output:

```
CurrentCulture: English (United States)
*** Filter and sort ***
SQLite database path: C:\cs13net9\Chapter11\LinqWithEFCore\Northwind.db
Products that cost less than $10:
41: Jack's New England Clam Chowder costs $9.65
45: Rogede sild costs $9.50
47: Zaanse koeken costs $9.50
19: Teatime Chocolate Biscuits costs $9.20
23: Tunnbröd costs $9.00
75: Rhönbräu Klosterbier costs $7.75
54: Tourtière costs $7.45
52: Filo Mix costs $7.00
13: Konbu costs $6.00
24: Guaraná Fantástica costs $4.50
33: Geitost costs $2.50
```

Although this query outputs the information we want, it does so inefficiently because it gets all the columns from the `Products` table, instead of just the three columns we need. Let's log the generated SQL.

7. In the `FilterAndSort` method, before enumerating the results using `foreach`, add a statement to output the SQL, as shown highlighted in the following code:

```
WriteLine("Products that cost less than $10:");
WriteLine(sortedAndFilteredProducts.ToQueryString());
```

8. Run the code, and view the result that shows the SQL executed before the product details, as shown in the following partial output:

```
Products that cost less than $10:
SELECT "p"."ProductId", "p"."CategoryId", "p"."Discontinued",
"p"."ProductName", "p"."QuantityPerUnit", "p"."ReorderLevel",
"p"."SupplierId", "p"."UnitPrice", "p"."UnitsInStock", "p"."UnitsOnOrder"
FROM "Products" AS "p"
WHERE "p"."UnitPrice" < 10.0
ORDER BY "p"."UnitPrice" DESC
41: Jack's New England Clam Chowder costs $9.65
...
```

Projecting sequences into new types

Before we look at **projection**, we should review object initialization syntax. If you have a class defined, then you can instantiate an object using the class name, `new()`, and curly braces to set initial values for fields and properties, as shown in the following code:

```
// Person.cs
public class Person
{
  public string Name { get; set; }
  public DateTime DateOfBirth { get; set; }
}

// Program.cs
Person knownTypeObject = new()
{
  Name = "Boris Johnson",
  DateOfBirth = new(year: 1964, month: 6, day: 19)
};
```

C# 3 and later allow instances of **anonymous types** to be instantiated using the var keyword, as shown in the following code:

```
var anonymouslyTypedObject = new
{
  Name = "Boris Johnson",
  DateOfBirth = new DateTime(year: 1964, month: 6, day: 19)
};
```

Although we did not specify a type, the compiler can infer an anonymous type from the setting of two properties, named Name and DateOfBirth. The compiler can infer the types of the two properties from the values assigned: a literal string and a new instance of a date/time value.

This capability is especially useful when writing LINQ queries to project an existing type into a new type, without having to explicitly define the new type. Since the type is anonymous, this can only work with var-declared local variables.

Let's make the SQL command executed against the database table more efficient by adding a call to the Select method, projecting instances of the Product class into instances of a new anonymous type with only three properties:

1. In Program.Functions.cs, in the FilterAndSort method, add a statement to extend the LINQ query to use the Select method to return only the three properties (that is, table columns) that we need, modify the call to ToQueryString to use the new projectedProducts query, and modify the foreach statement to use the var keyword and the new projectedProducts query, as shown highlighted in the following code:

```
IOrderedQueryable<Product> sortedAndFilteredProducts =
  filteredProducts.OrderByDescending(product => product.UnitPrice);

var projectedProducts = sortedAndFilteredProducts
  .Select(product => new // Anonymous type.
  {
    product.ProductId,
    product.ProductName,
    product.UnitPrice
  });

WriteLine("Products that cost less than $10:");
WriteLine(projectedProducts.ToQueryString());

foreach (var p in projectedProducts)
{
```

2. Hover your mouse over the `new` keyword in the `Select` method call, or the `var` keyword in the `foreach` statement, and note that it is an anonymous type, as shown in *Figure 11.5*:

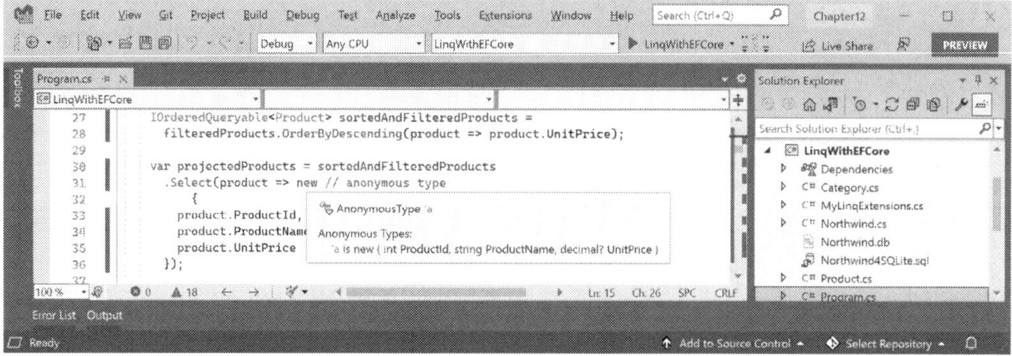

Figure 11.5: An anonymous type used during LINQ projection

3. Run the project, and confirm that the output is the same as before and the generated SQL is more efficient, as shown in the following output:

```
SELECT "p"."ProductId", "p"."ProductName", "p"."UnitPrice"
FROM "Products" AS "p"
WHERE "p"."UnitPrice" < 10.0
ORDER BY "p"."UnitPrice" DESC
```

More Information: You can learn more about projection using the `Select` method at the following link: `https://learn.microsoft.com/en-us/dotnet/csharp/programming-guide/concepts/linq/projection-operations`.

Let's continue to look at common LINQ queries by learning how to join, group, and perform lookups.

Joining, grouping, and lookups

There are three extension methods for joining, grouping, and creating grouped lookups:

- `Join`: This method has four parameters: the sequence that you want to join with, the property or properties on the *left* sequence to match on, the property or properties on the *right* sequence to match on, and a projection.
- `GroupJoin`: This method has the same parameters, but it combines the matches into a group object with a `Key` property for the matching value and an `IEnumerable<T>` type for the multiple matches.
- `ToLookup`: This method creates a new data structure with the sequence grouped by a key.

Joining sequences

Let's explore these methods when working with two tables, `Categories` and `Products`:

1. In `Program.Functions.cs`, add a method to select categories and products, join them, and output them, as shown in the following code:

```
private static void JoinCategoriesAndProducts()
{
  SectionTitle("Join categories and products");

  using NorthwindDb db = new();

  // Join every product to its category to return 77 matches.
  var queryJoin = db.Categories.Join(
    inner: db.Products,
    outerKeySelector: category => category.CategoryId,
    innerKeySelector: product => product.CategoryId,
    resultSelector: (c, p) =>
      new { c.CategoryName, p.ProductName, p.ProductId });

  foreach (var p in queryJoin)
  {
    WriteLine($"{p.ProductId}: {p.ProductName} in {p.CategoryName}.");
  }
}
```

In a join, there are two sequences, `outer` and `inner`. In the preceding example, `categories` is the outer sequence and `products` is the inner sequence.

2. In `Program.cs`, comment out the call to `FilterAndSort`, and then call the `JoinCategoriesAndProducts` method, as shown highlighted in the following code:

```
ConfigureConsole(); // Sets US English by default.
// FilterAndSort();
JoinCategoriesAndProducts();
```

3. Run the code and view the results. Note that there is a single line of output for each of the 77 products, as shown in the following output (edited to only include the first four items):

```
1: Chai in Beverages.
2: Chang in Beverages.
3: Aniseed Syrup in Condiments.
```

```
4: Chef Anton's Cajun Seasoning in Condiments.
...
```

4. In `Program.Functions.cs`, in the `JoinCategoriesAndProducts` method, at the end of the existing query, call the `OrderBy` method to sort by `CategoryName`, as shown highlighted in the following code:

```
var queryJoin = db.Categories.Join(
  inner: db.Products,
  outerKeySelector: category => category.CategoryId,
  innerKeySelector: product => product.CategoryId,
  resultSelector: (c, p) =>
    new { c.CategoryName, p.ProductName, p.ProductId })
  .OrderBy(cp => cp.CategoryName);
```

5. Run the code and view the results. Note that there is a single line of output for each of the 77 products, and the results show all products in the `Beverages` category first, then the `Condiments` category, and so on, as shown in the following partial output:

```
1: Chai in Beverages.
2: Chang in Beverages.
24: Guaraná Fantástica in Beverages.
34: Sasquatch Ale in Beverages.
...
```

Group-joining sequences

Let's explore group joining when working with the same two tables that we used to explore joining, `Categories` and `Products`, so that we can compare the subtle differences:

1. In `Program.Functions.cs`, add a method to group and join, show the group name, and then show all the items within each group, as shown in the following code:

```
private static void GroupJoinCategoriesAndProducts()
{
  SectionTitle("Group join categories and products");

  using NorthwindDb db = new();

  // Group all products by their category to return 8 matches.
  var queryGroup = db.Categories.AsEnumerable().GroupJoin(
    inner: db.Products,
    outerKeySelector: category => category.CategoryId,
    innerKeySelector: product => product.CategoryId,
    resultSelector: (c, matchingProducts) => new
```

```
    {
      c.CategoryName,
      Products = matchingProducts.OrderBy(p => p.ProductName)
    });

  foreach (var c in queryGroup)
  {
    WriteLine($"{c.CategoryName} has {c.Products.Count()} products.");

    foreach (var product in c.Products)
    {
      WriteLine($"  {product.ProductName}");
    }
  }
}
```

If we had not called the `AsEnumerable` method, then a runtime exception would have been thrown, as shown in the following output:

```
Unhandled exception. System.ArgumentException:  Argument type 'System.
Linq.IOrderedQueryable`1[Packt.Shared.Product]' does not match the
corresponding member type 'System.Linq.IOrderedEnumerable`1[Packt.Shared.
Product]' (Parameter 'arguments[1]')
```

This is because not all LINQ extension methods can be converted from expression trees into some other query syntax, like SQL. In these cases, we can convert from `IQueryable<T>` to `IEnumerable<T>` by calling the `AsEnumerable` method, which forces query processing to use LINQ to EF Core only to bring the data into the application, and then LINQ to Objects to execute more complex processing in memory. But, often, this is less efficient.

2. In `Program.cs`, call the `GroupJoinCategoriesAndProducts` method.
3. Run the code, view the results, and note that the products inside each category have been sorted by their name, as defined in the query and as shown in the following partial output:

```
Beverages has 12 products.
  Chai
  Chang
  ...
Condiments has 12 products.
  Aniseed Syrup
  Chef Anton's Cajun Seasoning
  ...
```

Grouping for lookups

Instead of writing a LINQ query expression to join and group and running it once, you might want to use a LINQ extension method to create, and then store a reusable in-memory collection that has entities that have been grouped.

We have a table named `Products` in the Northwind database that includes a column for the categories that they reside in, as partially shown in *Table 11.4*:

ProductName	CategoryID
Chai	1
Chang	1
Aniseed Syrup	2
Chef Anton's Cajun Seasoning	2
Chef Anton's Gumbo Mix	2
...	...

Table 11.4: The first five rows of the Products table

You might want to create a data structure in memory that can group the `Product` entities by their category, and then provide a quick way to look up all the products in a specific category.

You can create this using the `ToLookup` LINQ method, as shown in the following code:

```
ILookup<int, Product>? productsByCategoryId =
  db.Products.ToLookup(keySelector: category => category.CategoryId);
```

When you call the `ToLookup` method, you must specify a **key selector** to choose what value you want to group by. This value can then later be used to look up the group and its items.

The `ToLookup` method creates a dictionary-like data structure of key-value pairs in memory that has unique category IDs for the key and a collection of `Product` objects for the value, as partially shown in *Table 11.5*:

Key	Value (each one is a collection of Product objects)
1	[Chai], [Chang], and so on
2	[Aniseed Syrup], [Chef Anton's Cajun Seasoning], [Chef Anton's Gumbo Mix], and so on
...	...

Table 11.5: The first two rows in the lookup

Note that the product names in square brackets, such as `[Chai]`, represent an entire `Product` object.

Instead of using the `CategoryId` values as the key to the lookup, we could use the category names from the related categories table.

Let's do this in a code example:

1. In `Program.Functions.cs`, add a method to join products to category names, and then convert them into a lookup, enumerate through the whole lookup using an `IGrouping<string, Product>` to represent each row in the lookup dictionary, and look up an individual collection of products for a specific category, as shown in the following code:

```
private static void ProductsLookup()
{
  SectionTitle("Products lookup");

  using NorthwindDb db = new();

  // Join all products to their category to return 77 matches.
  var productQuery = db.Categories.Join(
    inner: db.Products,
    outerKeySelector: category => category.CategoryId,
    innerKeySelector: product => product.CategoryId,
    resultSelector: (c, p) => new { c.CategoryName, Product = p });

  ILookup<string, Product> productLookup = productQuery.ToLookup(
    keySelector: cp => cp.CategoryName,
    elementSelector: cp => cp.Product);

  foreach (IGrouping<string, Product> group in productLookup)
  {
    // Key is Beverages, Condiments, and so on.
    WriteLine($"{group.Key} has {group.Count()} products.");

    foreach (Product product in group)
    {
      WriteLine($" {product.ProductName}");
    }
  }

  // We can look up the products by a category name.
  Write("Enter a category name: ");
  string categoryName = ReadLine()!;
  WriteLine();
  WriteLine($"Products in {categoryName}:");
  IEnumerable<Product> productsInCategory = productLookup[categoryName];
  foreach (Product product in productsInCategory)
```

```
    {
      WriteLine($"  {product.ProductName}");
    }
  }
```

Selector parameters are lambda expressions that select sub-elements for different purposes. For example, `ToLookup` has a `keySelector` to select the part of each item that will be the key and an `elementSelector` to select the part of each item that will be the value. You can learn more at the following link: `https://learn.microsoft.com/en-us/dotnet/api/system.linq.enumerable.tolookup`.

2. In `Program.cs`, call the `ProductsLookup` method.
3. Run the code, view the results, enter a category name like `Seafoods`, and note that the products have been looked up and listed for that category, as shown in the following partial output:

```
Beverages has 12 products.
  Chai
  Chang
  ...
Condiments has 12 products.
  Aniseed Syrup
  Chef Anton's Cajun Seasoning
  ...
Enter a category name: Seafood

Products in Seafood:
  Ikura
  Konbu
  Carnarvon Tigers
  Nord-Ost Matjeshering
  Inlagd Sill
  Gravad lax
  Boston Crab Meat
  Jack's New England Clam Chowder
  Rogede sild
  Spegesild
  Escargots de Bourgogne
  Röd Kaviar
```

There is a lot more to LINQ, and in the final section, you will have the opportunity to explore further with some online material to walk through.

Practicing and exploring

Test your knowledge and understanding by answering some questions, getting some hands-on practice, and exploring with deeper research the topics covered in this chapter.

Exercise 11.1 – Online material

Online materials can be extra content written by me for this book, or it can be references to content created by Microsoft or third parties.

Aggregating and paging sequences

You can learn how to use LINQ aggregate methods and implement paging for data at the following link:

`https://github.com/markjprice/cs13net9/blob/main/docs/ch11-aggregating.md`

Using multiple threads with parallel LINQ

You can improve performance and scalability by using multiple threads to run LINQ queries. Learn how by completing the online-only section found at the following link:

`https://github.com/markjprice/cs13net9/blob/main/docs/ch11-plinq.md`

Working with LINQ to XML

If you want to process or generate XML using LINQ, then you can learn the basics of how by completing the online-only section found at the following link:

`https://github.com/markjprice/cs13net9/blob/main/docs/ch11-linq-to-xml.md`

Creating your own LINQ extension methods

If you want to create your own LINQ extension methods, then you can learn the basics of how by completing the online-only section found at the following link:

`https://github.com/markjprice/cs13net9/blob/main/docs/ch11-custom-linq-methods.md`

Design of the new LINQ methods in .NET 9

You can read about the design of the new LINQ extension methods introduced in .NET 9 at the following links:

- `AggregateBy`: `https://github.com/dotnet/runtime/issues/91533`.
- `CountBy`: `https://github.com/dotnet/runtime/issues/77716`.
- `Index`: `https://github.com/dotnet/runtime/issues/95563`.

Exercise 11.2 – Practice querying with LINQ

In the `Chapter11` solution, create a console application, named `Exercise_LinqQueries`, that prompts the user for a city and then lists the company names for Northwind customers in that city, as shown in the following output:

```
Enter the name of a city: London
There are 6 customers in London:
  Around the Horn
  B's Beverages
  Consolidated Holdings
  Eastern Connection
  North/South
  Seven Seas Imports
```

Then, enhance the application by displaying a list of all the unique cities that customers already reside in as a prompt to the user before they enter their preferred city, as shown in the following output:

```
Aachen, Albuquerque, Anchorage, Århus, Barcelona, Barquisimeto, Bergamo,
Berlin, Bern, Boise, Bräcke, Brandenburg, Bruxelles, Buenos Aires, Butte,
Campinas, Caracas, Charleroi, Cork, Cowes, Cunewalde, Elgin, Eugene, Frankfurt
a.M., Genève, Graz, Helsinki, I. de Margarita, Kirkland, Kobenhavn, Köln,
Lander, Leipzig, Lille, Lisboa, London, Luleå, Lyon, Madrid, Mannheim,
Marseille, México D.F., Montréal, München, Münster, Nantes, Oulu, Paris,
Portland, Reggio Emilia, Reims, Resende, Rio de Janeiro, Salzburg, San
Cristóbal, San Francisco, Sao Paulo, Seattle, Sevilla, Stavern, Strasbourg,
Stuttgart, Torino, Toulouse, Tsawassen, Vancouver, Versailles, Walla Walla,
Warszawa
```

Exercise 11.3 – Test your knowledge

Answer the following questions:

1. What are the two required parts of LINQ?
2. Which LINQ extension method would you use to return a subset of properties from a type?
3. Which LINQ extension method would you use to filter a sequence?
4. List five LINQ extension methods that perform aggregation.
5. What is the difference between the `Select` and `SelectMany` extension methods?
6. What is the difference between `IEnumerable<T>` and `IQueryable<T>`? How do you switch between them?
7. What does the last type parameter `T` in generic `Func` delegates like `Func<T1, T2, T>` represent?
8. What is the benefit of a LINQ extension method that ends with `OrDefault`?
9. Why is query comprehension syntax optional?
10. How can you create your own LINQ extension methods?

Exercise 11.4 – Explore topics

Use the links on the following page to learn more details about the topics covered in this chapter:

`https://github.com/markjprice/cs13net9/blob/main/docs/book-links.md#chapter-11---querying-and-manipulating-data-using-linq`

Summary

In this chapter, you learned how to write LINQ queries to perform common tasks like:

- Selecting just the properties of an item that you need.
- Filtering items based on conditions.
- Sorting items.
- Projecting items into different types.
- Joining and grouping items.

In the next chapter, you will be introduced to web development using ASP.NET Core. In the remaining chapters, you will learn how to implement modern features of ASP.NET Core, like Blazor and Minimal APIs.

12

Introducing Modern Web Development Using .NET

The third and final part of this book is about modern web development using .NET, which means ASP.NET Core, Blazor, and Minimal APIs. You will learn how to build cross-platform projects such as websites and web services.

Microsoft call platforms for building applications **app models** or **workloads.**

I recommend that you work through this and subsequent chapters sequentially because later chapters will reference projects in earlier chapters, and you will build up sufficient knowledge and skills to tackle the trickier problems in later chapters.

In this chapter, we will cover the following topics:

- Understanding ASP.NET Core
- New features in ASP.NET Core
- Structuring projects
- Building an entity model for use in the rest of the book
- Understanding web development

Understanding ASP.NET Core

Since this book is about C# and .NET, we will learn about app models that are used to build the practical applications that we will encounter in the remaining chapters of this book.

More Information: Microsoft has extensive guidance for implementing app models in its *.NET Architecture Guides* documentation, which you can read at the following link: `https://dotnet.microsoft.com/en-us/learn/dotnet/architecture-guides`.

ASP.NET Core is part of a history of evolving Microsoft technologies used to build websites and services that work with data:

- **ActiveX Data Objects (ADO)** was released in 1996 and was Microsoft's attempt to provide a single set of **Component Object Model** (**COM**) components to work with data. With the release of .NET, an equivalent was created named **ADO.NET**, which is still the faster method to work with data in .NET, with its core classes, `DbConnection`, `DbCommand`, and `DbDataReader`. **ORMs (Object-Relational Mappers)** like EF Core use ADO.NET internally.
- **Active Server Pages** (**ASP**) was released in 1996 and was Microsoft's first attempt at a platform for dynamic server-side execution of website code. ASP files contain a mix of HTML and code that executes on the server, written in the VBScript language.
- **ASP.NET Web Forms** was released in 2002 with the .NET Framework and was designed to enable non-web developers, such as those familiar with Visual Basic, to quickly create websites by dragging and dropping visual components and writing event-driven code in Visual Basic or C#. Web Forms should be avoided in new .NET Framework web projects in favor of ASP.NET MVC.
- **Windows Communication Foundation** (**WCF**) was released in 2006 and enables developers to build SOAP and REST services. SOAP is powerful but complex, so it should be avoided unless you need advanced features, such as distributed transactions and complex messaging topologies.
- **ASP.NET MVC** was released in 2009 to cleanly separate the concerns of web developers between **models**, which temporarily store the data; **views**, which present data using various formats in the UI; and **controllers**, which fetch a model and pass it to a view. This separation enables improved reuse and unit testing.
- **ASP.NET Web API** was released in 2012 and enables developers to create HTTP services (aka REST services) that are simpler and more scalable than SOAP services.
- **ASP.NET SignalR** was released in 2013 and enables real-time communication for websites by abstracting underlying technologies and techniques, such as WebSockets and long polling. This enables website features, such as live chat, and updates to time-sensitive data, such as stock prices, across a wide variety of web browsers, even when they do not support an underlying technology, such as WebSockets.
- **ASP.NET Core** was released in 2016 and combines modern implementations of .NET Framework technologies, such as MVC, Web API, and SignalR, with alternative technologies, such as Razor Pages, gRPC, and Blazor, all running on modern .NET. Therefore, ASP.NET Core can execute across platforms. ASP.NET Core has many project templates to get you started with its supported technologies.

Good Practice: Choose ASP.NET Core to develop websites and web services because it includes web-related technologies that are modern and cross-platform.

Classic ASP.NET versus modern ASP.NET Core

Until modern .NET, ASP.NET was built on top of a large assembly in .NET Framework, named `System.Web.dll`, and it was tightly coupled to Microsoft's Windows-only web server, named **Internet Information Services** (**IIS**). Over the years, this assembly has accumulated a lot of features, many of which are not suitable for modern cross-platform development.

ASP.NET Core is a major redesign of ASP.NET. It removes the dependency on the `System.Web.dll` assembly and IIS and is composed of modular lightweight packages, just like the rest of modern .NET. Using IIS as the web server is still supported by ASP.NET Core, but there is a better option.

You can develop and run ASP.NET Core applications across platforms on Windows, macOS, and Linux. Microsoft has even created a cross-platform, super-performant web server, named **Kestrel**, and the entire stack is open source.

ASP.NET Core 2.2 or later projects default to the new in-process hosting model. This gives a 400% performance improvement when hosting in Microsoft IIS, but Microsoft still recommends using Kestrel for even better performance.

Building websites using ASP.NET Core

Websites are made up of multiple web pages, loaded statically from the filesystem or generated dynamically by a server-side technology, such as ASP.NET Core. A web browser makes `GET` requests using **Unique Resource Locators** (**URLs**) that identify each page and can manipulate data stored on the server, using `POST`, `PUT`, and `DELETE` requests.

With many websites, the web browser is treated as a presentation layer, with almost all the processing performed on the server side. Some JavaScript might be used on the client side to implement form validation warnings and some presentation features, such as carousels.

ASP.NET Core provides multiple technologies to build the user interface for websites:

- **ASP.NET Core Razor Pages** is a simple way to dynamically generate HTML for simple websites. I recommend considering this as a legacy technology and using Blazor instead.
- **ASP.NET Core MVC** is an implementation of the **Model-View-Controller** (**MVC**) design pattern that is popular for developing complex websites.
- **Blazor** lets you build user interface components using C# and .NET instead of a JavaScript-based UI framework, like Angular, React, and Vue. Early versions of Blazor required a developer to choose a **hosting model**. The **Blazor WebAssembly** hosting model runs your code in the browser like a JavaScript-based framework would. The **Blazor Server** hosting model runs your code on the server and updates the web page dynamically. .NET 8 introduced a unified, full-stack hosting model that allows individual components to execute either on the server or client side, or even adapt dynamically at runtime. You will learn about Blazor in detail in *Chapter 14, Building Interactive Web Components Using Blazor*.

So, which should you choose?

> *"Blazor is now our recommended approach for building web UI with ASP.NET Core, but neither MVC nor Razor Pages are now obsolete. Both MVC & Razor Pages are mature, fully supported, and widely used frameworks that we plan to support for the foreseeable future. There is also no requirement or guidance to migrate existing MVC or Razor Pages apps to Blazor. For existing, well-established MVC-based projects, continuing to develop with MVC is a perfectly valid and reasonable approach."*
>
> *– Dan Roth*

You can see Dan's original comment post at the following link: `https://github.com/dotnet/aspnetcore/issues/51834#issuecomment-1913282747`.

Dan Roth is the principal product manager at ASP.NET, so he knows the future of ASP.NET Core better than anyone else: `https://devblogs.microsoft.com/dotnet/author/danroth27/`.

I agree with the quote by Dan Roth. For me, there are two main choices for web development:

- For websites or web services using modern web development: Choose Blazor for the web user interface and Minimal APIs for the web service. These technologies are covered in this book and its companion book, *Apps and Services with .NET 8*.
- For websites or web services using mature and proven web development: Choose controller-based ASP.NET Core MVC for the web user interface and Web API for the web service. For even more productivity, you can add third-party platforms on top of those, for example, a .NET CMS like Umbraco. These technologies are covered in my new companion book, *Real-World Web Development with .NET 9*.

Much of ASP.NET Core is shared across all these choices anyway, so you will only need to learn those shared components once, as shown in *Figure 12.1*:

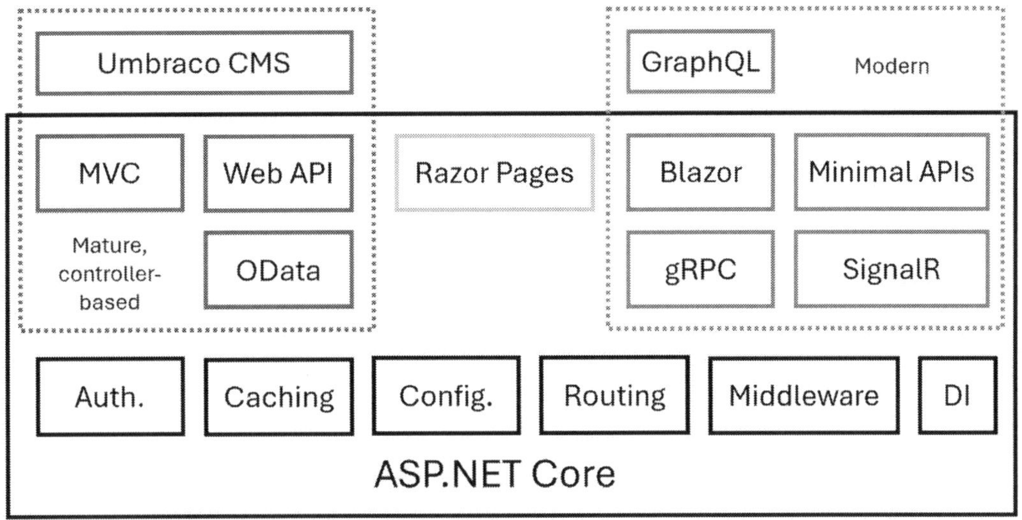

Figure 12.1: Modern or controller-based and shared ASP.NET Core components

Comparison of file types used in ASP.NET Core

It is useful to summarize the file types used by these technologies because they are similar but different. If you do not understand some subtle but important differences, it can cause much confusion when trying to implement your own projects. Please note the differences in *Table 12.1*:

Technology	Special filename	File extension	Directive
Razor Component (Blazor)		`.razor`	
Razor Component (Blazor with page routing)		`.razor`	`@page "<path>"`
Razor Component Imports (Blazor)	`_Imports`	`.razor`	
Razor Page		`.cshtml`	`@page`
Razor View (MVC)		`.cshtml`	
Razor Layout		`.cshtml`	
Razor View Start	`_ViewStart`	`.cshtml`	
Razor View Imports	`_ViewImports`	`.cshtml`	

Table 12.1: Comparison of file types used in ASP.NET Core

Directives like `@page` are added to the top of a file's contents.

If a file does not have a special filename, then it can be named anything. For example, you might create a Razor component for use in a Blazor project, named `Customer.razor`, or you might create a Razor layout for use in an MVC or Razor Pages project, named `_MobileLayout.cshtml`.

The naming convention for shared Razor files, like layouts and partial views, is to prefix with an underscore, `_`. For example, `_ViewStart.cshtml`, `_Layout.cshtml`, or `_Product.cshtml` (this might be a partial view for rendering a product).

A Razor layout file like `_MyCustomLayout.cshtml` is identical to a Razor view. What makes the file a layout is being set as the `Layout` property of another Razor file, as shown in the following code:

```
@{
  Layout = "_MyCustomLayout"; // File extension is not needed.
}
```

Warning! Be careful to use the correct file extension and directive at the top of the file; otherwise, you will get unexpected behavior.

Building websites using a content management system

Most websites have a lot of content, and if developers had to be involved every time some content needed to be changed, that would not scale well.

A **Content Management System** (**CMS**) enables developers to define content structure and templates to provide consistency and good design, while making it easy for a non-technical content owner to manage the actual content. They can create new pages or blocks of content and update existing content, knowing that it will look great for visitors with minimal effort.

There are a multitude of CMSs available for all web platforms, like WordPress for PHP or Django for Python. CMSs that support modern .NET include Optimizely Content Cloud, Umbraco, Piranha, and Orchard Core.

The key benefit of using a CMS is that it provides a friendly content management user interface. Content owners log in to the website and manage the content themselves. The content is then rendered and returned to visitors using ASP.NET Core MVC controllers and views, or via web service endpoints, known as a **headless CMS**, to provide that content to "heads" implemented as mobile or desktop apps, in-store touchpoints, or clients built with JavaScript frameworks or Blazor.

This book does not cover .NET CMSs, so I have included links where you can learn more about them in the GitHub repository: `https://github.com/markjprice/cs13net9/blob/main/docs/book-links.md#net-content-management-systems`. I also cover Umbraco CMS in my new book, *Real-World Web Development with .NET 9*.

Building web applications using SPA frameworks

Web applications are often built using technologies known as **Single-Page Application** (**SPA**) frameworks, such as Blazor, Angular, React, Vue, or a proprietary JavaScript library.

They can make requests to a backend web service to get more data when needed and post updated data using common serialization formats, such as XML and JSON. The canonical examples are Google web apps, like Gmail, Maps, and Docs.

With a web application, the client side uses JavaScript frameworks or Blazor to implement sophisticated user interactions, but most of the important processing and data access still happens on the server side, as the web browser has limited access to local system resources.

JavaScript is loosely typed and not designed for complex projects, so most JavaScript libraries these days use TypeScript, which adds strong typing to JavaScript and is designed with many modern language features to handle complex implementations.

The .NET SDK has project templates for JavaScript and TypeScript-based SPAs, but we will not spend any time learning how to build JavaScript and TypeScript-based SPAs in this book. Even though these SPAs are commonly used with ASP.NET Core as the backend, the focus of this book is on C# and not on other languages.

In summary, C# and .NET can be used on both the server side and the client side to build websites, as shown in *Figure 12.2*:

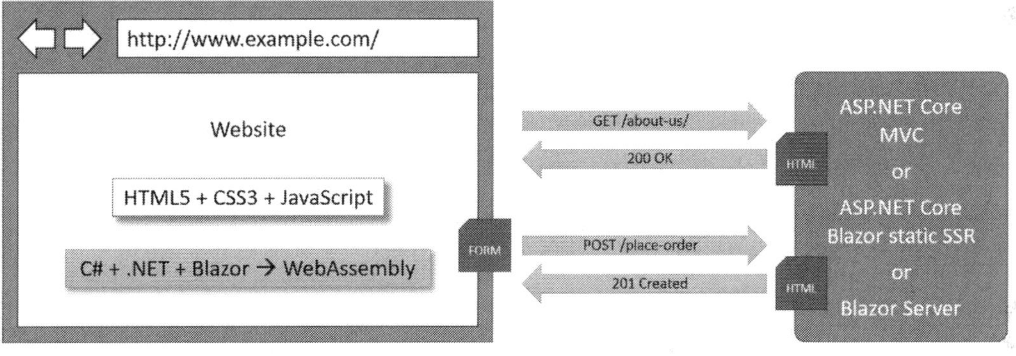

Figure 12.2: The use of C# and .NET to build websites on both the server side and client side

Building web and other services

Although we will not learn about JavaScript and TypeScript-based SPAs, we will learn how to build a web service using **ASP.NET Core Minimal API**, and then we will call that web service from Blazor components.

There are no formal definitions, but services are sometimes described based on their complexity:

- **Service**: All functionality needed by a client app in one monolithic service.
- **Microservice**: Multiple services that each focus on a smaller set of functionalities.
- **Nanoservice**: A single function provided as a service. Unlike services and microservices that are hosted 24/7/365, nanoservices are often inactive until called upon to reduce resources and costs.

At the start of the first part of this book, we briefly reviewed C# language features and in which versions they were introduced. At the start of the second part of this book, we briefly reviewed .NET library features and in which versions they were introduced. Now, in the third and final part of this book, we will briefly review ASP.NET Core features and in which versions they were introduced.

You can read this information in the GitHub repository at the following link: `https://github.com/markjprice/cs13net9/blob/main/docs/ch12-features.md`.

To summarize what's new in ASP.NET Core 9, let's end this section with another quote from Dan Roth:

> *"We're optimizing how the .NET runtime on WebAssembly initializes so that you start up faster and we're making Blazor's initialization logic more efficient by leveraging source generation for JSON serialization. We're also optimizing how static web assets are handled for all ASP.NET Core apps so that your files are pre-compressed as part of publishing your app. For API developers we're providing built-in support for OpenAPI document generation."*
>
> *– Dan Roth*

Building desktop and mobile apps

Since this book is about modern cross-platform development using C# and .NET, it does not include coverage of building desktop apps using **Windows Forms**, **Windows Presentation Foundation (WPF)**, or **WinUI 3** apps because they are Windows-only.

If you need to build apps for Windows, then the following links will be helpful:

- Official documentation to get started building apps for Windows: `https://learn.microsoft.com/en-us/windows/apps/get-started/`
- Is WPF Dead?: `https://avaloniaui.net/Blog/is-wpf-dead`
- How popular is WPF compared to WinUI , MAUI in 2024?: `https://twitter.com/DrAndrewBT/status/1759557538805108860`
- WinForms in a 64-Bit world – our strategy going forward: `https://devblogs.microsoft.com/dotnet/winforms-designer-64-bit-path-forward/`

Mobile app platforms

There are two major mobile platforms, Apple's iOS and Google's Android, each with its own programming languages and platform APIs. There are also two major desktop platforms, Apple's macOS and Microsoft's Windows, each with its own programming languages and platform APIs, as shown in the following list:

- **iOS**: Objective C or Swift and UIKit
- **Android**: Java or Kotlin and Android API
- **macOS**: Objective C or Swift and AppKit or Catalyst
- **Windows**: C, C++, or many other languages, and Win32 API or Windows App SDK

Since there are so many combinations to learn to do native mobile development, it would be useful if there were a single technology that could target all those mobile platforms.

.NET MAUI

Cross-platform mobile and desktop apps can be built once for the **.NET Multi-platform App User Interfaces (MAUI)** platform, and then they can run on many mobile and desktop platforms.

.NET MAUI makes it easy to develop those apps by sharing user interface components as well as business logic. They can target the same .NET APIs as used by console apps, websites, and web services. The apps will be executed by the Mono runtime on mobile devices and the CoreCLR runtime on desktop devices. The Mono runtime is better optimized for mobile devices compared to the normal .NET CoreCLR runtime. Blazor WebAssembly also uses the Mono runtime because, like a mobile app, it is resource-constrained.

The apps can exist on their own, but they usually call services to provide an experience that spans all your computing devices, from servers and laptops to phones and gaming systems.

I introduce .NET MAUI in my companion book, *Apps and Services with .NET 8*, and Packt has many other books that go much deeper into .NET MAUI, so if you're serious about learning MAUI, please take a look at the following Packt books:

- *.NET MAUI Cross-Platform Application Development*: `https://www.packtpub.com/en-us/product/net-maui-cross-platform-application-development-9781835080597`
- *The MVVM Pattern in .NET MAUI*: `https://www.packtpub.com/en-us/product/the-mvvm-pattern-in-net-maui-9781805125006`
- *.NET MAUI Projects*: `https://www.packtpub.com/en-us/product/net-maui-projects-9781837634910`

Before Microsoft created .NET MAUI, third parties created open source initiatives to enable .NET developers to build cross-platform apps using XAML, named **Uno** and **Avalonia**.

Warning! I have not tried Uno or Avalonia myself for any real-world projects, so I cannot give an evidence-based recommendation for either of them. I mention them in this book just to make you aware of them.

Uno platform

Uno is an *"Open-source platform for building single codebase native mobile, web, desktop and embedded apps quickly"*, as stated on their website, found at the following link: `https://platform.uno/`.

Developers can reuse 99% of the business logic and UI layer across native mobile, web, and desktop.

The Uno platform uses the Xamarin native platform but *not* Xamarin.Forms. For WebAssembly, Uno uses the Mono-WASM runtime. For Linux, Uno uses Skia to draw the user interface on canvas.

Avalonia

Avalonia is an *"open source framework for building beautiful, cross-platform applications from a single .NET codebase"*, as stated on their website, found at the following link: `https://avaloniaui.net/`.

You can think of Avalonia as a spiritual successor to WPF. WPF, Silverlight, and UWP developers familiar with WPF can continue to benefit from their years of pre-existing knowledge and skills.

It was used by JetBrains to modernize their WPF-based tools and make them cross-platform.

The Avalonia extension for Visual Studio and deep integration with Rider make development easier and more productive.

Structuring projects

How should you structure your projects? So far, we have mostly built small individual console apps to illustrate language or library features, with the occasional class library and unit test project to support them. In the rest of this book, we will build multiple projects using different technologies that work together to provide a single solution.

With large, complex solutions, it can be difficult to navigate through all the code. So the primary reason to structure your projects is to make it easier to find components. It is good to have an overall name for your solution that reflects the application or solution.

We will build multiple projects for a fictional company named **Northwind**. We will name the solution `ModernWeb` and use the name `Northwind` as a prefix for all the project names.

There are many ways to structure and name projects and solutions, for example, using a folder hierarchy as well as a naming convention. If you work in a team, make sure you know how your team does it.

Structuring projects in a solution

It is good to have a naming convention for your projects in a solution so that any developer can tell what each one does instantly. A common choice is to use the type of project, for example, class library, console app, website, and so on.

Since you might want to run multiple web projects at the same time and they will be hosted on a local web server, we need to differentiate each project by assigning different port numbers for their endpoints, for both HTTP and HTTPS.

Commonly assigned local port numbers are `5000` for HTTP and `5001` for HTTPS. We will use a numbering convention of `5<chapter>0` for HTTP and `5<chapter>1` for HTTPS. For example, for a website project we will create in *Chapter 13*, we will assign `5130` to HTTP and `5131` to HTTPS.

We will, therefore, use the following project names and port numbers, as shown in *Table 12.2*:

Name	Ports	Description
`Northwind.Common`	N/A	A class library project for common types, like interfaces, enums, classes, records, and structs, used across multiple projects.
`Northwind.EntityModels`	N/A	A class library project for common EF Core entity models. Entity models are often used on both the server side and client side, so it is best to separate dependencies on specific database providers.
`Northwind.DataContext`	N/A	A class library project for the EF Core database context, with dependencies on specific database providers.
`Northwind.UnitTests`	N/A	An xUnit test project for the solution.
`Northwind.Web`	`http` `5130` and `https` `5131`	An ASP.NET Core project for a simple website that uses a mixture of static HTML files and Blazor Static **Server-Side Rendering** (**SSR**).
`Northwind.Blazor`	`http` `5140` and `https` `5141`	An ASP.NET Core Blazor project.
`Northwind.WebApi`	`http` `5150` and `https` `5151`	An ASP.NET Core project for a web API, aka an HTTP service. A good choice for integrating with websites because it can use any JavaScript library or Blazor to interact with the service.

Table 12.2: Example project names for various project types

Central package management

In all previous projects in this book, if we needed to reference a NuGet package, we included the reference to the package name and version directly in the project file.

Central Package Management (**CPM**) is a feature that simplifies the management of NuGet package versions across multiple projects within a solution. This is particularly useful for large solutions with many projects, where managing package versions individually can become cumbersome and error-prone.

The key features and benefits of CPM include:

- **Centralized Control**: CPM allows you to define package versions in a single file, typically `Directory.Packages.props`, which is placed in the root directory of your solution. This file centralizes the version information for all NuGet packages used across the projects in your solution.
- **Consistency**: It ensures consistent package versions across multiple projects. By having a single source of truth for package versions, CPM eliminates discrepancies that can occur when different projects specify different versions of the same package.

- **Simplified Updates:** Updating a package version in a large solution becomes straightforward. You update the version in the central file, and all projects referencing that package automatically use the updated version. This significantly reduces the maintenance overhead.
- **Reduced Redundancy:** It removes the need to specify package versions in individual project files (`.csproj`). This makes project files cleaner and easier to manage, as they no longer contain repetitive version information.

Let's set up CPM for a solution that we will use throughout the rest of the chapters in this book:

1. In the `cs13net9` folder, create a new folder named `ModernWeb`.
2. In the `ModernWeb` folder, create a new file named `Directory.Packages.props`.
3. In `Directory.Packages.props`, modify its contents, as shown in the following markup:

```
<Project>

  <PropertyGroup>
    <ManagePackageVersionsCentrally>true</Man
agePackageVersionsCentrally>
  </PropertyGroup>

  <ItemGroup Label="For EF Core 9." >
    <PackageVersion
      Include="Microsoft.EntityFrameworkCore.Sqlite"
      Version="9.0.0" />
    <PackageVersion
      Include="Microsoft.EntityFrameworkCore.Design"
      Version="9.0.0" />
  </ItemGroup>

  <ItemGroup Label="For unit testing.">
    <PackageVersion Include="coverlet.collector"
      Version="6.0.2" />
    <PackageVersion Include="Microsoft.NET.Test.Sdk"
      Version="17.12.0" />
    <PackageVersion Include="xunit"
      Version="2.9.0" />
    <PackageVersion
      Include="xunit.runner.visualstudio"
      Version="3.0.0" />
  </ItemGroup>

  <ItemGroup Label="For Blazor.">
    <PackageVersion Include=
```

```
      "Microsoft.AspNetCore.Components.WebAssembly.Server"
      Version="9.0.0" />
    <PackageVersion Include=
      "Microsoft.AspNetCore.Components.WebAssembly"
      Version="9.0.0" />
    <PackageVersion Include=
"Microsoft.AspNetCore.Components.WebAssembly.DevServer"
      Version="9.0.0" />
  </ItemGroup>

  <ItemGroup Label="For web services.">
    <PackageVersion
      Include="Microsoft.AspNetCore.OpenApi"
      Version="9.0.0" />
    <PackageVersion
      Include="Microsoft.Extensions.Caching.Hybrid"
      Version="9.0.0" />
  </ItemGroup>

</Project>
```

Warning! The `<ManagePackageVersionsCentrally>` element and its `true` value must go all on one line. Also, you cannot use floating wildcard version numbers, like `10.0-*`, as you can in an individual project, to always get the latest patch version of EF Core 10 during previews automatically.

For any projects that we add underneath the folder containing this file, we can reference the packages without explicitly specifying the version, as shown in the following markup:

```
<ItemGroup>
  <PackageReference Include="Microsoft.EntityFrameworkCore.Sqlite" />
  <PackageReference Include="Microsoft.EntityFrameworkCore.Design" />
</ItemGroup>
```

You should regularly review and update the package versions in the `Directory.Packages.props` file to ensure that you are using the latest stable releases, with important bug fixes and performance improvements.

I recommend that you set a monthly event in your calendar for the second Wednesday of each month. This will occur after the second Tuesday of each month, which is Patch Tuesday, when Microsoft releases bug fixes and patches for .NET and related packages.

For example, in mid-December 2024, there are likely to be new versions, so you can go to the NuGet page for all your packages and then update the versions if necessary, as shown in the following markup:

```
<ItemGroup>
  <PackageVersion
    Include="Microsoft.EntityFrameworkCore.Sqlite"
    Version="9.0.1" />
  <PackageVersion
    Include="Microsoft.EntityFrameworkCore.Design"
    Version="9.0.1" />
</ItemGroup>
```

Before updating package versions, check for any breaking changes in the release notes of the packages. Test your solution thoroughly after updating to ensure compatibility.

Educate your team, and document the purpose and usage of the `Directory.Packages.props` file to ensure that everyone understands how to manage package versions centrally.

You can override an individual package version by using the `VersionOverride` attribute on a `<PackageReference />` element, as shown in the following markup:

```
<ItemGroup>
  <PackageReference Include="Microsoft.EntityFrameworkCore.Sqlite" />
  <PackageReference Include="Microsoft.EntityFrameworkCore.Design"
    VersionOverride="9.0.0" />
</ItemGroup>
```

This can be useful if a newer version introduces a regression bug.

More Information: You can learn more about CPM at the following link: `https://learn.microsoft.com/en-us/nuget/consume-packages/central-package-management`.

Building an entity model for use in the rest of the book

Websites and web services usually need to work with data in a relational database or another data store. In this section, we will define an entity data model for the Northwind database stored in SQL Server or SQLite. It will be used in most of the apps that we create in subsequent chapters.

Creating the Northwind database

The script files to create the Northwind database for SQLite and SQL Server are different. The script for SQL Server creates 13 tables as well as related views and stored procedures. The script for SQLite is a simplified version that only creates 10 tables because SQLite does not support as many features. The main projects in this book only need those 10 tables, so you can complete every task in this book with either database.

The SQL scripts are found at the following link: `https://github.com/markjprice/cs13net9/tree/main/scripts/sql-scripts`.

There are multiple SQL scripts to choose from, as described in the following list:

- `Northwind4Sqlite.sql` script: To use SQLite on a local Windows, macOS, or Linux computer. This script could probably also be used for other SQL systems, like PostgreSQL or MySQL, but has not been tested for use with those!
- `Northwind4SqlServer.sql` script: To use SQL Server on a local Windows computer. The script checks if the Northwind database already exists, and if the database does exists, then it drops (aka deletes) the database before recreating it.
- `Northwind4AzureSqlDatabaseCloud.sql` script: To use SQL Server with an Azure SQL Database resource created in the Azure cloud. These resources cost money as long as they exist! The script does not drop or create the Northwind database because you should manually create the Northwind database using the Azure portal user interface.
- `Northwind4AzureSqlEdgeDocker.sql` script: To use SQL Server on a local computer in Docker. The script creates the Northwind database. It does not drop it if it already exists because the Docker container should be empty anyway, as a fresh one will be spun up each time.

Instructions to install SQLite can be found in *Chapter 10, Working with Data Using Entity Framework Core*. In that chapter, you will also find instructions for installing the `dotnet-ef` tool, which you will use to scaffold an entity model from an existing database.

Instructions to install SQL Server Developer Edition (free) on your local Windows computer can be found in the GitHub repository for this book at the following link: `https://github.com/markjprice/cs13net9/blob/main/docs/sql-server/README.md`.

Instructions to set up Azure SQL Edge in Docker for Windows, macOS, or Linux can be found in the GitHub repository for this book at the following link: `https://github.com/markjprice/cs13net9/blob/main/docs/sql-server/sql-edge.md`.

Creating a class library for entity models using SQLite

You will now define entity data models in a class library so that they can be reused in other types of projects, including client-side app models.

Good Practice: You should create a separate class library project for your entity data models from the class library for your data context. This allows easier sharing of the entity models between backend web servers and frontend desktop, mobile, and Blazor clients, and only the backend needs to reference the data context class library.

We will automatically generate some entity models using the EF Core command-line tool:

1. Use your preferred code editor to create a new project and solution, as defined in the following list:
   - Project template: **Class Library** / `classlib`

- Project file and folder: `Northwind.EntityModels.Sqlite`
- Solution file and folder: `ModernWeb`

2. In the `Northwind.EntityModels.Sqlite` project, add package references for the SQLite database provider and EF Core design-time support, as shown in the following markup:

```
<ItemGroup>
  <PackageReference
    Include="Microsoft.EntityFrameworkCore.Sqlite" />
  <PackageReference
    Include="Microsoft.EntityFrameworkCore.Design">
    <PrivateAssets>all</PrivateAssets>
    <IncludeAssets>runtime; build; native; contentfiles; analyzers;
buildtransitive</IncludeAssets>
  </PackageReference>
</ItemGroup>
```

3. Delete the `Class1.cs` file.
4. Build the `Northwind.EntityModels.Sqlite` project to restore packages.
5. Copy the `Northwind4Sqlite.sql` file into the `ModernWeb` solution folder (not the project folder!).
6. At a command prompt or terminal in the `ModernWeb` folder, enter a command to create the `Northwind.db` file for SQLite, as shown in the following command:

```
sqlite3 Northwind.db -init Northwind4SQLite.sql
```

Be patient because this command might take a while to create the database structure.

7. To exit SQLite command mode, press *Ctrl* + *C* twice on Windows or *Cmd* + *D* on macOS or Linux.
8. At a command prompt or terminal in the `ModernWeb` folder, enter a command to list the files in the current directory, as shown in the following command:

```
dir
```

9. You should see that a new file named `Northwind.db` has been created, as shown in the following output:

```
    Directory: C:\cs13net9\ModernWeb

Length Name
------ ----
       Northwind.EntityModels.Sqlite
   382 Directory.Packages.props
```

```
  1193 ModernWeb.sln
557056 Northwind.db
480790 Northwind4SQLite.sql
```

10. Change to the project folder:

```
cd Northwind.EntityModels.Sqlite
```

11. At a command prompt or terminal in the `Northwind.EntityModels.Sqlite` project folder (the folder that contains the `.csproj` project file), generate entity class models for all tables, as shown in the following command:

```
dotnet ef dbcontext scaffold "Data Source=../Northwind.db" Microsoft.
EntityFrameworkCore.Sqlite --namespace Northwind.EntityModels --data-
annotations
```

Note the following:

- The command to perform: `dbcontext scaffold`
- The connection string refers to the database file in the solution folder, which is one folder up from the current project folder: `"Data Source=../Northwind.db"`
- The database provider: `Microsoft.EntityFrameworkCore.Sqlite`
- The namespace: `--namespace Northwind.EntityModels`
- To use data annotations as well as the Fluent API: `--data-annotations`

Warning! `dotnet-ef` commands must be entered all on one line and in a folder that contains a project; otherwise, you will see the following error: `No project was found. Change the current working directory or use the --project option.` Remember that all command lines can be found at and copied from the following link: `https://github.com/markjprice/cs13net9/blob/main/docs/command-lines.md`.

If you use SQLite, then you will see warnings about incompatible type mappings between the table columns and properties in the entity class models. For example, `The column 'BirthDate' on table 'Employees' should map to a property of type 'DateOnly', but its values are in an incompatible format. Using a different type.` This is due to SQLite using dynamic types. We will fix those issues in the next section.

Creating a class library for a database context using SQLite

You will now define a database context class library:

1. Add a new project to the solution, as defined in the following list:

    - Project template: **Class Library** / `classlib`
    - Project file and folder: `Northwind.DataContext.Sqlite`
    - Solution file and folder: `ModernWeb`

2. In the `Northwind.DataContext.Sqlite` project, statically and globally import the `Console` class, add a package reference to the EF Core data provider for SQLite, and add a project reference to the `Northwind.EntityModels.Sqlite` project, as shown in the following markup:

```
<ItemGroup Label="To simplify use of WriteLine.">
  <Using Include="System.Console" Static="true" />
</ItemGroup>

<ItemGroup Label="Versions are set at solution-level.">
  <PackageReference
    Include="Microsoft.EntityFrameworkCore.Sqlite" />
</ItemGroup>

<ItemGroup>
  <ProjectReference Include=
    "..\Northwind.EntityModels.Sqlite
\Northwind.EntityModels.Sqlite.csproj" />
</ItemGroup>
```

Warning! The path to the project reference should not have a line break in your project file.

3. In the `Northwind.DataContext.Sqlite` project, delete the `Class1.cs` file.
4. Build the `Northwind.DataContext.Sqlite` project to restore packages.
5. In the `Northwind.DataContext.Sqlite` project, add a class named `NorthwindContextLogger.cs`.
6. Modify its contents to define a static method, named `WriteLine`, that appends a string to the end of a text file, named `northwindlog-<date_time>.txt`, in a folder named `book-logs` on the desktop, as shown in the following code:

```
using static System.Environment;

namespace Northwind.EntityModels;

public class NorthwindContextLogger
{
  public static void WriteLine(string message)
  {
    string folder = Path.Combine(GetFolderPath(
      SpecialFolder.DesktopDirectory), "book-logs");
```

```
      if (!Directory.Exists(folder))
        Directory.CreateDirectory(folder);

      string dateTimeStamp = DateTime.Now.ToString(
        "yyyyMMdd_HHmmss");

      string path = Path.Combine(folder,
        $"northwindlog-{dateTimeStamp}.txt");

      StreamWriter textFile = File.AppendText(path);
      textFile.WriteLine(message);
      textFile.Close();
    }
  }
```

7. Move the `NorthwindContext.cs` file from the `Northwind.EntityModels.Sqlite` project/folder to the `Northwind.DataContext.Sqlite` project/folder.

In Visual Studio **Solution Explorer**, if you drag and drop a file between projects, it will be copied. If you hold down *Shift* while dragging and dropping, it will be moved. In VS Code **EXPLORER**, if you drag and drop a file between projects, it will be moved. If you hold down *Ctrl* while dragging and dropping, it will be copied.

8. In `NorthwindContext.cs`, note the second constructor can have `options` passed as a parameter, which allows us to override the default database connection string in any projects, such as websites that need to work with the Northwind database, as shown in the following code:

```
public NorthwindContext(DbContextOptions<NorthwindContext> options)
  : base(options)
{
}
```

9. In `NorthwindContext.cs`, in the `OnConfiguring` method, remove the compiler #warning about the connection string, and then add statements to check the end of the current directory to accommodate when running in Visual Studio compared to the command prompt with VS Code, as shown in the following code:

```
protected override void OnConfiguring(
  DbContextOptionsBuilder optionsBuilder)
{
  if (!optionsBuilder.IsConfigured)
  {
    string database = "Northwind.db";
```

```
    string dir = Environment.CurrentDirectory;
    string path = string.Empty;

    if (dir.EndsWith("net9.0"))
    {
      // In the <project>\bin\<Debug|Release>\net9.0 directory.
      path = Path.Combine("..", "..", "..", "..", database);
    }
    else
    {
      // In the <project> directory.
      path = Path.Combine("..", database);
    }

    path = Path.GetFullPath(path); // Convert to absolute path.
    try
    {
      NorthwindContextLogger.WriteLine($"Database path: {path}");
    }
    catch (Exception ex)
    {
      WriteLine(ex.Message);
    }

    if (!File.Exists(path))
    {
      throw new FileNotFoundException(
        message: $"{path} not found.", fileName: path);
    }

    optionsBuilder.UseSqlite($"Data Source={path}");

    optionsBuilder.LogTo(NorthwindContextLogger.WriteLine,
      new[] { Microsoft.EntityFrameworkCore
        .Diagnostics.RelationalEventId.CommandExecuting });
  }
}
```

The throwing of the exception is important because if the database file is missing, then the SQLite database provider will create an empty database file, and so if you test connecting to it, it works. But if you query it, then you will see an exception related to missing tables because it does not have any tables! After converting the relative path to an absolute path, you can set a breakpoint while debugging to more easily see where the database file is expected to be, or add a statement to log that path.

Customizing the model and defining an extension method

Now, we will simplify the `OnModelCreating` method. I will briefly explain the individual steps and then show the complete final method. You can either try to perform the individual steps or just use the final method code:

1. In the `OnModelCreating` method, remove all Fluent API statements that call the `ValueGeneratedNever` method, like the one shown in the following code. This will configure primary key properties, like `CategoryId`, to never generate a value automatically:

```
modelBuilder.Entity<Category>(entity =>
{
  entity.Property(e => e. CategoryId).ValueGeneratedNever();
});
```

If we do not remove the configuration like the statements above, then when we add new suppliers, the `CategoryId` value will always be `0`, and we will only be able to add one supplier with that value; all other attempts will throw an exception. You can compare your `NorthwindContext.cs` to the one in the GitHub repository at the following link: `https://github.com/markjprice/cs13net9/blob/main/code/ModernWeb/Northwind.DataContext.Sqlite/NorthwindContext.cs`.

2. In the `OnModelCreating` method, for the `Product` entity, tell SQLite that the `UnitPrice` can be converted from `decimal` to `double`, as shown in the following code:

```
entity.Property(product => product.UnitPrice)
  .HasConversion<double>();
```

3. In the `OnModelCreating` method, for the `Order` entity, pass a decimal value `0.0M` to the `HasDefaultValue` method, as shown in the following code:

```
modelBuilder.Entity<Order>(entity =>
{
  entity.Property(e => e.Freight).HasDefaultValue(0.0M);
});
```

4. In the `OnModelCreating` method, for the `Product` entity, pass a decimal value `0.0M` to the `HasDefaultValue` method, as shown in the following code:

```
modelBuilder.Entity<Product>(entity =>
{
  ...
  entity.Property(e => e.UnitPrice).HasDefaultValue(0.0M);
```

5. The `OnModelCreating` method should now be simpler, as shown in the following code:

```
protected override void OnModelCreating(
  ModelBuilder modelBuilder)
{
  modelBuilder.Entity<Order>(entity =>
  {
    entity.Property(e => e.Freight).HasDefaultValue(0.0M);
  });

  modelBuilder.Entity<OrderDetail>(entity =>
  {
    entity.Property(e =>
      e.Quantity).HasDefaultValue((short)1);

    entity.HasOne(d => d.Order)
      .WithMany(p => p.OrderDetails)
      .OnDelete(DeleteBehavior.ClientSetNull);

    entity.HasOne(d => d.Product)
      .WithMany(p => p.OrderDetails)
      .OnDelete(DeleteBehavior.ClientSetNull);
  });

  modelBuilder.Entity<Product>(entity =>
  {
    entity.Property(e => e.Discontinued)
      .HasDefaultValue((short)0);
    entity.Property(e => e.ReorderLevel)
      .HasDefaultValue((short)0);
    entity.Property(e => e.UnitPrice)
      .HasDefaultValue(0.0M);
    entity.Property(e => e.UnitsInStock)
      .HasDefaultValue((short)0);
    entity.Property(e => e.UnitsOnOrder)
```

```
      .HasDefaultValue((short)0);

    entity.Property(product => product.UnitPrice)
      .HasConversion<double>();
  });

  OnModelCreatingPartial(modelBuilder);
}
```

6. In the `Northwind.DataContext.Sqlite` project, add a class named `NorthwindContextExtensions.cs`. Modify its contents to define an extension method that adds the Northwind database context to a collection of dependency services, as shown in the following code:

```
using Microsoft.EntityFrameworkCore; // To use UseSqlite.
using Microsoft.Extensions.DependencyInjection; // To use
IServiceCollection.

namespace Northwind.EntityModels;

public static class NorthwindContextExtensions
{
  /// <summary>
  /// Adds NorthwindContext to the specified IServiceCollection. Uses the
Sqlite database provider.
  /// </summary>
  /// <param name="services">The service collection.</param>
  /// <param name="relativePath">Default is ".."</param>
  /// <param name="databaseName">Default is "Northwind.db"</param>
  /// <returns>An IServiceCollection that can be used to add more
services.</returns>
  public static IServiceCollection AddNorthwindContext(
    this IServiceCollection services, // The type to extend.
    string relativePath = "..",
    string databaseName = "Northwind.db")
  {
    string path = Path.Combine(relativePath, databaseName);
    path = Path.GetFullPath(path);
    NorthwindContextLogger.WriteLine($"Database path: {path}");

    if (!File.Exists(path))
    {
      throw new FileNotFoundException(
        message: $"{path} not found.", fileName: path);
```

```
      }

      services.AddDbContext<NorthwindContext>(options =>
      {
        // Data Source is the modern equivalent of Filename.
        options.UseSqlite($"Data Source={path}");

        options.LogTo(NorthwindContextLogger.WriteLine,
          new[] { Microsoft.EntityFrameworkCore
            .Diagnostics.RelationalEventId.CommandExecuting });
      },
      // Register with a transient lifetime to avoid concurrency
      // issues in Blazor server-side projects.
      contextLifetime: ServiceLifetime.Transient,
      optionsLifetime: ServiceLifetime.Transient);

      return services;
    }
  }
```

7. Build the two class libraries and fix any compiler errors.

HasDefaultValue and HasDefaultValueSql

What's the difference between these two methods?

You should use `HasDefaultValue()` when you need a constant, static value as a default for a column, and the value does not depend on any conditions or need to be dynamically calculated at the time of insertion. This constant value is set at the model level and is used by EF Core to insert into the database if no other value is provided.

For the equivalent to the preceding example, you would use `entity.Property(e => e.Freight).HasDefaultValue(0M);` because `0M` uses the decimal suffix `M`. Think of it as setting a default value on the client side.

You should use `HasDefaultValueSql()` when the default value should be calculated by the database at the time of insertion, especially if it involves SQL functions or dynamic data that the database should evaluate. The default is a string `"0"` because it will be concatenated to the SQL statement, as shown in the following code:

```
CREATE TABLE "Orders" ( ... "Freight" "money" NULL CONSTRAINT "DF_Orders_
Freight" DEFAULT (0), ... );
```

Think of this as configuring the database to set a default value on the server side.

EF Core 8 and earlier database reverse-engineering for SQLite used HasDefaultValueSql. EF Core 9 database reverse-engineering uses HasDefaultValue.

Registering the scope of a dependency service

By default, a DbContext class is registered using the Scope lifetime, meaning that multiple threads can share the same instance. But DbContext does not support multiple threads. If more than one thread attempts to use the same NorthwindContext class instance at the same time, then you will see the following runtime exception thrown: `A second operation started on this context before a previous operation completed. This is usually caused by different threads using the same instance of a DbContext, however instance members are not guaranteed to be thread safe.`

This happens in Blazor projects with components set to run on the server side because, whenever interactions on the client side happen, a SignalR call is made back to the server, where a single instance of the database context is shared between multiple clients. This issue does not occur if a component is set to run on the client side.

Creating class libraries for entity models using SQL Server

If you would like to use SQL Server instead of SQLite, then there are instructions at the following link:

https://github.com/markjprice/cs13net9/blob/main/docs/sql-server/README.md#chapter-12---introducing-web-development-using-aspnet-core

Improving the class-to-table mapping

The dotnet-ef command-line tool generates different code for SQL Server and SQLite because they support different levels of functionality, and SQLite uses dynamic typing. For example, with EF Core 7, all integer columns in SQLite were mapped to nullable long properties for maximum flexibility.

With EF Core 8 and later, the actual stored values are checked, and if they are all storable in an int, EF Core 8 and later will declare the mapped property as an int. If the stored values are all storable in a short, EF Core 8 and later will declare the mapped property as a short.

In this edition, we need to do less work to improve the mapping. Hooray!

As another example, SQL Server text columns can have limits on the number of characters. SQLite does not support this. So dotnet-ef will generate validation attributes to ensure that string properties are limited to a specified number of characters for SQL Server but not for SQLite, as shown in the following code:

```
// SQLite database provider-generated code.
[Column(TypeName = "nvarchar (15)")]
public string CategoryName { get; set; } = null!;

// SQL Server database provider-generated code.
[StringLength(15)]
public string CategoryName { get; set; } = null!;
```

We will make some small changes to improve the entity model mapping and validation rules for SQLite. Similar ones for SQL Server are in the online-only instructions.

Remember that all code is available in the GitHub repository for the book. Although you will learn more by typing the code yourself, you never have to. Go to the following link and press `.` to get a live code editor in your browser: `https://github.com/markjprice/cs13net9`.

First, we will add a regular expression to validate that a `CustomerId` value is exactly five uppercase letters. Second, we will add string length requirements to validate that multiple properties throughout the entity models know the maximum length allowed for their text values:

1. Activate your code editor's **Find and Replace** feature:
   - In Visual Studio, navigate to **Edit** | **Find and Replace** | **Quick Replace**, and then toggle on **Use Regular Expressions.**
2. Type a regular expression in the **Find** box, as shown in *Figure 12.3* and the following expression:

```
\[Column\(TypeName = "(nchar|nvarchar) \((.*)\)"\)\]
```

3. In the **Replace** box, type a replacement regular expression, as shown in the following expression:

```
$0\n    [StringLength($2)]
```

After the newline character, `\n`, I have included four space characters to indent correctly on my system, which uses two space characters per indentation level. You can insert as many as you wish.

4. Set **Find and Replace** to search files in the **Current project**.
5. Execute **Find and Replace** to replace all files, as shown in *Figure 12.3*:

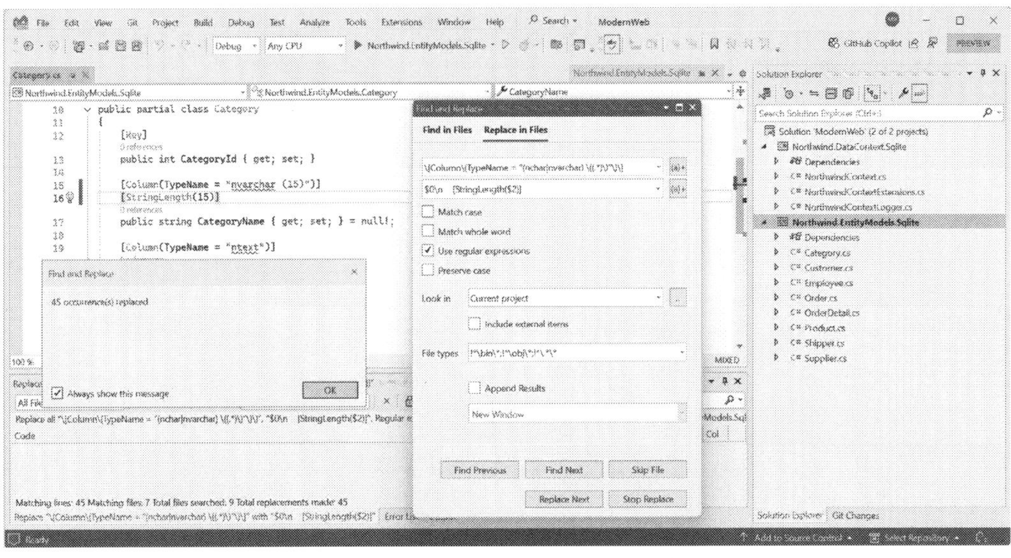

Figure 12.3: Find and replace all matches using regular expressions in Visual Studio

6. Change any date/time columns, for example, in `Employee.cs`, to use a nullable `DateTime` instead of a string, as shown in the following code:

```
// Before:
[Column(TypeName = "datetime")]
public string? BirthDate { get; set; }

// After:
[Column(TypeName = "datetime")]
public DateTime? BirthDate { get; set; }
```

Use your code editor's **Find** feature to search for `"datetime"` to find all the properties that need changing. There should be two in `Employee.cs` and three in `Order.cs`.

7. Change any money columns, for example, in Order.cs, to use a nullable decimal instead of a double, as shown in the following code:

```
// Before:
[Column(TypeName =  "money")]
public double? Freight { get; set; }

// After:
[Column(TypeName = "money")]
public decimal? Freight { get; set; }
```

Use your code editor's **Find** feature to search for "money" to find all the properties that need changing. There should be one in Order.cs, one in Orderdetail.cs, and one in Product.cs.

8. In Category.cs, make the CategoryName property required, as shown highlighted in the following code:

```
[Required]
[Column(TypeName = "nvarchar (15)")]
[StringLength(15)]
public string CategoryName { get; set; }
```

9. In Customer.cs, add a regular expression to validate its primary key CustomerId, only allowing uppercase Western characters and making the CompanyName property required, as shown highlighted in the following code:

```
[Key]
[Column(TypeName = "nchar (5)")]
[StringLength(5)]
[RegularExpression("[A-Z]{5}")]
public string CustomerId { get; set; } = null!;

[Required]
[Column(TypeName = "nvarchar (40)")]
[StringLength(40)]
public string CompanyName { get; set; }
```

10. In Order.cs, decorate the CustomerId property with a regular expression to enforce five uppercase characters.
11. In Employee.cs, make the FirstName and LastName properties required.
12. In Product.cs, make the ProductName property required.
13. In Shipper.cs, make the CompanyName property required.

14. In `Supplier.cs`, make the `CompanyName` property required.

Testing the class libraries

Now, let's build some unit tests to ensure that the class libraries are working correctly.

Warning! If you are using the SQLite database provider, then when you call the `CanConnect` method with a wrong or missing database file, the provider creates a `Northwind.db` with 0 bytes! This is why it is so important that, in our `NorthwindContext` class, we explicitly check if the database file exists and throw an exception when it is instantiated if it does not exist, preventing this behavior.

Let's write the tests:

1. Use your preferred coding tool to add a new **xUnit Test Project [C#]** / `xunit` project, named `Northwind.UnitTests`, to the `ModernWeb` solution.
2. In the `Northwind.UnitTests` project, add a project reference to the `Northwind.DataContext` project for either SQLite or SQL Server, as shown highlighted in the following configuration:

```
<ItemGroup>
  <!-- change Sqlite to SqlServer if you prefer -->
  <ProjectReference Include="..\Northwind.DataContext
.Sqlite\Northwind.DataContext.Sqlite.csproj" />
</ItemGroup>
```

Warning! The project reference must go all on one line with no line break.

3. If necessary, delete the version numbers specified for the testing packages in the project file. (Visual Studio and other code editors will give errors if you have projects that should use CPM but specify their own package versions, without using the `VersionOverride` attribute.)
4. Build the `Northwind.UnitTests` project to build referenced projects.
5. Rename `UnitTest1.cs` to `EntityModelTests.cs`.
6. Modify the contents of the file to define two tests, the first to connect to the database and the second to confirm that there are eight categories in the database, as shown in the following code:

```
using Northwind.EntityModels; // To use NorthwindContext.

namespace Northwind.UnitTests
{
  public class EntityModelTests
  {
```

```
    [Fact]
    public void DatabaseConnectTest()
    {
      using NorthwindContext db = new();
      Assert.True(db.Database.CanConnect());
    }

    [Fact]
    public void CategoryCountTest()
    {
      using NorthwindContext db = new();

      int expected = 8;
      int actual = db.Categories.Count();

      Assert.Equal(expected, actual);
    }

    [Fact]
    public void ProductId1IsChaiTest()
    {
      using NorthwindContext db = new();

      string expected = "Chai";

      Product? product = db.Products.Find(keyValues: 1);
      string actual = product?.ProductName ?? string.Empty;

      Assert.Equal(expected, actual);
    }
  }
}
```

7. Run the unit tests:

    - If you are using Visual Studio, navigate to **Test** | **Run All Tests**, and then view the results in **Test Explorer**.
    - If you are using VS Code, in the `Northwind.UnitTests` project's **TERMINAL** window, run the tests with the following command: `dotnet test`. Alternatively, use the **TESTING** window if you have installed C# Dev Kit.

8. Note that the results should indicate that three tests ran and all passed, as shown in *Figure 12.4*:

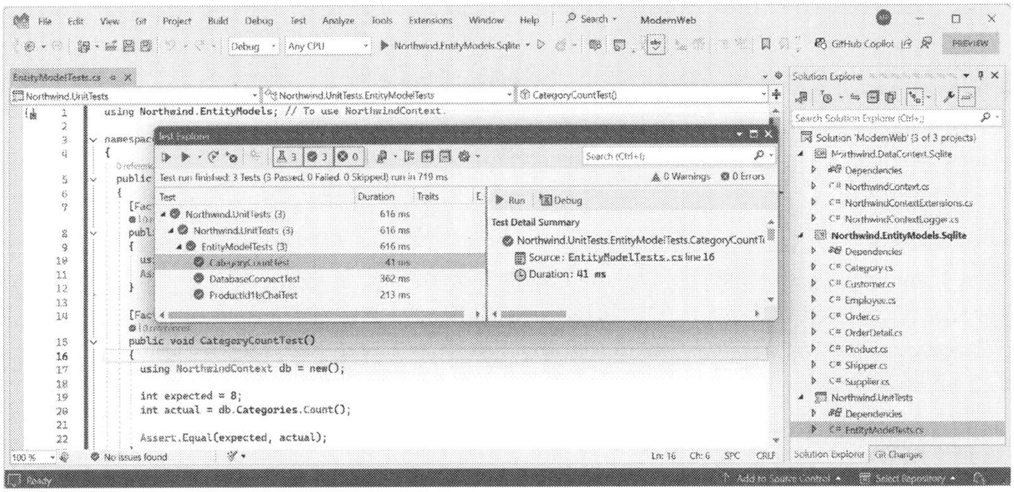

Figure 12.4: Three successful unit tests ran

If any of the tests fail, then fix the issue. For example, if you use SQLite, then check that the `Northwind.db` file is in the solution directory (one up from the project directories). Check the database path in the `northwindlog-<date_time>.txt` file in the `book-logs` folder on your desktop, which should output the database path it used three times for the three tests, as shown in the following log:

```
Database path: C:\cs13net9\ModernWeb\Northwind.db
Database path: C:\cs13net9\ModernWeb\Northwind.db
dbug: 18/09/2024 14:20:16.712 RelationalEventId.CommandExecuting[20100]
(Microsoft.EntityFrameworkCore.Database.Command)
      Executing DbCommand [Parameters=[@__p_0='?' (DbType = Int32)],
CommandType='Text', CommandTimeout='30']
      SELECT "p"."ProductId", "p"."CategoryId", "p"."Discontinued",
"p"."ProductName", "p"."QuantityPerUnit", "p"."ReorderLevel", "p"."SupplierId",
"p"."UnitPrice", "p"."UnitsInStock", "p"."UnitsOnOrder"
      FROM "Products" AS "p"
      WHERE "p"."ProductId" = @__p_0
      LIMIT 1
Database path: C:\cs13net9\ModernWeb\Northwind.db
dbug: 18/09/2024 14:20:16.832 RelationalEventId.CommandExecuting[20100]
(Microsoft.EntityFrameworkCore.Database.Command)
      Executing DbCommand [Parameters=[], CommandType='Text',
CommandTimeout='30']
      SELECT COUNT(*)
      FROM "Categories" AS "c"
```

Finally, in this chapter, let's review some key concepts about web development, enabling us to be better prepared to dive into ASP.NET Core Razor Pages in the next chapter.

Understanding web development

Developing for the web means developing with the **Hypertext Transfer Protocol** (**HTTP**), so we will start by reviewing this important foundational technology.

Understanding the Hypertext Transfer Protocol

To communicate with a web server, the client, also known as the **user agent**, makes calls over the network using HTTP. As such, HTTP is the technical underpinning of the web. So when we talk about websites and web services, we mean that they use HTTP to communicate between a client (often a web browser) and a server.

A client makes an HTTP request to a resource, such as a page, uniquely identified by a URL, and the server sends back an HTTP response, as shown in *Figure 12.5*:

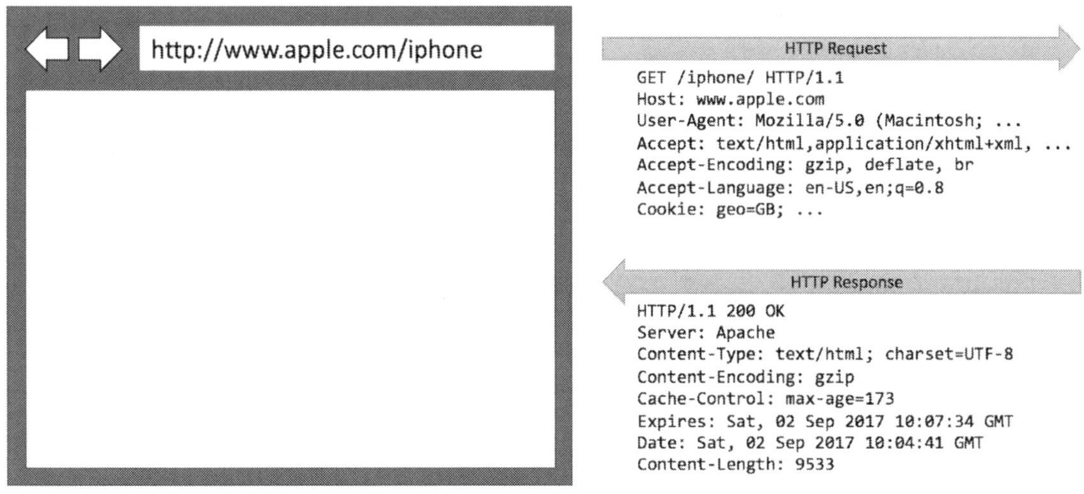

Figure 12.5: An HTTP request and response

You can use Google Chrome and other browsers to record requests and responses.

Good Practice: Google Chrome is currently used by about two-thirds of website visitors worldwide, and it has powerful, built-in developer tools, so it is a good first choice for trying out your websites. Try out your websites with Chrome and at least two other browsers, for example, Firefox and Safari for macOS and iPhone, respectively. Microsoft Edge switched from using Microsoft's own rendering engine to using Chromium in 2019, so it is less important to try out with it, although some say Edge has the best developer tools. If Microsoft's Internet Explorer is used at all, it tends to mostly be inside organizations for intranets.

Understanding the components of a URL

A **URL** is made up of several components:

- **Scheme:** `http` (clear text) or `https` (encrypted).
- **Domain:** For a production website or service, the **top-level domain** (**TLD**) might be `example.com`. You might have subdomains such as `www`, `jobs`, or `extranet`. During development, you typically use `localhost` for all websites and services.
- **Port number:** For a production website or service, use `80` for `http` and `443` for `https`. These port numbers are usually inferred from the scheme. During development, other port numbers are commonly used, such as `5000`, `5001`, and so on, to differentiate between websites and services that all use the shared domain `localhost`.
- **Path:** A relative path to a resource, for example, `/customers/germany`.
- **Query string:** A way to pass parameter values, for example, `?country=Germany&searchtext=shoes`.
- **Fragment:** A reference to an element on a web page using its `id`, for example, `#toc`.

A URL is a subset of a **Uniform Resource Identifier** (**URI**). A URL specifies where a resource is located and how to get it. A URI identifies a resource either by the URL or **URN** (**Uniform Resource Name**).

Using Google Chrome to make HTTP requests

Let's explore how to use Google Chrome to make HTTP requests:

1. Start Google Chrome.
2. Navigate to **More tools** | **Developer tools**.

3. Click the **Network** tab, and Chrome should immediately start recording the network traffic between your browser and any web servers (note the red circle), as shown in *Figure 12.6*:

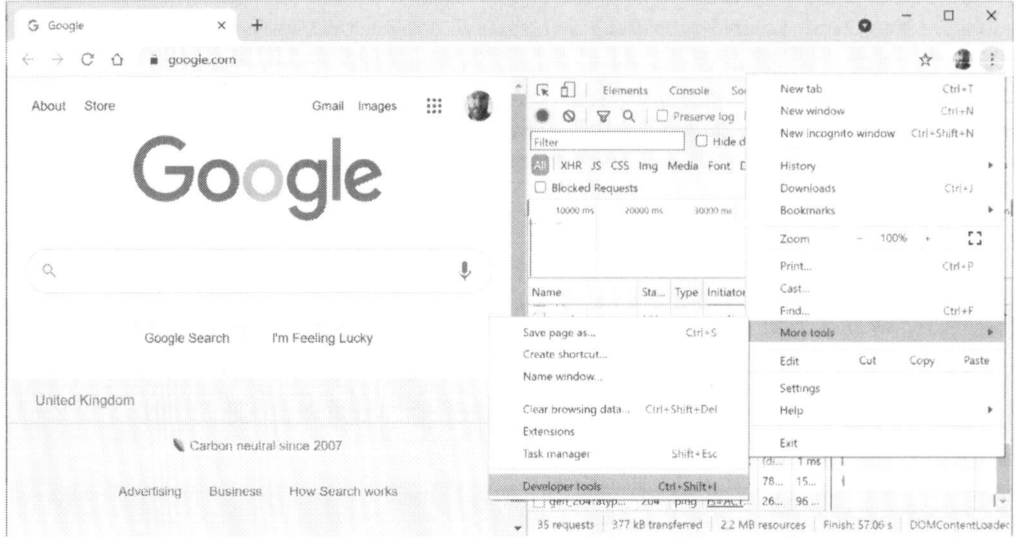

Figure 12.6: Chrome Developer tools recording network traffic

4. In Chrome's address box, enter the address of Microsoft's website for learning ASP.NET, which is the following URL:

```
https://dotnet.microsoft.com/en-us/learn/aspnet
```

5. In **Developer Tools**, in the list of recorded requests, scroll to the top and click on the first entry, the row where the **Type** is **document**, as shown in *Figure 12.7*:

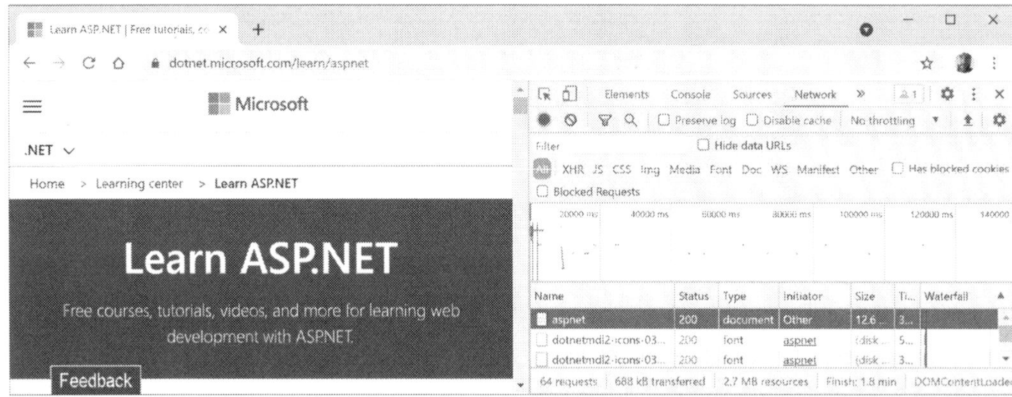

Figure 12.7: Recorded requests in Developer Tools

6. On the right-hand side, click on the **Headers** tab, and you will see details about **Request Headers** and **Response Headers**, as shown in *Figure 12.8*:

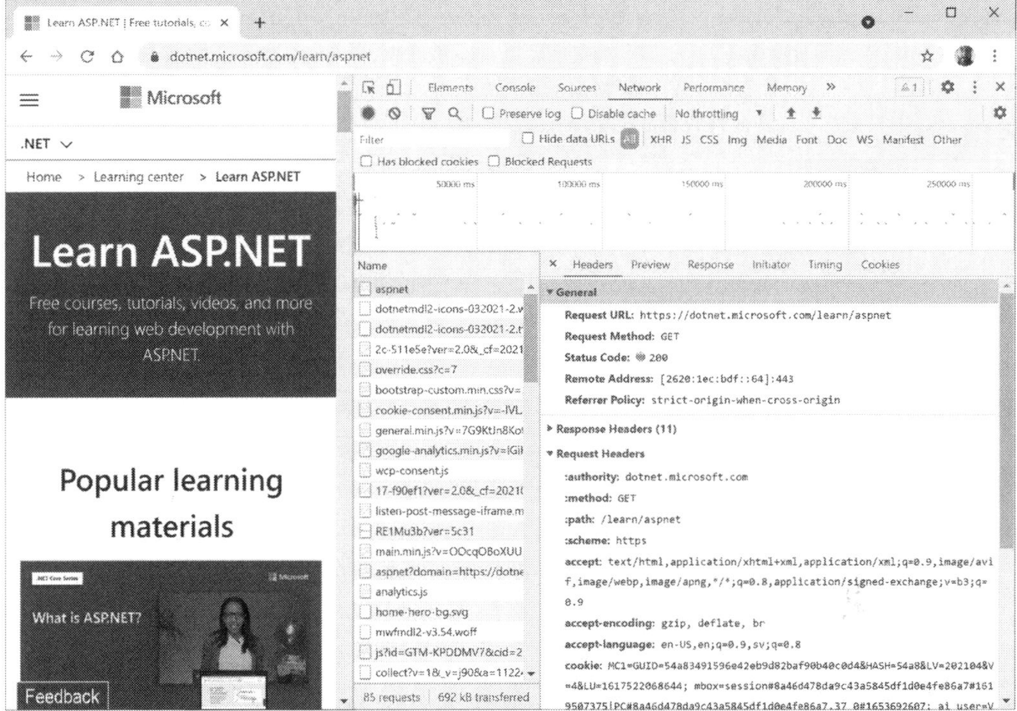

Figure 12.8: Request and response headers

Note the following aspects:

- **Request Method** is `GET`. Other HTTP methods that you could see here include `POST`, `PUT`, `DELETE`, `HEAD`, and `PATCH`.
- **Status Code** is `200` OK. This means that the server found the resource that the browser requested and has returned it in the body of the response. Other status codes that you might see in response to a `GET` request include `301 Moved Permanently`, `400 Bad Request`, `401 Unauthorized`, and `404 Not Found`.
- **Request Headers** sent by the browser to the web server include:
    - `accept`, which lists what formats the browser accepts. In this case, the browser is saying it understands HTML, XHTML, XML, and some image formats, but it will accept all other files (`*/*`). Default weightings, also known as quality values, are `1.0`. XML is specified with a quality value of `0.9`, so it is less preferable than HTML or XHTML. All other file types are given a quality value of `0.8` so are least preferred.
    - `accept-encoding`, which lists what compression algorithms the browser understands – in this case, GZIP, DEFLATE, and Brotli.

    - `accept-language`, which lists the human languages it would prefer the content to use – in this case, US English, which has a default quality value of `1.0`; any dialect of English, which has an explicitly specified quality value of `0.9`; and then any dialect of Swedish, which has an explicitly specified quality value of `0.8`.
  - **Response Headers** (`content-encoding`), which tells me that the server has sent back the HTML web page response compressed using the `gzip` algorithm, as it knows that the client can decompress that format. (This is not visible in *Figure 12.8* because there is not enough space to expand the **Response Headers** section.)

7. Close Chrome.

Understanding client-side web development technologies

When building websites, a developer needs to know more than just C# and .NET. On the client (that is, in the web browser), you will use a combination of the following technologies:

- **HTML5**: This is used for the content and structure of a web page.
- **CSS3**: This is used for the styles applied to elements on the web page.
- **JavaScript**: This is used to code any business logic needed on the web page, for example, validating form input or making calls to a web service to fetch more data needed by the web page.

Although HTML5, CSS3, and JavaScript are the fundamental components of frontend web development, there are many additional technologies that can make frontend web development more productive, including:

- **Bootstrap**, the world's most popular frontend open source toolkit
- **SASS** and **LESS**, CSS preprocessors for styling
- Microsoft's **TypeScript** language for writing more robust code
- JavaScript libraries such as **Angular**, **jQuery**, **React**, and **Vue**

All these higher-level technologies ultimately translate or compile to the underlying three core technologies, so they work across all modern browsers.

As part of the build and deploy process, you will likely use technologies such as:

- **Node.js**, a framework for server-side development using JavaScript
- **Node Package Manager** (**npm**) and **Yarn**, both client-side package managers
- **webpack**, a popular module bundler and a tool for compiling, transforming, and bundling website source files

Practicing and exploring

Test your knowledge and understanding by answering some questions and exploring this chapter's topics with deeper research.

Exercise 12.1 – Online material

Online materials can be extra content written by me for this book, or it can be references to content created by Microsoft or third parties.

One of the best sites for learning client-side web development is W3Schools, found at the following link: `https://www.w3schools.com/`.

A summary of what's new with ASP.NET Core 9 can be found at the following link:

`https://learn.microsoft.com/en-us/aspnet/core/release-notes/aspnetcore-9.0`

Microsoft's official recommendation for how to choose an ASP.NET Core web UI can be found here:

`https://learn.microsoft.com/en-us/aspnet/core/tutorials/choose-web-ui`

Microsoft's official recommendation for ASP.NET Core best practices can be found here:

`https://learn.microsoft.com/en-us/aspnet/core/fundamentals/best-practices`

Exercise 12.2 – Practice exercises

Practice exercises go deeper into the topics for this chapter.

Troubleshooting web development

It is common to have temporary issues with web development because there are so many moving parts. Sometimes, variations of the classic "turn it off and on again" can fix these!

1. Delete the project's `bin` and `release` folders.
2. Restart the web server to clear its caches.
3. Reboot the computer.

Exercise 12.3 – Test your knowledge

Answer the following questions:

1. What was the name of Microsoft's first dynamic server-side-executed web page technology, and why is it still useful to know this history today?
2. What are the names of two Microsoft web servers?
3. What are some differences between a microservice and a nanoservice?
4. What is Blazor?
5. What was the first version of ASP.NET Core that could not be hosted on .NET Framework?
6. What is a user agent?
7. What impact does the HTTP request-response communication model have on web developers?
8. Name and describe four components of a URL.
9. What capabilities do developer tools give you?
10. What are the three main client-side web development technologies, and what do they do?

Know your webbreviations

What do the following web abbreviations stand for, and what do they do?

1. URI
2. URL
3. WCF
4. TLD
5. API
6. SPA
7. CMS
8. Wasm
9. SASS
10. REST

Exercise 12.4 – Explore topics

Use the links on the following page to learn more details about the topics covered in this chapter:

`https://github.com/markjprice/cs13net9/blob/main/docs/book-links.md#chapter-12---introducing-web-development-using-aspnet-core`

Summary

In this chapter, you have:

- Been introduced to some of the app models that you can use to build websites and web services using C# and .NET
- Created class libraries to define an entity data model for working with the Northwind database, using SQLite, SQL Server, or both

In the following chapters, you will learn the details of how to build the following:

- Simple websites using static HTML pages and dynamically generated Blazor static SSR pages
- Blazor user interface components that can be hosted on a web server, in the browser, or on hybrid web-native mobile and desktop apps
- Web services that can be called by any platform that can make an HTTP request, as well as client websites that call those web services

Learn more on Discord

To join the Discord community for this book – where you can share feedback, ask questions to the author, and learn about new releases – follow the QR code below:

`https://packt.link/csharp13dotnet9`

Leave a Review!

Thank you for purchasing this book from Packt Publishing—we hope you enjoy it! Your feedback is invaluable and helps us improve and grow. Once you've completed reading it, please take a moment to leave an Amazon review; it will only take a minute, but it makes a big difference for readers like you.

Scan the QR or visit the link to receive a free ebook of your choice.

`https://packt.link/NzOWQ`

13

Building Websites Using ASP.NET Core

This chapter is about building websites with a modern HTTP architecture on the server side using ASP.NET Core.

ASP.NET Core is a free, open-source, and cross-platform web framework developed by Microsoft for building modern, cloud-based, and internet-connected applications. It is a significant redesign of the previous ASP.NET framework, focusing on flexibility, modularity, and performance.

ASP.NET Core provides a unified programming model for building web **user interface** (**UI**) and web APIs, which means you can use the same framework for both server-side and client-side logic.

ASP.NET Core is optimized for high performance and can handle a significant number of requests per second. It has a lightweight, modular architecture that allows developers to include only the necessary components, reducing the overall footprint.

You will learn about building simple websites using the Blazor static **Server-Side Rendering** (**SSR**) feature introduced with ASP.NET Core 8.

This chapter covers the following topics:

- Exploring ASP.NET Core
- History of Blazor
- Exploring Blazor static SSR
- Using Entity Framework Core with ASP.NET Core

Exploring ASP.NET Core

Before we explore an example ASP.NET Core project, let's review its architecture.

Architecture of ASP.NET Core

The architecture of ASP.NET Core is composed of several key components:

- **Middleware**: Middleware components are used to handle requests and responses. They are chained together in a pipeline, where each piece of middleware processes the request and passes it to the next component in the chain. In an optional online section for this chapter, you can learn how to customize this pipeline.
- **Routing**: The routing component is responsible for mapping incoming requests to appropriate endpoints and Blazor components.
- **Dependency Injection (DI) container**: This is built-in and used for managing service object lifetimes and dependencies.
- **Configuration**: ASP.NET Core uses a flexible configuration system that supports various sources like JSON files, environment variables, and command-line arguments.
- **Logging**: The framework has a logging API that supports various logging providers, allowing you to capture and store log data in a consistent manner.

Now we will create an empty ASP.NET Core project and explore how to enable it to serve simple web pages.

Creating an empty ASP.NET Core project

We will create an ASP.NET Core project that will show a list of suppliers from the `Northwind` database.

The `dotnet` tool has many project templates that do a lot of work for you, but it can be difficult to know which works best for a given situation, so we will start with the empty website project template and then add features step by step so that you can understand all the pieces:

1. Use your preferred code editor to open the `ModernWeb` solution and then add a new project, as defined in the following list:
   - **Project template: ASP.NET Core Empty [C#]** / `web`. For Rider, select the project template named **ASP.NET Core Web Application**, and then set **Type** to **Empty**.
   - **Project file and folder**: `Northwind.Web`.
   - **Solution file and folder**: `ModernWeb`.
   - For Visual Studio, leave all other options as their defaults.

     For example:

     - **Framework**: .NET 9.0 (Standard Term Support)
     - **Configure for HTTPS**: Selected
     - **Enable Docker**: Cleared
     - **Do not use top-level statements**: Cleared
   - For VS Code and the `dotnet new web` command, the defaults are the options we want. In future projects, if you want to change from top-level statements to the old `Program` class style, then specify the switch `--use-program-main`.

Summaries of Visual Studio and `dotnet new` options when creating new projects can be found in the GitHub repository at the following link: https://github.com/markjprice/cs13net9/blob/main/docs/ch01-project-options.md.

2. Build the `Northwind.Web` project.
3. In `Northwind.Web.csproj`, note that the project is like a class library except that the SDK is `Microsoft.NET.Sdk.Web`, as shown highlighted in the following markup:

```
<Project Sdk="Microsoft.NET.Sdk.Web">

  <PropertyGroup>
    <TargetFramework>net9.0</TargetFramework>
    <Nullable>enable</Nullable>
    <ImplicitUsings>enable</ImplicitUsings>
  </PropertyGroup>

</Project>
```

4. Add an element to import the `System.Console` class globally and statically, as shown in the following markup:

```
<ItemGroup Label="To simplify use of WriteLine.">
  <Using Include="System.Console" Static="true" />
</ItemGroup>
```

5. If you are using Visual Studio, in **Solution Explorer**, toggle **Show All Files**. If you are using Rider, then move the cursor over the **Solution** pane, and then click the "eyeball" icon.
6. Expand the `obj` folder, expand the `Debug` folder, expand the `net9.0` folder, and select the `Northwind.Web.GlobalUsings.g.cs` file. Note how the implicitly imported namespaces include all the ones for a console app or class library, as well as some ASP.NET Core ones, such as `Microsoft.AspNetCore.Builder`, as shown in the following code:

```
// <autogenerated />
global using global::Microsoft.AspNetCore.Builder;
global using global::Microsoft.AspNetCore.Hosting;
global using global::Microsoft.AspNetCore.Http;
global using global::Microsoft.AspNetCore.Routing;
global using global::Microsoft.Extensions.Configuration;
global using global::Microsoft.Extensions.DependencyInjection;
global using global::Microsoft.Extensions.Hosting;
global using global::Microsoft.Extensions.Logging;
global using global::System;
global using global::System.Collections.Generic;
```

```
global using global::System.IO;
global using global::System.Linq;
global using global::System.Net.Http;
global using global::System.Net.Http.Json;
global using global::System.Threading;
global using global::System.Threading.Tasks;
global using static global::System.Console;
```

7. Close the file and collapse the `obj` folder.
8. In the `Northwind.Web` project/folder, expand the folder named `Properties`, open the file named `launchSettings.json`, and note the profiles named `http` and `https`. They have randomly assigned port numbers that you will change in the next step so, for now, just note their locations, as shown highlighted in the following configuration:

```
{
  "$schema": "http://json.schemastore.org/launchsettings.json",
  "iisSettings": {
    "windowsAuthentication": false,
    "anonymousAuthentication": true,
    "iisExpress": {
      "applicationUrl": "http://localhost:14842",
      "sslPort": 44352
    }
  },
  "profiles": {
    "http": {
      "commandName": "Project",
      "dotnetRunMessages": true,
      "launchBrowser": true,
      "applicationUrl": "http://localhost:5122",
      "environmentVariables": {
        "ASPNETCORE_ENVIRONMENT": "Development"
      }
    },
    "https": {
      "commandName": "Project",
      "dotnetRunMessages": true,
      "launchBrowser": true,
      "applicationUrl": "https://localhost:7155;http://localhost:5122",
      "environmentVariables": {
        "ASPNETCORE_ENVIRONMENT": "Development"
      }
```

```
    },
    "IIS Express": {
      "commandName": "IISExpress",
      "launchBrowser": true,
      "environmentVariables": {
        "ASPNETCORE_ENVIRONMENT": "Development"
      }
    }
  }
}
```

The `launchSettings.json` file is only for use during development. It has no effect on the build process. It is not deployed with the compiled website project, so it has no effect on the production runtime. It is only processed by code editors like Visual Studio and Rider to set up environment variables and define URLs for the web server to listen on when the project is started by a code editor.

9. For the `https` profile, for its `applicationUrl`, change the assigned port numbers for `http` to `5130` and `https` to `5131`, and swap the order so `http` is first in the list and will be used by default, as shown highlighted in the following markup:

```
"applicationUrl": "http://localhost:5130;https://localhost:5131",
```

The `http` and `https` launch profiles have a `commandName` of `Project`, meaning they use the web server configured in the project to host the website, which is Kestrel by default. There is also a profile and settings for **IIS**, which is a Windows-only web server. In this book, we will only use Kestrel as the web server since it is cross-platform. To declutter your `launchSettings.json` file, you could even delete the `iisSettings` and `IIS Express` sections.

10. Save the changes to `launchSettings.json` and close the file.
11. In `Program.cs`, note the following:
    - An ASP.NET Core project is like a top-level console app, with a hidden `<Main>$` method as its entry point that has an argument passed using the name `args`.
    - It calls `WebApplication.CreateBuilder`, which creates a host for the website using defaults for a web host that is then built.
    - The website will respond to all HTTP `GET` requests to the root path of the website with this plain text: `Hello World!`.
    - The call to the `Run` method is a blocking call, so the hidden `<Main>$` method does not return until the web server stops running.

The contents of `Program.cs` are shown in the following code:

```
var builder = WebApplication.CreateBuilder(args);
var app = builder.Build();

app.MapGet("/", () => "Hello World!");

app.Run();
```

12. After the existing call to `MapGet`, add two calls to define an endpoint that returns an anonymous C# object as JSON, and an endpoint that returns a web page, as shown in the following code:

```
app.MapGet("/data", () => Results.Json(new
{
  firstName = "John",
  lastName = "Doe",
  age = 30
}));

app.MapGet("/welcome", () => Results.Content(
  content: $"""
  <!doctype html>
  <html lang="en">
  <head>
    <title>Welcome to Northwind Web!</title>
  </head>
  <body>
    <h1>Welcome to Northwind Web!</h1>
  </body>
  </html>
  """,
  contentType: "text/html"));
```

As well as `Json` and `Content`, the `Results` class has many more methods, like `BadRequest`, `Bytes`, `Created`, `File`, `NoContent`, `NotFound`, `Ok`, `Problem`, `Redirect`, and so on. You can learn more at the following link: `https://learn.microsoft.com/en-us/dotnet/api/microsoft.aspnetcore.http.results`.

13. At the bottom of `Program.cs`, add a comment to explain the `Run` method and a statement to write a message to the console after `Run` and, therefore, after the web server has stopped, as shown highlighted in the following code:

```
// Start the web server, host the website, and wait for requests.
app.Run(); // This is a thread-blocking call.
WriteLine("This executes after the web server has stopped!");
```

Testing and securing the website

We will now test the functionality of the ASP.NET Core Empty website project. We will also enable the encryption of all traffic between the browser and web server for privacy by switching from HTTP to HTTPS. HTTPS is the secure encrypted version of HTTP.

- For Visual Studio:
    1. In the toolbar, make sure that the **https** profile is selected (rather than **http, IIS Express**, or **WSL**), and then change **Web Browser** to **Google Chrome**, as shown in *Figure 13.1*:

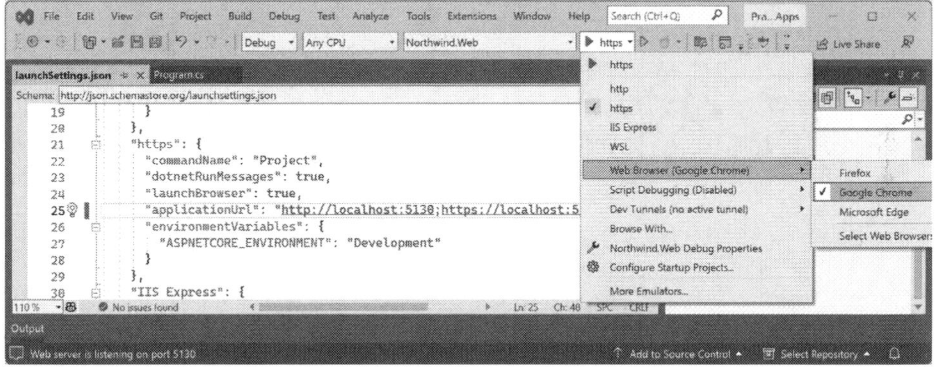

Figure 13.1: Selecting the https profile with its Kestrel web server in Visual Studio

    2. Navigate to **Debug | Start Without Debugging....**
    3. In Windows, if you see a **Windows Security Alert** saying **Windows Defender Firewall has blocked some features of this app**, then click the **Allow access** button.
    4. The first time you start a secure website, you might be notified that your project is configured to use SSL, and to avoid warnings in the browser, you can choose to trust the self-signed certificate that ASP.NET Core has generated. Click **Yes**. When you see the **Security Warning** dialog box, click **Yes** again.
- For VS Code, enter the command to start the project with the `https` profile, like this: `dotnet run --launch-profile https`. Then start Chrome.
- For Rider:
    1. Navigate to **Run | Edit Configurations....**
    2. In the **Run/Debug Configurations** dialog box, select **Northwind.Web: https**.
    3. At the bottom of the dialog box, to the right of the **After launch** check box, select **Chrome** and then click **OK**.
    4. Navigate to **Run | Run 'Northwind.Web: https'**.

1. In either Visual Studio's command prompt window or VS Code's terminal, note the following, as shown in the following output:

    - The web server has started listening on the ports we assigned for HTTP and HTTPS.
    - You can press *Ctrl* + *C* to shut down the Kestrel web server.
    - The hosting environment is `Development`.

```
info: Microsoft.Hosting.Lifetime[14]
  Now listening on: http://localhost:5130
info: Microsoft.Hosting.Lifetime[14]
  Now listening on: https://localhost:5131
info: Microsoft.Hosting.Lifetime[0]
  Application started. Press Ctrl+C to shut down.
info: Microsoft.Hosting.Lifetime[0]
  Hosting environment: Development
info: Microsoft.Hosting.Lifetime[0]
  Content root path: C:\cs13net9\ModernWeb\Northwind.Web
```

Visual Studio will also start your chosen browser automatically and navigate to the first URL. If you are using VS Code, you will have to start Chrome manually.

2. Leave the Kestrel web server running in the command prompt or terminal.
3. In Chrome, show **Developer Tools**, and click the **Network** tab.
4. Request the home page for the website project:

    - If you are using Visual Studio and Chrome launched automatically with the URL already entered for you, then click the **Reload this page** button or press *F5*.
    - If you are using VS Code and the command prompt or terminal, then in the Chrome address bar, manually enter the address `http://localhost:5130/`.

5. In the **Network** tab, click **localhost**, and note the response is **Hello World!** in plain text from the cross-platform Kestrel web server, as shown in *Figure 13.2*:

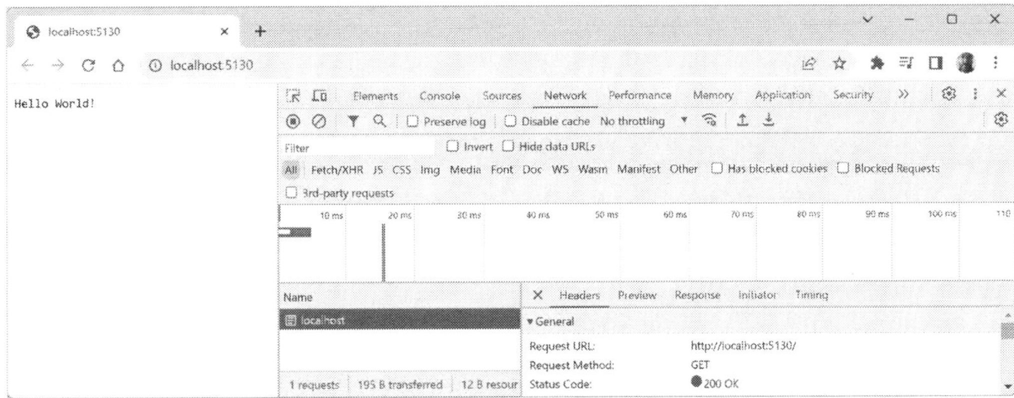

Figure 13.2: Plain text response from the website project

Browsers like Chrome might also request a `favicon.ico` file to show in their browser window or tab, but this file does not exist in our project, so it shows as a `404 Not Found` error. If this annoys you, then you can generate a `favicon.ico` file for free at the following link and put it in the project folder: `https://favicon.io/`. On a web page, you can also specify one in the meta tags, for example, a blank one using Base64 encoding, as shown in the following markup:

```
<link rel="icon" href="data:;base64,iVBORw0KGgo=">
```

6. Enter the address `https://localhost:5131/` and note that if you are not using Visual Studio or if you clicked **No** when prompted to trust the SSL certificate, then the response is a privacy error. You will see this error when you have not configured a certificate that the browser can trust to encrypt and decrypt HTTPS traffic (if you do not see this error, it is because you have already configured a certificate). In a production environment, you would want to pay a company such as Verisign for an SSL certificate because they provide liability protection and technical support. During development, you can tell your OS to trust a temporary development certificate provided by ASP.NET Core.

Visual Studio will also start your chosen browser automatically and navigate to the first URL. If you are using VS Code, you will have to start Chrome manually.

7. Enter the address `https://localhost:5131/data` and note that some JSON is returned.
8. Enter the address `https://localhost:5131/welcome` and note that a web page is returned.

9. At the command prompt or terminal, press *Ctrl* + *C* to shut down the web server, and note the message that is written, as shown highlighted in the following output:

```
info: Microsoft.Hosting.Lifetime[0]
      Application is shutting down...
This executes after the web server has stopped!
C:\cs13net9\ModernWeb\Northwind.Web\bin\Debug\net9.0\Northwind.Web.exe
(process 19888) exited with code 0.
```

10. If you need to trust a local self-signed SSL certificate, then at the command line or terminal, enter the following command: `dotnet dev-certs https --trust`.
11. Note the message **Trusting the HTTPS development certificate was requested.** You might be prompted to enter your password, and a valid HTTPS certificate may already be present.

Enabling stronger security and redirecting to a secure connection

It is good practice to enable stricter security and automatically redirect requests for HTTP to HTTPS.

Good Practice: An optional but recommended security enhancement is **HTTP Strict Transport Security (HSTS)**, which you should always enable. If a website specifies it and a browser supports it, then it forces all communication over HTTPS and prevents the visitor from using untrusted or invalid certificates.

Let's do that now:

1. In `Program.cs`, after the statement that builds the `app`, add a region and an `if` statement to enable HSTS when *not* in development, and redirect HTTP requests to HTTPS, as shown highlighted in the following code:

```
var builder = WebApplication.CreateBuilder(args);
var app = builder.Build();

#region Configure the HTTP pipeline and routes

if (!app.Environment.IsDevelopment())
{
  app.UseHsts();
}

app.UseHttpsRedirection();

app.MapGet("/", () => "Hello World!");

// Other MapGet calls.
```

```
#endregion

// Start the web server, host the website, and wait for requests.
app.Run(); // This is a thread-blocking call.
WriteLine("This executes after the web server has stopped!");
```

2. Start the `Northwind.Web` website project without debugging using the `https` launch profile.
3. If Chrome is still running, close and restart it.
4. In Chrome, show **Developer Tools**, and click the **Network** tab.
5. Enter the address `http://localhost:5130/`, and note how the server responds with a **307 Temporary Redirect** to `https://localhost:5131/`, and that the certificate is valid and trusted, as shown in *Figure 13.3*:

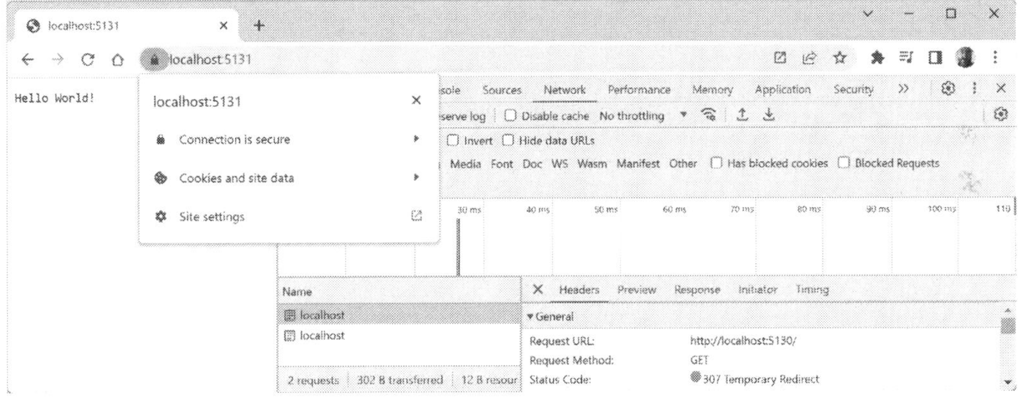

Figure 13.3: The connection is now secured using a valid certificate and a 307 redirect

6. Close Chrome and shut down the web server.

Good Practice: Remember to shut down the Kestrel web server by switching to the command prompt or terminal and pressing *Ctrl* + *C* whenever you have finished testing a website.

Controlling the hosting environment

In ASP.NET Core 5 and earlier, the project template sets a rule to say that while in development mode, any unhandled exceptions will be shown in the browser window for the developer to see the details of the exception, as shown in the following code:

```
if (app.Environment.IsDevelopment())
{
  app.UseDeveloperExceptionPage();
}
```

With ASP.NET Core 6 and later, this code is configured by default by the `WebApplication` builder and therefore executed automatically so it is no longer needed in the project template `Program.cs` source code.

How does ASP.NET Core know when we are running in development mode so that the `IsDevelopment` method returns `true`, and this extra code executes to set up the developer exception page? Let's find out.

ASP.NET Core can read from settings files and environment variables to determine what hosting environment to use, for example, `DOTNET_ENVIRONMENT` or `ASPNETCORE_ENVIRONMENT`.

You can override these settings during local development:

1. In the `Northwind.Web` folder, expand the folder named `Properties`, and open the file named `launchSettings.json`. Note that the `https` launch profile sets the environment variable for the hosting environment to `Development`, as shown highlighted in the following configuration:

```
"https": {
  "commandName": "Project",
  "dotnetRunMessages": true,
  "launchBrowser": true,
  "applicationUrl": "https://localhost:5131;http://localhost:5130",
  "environmentVariables": {
    "ASPNETCORE_ENVIRONMENT": "Development"
  }
},
```

2. Change the `ASPNETCORE_ENVIRONMENT` environment variable from `Development` to `Production`.
3. If you are using Visual Studio, optionally, change `launchBrowser` to `false` to prevent Visual Studio from automatically launching a browser. This setting is ignored when you start a website project using `dotnet run` or Rider.
4. In `Program.cs`, modify the first `MapGet` statement to return the name of the environment instead of `Hello World`, as shown highlighted in the following code:

```
app.MapGet("/", () =>
  $"Environment is {app.Environment.EnvironmentName}");
```

5. Start the website project using the `https` launch profile and note that the hosting environment is `Production`, as shown in the following output:

```
info: Microsoft.Hosting.Lifetime[0]
  Hosting environment: Production
```

6. In Chrome, note that the plain text is `Environment is Production`.
7. Shut down the web server.
8. In `launchSettings.json`, change the environment variable back to `Development`, and save the changes to the file.

More Information: You can learn more about environments at the following link: https://learn.microsoft.com/en-us/aspnet/core/fundamentals/environments.

Enabling a website to serve static content

A website that only ever returns a single plain text message isn't very useful!

At a minimum, it ought to return static HTML pages, CSS that the web pages will use for styling, and any other static resources, such as images and videos.

By convention, these files should be stored in a directory named `wwwroot` to keep them separate from the dynamically executing parts of your website project.

Creating a folder for static files and a web page

You will now create a folder for your static website resources and a basic index page that uses Bootstrap for styling:

1. In the `Northwind.Web` project/folder, create a folder named `wwwroot`. Note that Visual Studio recognizes it as a special type of folder by giving it a globe icon, 🌐.
2. In the `wwwroot` folder, add a new file named `index.html`. (In Visual Studio, the project item template is named **HTML Page**.)
3. In `index.html`, modify its markup to link to CDN-hosted Bootstrap for styling, and use modern good practices such as setting the viewport, as shown in the following markup:

```
<!doctype html>
<html lang="en">
<head>
  <!-- Required meta tags -->
  <meta charset="utf-8" />
  <meta name="viewport" content=
    "width=device-width, initial-scale=1, shrink-to-fit=no" />
  <!-- Bootstrap CSS -->
  <link href="https://cdn.jsdelivr.net/npm/bootstrap@5.3.3/dist/css/
bootstrap.min.css" rel="stylesheet" integrity="sha384-QWTKZyjpPEjISv5WaRU
9OFeRpok6YctnYmDr5pNlyT2bRjXh0JMhjY6hW+ALEwIH" crossorigin="anonymous">
  <title>Welcome ASP.NET Core!</title>
</head>
<body>
  <div class="container">
    <div class="jumbotron">
      <h1 class="display-3">Welcome to Northwind B2B</h1>
      <p class="lead">We supply products to our customers.</p>
```

```
      <hr />
      <h2>This is a static HTML page.</h2>
      <p>Our customers include restaurants, hotels, and cruise lines.</p>
      <p>
        <a class="btn btn-primary"
          href="https://www.asp.net/">Learn more</a>
      </p>
    </div>
  </div>
</body>
</html>
```

Good Practice: Check out the latest version at the following link: `https://getbootstrap.com/docs/versions/`. Click the latest version to go to its **Get started with Bootstrap** page. Scroll down the page to step 2 to find the latest `<link>` element (and later in this chapter, the `<script>` element), which you can then copy and paste.

Bootstrap is the world's most popular framework for building responsive, mobile-first websites. You can read an online-only section introducing some of Bootstrap's most important features at the following link: `https://github.com/markjprice/cs13net9/blob/main/docs/ch13-bootstrap.md`.

4. In the `wwwroot` folder, add a file named `site.css` and modify its content, as shown in the following markup:

```
h1 {
  color: darkcyan;
  font-style: italic;
}
```

5. In the `wwwroot` folder, add a file named `categories.jpeg`. You can download it from the following link: `https://github.com/markjprice/cs13net9/blob/main/code/images/Categories/categories.jpeg`.
6. In the `wwwroot` folder, add a file named `about.html` and modify its content, as shown in the following markup:

```
<!doctype html>
<html lang="en">
<head>
  <meta charset="utf-8" />
  <meta name="viewport" content="width=device-width, initial-scale=1,
shrink-to-fit=no" />
```

```
  <link href="https://cdn.jsdelivr.net/npm/bootstrap@5.3.3/dist/css/
bootstrap.min.css" rel="stylesheet" integrity="sha384-QWTKZyjpPEjISv5WaRU
9OFeRpok6YctnYmDr5pNlyT2bRjXh0JMhjY6hW+ALEwIH" crossorigin="anonymous">
  <link href="site.css" rel="stylesheet" />
  <title>About Northwind Web</title>
</head>
<body>
  <div class="container">
    <div class="jumbotron">
      <h1 class="display-3">About Northwind Web</h1>
      <p class="lead">We supply products to our customers.</p>
      <img src="categories.jpeg" />
    </div>
  </div>
</body>
</html>
```

Enabling static and default files

If you were to start the website now and enter `http://localhost:5130/index.html` or `https://localhost:5131/index.html` in the address box, the website would return a `404 Not Found` error saying no web page was found. To enable the website to return static files such as `index.html`, we must explicitly configure that feature.

Even if we enable static files stored in `wwwroot`, if you were to start the website and enter `http://localhost:5130/` or `https://localhost:5131/` in the address box, the website would still return a `404 Not Found` error because the web server does not know what to return by default if no named file is requested.

You will now enable static files, explicitly configure default files like `index.html`, and change the URL path registered that returns the plain text response:

1. In `Program.cs`, add statements after enabling HTTPS redirection to enable static files and default files. Also, modify the statement that maps a `GET` request to return the plain text response containing the environment name to only respond to the URL path `/env`, as shown highlighted in the following code:

```
app.UseDefaultFiles(); // index.html, default.html, and so on.
app.MapStaticAssets(); // .NET 9 or later.
// app.UseStaticFiles(); // .NET 8 or earlier.

app.MapGet("/env", () =>
  $"Environment is {app.Environment.EnvironmentName}");
```

ASP.NET Core 9 introduced the `MapStaticAssets` method that automatically compresses the static files, which reduces bandwidth requirements. For ASP.NET Core 8 and earlier, you must call the `UseStaticFiles` method instead. You can learn more at the following link: `https://learn.microsoft.com/en-us/aspnet/core/release-notes/aspnetcore-9.0#optimizing-static-web-asset-delivery`.

Warning! The call to `UseDefaultFiles` must come before the call to `MapStaticAssets` or `UseStaticFiles`, or it will not work! You will learn more about the ordering of middleware and endpoint routing in an online section linked at the end of this chapter.

2. Start the website.
3. Start Chrome and show **Developer Tools**.
4. In Chrome, enter `http://localhost:5130/`. Note that you are redirected to the HTTPS address on port `5131`, and the `index.html` file is now returned over that secure connection because it is one of the possible default files for this website and it was the first match found in the `wwwroot` folder.

Warning! If you still see the environment name as plain text, make sure that you mapped the relative path `/env` to that endpoint instead of just `/`, otherwise that will override the default request for the `index.html` file.

5. In Chrome, enter `http://localhost:5131/about.html` and note that the web page is returned, and it requests the `site.css` file, which applies additional styles.
6. In **Developer Tools**, note the request for the Bootstrap stylesheet.
7. In Chrome, enter `http://localhost:5130/env` and note that it returns the plain text environment name as before.
8. Close Chrome and shut down the web server.

Understanding MapStaticAssets

Introduced in .NET 9, `MapStaticAssets` works by integrating build and publish-time processes to gather data about all the static resources in an application. This data is then used by the runtime library to serve these files efficiently to the browser.

While `MapStaticAssets` can often directly replace `UseStaticFiles`, it is specifically optimized for serving assets known to the app at build and publish time. For assets served from other locations, such as disk or embedded resources, `UseStaticFiles` should still be used.

`MapStaticAssets` provides the following benefits compared to `UseStaticFiles`:

- Build time compression for all the assets in the app:
    - `gzip` during development and `gzip` + `brotli` during publishing.
    - All assets are compressed with the goal of reducing the size of the assets to the minimum.
- The ETags for each resource are the Base64-encoded string of the SHA-256 hash of the content. This ensures that the browser only redownloads a file if its contents have changed.

As an example, *Table 13.1* shows the original and compressed sizes using the Fluent UI Blazor components library, with a total of 478 KB uncompressed to 84 KB compressed:

File	Original	Compressed	% Reduction
`fluent.js`	384	73	80.99%
`fluent.css`	94	11	88.30%
Total	478	84	82.43%

Table 13.1: How MapStaticAssets compresses Fluent UI Blazor components

If all web pages are static (that is, they only get changed manually by a web editor), then our website programming work is complete. But almost all websites need dynamic content, which means a web page that is generated at runtime by executing code.

The easiest way to do that is to use a feature of ASP.NET Core named Blazor static SSR. But before that, let's understand why you might see additional requests in tools like **Developer Tools** that you don't expect.

Understanding browser requests during development

In **Developer Tools**, we can see all the requests made by the browser. Some will be requests that you expect, for example:

- `localhost`: This is the request for the home page in the website project. For our current project, the address will be `http://localhost:5130/` or `https://localhost:5131/`.
- `bootstrap.min.css`: This is the request for Bootstrap's styles. We added a reference to this on the home page, so the browser then made this request for the stylesheet.

Some of the requests are made only during development and are determined by the code editor that you use. You can usually ignore them if you see them in **Developer Tools**. For example:

- `browserLink` and `aspnetcore-browser-refresh.js`: These are requests made by Visual Studio to connect the browser to Visual Studio for debugging and Hot Reload. For example, `https://localhost:5131/_vs/browserLink` and `https://localhost:5131/_framework/aspnetcore-browser-refresh.js`.
- `negotiate?requestUrl`, `connect?transport`, `abort?Transport`, and so on: These are additional requests used to connect Visual Studio with the browser.
- `Northwind.Web/`: This is a secure WebSockets request related to SignalR used to connect Visual Studio with the browser: `wss://localhost:44396/Northwind.Web/`.

Now that you have seen how to set up a basic website with support for static files like HTML web pages and CSS, let's make it more interesting by adding support for dynamically generated static web pages (once the web page arrives at the browser client, it is static).

ASP.NET Core has multiple technologies for dynamically generating static web pages, including Razor Pages and **Model View Controller** (**MVC**) Razor views. The most modern technology for dynamic web pages is Blazor static SSR pages. But let's first review Blazor and why it was created.

History of Blazor

Blazor lets you build interactive web UI components using C# instead of JavaScript. Blazor is supported on all modern browsers.

JavaScript and friends

Traditionally, any code that needs to be executed in a web browser must be written using the JavaScript programming language or a higher-level technology that **transpiles** (transforms or compiles) into JavaScript. This is because all browsers have supported JavaScript for over two decades, so it is the lowest common denominator for implementing business logic in the client.

JavaScript does have some issues, however. Although it has superficial similarities to C-style languages like C# and Java, it is actually very different once you dig beneath the surface. It is a dynamically typed pseudo-functional language that uses prototypes instead of class inheritance for object reuse. It might look human, but you will get a surprise when it's revealed to be a Skrull.

It'd be great if we could use the same language and libraries in a browser as we do on the server.

Even Blazor cannot replace JavaScript completely. For example, some parts of the browser are only accessible to JavaScript. Blazor provides an interop service so that your C# code can call JavaScript code, and vice versa. You will see this in the online-only *Interop with JavaScript* section in *Chapter 14, Building Interactive Web Components Using Blazor*.

Silverlight – C# and .NET using a plugin

Microsoft made a previous attempt at achieving this goal with a technology named Silverlight. When Silverlight 2 was released in 2008, a C# and .NET developer could use their skills to build libraries and visual components that were executed in the web browser by the Silverlight plugin.

By 2011 and Silverlight 5, Apple's success with the iPhone and Steve Jobs' hatred of browser plugins like Flash eventually led to Microsoft abandoning Silverlight since, like Flash, Silverlight is banned from iPhones and iPads.

WebAssembly – a target for Blazor

Another development in web browsers has given Microsoft the opportunity to make another attempt. In 2017, the **WebAssembly Consensus** was completed, and all major browsers now support it: Chromium (Chrome, Edge, Opera, and Brave), Firefox, and WebKit (Safari).

WebAssembly (**Wasm**) is a binary instruction format for a virtual machine that provides a way to run code written in multiple languages on the web at near-native speed. Wasm is designed as a portable target for the compilation of high-level languages like C#.

Blazor hosting models in .NET 7 and earlier

Blazor is a single programming or app model. For .NET 7 and earlier, a developer had to choose one hosting model for each project:

- A **Blazor Server** project runs on the server side, so the C# code has full access to all resources that your business logic might need without needing to supply credentials to authenticate. It uses SignalR to communicate UI updates to the client side. The server must keep a live SignalR connection to each client and track the current state of every client. This means that Blazor Server does not scale well if you need to support lots of clients. It first shipped as part of ASP.NET Core 3 in September 2019.
- A **Blazor Wasm** project runs on the client side, so the C# code only has access to resources in the browser. It must make HTTP calls (which might require authentication) before it can access resources on the server. It first shipped as an extension to ASP.NET Core 3.1 in May 2020 and was versioned 3.2 because it was a current release and therefore not covered by ASP.NET Core 3.1's Long-Term Support. The Blazor Wasm 3.2 version used the Mono runtime and Mono libraries. .NET 5 and later use the Mono runtime and the .NET libraries.
- A **.NET MAUI Blazor app**, aka **Blazor Hybrid**, project renders its web UI to a web view control using a local interop channel and is hosted in a .NET MAUI app. It is conceptually like an Electron app.

Unification of Blazor hosting models in .NET 8

With .NET 8, the Blazor team has created a unified hosting model where each individual component can be set to execute using a different rendering model:

- **SSR**: Executes code on the server side like Razor Pages and MVC do. The complete response is then sent to the browser to display to the visitor and there is no further interaction between the server and client until the browser makes a new HTTP request. As far as the browser is concerned, the web page is static just like any other HTML file.
- **Streaming rendering**: Executes code on the server side. HTML markup can be returned and displayed in the browser, and while the connection is still open, any asynchronous operations can continue to execute. When all asynchronous operations are complete, the final markup is sent by the server to update the contents of the page. This improves the experience for the visitor because they see some content like a "Loading..." message while waiting for the rest.
- **Interactive server rendering**: Executes code on the server side during live interactions, which means the code has full and easy access to server-side resources like databases. This can simplify implementing functionality. Interactive requests are made using SignalR, which is more efficient than a full request. A permanent connection is needed between the browser and server, which limits scalability. This is a good choice for intranet websites where there are a limited number of clients and high bandwidth networking.

- **Interactive Wasm rendering:** Executes code on the client side, which means the code only has access to resources within the browser. This can complicate the implementation because a callback to the server must be made whenever new data is required. A good choice for public websites where there are potentially a large number of clients and low bandwidth connections for some of them.
- **Interactive automatic rendering:** Starts by rendering on the server for faster initial display, downloads Wasm components in the background, and then switches to Wasm for subsequent interactivity.

This unified model means that, with careful planning, a developer can write Blazor components once and then choose to run them on the web server side, or the web client side, or dynamically switch. This gives the best of all worlds.

Understanding Blazor components

It is important to understand that Blazor is used to create **UI components**. Components define how to render the UI and react to user events, and can be composed, nested, and compiled into a Razor class library for packaging and distribution.

For example, to provide a UI for star ratings of products on a commerce site, you might create a component named `Rating.razor`, as shown in the following markup:

```
<div>
@for (int i = 0; i < Maximum; i++)
{
  if (i < Value)
  {
    <span class="oi oi-star-filled" />
  }
  else
  {
    <span class="oi oi-star-empty" />
  }
}
</div>

@code {
  [Parameter]
  public byte Maximum { get; set; }
  [Parameter]
  public byte Value { get; set; }
}
```

You could then use the component on a web page, as shown in the following markup:

```
<h1>Review</h1>
<Rating id="rating" Maximum="5" Value="3" />
<textarea id="comment" />
```

The markup for creating an instance of a component looks like an HTML tag, where the name of the tag is the component type. Components can be embedded in a web page using an element, for example, `<Rating Value="5" />`, or they can be routed to, like a mapped endpoint.

Instead of a single file with both markup and an `@code` block, the code can be stored in a separate code-behind file named `Rating.razor.cs`. The class in this file must be `partial` and have the same name as the component.

There are many built-in Blazor components, including ones to set elements like `<title>` in the `<head>` section of a web page, and plenty of third parties who will sell you components for common purposes.

What is the difference between Blazor and Razor?

You might wonder why Blazor components use `.razor` as their file extension. Razor is a template markup syntax that allows the mixing of HTML and C#. Older technologies that support Razor syntax use the `.cshtml` file extension to indicate the mix of C# and HTML.

Razor syntax is used for:

- ASP.NET Core MVC **views** and **partial views** that use the `.cshtml` file extension. The business logic is separated into a controller class that treats the view as a template to push the view model to, which then outputs it to a web page.
- **Razor Pages** that use the `.cshtml` file extension. The business logic can be embedded or separated into a file that uses the `.cshtml.cs` file extension. The output is a web page.
- **Blazor components** that use the `.razor` file extension. The output is rendered as part of a web page, although layouts can be used to wrap a component so it outputs as a web page, and the `@page` directive can be used to assign a route that defines the URL path to retrieve the component as a page.

Now that you understand the background of Blazor, let's see something more practical: how to add Blazor support to an existing ASP.NET Core project.

Exploring Blazor static SSR

ASP.NET Core Blazor static SSR allows a developer to easily mix C# code statements with HTML markup to make the generated web page dynamic.

Enabling Blazor static SSR

Here is a summary of the tasks that you must complete to enable Blazor and its static SSR capability in an existing ASP.NET Core project:

1. Create a `Components` folder to contain your Blazor components.
2. Create a `Components\Pages` folder to contain your Blazor page components.

3. In the Components folder, create three .razor files:
    - _Imports.razor: This file imports namespaces for all .razor files so that you do not need to import them at the top of every .razor file. At a minimum, you will want to import the namespace for Blazor routing and your local project Blazor components.
    - App.razor: This file contains HTML for the web page that will contain all your Blazor components. It also needs to reference your Blazor Routes component somewhere in the <body> of the web page.
    - Routes.razor: This file defines a <Router> component that scans the current assembly for page components and their registered routes.
4. In Components\Pages, create an Index.razor file, a Blazor page component that will be your home page shown by default by the Blazor router. It will need a directive at the top of the file to define a route for the root path: @page "/".
5. In Program.cs, you must call AddRazorComponents() to register Blazor aka Razor components (*.razor) files with ASP.NET Core's dependency services collection, and then call MapRazorComponents<App>() to map endpoints for all the found Blazor routes. You must also call UseAntiforgery() because Blazor components automatically check for anti-forgery tokens; therefore, the HTTP pipeline must enable middleware to support them.

You will now add and enable the Blazor static SSR service, and then copy and change the static HTML page into a Blazor static SSR file:

1. In the Northwind.Web project folder, create a folder named Components.
2. In the Components folder, create a folder named Pages.
3. In the Components folder, create a file named _Imports.razor.
4. In _Imports.razor, add statements to import the namespaces for Blazor components routing, and for your Northwind.Web project and its components, as shown in the following markup:

```
@using Microsoft.AspNetCore.Components.Routing
@using Northwind.Web
@using Northwind.Web.Components
```

5. In the Components folder, create a file named Routes.razor.
6. In Routes.razor, add statements to define a router that scans the current assembly for Blazor page components that have been registered with a route, as shown in the following markup:

```
<Router AppAssembly="typeof(Program).Assembly">
  <Found Context="routeData">
    <RouteView RouteData="routeData" />
  </Found>
</Router>
```

7. Copy the index.html file into the Components\Pages folder. (In Visual Studio or Rider, hold down *Ctrl* while dragging and dropping.)

8. For the file in the `Components\Pages` folder (not the original file), rename the file extension from `index.html` to `Index.razor`. Make sure the "I" is uppercase. (Blazor components must start with an uppercase letter or you will get a compiler error!)
9. In the `Components` folder, create a file named `App.razor`.
10. In `App.razor`, cut and paste the root markup, including head and body elements from `Index.razor`, and then add elements to insert a page title using `<HeadOutlet />` and for your `<Routes>` component, as shown in the following markup:

```
<!doctype html>
<html lang="en">

<head>
  <!-- Required meta tags -->
  <meta charset="utf-8" />
  <meta name="viewport" content="width=device-width, initial-scale=1,
shrink-to-fit=no" />
  <!-- Bootstrap CSS -->
  <link href="https://cdn.jsdelivr.net/npm/bootstrap@5.3.3/dist/css/
bootstrap.min.css" rel="stylesheet" integrity="sha384-QWTKZyjpPEjISv5WaRU
9OFeRpok6YctnYmDr5pNlyT2bRjXh0JMhjY6hW+ALEwIH" crossorigin="anonymous">
  <HeadOutlet />
</head>

<body>
  <Routes />
</body>

</html>
```

11. In `Index.razor`, add the `@page` directive to the top of the file and set its route to `/`, add a `<PageTitle>` component to set the web page `<title>`, remove the `<h2>` element that says that this is a static HTML page, and note the result, as shown in the following markup:

```
@page "/"
<PageTitle>Welcome to Northwind B2B</PageTitle>
<div class="container">
  <div class="jumbotron">
    <h1 class="display-3">Welcome to Northwind B2B</h1>
    <p class="lead">We supply products to our customers.</p>
    <hr />
    <p>Our customers include restaurants, hotels, and cruise lines.</p>
    <p>
      <a class="btn btn-primary"
```

```
        href="https://www.asp.net/">Learn more</a>
    </p>
  </div>
</div>
```

Warning! If you have installed ReSharper for Visual Studio, or you use Rider, then they might give warnings like "Cannot resolve symbol" in your Razor Pages, Razor views, and Blazor components. This does not always mean there is an actual problem. If the file compiles, then you can ignore their errors. Sometimes these tools get confused and needlessly worry developers.

12. In `Program.cs`, after the statement that creates the `builder`, add a statement to add ASP.NET Core Blazor components and their related services, and optionally define a `#region`, as shown in the following code:

```
#region Configure the web server host and services.

var builder = WebApplication.CreateBuilder(args);

builder.Services.AddRazorComponents();

var app = builder.Build();

#endregion
```

13. In `Program.cs`, after calling the method to use HTTPS redirection, add a statement to use anti-forgery middleware, as shown highlighted in the following code:

```
app.UseHttpsRedirection();

app.UseAntiforgery();
```

14. In `Program.cs`, at the top of the file, add a statement to import your project components, as shown in the following code:

```
using Northwind.Web.Components; // To use App.
```

15. In `Program.cs`, before the statement that maps an HTTP `GET` request for the path `/env`, add a statement to call the `MapRazorComponents` method, as shown highlighted in the following code:

```
app.MapRazorComponents<App>();

app.MapGet("/env", () =>
  $"Environment is {app.Environment.EnvironmentName}");
```

16. Start the website project using the `https` launch profile.
17. In Chrome, enter `https://localhost:5131/` and note that the element that says this is a static HTML page is gone. If it is still there, then you might have to empty the browser cache. View **Developer Tools**, click and hold on the **Reload this page** button, and then select **Empty cache and hard reload**, as shown in *Figure 13.4*:

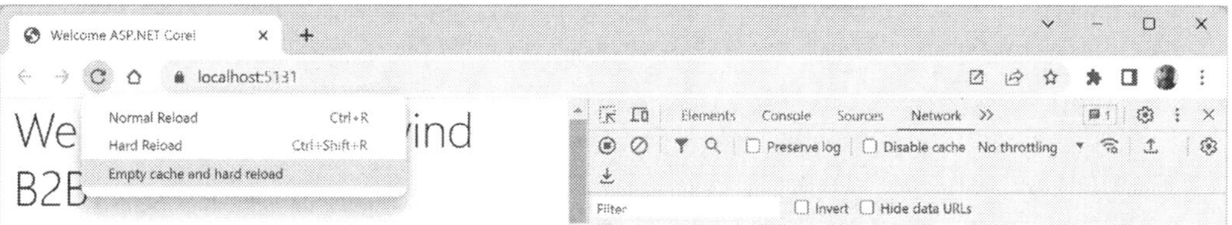

Figure 13.4: View Developer Tools, then click and hold the Reload this page button to see more commands

Adding code to a Blazor static SSR page

In the HTML markup of a web page, Razor syntax is indicated by the @ symbol. Blazor static SSR pages can be described as follows:

- Blazor static SSR pages require the `@page` directive at the top of the file with a route specified as a string value like `"/"` or `"/customers"`.

Warning! *Blazor static SSR pages* are different from *Razor Pages* and *Razor views* (used in ASP.NET Core MVC) but they share the same Razor syntax. Blazor components have the `.razor` file extension. *Razor Pages* and *Razor views* have the `.cshtml` file extension. *Razor Pages* must have the `@page` directive. *Razor views* must **not** use the `@page` directive.

- Blazor static SSR pages can optionally have an `@code` section that defines properties for storing data values, like in a class definition. You can then get the property values in the HTML.
- Blazor static SSR pages' markup can have comments using `@*` and `*@`, as shown in the following code: `@* This is a comment. *@`.

Let's now add some dynamic content to the Blazor static SSR page using an `@code` block:

1. In the `Components/Pages` folder, in `Index.cshtml`, after the `@page` directive, add an `@code` statement block to define a property to store the name of the current day as a `string` value, as shown in the following code:

```
@page "/"
@code
{
  public string? DayName { get; set; }
    = DateTime.Now.ToString("dddd");
}
```

2. In the second HTML paragraph, `<p>`, render the day name, as shown highlighted in the following markup:

```
<p>It's @DayName! Our customers include restaurants, hotels, and cruise
lines.</p>
```

3. Start the website project using the `https` profile.
4. In Chrome, if necessary, enter `https://localhost:5131/`, and note that the current day name is output on the page, as shown in *Figure 13.5*:

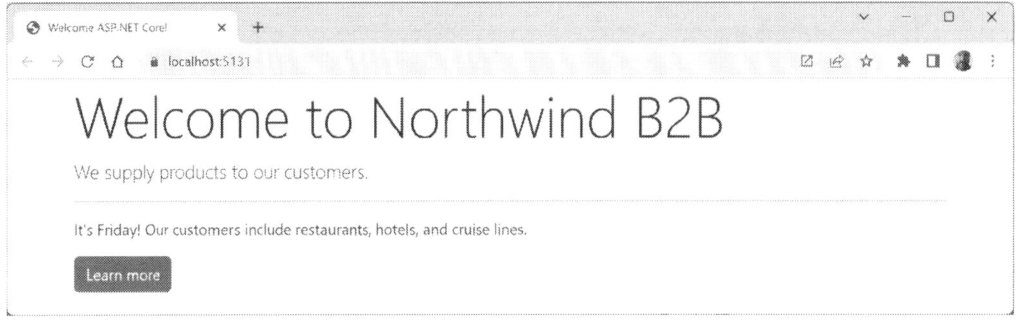

Figure 13.5: Welcome to the Northwind page showing the current day

5. In Chrome, enter `https://localhost:5131/index.html`, which exactly matches the static filename, and note that it returns the static HTML page as before.
6. In Chrome, enter `https://localhost:5131/env`, which exactly matches the endpoint route that returns plain text, and note that it returns the plain text as before.
7. Close Chrome and shut down the web server.

Using shared layouts with Blazor static SSR pages

Most websites have more than one page. The boilerplate markup for the body of the web page that is currently in `Index.razor` is minimal. Usually, there will be common markup in the body that needs to be visible on every page. If each page had to duplicate that markup, then it would become a pain to manage. So, ASP.NET Core has a feature named **layouts**.

To use layouts, we must create a Razor file to define the default layout for all Blazor components and specify it in the `Routes` component. By convention, we store it in a `Layout` folder. The name of this file can be anything, as we will specify it, but `MainLayout.razor` is good practice.

Typically, your main layout will include a navigation menu with links to all the other Blazor page components. The naming convention for this file is `NavMenu.razor`.

Let's see layouts in action:

1. In the `Components` folder, add a new folder named `Layout`.
2. In the `Layout` folder, add a file named `NavMenu.razor`.

3. In `NavMenu.razor`, modify the content, as shown in the following markup:

```
<nav class="navbar navbar-expand-lg bg-body-tertiary">
  <div class="container-fluid">
    <a class="navbar-brand" href="#">Northwind Web</a>
    <button class="navbar-toggler" type="button"
            data-bs-toggle="collapse"
            data-bs-target="#navbarNavAltMarkup"
            aria-controls="navbarNavAltMarkup"
            aria-expanded="false"
            aria-label="Toggle navigation">
      <span class="navbar-toggler-icon"></span>
    </button>
    <div class="collapse navbar-collapse"
         id="navbarNavAltMarkup">
      <div class="navbar-nav">
        <NavLink class="nav-link" href="/">
          Home
        </NavLink>
        <NavLink class="nav-link" href="suppliers">
          Suppliers
        </NavLink>
      </div>
    </div>
  </div>
</nav>
```

More Information: This markup uses a Bootstrap feature called `navbar`, which you can read about at the following link: `https://getbootstrap.com/docs/5.3/components/navbar/`.

4. In the `Layout` folder, add a file named `MainLayout.razor`.
5. In `MainLayout.razor`, modify the content, as shown in the following markup:

```
@inherits LayoutComponentBase
<div class="page">
  <div class="sidebar">
    <NavMenu />
  </div>
  <main>
    <article class="content px-4">
```

```
        @Body
      </article>
    </main>
    <!-- JavaScript to enable features like carousel -->
    <script src="https://cdn.jsdelivr.net/npm/bootstrap@5.3.0/dist/js/
bootstrap.bundle.min.js" integrity="sha384-geWF76RCwLtnZ8qwWowPQNguL3RmwH
VBC9FhGdlKrxdiJJigb/j/68SIy3Te4Bkz" crossorigin="anonymous"></script>
  </div>
```

While reviewing the preceding markup, note the following:

- `@Body` marks the insertion point for the Blazor page component being requested.
- At the bottom of the layout is a script to implement some cool features of Bootstrap that we can use later, such as a carousel of images.

6. In `Routes.razor`, add an attribute to set the class name for the default layout, as shown highlighted in the following markup:

```
<Router AppAssembly="typeof(Program).Assembly">
  <Found Context="routeData">
    <RouteView RouteData="routeData"
      DefaultLayout="typeof(Layout.MainLayout)" />
  </Found>
</Router>
```

7. Start the website using the `https` launch profile.
8. Visit it with Chrome and note that it now has a navigation menu at the top of every page, as shown in *Figure 13.6*:

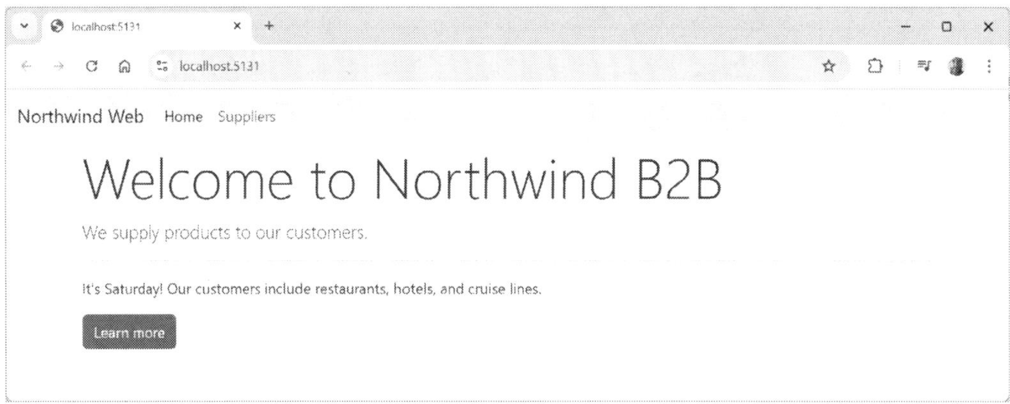

Figure 13.6: Navigation menu in the shared layout component

Warning! Clicking the navigation menu for suppliers will give a `404 Not Found` error because we have not created that Blazor page component yet.

9. Close Chrome and shut down the web server.

Creating a suppliers page

You will now create a Blazor static SSR page component that shows a list of suppliers. In the next topic, we will load the list of suppliers from a database, but for now, we will simulate that with a hardcoded array of `string` values:

1. In the `Pages` folder, add a new Blazor page component named `Suppliers.razor`. If you are using Visual Studio or Rider, then the project item template is named **Razor Component**.
2. In `Suppliers.razor`, replace the existing contents with markup to render an HTML table containing supplier company names, as shown in the following markup:

```
@page "/suppliers"
@code {
  public IEnumerable<string>? Companies { get; set; }
    = new[] { "Alpha Co", "Beta Limited", "Gamma Corp" };
}
<div class="row">
  <h1 class="display-2">Suppliers</h1>
  <table class="table">
    <thead class="thead-inverse">
      <tr>
        <th>Company Name</th>
      </tr>
    </thead>
    <tbody>
    @if (Companies is not null)
    {
      @foreach(string name in Companies)
      {
        <tr>
          <td>@name</td>
        </tr>
      }
    }
    </tbody>
  </table>
</div>
```

While reviewing the preceding markup, note the following:

- The `Suppliers` property is populated with some example supplier company names from an array of `string` values. Later, we will populate this from the `Northwind` database.
- The page outputs an HTML table with Bootstrap styles.
- The page uses Razor syntax `@if` and `@for` statements to embed C# code in HTML.
- The data rows in the table are generated by looping through the `Suppliers` property if it is not `null`.

3. Start the website using the `https` launch profile and visit it using Chrome.
4. Click on the button to learn more about suppliers, and note the table of suppliers, as shown in *Figure 13.7*:

Figure 13.7: The table of suppliers loaded from an array of strings

Configuring files included in an ASP.NET Core project

Until now, most of our projects have been simple console apps and class libraries with a few C# class files. By default, when we compiled those projects, all `.cs` files in the project folder or subfolders were automatically included in the build at compile time.

ASP.NET Core projects get more complicated. There are many more file types; some of them can be compiled at runtime instead of compile time, and some of them are just content that does not need to be compiled but does need to be deployed along with the compiled assemblies.

You can control how files are processed during a build, and which are included or excluded from a deployment, by putting elements in the project file. These are processed by **MS Build** and other tools during builds and deployments.

You declare items in the project file as child elements of an `<ItemGroup>` element. For example:

```
<--Include the greet.proto file in the build process.-->
<ItemGroup>
  <Protobuf Include="Protos\greet.proto" GrpcServices="Server" />
</ItemGroup>
```

```
<--Remove the stylecop.json file from the build process.-->
<ItemGroup>
  <None Remove="stylecop.json" />
</ItemGroup>

<--Include the stylecop.json file in the deployment.-->
<ItemGroup>
  <AdditionalFiles Include="stylecop.json" />
</ItemGroup>
```

You can have as many `<ItemGroup>` elements as you want, so it is good practice to use them to logically divide elements by type. They are merged automatically by build tools.

Usually, you manually add these elements when you know you need to use them, but unfortunately, Visual Studio and other code editors sometimes mess things up by trying to be helpful.

In one scenario, you might have added a new Blazor component in the `Components\Pages` folder named `Customers.razor`. You start the web server, but the page does not appear. Or, you are working on a GraphQL service, and you add a file named `seafoodProducts.graphql`, but when you run the GraphQL tool to auto-generate client-side proxies, it fails.

These are both common indications that your code editor has decided that the new file should not be part of the project. It has automatically added an element to the project file to remove the file from the build process without telling you.

To solve this type of problem, review the project file for unexpected entries like the following, and delete them:

```
<ItemGroup>
  <Content Remove="Components\Pages\Customers.razor" />
</ItemGroup>

<ItemGroup>
  <GraphQL Remove="seafoodProducts.graphql" />
</ItemGroup>
```

Good Practice: When using tools that automatically "fix" problems without telling you, review your project file for unexpected elements when unexpected results happen.

More Information: You can read more about managing MS Build items at the following link: `https://learn.microsoft.com/en-us/visualstudio/msbuild/msbuild-items`.

Project file build actions

As we have just seen, it is important that ASP.NET Core developers understand how project build actions affect compilation.

All files in a .NET SDK project have a build action. Most are set implicitly based on their file extension. You can override the default behavior by explicitly setting a build action. You can do this either by directly editing the `.csproj` project file or by using your code editor's **Properties** window.

Common build actions for ASP.NET Core project files are shown in *Table 13.2*:

Build action	Description
Compile or **C# compiler**	Passed to the compiler as a source file.
Content	Included as part of the website when it's deployed.
Embedded Resource	Passed to the compiler as a resource to be embedded in the assembly.
None	Not part of the build. This value can be used for documentation and other files that should not be deployed with the website.

Table 13.2: Common build actions for ASP.NET Core project files

More Information: You can learn more about **build actions** and `.csproj` entries at the following link: `https://learn.microsoft.com/en-us/visualstudio/ide/build-actions`.

Using Entity Framework Core with ASP.NET Core

Entity Framework Core is a natural way to get real data onto a website. In *Chapter 12, Introducing Modern Web Development Using .NET*, you created a pair of class libraries (one for the entity models and one for the `Northwind` database context) using SQLite as the data provider (or SQL Server). You will now use them in your website project.

Configuring Entity Framework Core as a service

Functionality that is needed by an ASP.NET Core project, such as Entity Framework Core database contexts, should be registered as a dependency service during website startup. The code in the GitHub repository solution and below uses SQLite, but you can easily use SQL Server if you prefer.

Let's see how:

1. In the `Northwind.Web` project, add a project reference to the `Northwind.DataContext` project for either SQLite or SQL Server, as shown in the following markup:

```
<!-- Change Sqlite to SqlServer if you prefer. -->
<ItemGroup>
  <ProjectReference Include="..\Northwind.DataContext.Sqlite\
Northwind.DataContext.Sqlite.csproj" />
</ItemGroup>
```

Warning! The project reference must go all on one line with no line break.

2. Build the `Northwind.Web` project.
3. In `Program.cs`, import the namespace to work with your entity model types, as shown in the following code:

```
using Northwind.EntityModels; // To use AddNorthwindContext method.
```

4. In `Program.cs`, after the statement that adds Razor components to the registered services, add a statement to register the `Northwind` database context class, as shown in the following code:

```
builder.Services.AddNorthwindContext();
```

5. In the `Components` folder, in `_Imports.razor`, import the namespace for our database context, as shown in the following code:

```
@using Northwind.EntityModels @* To use NorthwindContext. *@
```

6. In the `Components\Pages` folder, in `Suppliers.razor`, inject an instance of the Northwind context service, as shown in the following code:

```
@inject NorthwindContext _db
```

7. Change the `Companies` property to be declared as a sequence of `Supplier` objects instead of `string` values, as shown highlighted in the following code:

```
public IEnumerable<Supplier>? Companies { get; set; }
```

8. Define an `OnInitialized` method to set the `Companies` property using data from the Northwind `Suppliers` tables, as shown in the following code:

```
protected override void OnInitialized()
{
  Companies = _db.Suppliers
    .OrderBy(c => c.Country)
    .ThenBy(c => c.CompanyName);
}
```

9. Modify the markup to render multiple columns for each supplier, as shown highlighted in the following markup:

```
  <table class="table">
    <thead class="thead-inverse">
      <tr>
        <th>Company Name</th>
```

```
            <th>Country</th>
            <th>Phone</th>
          </tr>
        </thead>
        <tbody>
        @if (Model.Suppliers is not null)
        {
          @foreach(Supplier s in Companies)
          {
            <tr>
              <td>@s.CompanyName</td>
              <td>@s.Country</td>
              <td>@s.Phone</td>
            </tr>
          }
        }
        </tbody>
      </table>
    </div>
```

10. Start the website using the `https` launch profile and go to the website home page.
11. Click **Learn more about our suppliers** and note that the `Suppliers` table now loads from the database and the data is sorted first by country and then by company name, as shown in *Figure 13.8*:

Figure 13.8: The Suppliers table loaded from the Northwind database

In the next chapter, we will implement a Blazor website with interactivity so we can easily modify data as well as display it.

Practicing and exploring

Test your knowledge and understanding by answering some questions, getting some hands-on practice, and exploring this chapter's topics with deeper research.

Exercise 13.1 – online material

Online material could be material created by Microsoft or third parties, or extra content for this book.

Configuring services and the HTTP request pipeline

This online section reviews how services and the HTTP request pipeline work in more detail:

`https://github.com/markjprice/cs13net9/blob/main/docs/ch13-http-pipeline.md`

Enabling HTTP/3 and request decompression support

HTTP/3 brings benefits to all internet-connected apps, but especially mobile. I have written an online-only section introducing HTTP/3 and showing how to enable it in an ASP.NET Core project like `Northwind.Web` when targeting .NET 7 or later.

In previews of .NET 8, HTTP/3 was enabled by default, but the Microsoft team decided to revert to disabling HTTP/3 by default. They did this due to a bad experience caused by some antivirus software. Hopefully, in ASP.NET Core 9, they will resolve this issue and re-enable HTTP/3 by default. You can read more about their decision at the following link:

`https://devblogs.microsoft.com/dotnet/asp-net-core-updates-in-dotnet-8-rc-1/#http-3-disabled-by-default`

The page also includes a section about enabling request decompression support. You can read the page at the following link:

`https://github.com/markjprice/cs13net9/blob/main/docs/ch13-enabling-http3.md`

Introducing Bootstrap

Bootstrap is the world's most popular framework for building responsive, mobile-first websites. You can read an online-only section introducing some of Bootstrap's most important features at the following link:

`https://github.com/markjprice/cs13net9/blob/main/docs/ch13-bootstrap.md`

Exercise 13.2 – practice exercises

Practice exercises go deeper into the topics for this chapter.

Build a data-driven web page

Add a Razor Page to the `Northwind.Web` website that enables the user to see a list of customers grouped by country. When the user clicks on a customer record, they should then see a page showing the full contact details of that customer and a list of their orders.

My suggested solution can be found at the following links:

- `https://github.com/markjprice/cs13net9/blob/main/code/ModernWeb/Northwind.Web/Components/Pages/Customers.razor`
- `https://github.com/markjprice/cs13net9/blob/main/code/ModernWeb/Northwind.Web/Components/Pages/CustomerOrders.razor`

Build web pages for functions

Reimplement some of the console apps from earlier chapters as Razor Pages; for example, from *Chapter 4, Writing, Debugging, and Testing Functions*, provide a web UI to output times tables, calculate tax, and generate factorials and the Fibonacci sequence.

My suggested solution can be found at the following links:

- `https://github.com/markjprice/cs13net9/blob/main/code/ModernWeb/Northwind.Web/Components/Pages/Functions.razor`

Exercise 13.3 – test your knowledge

Answer the following questions:

1. List six method names that can be specified in an HTTP request.
2. List six status codes and their descriptions that can be returned in an HTTP response.
3. In ASP.NET Core, what is the `Program` class used for?
4. What does the acronym HSTS stand for and what does it do?
5. How do you enable static HTML pages for a website?
6. How do you mix C# code into the middle of HTML to create a dynamic page?
7. How can you define shared layouts for Blazor components?
8. In a Blazor project, what three files do you typically create in a `Components` folder and what do they do?
9. How do you configure an Entity Framework Core data context for use with an ASP.NET Core website?
10. What are the tasks that you must complete to enable Blazor and its static SSR capability in an existing ASP.NET Core Empty project?

Exercise 13.4 – explore topics

Use the links on the following page to learn more about the topics covered in this chapter:

`https://github.com/markjprice/cs13net9/blob/main/docs/book-links.md#chapter-13---building-websites-using-aspnet-core`

Summary

In this chapter, you learned:

- About the foundations of web development using HTTP
- How to build a simple website that returns static files
- How to use ASP.NET Core with Entity Framework Core to create web pages that are dynamically generated from information in a database

In the next chapter, you will learn how to build interactive UI using Blazor.

14

Building Interactive Web Components Using Blazor

This chapter is about using Blazor to build interactive web user interface components. You will learn how to build Blazor components that can execute their C# and .NET code on the web server or in the web browser.

When components execute on the server, Blazor uses SignalR to communicate necessary updates to the user interface in the browser.

When components execute in the browser using WebAssembly, they must make HTTP calls to interact with data on the server. You will learn more about this in *Chapter 15, Building and Consuming Web Services*.

In this chapter, we will cover the following topics:

- Reviewing the Blazor Web App project template
- Building components using Blazor
- Defining forms using the EditForm component

Reviewing the Blazor Web App project template

Before .NET 8, there were separate project templates for the different hosting models, for example, **Blazor Server App**, **Blazor WebAssembly App**, and **Blazor WebAssembly App Empty**. .NET 8 introduced a unified project template, named **Blazor Web App**, and a client-only project template, renamed **Blazor WebAssembly Standalone App**. Avoid using the other legacy project templates unless you must use older .NET SDKs.

Creating a Blazor Web App project

Let's look at the default template for a Blazor Web App project. Mostly, you will see that it is the same as an ASP.NET Core Empty template, with a few key additions:

1. Use your preferred code editor to open the `ModernWeb` solution, and then add a new project, as defined in the following list:
    - Project template: **Blazor Web App** / `blazor --interactivity Auto`
    - Solution file and folder: `ModernWeb`
    - Project file and folder: `Northwind.Blazor`
    - **Authentication type**: None
    - **Configure for HTTPS**: Selected
    - **Interactive render mode**: Auto (Server and WebAssembly)
    - **Interactivity location:** Per page/component
    - **Include sample pages**: Selected
    - **Do not use top-level statements:** Cleared

If you are using VS Code or Rider, then enter the following command at the command prompt or terminal: `dotnet new blazor --interactivity Auto -o Northwind.Blazor`

Good Practice: The default interactive render moder is **Server.** We have explicitly chosen **Auto** so that we can see both render modes in this project. We have also selected to include sample pages, which you will likely not want in a real-world project.

2. Note that two projects are created:
    - `Northwind.Blazor`: This is the main ASP.NET Core project that defines and runs static SSR, streaming, and server-side Blazor components. It also references and hosts your client-side WebAssembly Blazor components.
    - `Northwind.Blazor.Client`: This is a Blazor WebAssembly project for any client-side components that you define. In the future, it might not need to be in a separate project, but for .NET 8 and .NET 9, it still does.
3. In the `ModernWeb` folder, in `Directory.Packages.props`, add an `<ItemGroup>` to set version numbers for server-side hosting and defining Blazor WebAssembly packages, as shown in the following markup:

```
<ItemGroup Label="For Blazor.">
  <PackageVersion Include=
    "Microsoft.AspNetCore.Components.WebAssembly.Server"
    Version="9.0.0" />
  <PackageVersion Include=
    "Microsoft.AspNetCore.Components.WebAssembly"
```

```
  Version="9.0.0" />
</ItemGroup>
```

4. In `Northwind.Blazor.csproj`, note that it is identical to an ASP.NET Core project that uses the Web SDK and targets .NET 9. Note also that it references the client-side project.
5. In `Northwind.Blazor.csproj`, delete the version number for the `Microsoft.AspNetCore.Components.WebAssembly.Server` package that allows this project to host WebAssembly components, as shown in the following markup:

```
<Project Sdk="Microsoft.NET.Sdk.Web">

  <PropertyGroup>
    <TargetFramework>net9.0</TargetFramework>
    <Nullable>enable</Nullable>
    <ImplicitUsings>enable</ImplicitUsings>
  </PropertyGroup>

  <ItemGroup>
    <ProjectReference Include="..\Northwind.Blazor.Client\Northwind.
Blazor.Client.csproj" />
    <PackageReference Include="Microsoft.AspNetCore.Components.
WebAssembly.Server" />
  </ItemGroup>

</Project>
```

6. In `Northwind.Blazor.Client.csproj`, note that it is similar to an ASP.NET Core project but uses the Blazor WebAssembly SDK.
7. In `Northwind.Blazor.Client.csproj`, delete the version number for the `Microsoft.AspNetCore.Components.WebAssembly` package that allows this project to define WebAssembly components, as shown in the following markup:

```
<Project Sdk="Microsoft.NET.Sdk.BlazorWebAssembly">

  <PropertyGroup>
    <TargetFramework>net9.0</TargetFramework>
    <ImplicitUsings>enable</ImplicitUsings>
    <Nullable>enable</Nullable>
    <NoDefaultLaunchSettingsFile>true</NoDefaultLaunchSettingsFile>
    <StaticWebAssetProjectMode>Default</StaticWebAssetProjectMode>
  </PropertyGroup>

  <ItemGroup>
```

```
    <PackageReference Include="Microsoft.AspNetCore.Components.
WebAssembly" />
  </ItemGroup>

</Project>
```

8. Build the `Northwind.Blazor` and `Northwind.Blazor.Client` projects.
9. In `Northwind.Blazor`, note that `Program.cs` is almost identical to an ASP.NET Core project. One difference is the section that configures services, with its call to the `AddRazorComponents` method, which we had in our `Northwind.Web` project. This section also calls to enable interactivity on the server and client sides, as shown highlighted in the following code:

```
using Northwind.Blazor.Client.Pages;
using Northwind.Blazor.Components;

var builder = WebApplication.CreateBuilder(args);

// Add services to the container.
builder.Services.AddRazorComponents()
  .AddInteractiveServerComponents()
  .AddInteractiveWebAssemblyComponents();

var app = builder.Build();
```

10. Also note the section for configuring the HTTP pipeline, which calls the `MapRazorComponents<App>` method. This configures a root application component that will be named `App.razor`, as shown highlighted in the following code:

```
// Configure the HTTP request pipeline.
if (app.Environment.IsDevelopment())
{
  app.UseWebAssemblyDebugging();
}
else
{
  app.UseExceptionHandler(
    "/Error", createScopeForErrors: true);
  // The default HSTS value is 30 days. You may want to change this for
production scenarios, see https://aka.ms/aspnetcore-hsts.
  app.UseHsts();
}

app.UseHttpsRedirection();
```

```
app.UseAntiforgery();

app.MapStaticAssets();
app.MapRazorComponents<App>()
  .AddInteractiveServerRenderMode()
  .AddInteractiveWebAssemblyRenderMode()
  .AddAdditionalAssemblies(
    typeof(Northwind.Blazor.Client._Imports).Assembly);

app.Run();
```

11. In `Northwind.Blazor`, note that the `Components` folder and its subfolders, like `Layout` and `Pages`, use the same naming conventions that you used in your `Northwind.Web` project when enabling Blazor components.
12. In `Northwind.Blazor.Client`, in `Program.cs`, note that it creates a `WebAssemblyHostBuilder` instead of the normal `WebApplication` builder, as shown highlighted in the following code:

```
using Microsoft.AspNetCore.Components.WebAssembly.Hosting;

var builder = WebAssemblyHostBuilder.CreateDefault(args);

await builder.Build().RunAsync();
```

13. In `Northwind.Blazor.Client`, in the `Pages` folder, note that there is one Blazor component named `Counter.razor`.

Reviewing Blazor routing, layouts, and navigation

Let's review how routing is configured for this Blazor project, the layouts, and the navigation menu:

1. In the `Northwind.Blazor` project folder, in the `Components` folder, in `App.razor`, note that it defines basic HTML page markup that references a local copy of Bootstrap for styling, as well as a few Blazor-specific elements, as shown highlighted in the following markup and noted in the list after the markup:

```
<!DOCTYPE html>
<html lang="en">

<head>
  <meta charset="utf-8" />
  <meta name="viewport"
        content="width=device-width, initial-scale=1.0 " />
  <base href="/" />
  <link rel="stylesheet"
        href="@Assets["bootstrap/bootstrap.min.css"]" />
```

```
    <link rel="stylesheet" href="@Assets["app.css"]" />
    <link rel="stylesheet"
          href="@Assets["Northwind.Blazor.styles.css"]" />
    <ImportMap />
    <link rel="icon" type="image/png" href="favicon.png" />
    <HeadOutlet />
  </head>

  <body>
    <Routes />
    <script src="_framework/blazor.web.js"></script>
  </body>

  </html>
```

While reviewing the preceding markup, note the following:

- Assets are referenced using the `ComponentBase.Assets` property, which resolves the fingerprinted URL for a given asset. This should be used when you use `MapStaticAssets` middleware in `Program.cs`.
- A `<ImportMap />` Blazor component to represent an import map element (`<script type="importmap"></script>`) that defines the import map for module scripts. You can learn about import maps at the following link: `https://developer.mozilla.org/en-US/docs/Web/HTML/Element/script/type/importmap`.
- A `<HeadOutlet />` Blazor component to inject additional content into the `<head>` section. This is one of the built-in components available in all Blazor projects. For example, in a Blazor page component, use the `<PageTitle>` component to set the `<title>` of a web page.
- A `<Routes />` Blazor component to define the custom routes in this project. This component can be completely customized by the developer because it is part of the current project, in a file named `Routes.razor`.
- A script block for `blazor.web.js` that manages communication back to the server for Blazor's dynamic features, like downloading WebAssembly components in the background and later switching from server-side to client-side component execution.

2. In the `Components` folder, in `Routes.razor`, note that a `<Router>` enables routing for all Blazor components found in the current project assembly or the `Northwind.Blazor.Client` project assembly (for WebAssembly components), and if a matching route is found, then `RouteView` is executed, which sets the default layout for the component to `MainLayout` and passes any route data parameters to the component. For that component, the first `<h1>` element in it will be focused on, as shown in the following code:

```
<Router AppAssembly="@typeof(Program).Assembly"
  AdditionalAssemblies="new[] {
```

```
    typeof(Client._Imports).Assembly }">
  <Found Context="routeData">
    <RouteView RouteData="@routeData"
               DefaultLayout="@typeof(Layout.MainLayout)" />
    <FocusOnNavigate RouteData="@routeData" Selector="h1" />
  </Found>
</Router>
```

3. In the `Components` folder, in `_Imports.razor`, note that this file imports some useful namespaces for use in all your custom Blazor components, as shown in the following code:

```
@using System.Net.Http
@using System.Net.Http.Json
@using Microsoft.AspNetCore.Components.Forms
@using Microsoft.AspNetCore.Components.Routing
@using Microsoft.AspNetCore.Components.Web
@using static Microsoft.AspNetCore.Components.Web.RenderMode
@using Microsoft.AspNetCore.Components.Web.Virtualization
@using Microsoft.JSInterop
@using Northwind.Blazor
@using Northwind.Blazor.Client
@using Northwind.Blazor.Components
```

4. In the `Components\Layout` folder, in `MainLayout.razor`, note that it defines `<div>` for a sidebar containing a navigation menu that is implemented by the `NavMenu.razor` component file in this project, as well as HTML5 elements like `<main>` and `<article>` for the content, with a `<div>` to show unhandled errors, as shown in the following code:

```
@inherits LayoutComponentBase

<div class="page">
  <div class="sidebar">
    <NavMenu />
  </div>

  <main>
    <div class="top-row px-4">
      <a href="https://learn.microsoft.com/aspnet/core/"
         target="_blank">About</a>
    </div>

    <article class="content px-4">
      @Body
```

```
      </article>
    </main>
  </div>

  <div id="blazor-error-ui" data-nosnippet>
      An unhandled error has occurred.
      <a href="." class="reload">Reload</a>
      <span class="dismiss">✕</span>
  </div>
```

5. In the `Components\Layout` folder, in `MainLayout.razor.css`, note that it contains isolated CSS styles for the component. Due to the naming convention, styles defined in this file take priority over others defined elsewhere that might affect the component.

Blazor components often need to provide their own CSS to apply styling or JavaScript for activities that cannot be performed purely in C#, like access to browser APIs. To ensure that this does not conflict with site-level CSS and JavaScript, Blazor supports CSS and JavaScript isolation. If you have a component named `Home.razor`, simply create a CSS file named `Home.razor.css`. The styles defined within this file will override any other styles in the project.

6. In the `Components\Layout` folder, in `NavMenu.razor`, note that it has three menu items, **Home**, **Counter**, and **Weather**. These menu links are created by using a component named `NavLink`, as shown in the following markup:

```
  <div class="top-row ps-3 navbar navbar-dark">
    <div class="container-fluid">
      <a class="navbar-brand" href="">Northwind.Blazor</a>
    </div>
  </div>

  <input type="checkbox" title="Navigation menu"
         class="navbar-toggler" />

  <div class="nav-scrollable" onclick=
       "document.querySelector('.navbar-toggler').click()">
    <nav class="flex-column">
      <div class="nav-item px-3">
        <NavLink class="nav-link" href=""
                 Match="NavLinkMatch.All">
          <span class="bi bi-house-door-fill-nav-menu"
                aria-hidden="true"></span> Home
        </NavLink>
```

```
    </div>

    <div class="nav-item px-3">
      <NavLink class="nav-link" href="counter">
        <span class="bi bi-plus-square-fill-nav-menu"
              aria-hidden="true"></span> Counter
      </NavLink>
    </div>

    <div class="nav-item px-3">
      <NavLink class="nav-link" href="weather">
        <span class="bi bi-list-nested-nav-menu"
              aria-hidden="true"></span> Weather
      </NavLink>
    </div>
  </nav>
</div>
```

7. Note that `NavMenu.razor` has its own isolated stylesheet named `NavMenu.razor.css`.
8. In the `Components\Pages` folder, in `Home.razor`, note that it defines a component that sets the page title, and then it renders a heading and a welcome message, as shown in the following code:

```
@page "/"

<PageTitle>Home</PageTitle>

<h1>Hello, world!</h1>

Welcome to your new app.
```

9. In the `Components\Pages` folder, in `Weather.razor`, note that it defines a component that fetches weather forecasts from an injected dependency weather service and then renders them in a table, as shown in the following code:

```
@page "/weather"
@attribute [StreamRendering]

<PageTitle>Weather</PageTitle>

<h1>Weather</h1>

<p>This component demonstrates showing data.</p>
```

```
@if (forecasts == null)
{
  <p><em>Loading...</em></p>
}
else
{
  <table class="table">
    <thead>
      <tr>
        <th>Date</th>
        <th>Temp. (C)</th>
        <th>Temp. (F)</th>
        <th>Summary</th>
      </tr>
    </thead>
    <tbody>
    @foreach (var forecast in forecasts)
    {
      <tr>
        <td>@forecast.Date.ToShortDateString()</td>
        <td>@forecast.TemperatureC</td>
        <td>@forecast.TemperatureF</td>
        <td>@forecast.Summary</td>
       </tr>
    }
    </tbody>
  </table>
}

@code {
  private WeatherForecast[]? forecasts;

  protected override async Task OnInitializedAsync()
  {
    // Simulate asynchronous loading to demonstrate streaming rendering
    await Task.Delay(500);

    var startDate = DateOnly.FromDateTime(DateTime.Now);
    var summaries = new[] { "Freezing", "Bracing", "Chilly", "Cool",
      "Mild", "Warm", "Balmy", "Hot", "Sweltering", "Scorching" };
    forecasts = Enumerable.Range(1, 5).Select(index =>
```

```
        new WeatherForecast
        {
          Date = startDate.AddDays(index),
          TemperatureC = Random.Shared.Next(-20, 55),
          Summary = summaries[Random.Shared.Next(summaries.Length)]
        }).ToArray();
    }

    private class WeatherForecast
    {
      public DateOnly Date { get; set; }
      public int TemperatureC { get; set; }
      public string? Summary { get; set; }
      public int TemperatureF => 32 + (int)(TemperatureC / 0.5556);
    }
  }
```

10. In the `Northwind.Blazor.Client` project, in the `Pages` folder, in `Counter.razor`, note that a Blazor page component is defined with a route of `/counter`, a render mode that will automatically switch between Server and WebAssembly, which has a private field named `currentCount` that is incremented on each click of a button, as shown in the following markup:

```
@page "/counter"
@rendermode InteractiveAuto

<PageTitle>Counter</PageTitle>

<h1>Counter</h1>

<p role="status">Current count: @currentCount</p>

<button class="btn btn-primary"
  @onclick="IncrementCount">Click me</button>

@code {
  private int currentCount = 0;

  private void IncrementCount()
  {
    currentCount++;
  }
}
```

How to define a routable page component

To create a routable page component, add the @page directive to the top of a component's `.razor` file, as shown in the following markup:

```
@page "/customers"
```

The preceding code is the equivalent of a mapped endpoint in `Program.cs`, as shown in the following code:

```
app.MapGet("/customers", () => ...);
```

A page component can have multiple @page directives to register multiple routes, as shown in the following code:

```
@page "/weather"
@page "/forecast"
```

The `Router` component scans the assembly specifically in its `AppAssembly` parameter for Blazor components with the @page directive, registering their URL paths as endpoints.

At runtime, a page component is merged with any specific layout that you have specified in the `Routes.razor` file `<RouteView>` component. By default, the Blazor Web App project template defines `MainLayout.razor` as the layout for page components.

Good Practice: By convention, put routable page Blazor components in the `Components\Pages` folder.

How to navigate routes and pass route parameters

Microsoft provides a dependency service named `NavigationManager` that understands Blazor routing and the `NavLink` component. The `NavigateTo` method is used to go to the specified URL.

Blazor routes can include case-insensitive named parameters, and your code can most easily access the values passed by binding the parameter to a property in the code block, using the `[Parameter]` attribute, as shown in the following markup:

```
@page "/customers/{country}"

<div>Country parameter as the value: @Country</div>

@code {
  [Parameter]
  public string Country { get; set; }
}
```

The recommended way to handle a parameter that should have a default value when it is missing is to suffix the parameter with ? and use the null-coalescing operator in the `OnParametersSet` method, as shown in the following markup:

```
@page "/customers/{country?}"

<div>Country parameter as the value: @Country</div>

@code {
  [Parameter]
  public string Country { get; set; }

  protected override void OnParametersSet()
  {
    // If the automatically set property is null, then
    // set its value to USA.
    Country = Country ?? "USA";
  }
}
```

How to use the navigation link component with routes

In HTML, you use the `<a>` element to define navigation links, as shown in the following markup:

```
<a href="/customers">Customers</a>
```

In Blazor, use the `<NavLink>` component, as shown in the following markup:

```
<NavLink href="/customers">Customers</NavLink>
```

The `NavLink` component is better than an anchor element because it automatically sets its class to `active` if its `href` is a match with the current location URL. If your CSS uses a different class name, then you can set the class name in the `NavLink.ActiveClass` property.

By default, in the matching algorithm, the `href` is a path *prefix*, so if `NavLink` has an `href` of `/customers`, as shown in the preceding code example, then it would match all the following paths and set them all to have the `active` class style:

```
/customers
/customers/USA
/customers/Germany/Berlin
```

To ensure that the matching algorithm only performs matches on *all* of the text in the path (in other words, there is only a match when the whole complete text matches and not when just part of the path matches), then set the `Match` parameter to `NavLinkMatch.All`, as shown in the following code:

```
<NavLink href="/customers" Match="NavLinkMatch.All">Customers</NavLink>
```

If you set other attributes, such as `target`, they are passed through to the underlying `<a>` element that is generated.

Understanding base component classes

The `OnParametersSet` method is defined by the base class that components inherit from by default, named `ComponentBase`, as shown in the following code:

```
using Microsoft.AspNetCore.Components;

public abstract class ComponentBase
  : IComponent, IHandleAfterRender, IHandleEvent
{
  // Members not shown.
}
```

`ComponentBase` has some useful methods that you can call and override, as shown in *Table 14.1*:

Method(s)	Description
`InvokeAsync`	Call this method to execute a function on the associated renderer's synchronization context.
`OnAfterRender`, `OnAfterRenderAsync`	Override these methods to invoke code after each time the component has been rendered.
`OnInitialized`, `OnInitializedAsync`	Override these methods to invoke code after the component has received its initial parameters from its parent in the render tree.
`OnParametersSet`, `OnParametersSetAsync`	Override these methods to invoke code after the component has received parameters and the values have been assigned to properties.
`ShouldRender`	Override this method to indicate if the component should render.
`StateHasChanged`	Call this method to cause the component to re-render.

Table 14.1: Useful methods to override in ComponentBase

As you saw in *Chapter 13, Building Websites Using ASP.NET Core*, Blazor components can have shared layouts. You would create a `.razor` component file and make it explicitly inherit from `LayoutComponentBase`, as shown in the following markup:

```
@inherits LayoutComponentBase

<div>
  ...
```

```
    @Body
  ...
</div>
```

The base class has a property named `Body` that you can render in the markup at the correct place within the layout.

You can set a default layout for components in the `App.razor` file and its `Router` component. To explicitly set a layout for a component, use the `@layout` directive, as shown in the following markup:

```
@page "/customers"

@layout AlternativeLayout

<div>
  ...
</div>
```

Running the Blazor Web App project template

Now that we have reviewed the project template and the important parts that are specific to the Blazor server, we can start the website and review its behavior:

1. In the `Northwind.Blazor` project, in the `Properties` folder, in `launchSettings.json`, for the `https` profile, modify the `applicationUrl` to use port `5141` for HTTPS and port `5140` for HTTP, as shown highlighted in the following markup:

   ```
   "applicationUrl": "https://localhost:5141;http://localhost:5140",
   ```

2. Start the `Northwind.Blazor` project using the `https` launch profile.
3. Start Chrome and navigate to `https://localhost:5141/`.
4. In the left navigation menu, click **Weather** and then the **Click me** button three times, and note that the label shows 3, as shown in *Figure 14.1*:

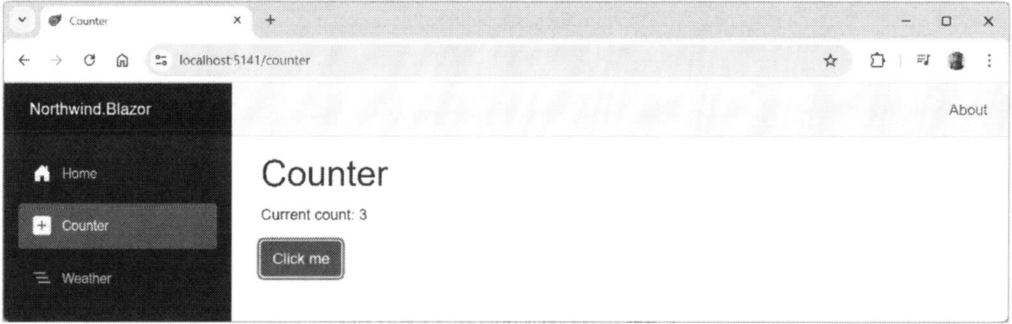

Figure 14.1: A Blazor Counter component using WebAssembly

5. In the left navigation menu, click **Weather**, and note that for half a second, a **Loading...** message appears, and then it is replaced by a table of five random weather forecasts, as shown in *Figure 14.2*:

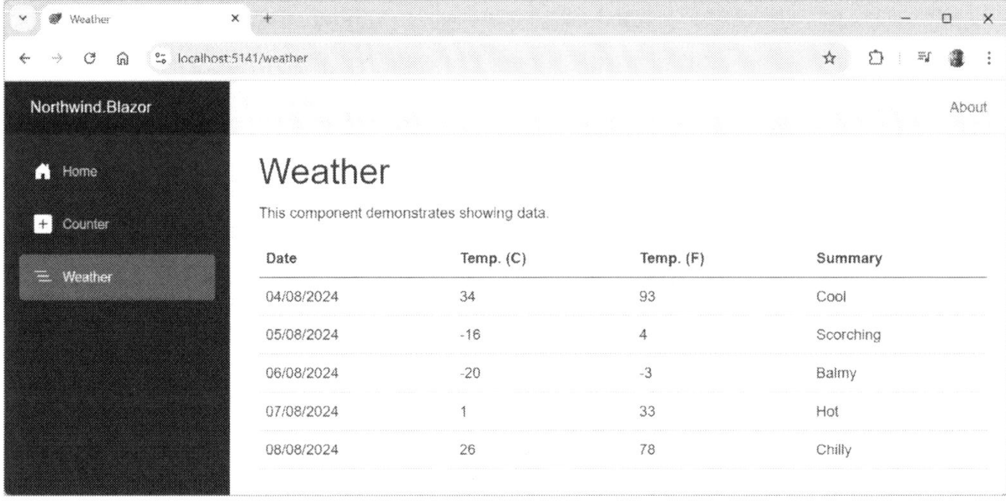

Figure 14.2: Fetching weather data into a Blazor Web App using streaming

6. Close Chrome and shut down the web server.

Now that you've reviewed the sample Blazor components in the Blazor Web App project template, let's dig in and build our own.

Building components using Blazor

In this section, we will build a component to list, create, and edit customers in the Northwind database.

We will build it over several steps:

1. Make a Blazor component that renders the name of a country set as a parameter.
2. Make it work as a routable page as well as a component.
3. Implement the functionality to perform **CRUD** (**create**, **read**, **update** and **delete**) operations on customers in a database.

Defining and testing a simple Blazor component

We will add the new component to the existing Blazor Web App project:

1. In the `Northwind.Blazor` project, in the `Components\Pages` folder, add a new file named `Customers.razor`. In Visual Studio, the project item template is named **Razor Component**. In Rider, the project item template is named **Blazor Component**.

Good Practice: Remember that Blazor component filenames must start with an uppercase letter; otherwise, you will have compile errors!

2. Add statements to output a heading for the `Customers` component, and define a code block that defines a property to store the name of a country, as shown highlighted in the following markup:

```
<h3>
  Customers @(string.IsNullOrWhiteSpace(Country)
    ? "Worldwide" : "in " + Country)
</h3>

@code {
  [Parameter]
  public string? Country { get; set; }
}
```

The @code block can be anywhere in the file. Some developers prefer it at the top so that any properties defined in it are available while typing the markup below, and they can more easily reference them.

3. In the `Components\Pages` folder, in `Home.razor`, add statements to the bottom of the file to instantiate the `Customers` component twice, once with `Germany` set as the `Country` parameter and once without setting the country, as shown in the following markup:

```
<Customers Country="Germany" />
<Customers />
```

4. Start the `Northwind.Blazor` project using the `https` launch profile.
5. Start Chrome, navigate to `https://localhost:5141/`, and note the `Customers` components, as shown in *Figure 14.3*:

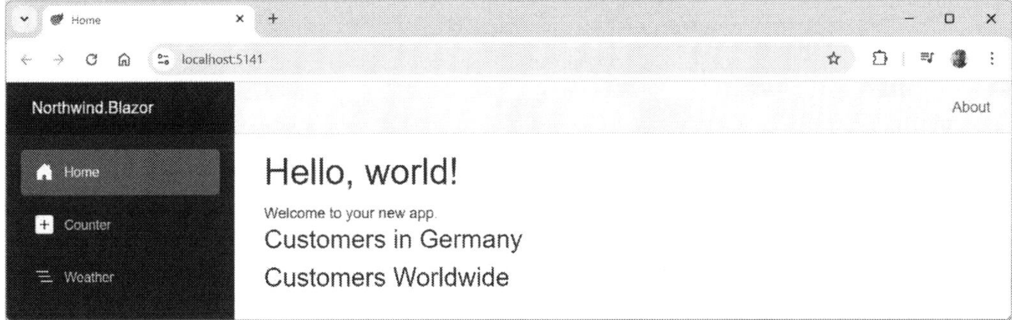

Figure 14.3: The Customers components with the Country parameter set to Germany, and not set

6. Close Chrome and shut down the web server.

Using Bootstrap icons

In the older Blazor project templates with .NET 7 and earlier, they included all Bootstrap icons. In the new project templates with .NET 8 and later, only three icons are defined using SVG. Let's see how the Blazor team defined those icons, and then we'll add some more for our own use:

1. In the `Components\Layout` folder, in `NavMenu.razor.css`, find the text `bi-house`, and note the three icons defined using SVG, as partially shown in the following code:

```
.bi-house-door-fill-nav-menu {
    background-image: url("data:image/svg+xml,...");
}

.bi-plus-square-fill-nav-menu {
    background-image: url("data:image/svg+xml,...");
}

.bi-list-nested-nav-menu {
    background-image: url("data:image/svg+xml,...");
}
```

2. In your favorite browser, navigate to `https://icon-sets.iconify.design/bi/`, and note that **Bootstrap Icons** have an MIT license and contain more than 2,000 icons.
3. There are two input boxes on the web page, one labeled **Search Icons** and one labeled **Filter Icons**. In the **Filter Icons** box, enter `globe`, and note that six globe icons are found.
4. Click the first globe, in the **CSS** section, click the **CSS** button, and note that you could click the **Copy to Clipboard** button to copy and paste the definition of this icon for use in the CSS stylesheet; however, you do not need to because I have already created a CSS file for you to use, with five icons defined for you to use in your Blazor project.
5. In your favorite browser, navigate to `https://github.com/markjprice/cs13net9/blob/main/code/ModernWeb/Northwind.Blazor/Northwind.Blazor/wwwroot/icons.css`, download the file, and save it in your own project in its `wwwroot` folder.
6. In the `Components` folder, in the `App.razor` component, in the `<head>`, add a `<link>` element to reference the `icons.css` stylesheet, as shown in the following markup:

```
<link rel="stylesheet" href="@Assets["icons.css"]" />
```

7. Save and close the file.

Making the component a routable page component

It is simple to turn this component into a routable page component with a route parameter for the country:

1. In the `Components\Pages` folder, in the `Customers.razor` component, add a statement at the top of the file to register `/customers` as its route with an optional `country` route parameter, as shown in the following markup:

```
@page "/customers/{country?}"
```

2. In the `Components\Layout` folder, in `NavMenu.razor`, at the bottom of the existing list item elements, add two list item elements for our routable page component, showing customers worldwide and in Germany that both use an icon of people, as shown in the following markup:

```
<div class="nav-item px-3">
  <NavLink class="nav-link" href="customers"
           Match="NavLinkMatch.All">
    <span class="bi bi-globe" aria-hidden="true"></span>
    Customers Worldwide
  </NavLink>
</div>

<div class="nav-item px-3">
  <NavLink class="nav-link" href="customers/Germany">
    <span class="bi bi-globe-europe-africa"
          aria-hidden="true"></span>
    Customers in Germany
  </NavLink>
</div>
```

3. In the `Components\Pages` folder, in `Home.razor`, remove the two `<Customers>` components because we can test them using their navigation menu items from now on, and we want to keep the home page as simple as possible.
4. Start the `Northwind.Blazor` project using the `https` launch profile.
5. Start Chrome and navigate to `https://localhost:5141/`.

6. In the left navigation menu, click **Customers in Germany**. Note that the country name is correctly passed to the page component and that the component uses the same layout as the other page components, like `Home.razor`. Also note the URL, `https://localhost:5141/customers/Germany`, as shown in *Figure 14.4*:

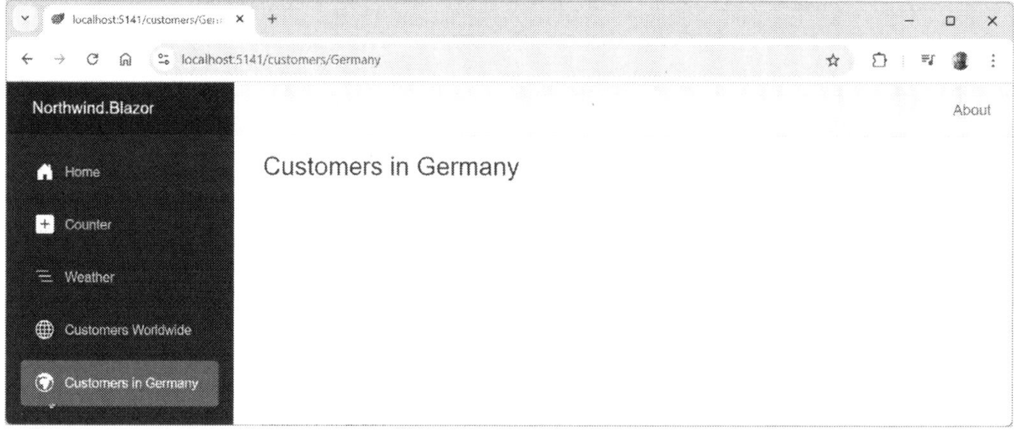

Figure 14.4: Customers in Germany

7. Close Chrome and shut down the web server.

Getting entities into a Blazor component

Now that you have seen the minimum implementation of a Blazor page component, we can add some useful functionality to it. In this case, we will use the Northwind database context to fetch customers from the database:

1. In `Northwind.Blazor.csproj`, add a reference to the Northwind database context project for either SQL Server or SQLite, and globally import the namespace to work with Northwind entities, as shown in the following markup:

```
<ItemGroup>
  <!-- change Sqlite to SqlServer if you prefer -->
  <ProjectReference Include="..\..\Northwind.DataContext.Sqlite
\Northwind.DataContext.Sqlite.csproj" />
</ItemGroup>

<ItemGroup>
  <Using Include="Northwind.EntityModels" />
</ItemGroup>
```

Warning! The relative path to the data context project is two directories up from the current project, so we must use `..\..\`.

2. Build the `Northwind.Blazor` project.
3. In `Program.cs`, before the call to `Build`, add a statement to register the Northwind database context in the dependency services collection, as shown in the following code:

```
builder.Services.AddNorthwindContext(
  relativePath: @"..\..");
```

We need to explicitly set the `relativePath` because the `Northwind.db` file is in the `ModernWeb` folder and the project runs in the `ModernWeb\Northwind.Blazor\Northwind.Blazor` folder.

Abstracting a service for a Blazor component

We could implement the Blazor component so that it directly calls the Northwind database context to fetch the customers using an entity model. This would work if the Blazor component executes on the server. However, if the component ran in the browser using WebAssembly, then it would not work.

We will now create a local dependency service to enable better reuse of the components:

1. Use your preferred coding tool to add a new project, as defined in the following list:
   - Project template: **Class Library** / `classlib`
   - Project file and folder: `Northwind.Blazor.Services`
   - Solution file and folder: `ModernWeb`
2. In the `Northwind.Blazor.Services.csproj` project file, add a project reference to the Northwind entity models library, as shown in the following markup:

```
<ItemGroup>
  <!-- change Sqlite to SqlServer if you prefer -->
  <ProjectReference Include="..\Northwind.EntityModels.Sqlite\
Northwind.EntityModels.Sqlite.csproj" />
</ItemGroup>
```

3. Build the `Northwind.Blazor.Services` project.
4. In the `Northwind.Blazor.Services` project, rename `Class1.cs` to `INorthwindService.cs`.
5. In `INorthwindService.cs`, define a contract for a local service that abstracts CRUD operations, as shown in the following code:

```
using Northwind.EntityModels; // To use Customer.

namespace Northwind.Blazor.Services;

public interface INorthwindService
{
```

```
  Task<List<Customer>> GetCustomersAsync();
  Task<List<Customer>> GetCustomersAsync(string country);
  Task<Customer?> GetCustomerAsync(string id);
  Task<Customer> CreateCustomerAsync(Customer c);
  Task<Customer> UpdateCustomerAsync(Customer c);
  Task DeleteCustomerAsync(string id);
}
```

6. In the `Northwind.Blazor.csproj` project file, add a project reference to the services class library, as shown highlighted in the following markup:

```
<ItemGroup>
  <!-- change Sqlite to SqlServer if you prefer -->
  <ProjectReference Include="..\Northwind.DataContext.Sqlite
\Northwind.DataContext.Sqlite.csproj" />
  <ProjectReference Include="..\..\Northwind.Blazor.Services\
Northwind.Blazor.Services.csproj" />
</ItemGroup>
```

7. Build the `Northwind.Blazor` project.
8. In the `Northwind.Blazor` project, add a new folder named `Services`.
9. In the `Services` folder, add a new file named `NorthwindServiceServerSide.cs`, and modify its contents to implement the `INorthwindService` interface by using the Northwind database context, as shown in the following code:

```
using Microsoft.EntityFrameworkCore; // To use ToListAsync<T>.

namespace Northwind.Blazor.Services;

public class NorthwindServiceServerSide : INorthwindService
{
  private readonly NorthwindContext _db;

  public NorthwindServiceServerSide(NorthwindContext db)
  {
    _db = db;
  }

  public Task<List<Customer>> GetCustomersAsync()
  {
    return _db.Customers.ToListAsync();
  }
```

```
  public Task<List<Customer>> GetCustomersAsync(string country)
  {
    return _db.Customers.Where(c => c.Country == country).ToListAsync();
  }

  public Task<Customer?> GetCustomerAsync(string id)
  {
    return _db.Customers.FirstOrDefaultAsync
      (c => c.CustomerId == id);
  }

  public Task<Customer> CreateCustomerAsync(Customer c)
  {
    _db.Customers.Add(c);
    _db.SaveChangesAsync();
    return Task.FromResult(c);
  }

  public Task<Customer> UpdateCustomerAsync(Customer c)
  {
    _db.Entry(c).State = EntityState.Modified;
    _db.SaveChangesAsync();
    return Task.FromResult(c);
  }

  public Task DeleteCustomerAsync(string id)
  {
    Customer? customer = _db.Customers.FirstOrDefaultAsync
      (c => c.CustomerId == id).Result;

    if (customer == null)
    {
      return Task.CompletedTask;
    }
    else
    {
      _db.Customers.Remove(customer);
      return _db.SaveChangesAsync();
    }
  }
}
```

10. In `Program.cs`, import the namespace for our service, as shown in the following code:

```
using Northwind.Blazor.Services; // To use INorthwindService.
```

11. In `Program.cs`, before the call to `Build`, add a statement to register `NorthwindServiceServerSide` as a transient service that implements the `INorthwindService` interface, as shown in the following code:

```
builder.Services.AddTransient<INorthwindService,
  NorthwindServiceServerSide>();
```

A transient service is one that creates a new instance for each request. You can read more about the different lifetimes for services at the following link: https://learn.microsoft.com/en-us/dotnet/core/extensions/dependency-injection#service-lifetimes.

12. In the `Components` folder, in `_Imports.razor`, import the namespace to work with the Northwind entities and our service so that Blazor components that we build do not need to import the namespaces individually, as shown in the following markup:

```
@using Northwind.Blazor.Services @* To use INorthwindService. *@
@using Northwind.EntityModels @* To use Northwind entities. *@
```

The `_Imports.razor` file only applies to `.razor` files. If you use code-behind `.cs` files to implement component code, then they must have namespaces imported separately or use global `usings` to implicitly import the namespaces.

13. In the `Components\Pages` folder, in `Customers.razor`, add statements to inject the service, and then use it to output a table of all customers using synchronous database operations, as shown highlighted in the following code:

```
@page "/customers/{country?}"
@inject INorthwindService _service
<h3>
  Customers @(string.IsNullOrWhiteSpace(Country)
    ? "Worldwide" : "in " + Country)
</h3>
@if (customers is null)
{
<p><em>Loading...</em></p>
}
else
{
<table class="table">
```

```
  <thead>
    <tr>
      <th>Id</th>
      <th>Company Name</th>
      <th>Address</th>
      <th>Phone</th>
      <th></th>
    </tr>
  </thead>
  <tbody>
  @foreach (Customer c in customers)
  {
    <tr>
      <td>@c.CustomerId</td>
      <td>@c.CompanyName</td>
      <td>
        @c.Address<br/>
        @c.City<br/>
        @c.PostalCode<br/>
        @c.Country
      </td>
      <td>@c.Phone</td>
      <td>
        <a class="btn btn-info" href="editcustomer/@c.CustomerId">
          <i class="bi bi-pencil"></i></a>
        <a class="btn btn-danger"
           href="deletecustomer/@c.CustomerId">
          <i class="bi bi-trash"></i></a>
      </td>
    </tr>
  }
  </tbody>
</table>
}
@code {
  [Parameter]
  public string? Country { get; set; }

  private IEnumerable<Customer>? customers;

  protected override async Task OnParametersSetAsync()
```

```
  {
    if (string.IsNullOrWhiteSpace(Country))
    {
      customers = await _service.GetCustomersAsync();
    }
    else
    {
      customers = await _service.GetCustomersAsync(Country);
    }
  }
}
```

14. Start the `Northwind.Blazor` project using the `https` launch profile.
15. Start Chrome and navigate to `https://localhost:5141/`.
16. In the left navigation menu, click **Customers in Germany**, and note that the table of customers loads from the database and renders in the web page, as shown in *Figure 14.5*:

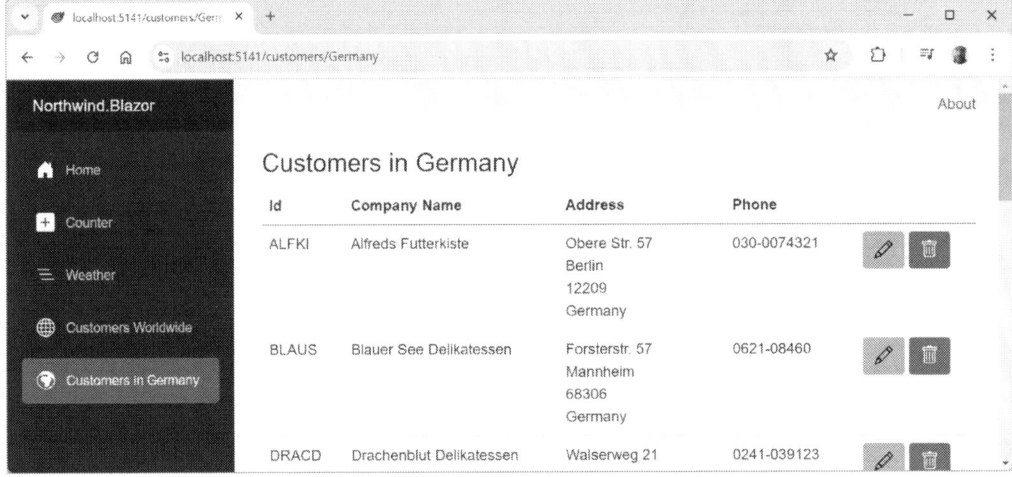

Figure 14.5: The list of customers in Germany

17. In the browser address bar, change `Germany` to `UK`, and note that the table of customers is filtered to only show UK customers.
18. In the left navigation menu, click **Customers Worldwide**, and note that the table of customers is unfiltered by country.
19. Click any of the edit or delete buttons, and note that they return a message saying **Error: 404** because we have not yet implemented that functionality. Also, note the link to edit the customer identified by the five-character identifier `ALFKI`, as shown in the following link: `https://localhost:5141/editcustomer/ALFKI`.
20. Close Chrome and shut down the web server.

Enabling streaming rendering

Now, let's improve the rendering of the table of customers by making it happen after the page has appeared to the visitor. We are already using an asynchronous operation to fetch the data, but this operation must finish before the web page response is sent back to the browser. This is why we never see the **Loading...** message on the page. To see it, we must enable streaming rendering. But if you are fetching data from a local database, it might still happen too quickly. So to make sure we see it, we will also slow down the fetching of the data by adding a delay:

1. In the `Components\Pages` folder, at the top of `Customers.razor`, add an attribute to enable streaming rendering, as shown in the following code:

```
@attribute [StreamRendering]
```

2. In `Customers.razor`, in the `OnParametersSetAsync` method, add a statement to asynchronously delay for one second, as shown highlighted in the following code:

```
protected override async Task OnParametersSetAsync()
{
  await Task.Delay(1000); // Delay for one second.
  ...
}
```

3. Start the `Northwind.Blazor` project using the `https` launch profile.
4. Start Chrome and navigate to `https://localhost:5141/`.
5. In the left navigation menu, click **Customers in Germany**, and note that the **Loading...** message appears for a second and is then replaced by the table of customers.

So far, the component provides only a read-only table of customers. Now, we will extend it with full CRUD operations.

Defining forms using the EditForm component

Microsoft provides ready-made components to build forms. We will use them to provide create, edit, and delete functionality for customers.

Microsoft provides the `EditForm` component and several form elements, such as `InputText`, to make it easier to use forms with Blazor.

`EditForm` can have a model set to bind it to an object with properties and event handlers for custom validation, as well as to recognize standard Microsoft validation attributes on the model class, as shown in the following code:

```
<EditForm Model="@customer" OnSubmit="ExtraValidation">
  <DataAnnotationsValidator />
  <ValidationSummary />
  <InputText id="name" @bind-Value="customer.CompanyName" />
  <button type="submit">Submit</button>
```

```
</EditForm>

@code {
  private Customer customer = new();

  private void ExtraValidation()
  {
    // Perform any extra validation you want.
  }
}
```

As an alternative to a `ValidationSummary` component, you can use the `ValidationMessage` component to show a message next to an individual form element. To bind the validation message to a property, you use a lambda expression to select the property, as shown in the following code:

```
<ValidationMessage For="@(() => Customer.CompanyName)" />
```

Building a customer detail component

We will create a component to show the details of a customer. This will only be a component, never a page, so it will not have a @page directive:

1. In the `Northwind.Blazor` project, in the `Components` folder, create a new file named `CustomerDetail.razor`. (The Visual Studio project item template is named **Razor Component**; the Rider project item template is named **Blazor Component**.)
2. Modify its contents to define a form to edit the properties of a customer, as shown in the following markup:

```
<EditForm Model="@Customer" OnValidSubmit="@OnValidSubmit">
  <DataAnnotationsValidator />
  <div>
    <label class="form-label">Customer Id</label>
    <div>
      <InputText class="form-control"
                 @bind-Value="@Customer.CustomerId" />
      <ValidationMessage For="@(() => Customer.CustomerId)" />
    </div>
  </div>
  <div>
    <label class="form-label">Company Name</label>
    <div>
      <InputText class="form-control"
                 @bind-Value="@Customer.CompanyName" />
      <ValidationMessage For="@(() => Customer.CompanyName)" />
```

```
    </div>
  </div>
  <div>
    <label class="form-label">Address</label>
    <div>
      <InputText class="form-control"
                 @bind-Value="@Customer.Address" />
      <ValidationMessage For="@(() => Customer.Address)" />
    </div>
  </div>
  <div>
    <label class="form-label">Country</label>
    <div>
      <InputText class="form-control"
                 @bind-Value="@Customer.Country" />
      <ValidationMessage For="@(() => Customer.Country)" />
    </div>
  </div>
  <button type="submit" class="btn btn-@ButtonStyle">
    @ButtonText
  </button>
</EditForm>

@code {
  [Parameter]
  public Customer Customer { get; set; } = null!;

  [Parameter]
  public string ButtonText { get; set; } = "Save Changes";

  [Parameter]
  public string ButtonStyle { get; set; } = "info";

  [Parameter]
  public EventCallback OnValidSubmit { get; set; }
}
```

Building customer create, edit, and delete components

Now, we can create three routable page components that use the component:

1. In the `Components\Pages` folder, create a new file named `CreateCustomer.razor`.

2. In `CreateCustomer.razor`, modify its contents to use the customer detail component to create a new customer, as shown in the following code:

```
@rendermode RenderMode.InteractiveServer
@page "/createcustomer"
@inject INorthwindService _service
@inject NavigationManager _navigation

<h3>Create Customer</h3>

<CustomerDetail ButtonText="Create Customer"
                Customer="@customer"
                OnValidSubmit="@Create" />

@code {
  private Customer customer = new();

  private async Task Create()
  {
    await _service.CreateCustomerAsync(customer);
    _navigation.NavigateTo("customers");
  }
}
```

3. In the `Components\Pages` folder, in `Customers.razor`, after the `<h3>` element, add a `<div>` element with a button to navigate to the create customer page component, as shown in the following markup:

```
<div>
  <a class="btn btn-info" href="createcustomer">
  <i class="bi bi-plus-square"></i> Create New</a>
</div>
```

4. In the `Components\Pages` folder, create a new file named `EditCustomer.razor`, and modify its contents to use the customer detail component to edit and save changes to an existing customer, as shown in the following code:

```
@rendermode RenderMode.InteractiveServer
@page "/editcustomer/{customerid}"
@inject INorthwindService _service
@inject NavigationManager _navigation

<h3>Edit Customer</h3>
```

```
<CustomerDetail ButtonText="Update"
                Customer="@customer"
                OnValidSubmit="@Update" />

@code {
  [Parameter]
  public string CustomerId { get; set; } = null!;

  private Customer? customer = new();

  protected override async Task OnParametersSetAsync()
  {
    customer = await _service.GetCustomerAsync(CustomerId);
  }

  private async Task Update()
  {
    if (customer is not null)
    {
      await _service.UpdateCustomerAsync(customer);
    }

    _navigation.NavigateTo("customers");
  }
}
```

5. In the `Components\Pages` folder, create a new file named `DeleteCustomer.razor`, and modify its contents to use the customer detail component to show the customer that is about to be deleted, as shown in the following code:

```
@rendermode RenderMode.InteractiveServer
@page "/deletecustomer/{customerid}"
@inject INorthwindService _service
@inject NavigationManager _navigation

<h3>Delete Customer</h3>

<div class="alert alert-danger">
  Warning! This action cannot be undone!
</div>

<CustomerDetail ButtonText="Delete Customer"
```

```
                  ButtonStyle="danger"
                  Customer="@customer"
                  OnValidSubmit="@Delete" />

@code {
  [Parameter]
  public string CustomerId { get; set; } = null!;

  private Customer? customer = new();

  protected override async Task OnParametersSetAsync()
  {
    customer = await _service.GetCustomerAsync(CustomerId);
  }

  private async Task Delete()
  {
    if (customer is not null)
    {
      await _service.DeleteCustomerAsync(CustomerId);
    }

    _navigation.NavigateTo("customers");
  }
}
```

Testing the customer components

Now, we can test the customer components and how to use them to create, edit, and delete customers:

1. Start the `Northwind.Blazor` project using the `https` launch profile.
2. Start Chrome and navigate to `https://localhost:5141/`.
3. Navigate to **Customers Worldwide** or **Customers in Germany** and click the **+ Create New** button.
4. Enter an invalid **Customer Id** like `ABCDEF`, leave the textbox, and note the validation message, as shown in *Figure 14.6*:

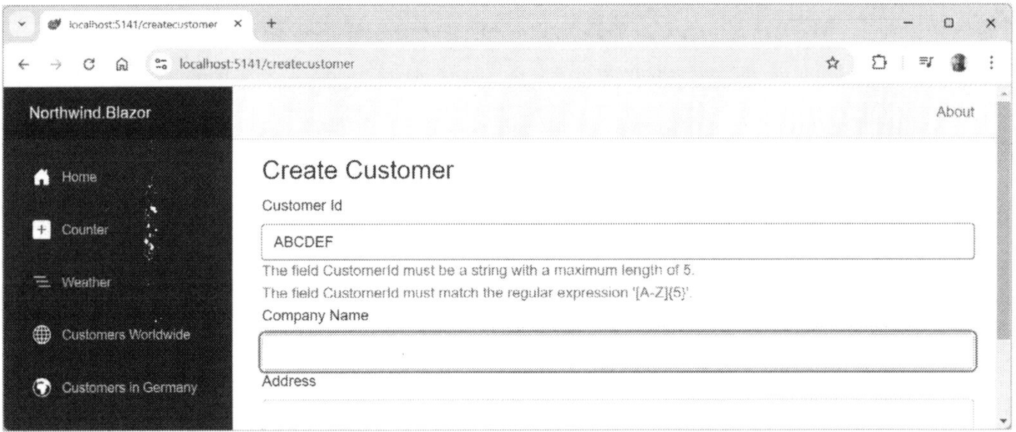

Figure 14.6: Creating a new customer and entering an invalid customer ID

5. Change **Customer Id** to `ABCDE`, enter values for the other textboxes like `Alpha Corp`, `Main Street`, and `Germany`, and then click the **Create Customer** button.
6. When the list of customers appears, click **Customers in Germany**, and scroll down to the bottom of the page to see the new customer.
7. On the **ABCDE** customer row, click the **Edit** icon button, change the address to something like `Upper Avenue`, click the **Update** button, and note that the customer record has been updated.
8. On the **ABCDE** customer row, click the **Delete** icon button, note the warning, click the **Delete Customer** button, and note that the customer record has been deleted.
9. Close Chrome and shut down the web server.

Practicing and exploring

Test your knowledge and understanding by answering some questions, getting some hands-on practice, and exploring this chapter's topics with deeper research.

Exercise 14.1 – Online material

Online material can be created by Microsoft or third parties, or it can be found in extra content for this book.

Enhancing Blazor apps

To learn how to enhance Blazor apps using AOT native publish and interop with JavaScript, as well as handle location-changing events, you can read an optional online-only section, found at the following link: `https://github.com/markjprice/cs13net9/blob/main/docs/ch14-enhanced-blazor.md`.

Leveraging open source Blazor component libraries

To learn how to use some common Blazor open source components, I have written an online-only section for my *Apps and Services with .NET 8* companion book, found at the following link: `https://github.com/markjprice/apps-services-net8/blob/main/docs/ch15-blazor-libraries.md`.

Exercise 14.2 – Practice exercises

Practice exercises go deeper into the topics for this chapter. First, you will create a times table component, and then you will create a country navigation component.

Creating a times table component

In the `Northwind.Blazor` project, create a routable page component that renders a times table, based on a parameter named `Number`, and then test your component in two ways.

First, do so by adding an instance of your component to the `Home.razor` file, as shown in the following markup, to generate the 6 times table with a default size of 12 rows, or the 7 times table with a size of 10 rows:

```
<TimesTable Number="6" />
<TimesTable Number="7" Size="10" />
```

Second, do so by entering a path in the browser address bar, as shown in the following links:

- `https://localhost:5141/timestable/6`
- `https://localhost:5141/timestable/7/10`

Creating a country navigation item

In the `Northwind.Blazor` project, in the `NavMenu` component, call the customer's web service to get the list of country names and loop through them, creating a menu item for each country.

Here's an example:

1. In the `Northwind.Blazor` project, in `INorthwindService.cs`, add the following code:

   ```
   List<string?> GetCountries();
   ```

2. In `NorthwindServiceServerSide.cs`, add the following code:

   ```
   public List<string?> GetCountries()
   {
     return _db.Customers.Select(c => c.Country)
       .Distinct().OrderBy(country => country).ToList();
   }
   ```

3. In NavMenu.razor, add the following markup:

```
@inject INorthwindService _service

...

@foreach(string? country in _service.GetCountries())
{
    string countryLink = "customers/" + country;

    <div class="nav-item px-3">
        <NavLink class="nav-link" href="@countryLink">
        <span class="oi oi-people" aria-hidden="true"></span>
        Customers in @country
        </NavLink>
    </div>
}
```

You cannot use `<NavLink class="nav-link" href="customers/@c">` because Blazor does not allow combined text and @ Razor expressions in components. That is why the code above creates a local variable to combine the text to make the country URL.

Exercise 14.3 – Test your knowledge

Answer the following questions:

1. What are the four Blazor render modes, and how are they different?
2. In a Blazor Web App project, compared to an ASP.NET Core MVC project, what extra configuration is required?
3. Why should you avoid the Blazor Server and Blazor Server Empty project templates?
4. In a Blazor Web App project, what does the App.razor file do?
5. What is the main benefit of using the <NavLink> component?
6. How can you pass a value into a component?
7. What is the main benefit of using the <EditForm> component?
8. How can you execute some statements when parameters are set?
9. How can you execute some statements when a component appears?
10. One of the benefits of Blazor is being able to implement client-side components using C# and .NET instead of JavaScript. Does a Blazor component need any JavaScript?

Exercise 14.4 – Explore topics

Use the links on the following page to learn more details about the topics covered in this chapter:

```
https://github.com/markjprice/cs13net9/blob/main/docs/book-links.md#chapter-15---
building-interactive-web-components-using-blazor
```

Summary

In this chapter, you learned:

- About the concepts of Blazor components.
- How to build Blazor components that execute on the server side, using SignalR to update the user interface during interactivity.

In the next chapter, you will learn how to build and consume services that use HTTP as the communication layer, aka web services. You will also learn how to build Blazor components that execute on the client side, using WebAssembly.

15

Building and Consuming Web Services

This chapter is about learning how to build web services (aka HTTP or **Representational State Transfer (REST)** services) using ASP.NET Core Minimal APIs. You will then learn how to consume web services using HTTP clients, which could be any other type of .NET app, including a website or a mobile or desktop app. We will create a Blazor WebAssembly client.

This chapter requires the knowledge and skills that you gained in *Chapter 10, Working with Data Using Entity Framework Core,* and *Chapters 12* to *14*, about building websites using ASP.NET Core and Blazor.

In this chapter, we will cover the following topics:

- Building web services using ASP.NET Core
- Creating a web service for the Northwind database
- Documenting and trying out web services
- Consuming web services using HTTP clients

Building web services using ASP.NET Core

Before we build a modern web service, we need to cover some background to set the context for this chapter.

Understanding web service acronyms

Although HTTP was originally designed to request and respond with HTML and other resources for humans to look at, it is also good for building services.

Roy Fielding stated in his doctoral dissertation, describing the **REST** architectural style, that the HTTP standard would be good for building services because it defines the following:

- URIs to uniquely identify resources, like `https://localhost:5151/products/23`.
- Methods for performing common tasks on those resources, like `GET`, `POST`, `PUT`, and `DELETE`.

- The ability to negotiate the media type of content exchanged in requests and responses, such as XML and JSON. Content negotiation happens when the client specifies a request header like `Accept: application/xml,*/*;q=0.8`. The default response format used by the ASP.NET Core web services is JSON, which means one of the response headers would be `Content-Type: application/json; charset=utf-8`.

Web services use the HTTP communication standard, so they are sometimes called **HTTP services** or **RESTful services**.

Understanding HTTP requests and responses

HTTP defines standard types of requests and standard codes to indicate a type of response. Most of them can be used to implement web services.

The most common type of request is `GET`, to retrieve a resource identified by a unique path, with additional options like what media type is acceptable to set as a request header, like `Accept`, as shown in the following example:

```
GET /path/to/resource
Accept: application/json
```

Common responses include success and multiple types of failure, as shown in *Table 15.1*:

Status code	Description
`101 Switching Protocols`	The requester has asked the server to switch protocols and the server has agreed to do so. For example, it is common to switch from HTTP to **WebSockets** (**WS**) for more efficient communication.
`103 Early Hints`	Used to convey hints that help a client make preparations to process the final response. For example, the server might send the following response before then sending a normal `200 OK` response for a web page that uses a stylesheet and JavaScript file: HTTP/1.1 103 Early Hints Link: </style.css>; rel=preload; as=style Link: </script.js>; rel=preload; as=script
`200 OK`	The path was correctly formed and the resource was successfully found, serialized into an acceptable media type, and then returned in the response body. The response headers specify the `Content-Type`, `Content-Length`, and `Content-Encoding`, for example, `GZIP`.
`301 Moved Permanently`	Over time, a web service may change its resource model, including the path used to identify an existing resource. The web service can indicate the new path by returning this status code and a response header named `Location` that has the new path.
`302 Found`	This is the same as `301`.

`304 Not Modified`	If the request includes the `If-Modified-Since` header, then the web service can respond with this status code. The response body is empty because the client should use its cached copy of the resource.
`307 Temporary Redirect`	The requested resource has been temporarily moved to the URL in the `Location` header. The browser should make a new request using that URL. For example, this is what happens if you enable `UseHttpsRedirection` and a client makes an HTTP request.
`400 Bad Request`	The request was invalid, for example, it used a path for a product using an integer ID where the ID value is missing.
`401 Unauthorized`	The request was valid and the resource was found, but the client did not supply credentials or is not authorized to access that resource. Re-authenticating may enable access, for example, by adding or changing the `Authorization` request header.
`403 Forbidden`	The request was valid and the resource was found, but the client is not authorized to access that resource. Re-authenticating will not fix the issue.
`404 Not Found`	The request was valid, but the resource was not found. The resource may be found if the request is repeated later. To indicate that a resource will never be found, return `410 Gone`.
`406 Not Acceptable`	If the request has an `Accept` header that only lists media types that the web service does not support. For example, if the client requests JSON but the web service can only return XML.
`451 Unavailable for Legal Reasons`	A website hosted in the USA might return this for requests coming from Europe to avoid having to comply with the **General Data Protection Regulation** (**GDPR**). The number was chosen as a reference to the novel Fahrenheit 451, in which books are banned and burned.
`500 Server Error`	The request was valid, but something went wrong on the server side while processing the request. Retrying again later might work.
`503 Service Unavailable`	The web service is busy and cannot handle the request. Trying again later might work.

Table 15.1: Common HTTP status code responses to the GET method

Other common types of HTTP requests include `POST`, `PUT`, `PATCH`, or `DELETE`, which create, modify, or delete resources.

To create a new resource, you might make a `POST` request with a body that contains the new resource, as shown in the following code:

```
POST /path/to/resource
Content-Length: 123
Content-Type: application/json
```

To create a new resource or update an existing resource, you might make a `PUT` request with a body that contains a whole new version of the existing resource, and if the resource does not exist, it is created, or if it does exist, it is replaced (sometimes called an **upsert** operation), as shown in the following code:

```
PUT /path/to/resource
Content-Length: 123
Content-Type: application/json
```

To update an existing resource more efficiently, you might make a `PATCH` request with a body that contains an object with only the properties that need changing, as shown in the following code:

```
PATCH /path/to/resource
Content-Length: 123
Content-Type: application/json
```

To delete an existing resource, you might make a `DELETE` request, as shown in the following code:

```
DELETE /path/to/resource
```

As well as the responses shown in the table above for a `GET` request, all the types of requests that create, modify, or delete a resource have additional possible common responses, as shown in *Table 15.2*:

Status code	Description
`201 Created`	The new resource was created successfully, the response header named `Location` contains its path, and the response body contains the newly created resource. Immediately `GET`-ing the resource should return `200`.
`202 Accepted`	The new resource cannot be created immediately, so the request is queued for later processing, and immediately `GET`-ing the resource might return `404`. The body can contain a resource that points to some form of status checker or an estimate of when the resource will become available.
`204 No Content`	Commonly used in response to a `DELETE` request since returning the resource in the body after deleting it does not usually make sense! Sometimes used in response to `POST`, `PUT`, or `PATCH` requests if the client does not need to confirm that the request was processed correctly.
`405 Method Not Allowed`	Returned when the request used a method that is not supported. For example, a web service designed to be read-only may explicitly disallow `PUT`, `DELETE`, and so on.
`415 Unsupported Media Type`	Returned when the resource in the request body uses a media type that the web service cannot handle. For example, if the body contains a resource in XML format but the web service can only process JSON.

Table 15.2: Common HTTP status code responses to other methods like POST and PUT

ASP.NET Core Minimal APIs projects

We will build a web service that provides a way to work with data in the Northwind database using ASP.NET Core so that the data can be used by any client application on any platform that can make HTTP requests and receive HTTP responses.

Traditionally, you use the **ASP.NET Core Web API** / `dotnet new webapi` project template. This allows the creation of a web service implemented using either controllers or the newer Minimal APIs.

Warning! With .NET 6 and .NET 7, the `dotnet new webapi` command creates a service implemented using controllers. With .NET 6 and .NET 7, to implement the service using Minimal APIs, you need to add the `--use-minimal-apis` switch to the command. Using .NET 8 or later, the `dotnet new webapi` command creates a service implemented using Minimal APIs. To implement the service using controllers, you need to add the `--use-controllers` switch.

Minimal APIs web service and native AOT compilation

.NET 8 introduced the **ASP.NET Core Web API (native AOT)** / `dotnet new webapiaot` project template, which only uses Minimal APIs and supports native AOT publishing. More components of .NET will support AOT over time, as you can read in the following quote:

> *"We expect to make progress investigating Native AOT support for MVC & Blazor in the .NET 9 timeframe, but we don't expect to deliver production ready Native AOT support for .NET 9 given the large amount of work involved."* – Dan Roth

`https://github.com/dotnet/aspnetcore/issues/51834#issuecomment-1913300365`

Good Practice: Minimal APIs work especially well with **Vertical Slice Architecture (VSA)**. A major benefit of Minimal APIs over a controller-based Web API is that each Minimal API endpoint only needs to instantiate the **dependency injection** (**DI**) services that it needs. With controllers, to execute any action method within that controller, all DI services used in any of the action methods must be instantiated for every call. This is a waste of time and resources!

Creating an ASP.NET Core Minimal API project

Let's go:

1. Use your preferred code editor to open the `ModernWeb` solution and then add a new project, as defined in the following list:
    - Project template: **ASP.NET Core Web API** / `webapi`
    - Solution file and folder: `ModernWeb`
    - Project file and folder: `Northwind.WebApi`

2. If you are using Visual Studio, then confirm the following defaults have been chosen:
    - **Authentication type**: None
    - **Configure for HTTPS**: Selected
    - **Enable container support**: Cleared
    - **Enable OpenAPI support**: Selected
    - **Do not use top-level statements:** Cleared
    - **Use controllers:** Cleared

Make sure to clear the **Use controllers** check box, or your code will look very different to what you will see in this book!

3. If you are using VS Code or Rider, then in the `ModernWeb` directory, at the command prompt or terminal, enter the following:

```
dotnet new webapi -o Northwind.WebApi
```

4. Build the `Northwind.WebApi` project.
5. In the project file, remove the version number for the package that implements OpenAPI web service documentation because we are using CPM, as shown in the following markup:

```
<PackageReference Include="Microsoft.AspNetCore.OpenApi" />
```

6. In `Program.cs`, review the code, as shown in the following code:

```
var builder = WebApplication.CreateBuilder(args);

// Add services to the container.
// Learn more about configuring OpenAPI at https://aka.ms/aspnet/openapi
builder.Services.AddOpenApi();

var app = builder.Build();

// Configure the HTTP request pipeline.
if (app.Environment.IsDevelopment())
{
  app.MapOpenApi();
}

app.UseHttpsRedirection();
```

```
var summaries = new[]
{
    "Freezing", "Bracing", "Chilly", "Cool", "Mild", "Warm", "Balmy",
"Hot", "Sweltering", "Scorching"
};

app.MapGet("/weatherforecast", () =>
{
  var forecast = Enumerable.Range(1, 5).Select(index =>
    new WeatherForecast
    (
      DateOnly.FromDateTime(DateTime.Now.AddDays(index)),
      Random.Shared.Next(-20, 55),
      summaries[Random.Shared.Next(summaries.Length)]
    ))
    .ToArray();
  return forecast;
})
.WithName("GetWeatherForecast");

app.Run();

internal record WeatherForecast(DateOnly Date,
  int TemperatureC, string? Summary)
{
  public int TemperatureF => 32 +
    (int)(TemperatureC / 0.5556);
}
```

While reviewing the preceding code, note the following:

- The program starts with a similar configuration to any other ASP.NET Core project, with a call to `WebApplication.CreateBuilder`.
- The services collection has an OpenAPI service added. This is used to document a web service. In .NET 8 and earlier, the third-party Swashbuckle package was used to do this, but with .NET 9 and later, Microsoft has written their own implementation. You can read more about this at the following link: https://github.com/dotnet/aspnetcore/issues/54599. By default, OpenAPI document generation creates a document that is compliant with v3.0 of the OpenAPI specification: https://spec.openapis.org/oas/v3.0.0.

- During development, the OpenAPI documentation is mapped as endpoints so that other developers can use it to create clients easily. By default, the OpenAPI endpoint registered via a call to MapOpenApi exposes the document at the `/openapi/{documentName}.json` endpoint. By default, the `documentName` is `v1`. In the production environment, these endpoints are not mapped because they are no longer necessary.
- The `MapGet` call registers a relative path of `/weatherforecast` to respond to HTTP `GET` requests, and its implementation uses the shared `Random` object to return an array of `WeatherForecast` objects with random temperatures and summaries like `Bracing` or `Balmy` for the next five days of weather.

Now let's allow the HTTP request to specify how many days ahead the forecast should be. At the same time, we will implement good practices by putting the weather endpoint implementation in its own code file:

1. Add a new class file named `Program.Weather.cs`.
2. In `Program.Weather.cs`, add statements to extend the automatically generated `partial Program` class by moving (cut and paste the statements) the weather-related statements from `Program.cs` and making small adjustments like defining a `GetWeather` method with a `days` parameter to control how many weather forecasts to generate, as shown in the following code:

```
public partial class Program
{
  static string[] summaries = { "Freezing", "Bracing",
    "Chilly", "Cool", "Mild", "Warm", "Balmy", "Hot",
    "Sweltering", "Scorching" };

  internal static WeatherForecast[]? GetWeather(int days)
  {
    WeatherForecast[]? forecast = Enumerable.Range(1, days)
      .Select(index => new WeatherForecast
      (
        DateOnly.FromDateTime(DateTime.Now.AddDays(index)),
        Random.Shared.Next(-20, 55),
        summaries[Random.Shared.Next(summaries.Length)]
      ))
      .ToArray();

    return forecast;
  }

  internal record WeatherForecast(DateOnly Date,
    int TemperatureC, string? Summary)
  {
```

```
        public int TemperatureF => 32 +
          (int)(TemperatureC / 0.5556);
    }
}
```

3. In `Program.cs`, modify the `MapGet` call, as shown highlighted in the following code:

```
app.UseHttpsRedirection();

app.MapGet("/weatherforecast/{days:int?}",
  (int days = 5) => GetWeather(days))
  .WithName("GetWeatherForecast");

app.Run();
```

In the `MapGet` call, note the route template pattern `{days:int?}` constrains the `days` parameter to `int` values. The `?` makes the `days` parameter optional, and if missing it will default to `5`.

Reviewing the web service's functionality

Now, we will test the web service's functionality:

1. In the `Properties` folder, in `launchSettings.json`, note that by default, if you are using Visual Studio, the `https` profile will launch the browser and navigate to the `/weatherforecast` relative URL path, as shown highlighted in the following markup:

```
"https": {
  "commandName": "Project",
  "dotnetRunMessages": true,
  "launchBrowser": true,
  "launchUrl": "weatherforecast",
```

2. For the `https` profile, for its `applicationUrl`, change the random port number for HTTPS to `5151` and for HTTP to `5150`, as shown highlighted in the following markup:

```
"applicationUrl": "https://localhost:5151;http://localhost:5150",
```

3. Save changes to all modified files.
4. Start the `Northwind.WebApi` web service project using the `https` launch profile.
5. On Windows, if you see a **Windows Security Alert** dialog box saying **Windows Defender Firewall has blocked some features of this app**, then click the **Allow access** button.

6. Start Chrome, navigate to `https://localhost:5151/`, and note you will get a `404` status code response because we have not enabled static files and there is not an `index.html`. Remember that this project is not designed for a human to view and interact with, so this is expected behavior for a web service.
7. In Chrome, show **Developer Tools**.
8. Navigate to `https://localhost:5151/weatherforecast` and note the web service should return a JSON document with five random weather forecast objects in an array, as shown in *Figure 15.1*:

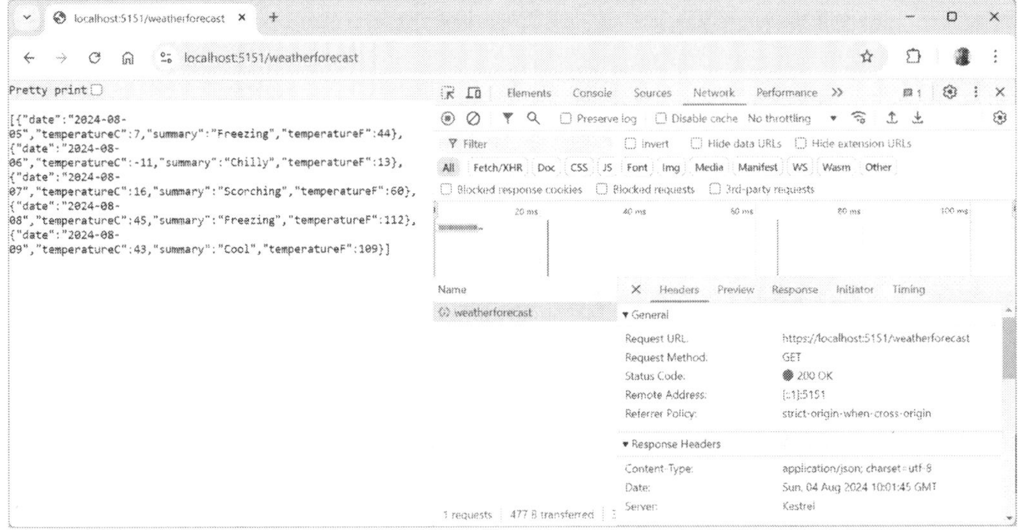

Figure 15.1: A request and response from a weather forecast web service

9. Close **Developer Tools**.
10. Navigate to `https://localhost:5151/weatherforecast/14` and note that the response when requesting a two-week weather forecast contains 14 forecasts.
11. Select the **Pretty print** checkbox, as shown in *Figure 15.1*, and note that recent versions of Chrome can now format JSON responses better for humans to read.
12. Close Chrome and shut down the web server.

Route constraints

To register the `/weatherforecast` route endpoint, we used a route constraint to limit acceptable values for the days parameter to integers, as shown highlighted in the following code:

```
app.MapGet("/weatherforecast/{days:int?}", ...
```

Route constraints allow us to control matches based on data types and other validation. They are summarized in *Table 15.3*:

Constraint	Example	Description
`required`	`{id:required}`	The parameter has been provided.
`int` and `long`	`{id:int}`	Any integer of the correct size.
`decimal`, `double`, and `float`	`{unitprice:decimal}`	Any real number of the correct size.
`bool`	`{discontinued:bool}`	Case-insensitive match on `true` or `false`.
`datetime`	`{hired:datetime}`	An invariant culture date/time.
`guid`	`{id:guid}`	A GUID value.
`minlength(n)`, `maxlength(n)`, `length(n)`, and `length(n, m)`	`{title:minlength(5)}`, `{title:length(5, 25)}`	The text must have the defined minimum and/or maximum length.
`min(n)`, `max(n)`, and `range(n, m)`	`{age:range(18, 65)}`	The integer must be within the defined minimum and/or maximum range.
`alpha`, `regex`	`{firstname:alpha}`, `{id:regex(^[A-Z]{{5}}$)}`	The parameter must match one or more alphabetic characters or the regular expression.

Table 15.3: Route constraints with examples and descriptions

Use colons to separate multiple constraints, as shown in the following example:

```
app.MapGet("/weatherforecast/{days:int:min(5)}", ...
```

For regular expressions, `RegexOptions.IgnoreCase | RegexOptions.Compiled | RegexOptions.CultureInvariant` is added automatically. Regular expression tokens must be escaped (replace \ with \\, { with {{, and } with }}) or use verbatim string literals.

More Information: You can create custom route constraints by defining a class that implements `IRouteConstraint`. This is beyond the scope of this book, but you can read about it at the following link: `https://learn.microsoft.com/en-us/aspnet/core/fundamentals/routing#custom-route-constraints`.

Short-circuit routes

When routing matches a request to an endpoint, it lets the rest of the middleware pipeline run before invoking the endpoint logic. That takes time, so in ASP.NET Core 8 and later, you can invoke the endpoint immediately and return the response.

You do this by calling the `ShortCircuit` method on a mapped endpoint route, as shown in the following code:

```
app.MapGet("/", () => "Hello World").ShortCircuit();
```

Alternatively, you can call the `MapShortCircuit` method to respond with a `404 Missing Resource` or other status code for resources that don't need further processing, as shown in the following code:

```
app.MapShortCircuit(404, "robots.txt", "favicon.ico");
```

Improved route tooling in ASP.NET Core 8 and later

Microsoft has improved the tooling for working with routes for all ASP.NET Core 8 and later technologies including Web APIs and Blazor. The features include the following:

- **Route syntax highlighting**: Different parts of routes are now highlighted in your code editor.
- **Autocompletion**: Parameter and route names and route constraints are autocompleted.
- **Route analyzers and fixers**: These address common problems that developers have when implementing their routes.

You can read about them in the blog article *ASP.NET Core Route Tooling Enhancements in .NET 8*, found at the following link: `https://devblogs.microsoft.com/dotnet/aspnet-core-route-tooling-dotnet-8/`.

Understanding endpoint route handler return types

An endpoint lambda expression can return .NET types like a single `string` value; complex objects defined by a `class`, `record`, or `struct`; or collections of complex objects. ASP.NET Core Minimal APIs will serialize them into JSON or plain text.

Consider the following endpoint route handler, which returns a `string` value, as shown in the following code:

```
app.MapGet("/hello", () => "Hello World");
```

ASP.NET Core Minimal APIs will return a `200` status code with a `Content-Type: text/plain; charset=utf-8` header and the following content in the body: `Hello World`.

Now consider the following endpoint route handler, which returns an anonymous type, as shown in the following code:

```
app.MapGet("/user", () => new {
  FirstName = "Bob",
  Age = 45 });
```

ASP.NET Core Minimal APIs will return a `200` status code with a `Content-Type: application/json; charset=utf-8` header and the following content in the body:

```
{"firstName":"Bob","age":45}
```

For more control over the response, there are helper methods that return an `IResult`, which defines a contract that represents the result of an HTTP endpoint. The static `Results` and `TypedResults` classes can be used to create various `IResult` objects that represent different types of responses.

Returning `TypedResults` rather than `Results` has the following advantages:

- `TypedResults` helpers return strongly typed objects, which can improve code readability and unit testing and reduce the chance of runtime errors.
- The implementation type automatically provides the response type metadata for OpenAPI to describe the endpoint.

The `TypedResults` class has a property named `Empty` that produces an empty result response, which when executed will do nothing.

The `TypedResults` class has methods to make it easy to return different responses, as shown in *Table 15.4*:

Method	**Description**
`Bytes`	Returns a `200` status code and writes byte-array content to the response.
`Content`	Returns a `200` status code and writes the content string to the HTTP response. Has an optional parameter to specify the media type header.
`File`	Returns a `200` status code and writes the specified `Stream` to the response.
`Json`	Returns a `200` status code and serializes the specified data object in JSON format to the response.
`Ok`	Returns a `200` status code and a resource converted into the client's preferred format, like JSON or XML. Commonly used in response to a `GET` request.
`Created`, `CreatedAtRoute`	Returns a `201` status code and the path to the new resource. Commonly used in response to a `POST` request to create a resource that can be created quickly.
`Accepted`, `AcceptedAtRoute`	Returns a `202` status code to indicate the request is being processed but has not been completed. Commonly used in response to a `POST`, `PUT`, `PATCH`, or `DELETE` request that triggers a background process that takes a long time to complete.
`Redirect`, `RedirectToRoute`	Returns a `301`, `307`, or `308` status code depending on a temporary or permanent redirect `bool` parameter, with the `url` to redirect to.
`Problem`	Produces a `ProblemDetails` response.
`NoContent`	Returns a `204` status code and an empty response body. Commonly used in response to a `PUT`, `PATCH`, or `DELETE` request when the response does not need to contain the affected resource.
`BadRequest`	Returns a `400` status code and an optional message string with more details.
`NotFound`	Returns a `404` status code and automatically populates the `ProblemDetails` body (requires a compatibility version of 2.2 or later).

Table 15.4: TypedResults helper methods that return a response

More Information: You can learn more about how to create responses in a Minimal APIs web service at the following link: `https://learn.microsoft.com/en-us/aspnet/core/fundamentals/minimal-apis/responses?view=aspnetcore-9.0`.

Creating a web service for the Northwind database

We will reference the Entity Framework Core entity data model for the Northwind database that you created in *Chapter 12, Introducing Modern Web Development Using .NET*:

1. In the `Northwind.WebApi` project, globally and statically import the `System.Console` class, and add a project reference to the Northwind data context class library for either SQLite or SQL Server, as shown in the following markup:

```
<ItemGroup Label="To simplify use of WriteLine.">
  <Using Include="System.Console" Static="true" />
</ItemGroup>

<ItemGroup>
  <!-- change Sqlite to SqlServer if you prefer -->
  <ProjectReference Include=
"..\Northwind.DataContext.Sqlite\Northwind.DataContext.Sqlite.csproj" />
</ItemGroup>
```

2. Build the `Northwind.WebApi` project and fix any compile errors in your code.
3. In `Program.cs`, import namespaces for working with the Northwind entity model, as shown in the following code:

```
using Northwind.EntityModels; // To use AddNorthwindContext method.
```

4. In `Program.cs`, add a statement before the call to `Build` to register the `Northwind` database context class (it will use either SQLite or SQL Server depending on which database provider you referenced in the project file), as shown in the following code:

```
builder.Services.AddNorthwindContext();
```

Registering dependency services

You can register dependency services with different lifetimes, as shown in the following list:

- **Transient**: These services are created each time they're requested. Transient services should be lightweight and stateless.
- **Scoped**: These services are created once per client request and are disposed of. Then, the response is returned to the client.
- **Singleton**: These services are usually created the first time they are requested and then are shared, although you can provide an instance at the time of registration too.

Introduced in .NET 8 is the ability to set a key for a dependency service. This allows multiple services to be registered with different keys and then retrieved later using that key.

```
builder.Services.AddKeyedsingleton<IMemoryCache, BigCache>("big");
builder.Services.AddKeyedSingleton<IMemoryCache, SmallCache>("small");
```

```
class BigCacheConsumer([FromKeyedServices("big")] IMemoryCache cache)
{
  public object? GetData() => cache.Get("data");
}

class SmallCacheConsumer(IKeyedServiceProvider keyedServiceProvider)
{
  public object? GetData() => keyedServiceProvider
    .GetRequiredKeyedService<IMemoryCache>("small");
}
```

In this book, you will use all three types of lifetimes, but we will not need to use keyed services.

In-memory, distributed, and hybrid caches

Now let's see an overview of in-memory, distributed, and hybrid caching.

In-memory caching

In-memory caching stores data in the memory of the web server where the application is running. This is useful for small to medium-sized applications where the caching needs are not too extensive and can be handled by a single server's memory.

The key points about in-memory caching are shown in the following list:

- **Performance**: Fast retrieval since the data is stored locally in RAM.
- **Simplicity**: Easy to implement and configure within the application.
- **Volatility**: Data is lost if the application restarts or the server goes down.
- **Scalability**: Limited to a single server's memory; not suitable for large-scale applications needing distributed caching.

To implement in-memory caching, add the memory cache service to the services collection in `Program.cs`, as shown in the following code:

```
services.AddMemoryCache();
```

Retrieve the service in an endpoint, as shown in the following code:

```
private readonly IMemoryCache _cache;
```

Set data in the cache, as shown in the following code:

```
_cache.Set(key, data);
```

Get data from the cache, as shown in the following code:

```
return _cache.TryGetValue(key, out var data) ? data : null;
```

Now let's compare in-memory caching to distributed caching.

Distributed caching

Distributed caching allows caching data across multiple servers, making it suitable for large-scale, distributed applications. This ensures data availability and consistency across different nodes in a web farm.

The key points about in-memory caching are shown in the following list:

- **Scalability**: Can handle large datasets and provide caching across multiple servers.
- **Persistence**: Depending on the provider, data can be persisted beyond application restarts.
- **Latency**: May have higher latency compared to in-memory caching due to network calls.
- **Providers**: Common providers include Redis, SQL Server, and NCache.

To implement in-memory caching, add it to the services collection in `Program.cs`, as shown in the following code:

```
services.AddStackExchangeRedisCache(options =>
  options.Configuration = "localhost:6379";
  options.InstanceName = "SampleInstance";
});
```

Retrieve the service in an endpoint, as shown in the following code:

```
private readonly IDistributedCache _cache;
```

Set data in the cache, as shown in the following code:

```
await _cache.SetStringAsync(key, value);
```

Get data from the cache, as shown in the following code:

```
return await _cache.GetStringAsync(key);
```

So, in-memory caching is fast and simple but limited to the server's memory and loses data on restart. Distributed caching is scalable and persistent, ideal for large applications, with various providers like Redis and SQL Server.

Both approaches help improve application performance by reducing the need to repeatedly fetch or compute data. The choice between them depends on the application's scale, performance needs, and architecture.

But what if we could get the best of both worlds?

Let's see a new option called hybrid caching.

Hybrid caching

The HybridCache API introduced with ASP.NET Core 9 addresses some limitations found in the `IDistributedCache` and `IMemoryCache` APIs. As an abstract class with a default implementation, `HybridCache` efficiently manages most tasks related to storing and retrieving data from the cache.

The key points about hybrid caching are shown in the following list:

- **Unified API**: Provides a single interface for both in-process and out-of-process caching. `HybridCache` can seamlessly replace any existing `IDistributedCache` and `IMemoryCache` usage. It always uses the in-memory cache initially, and when an `IDistributedCache` implementation is available, `HybridCache` leverages it for secondary caching. This dual-level caching approach combines the speed of in-memory caching with the durability of distributed or persistent caching.
- **Stampede Protection**: `HybridCache` prevents cache stampedes, which occur when a frequently used cache entry is invalidated, causing multiple requests to try to repopulate it simultaneously. `HybridCache` merges concurrent operations, ensuring all requests for the same response wait for the first request to be completed.
- **Configurable Serialization**: `HybridCache` allows for configurable serialization during service registration, supporting both type-specific and generalized serializers via the `WithSerializer` and `WithSerializerFactory` methods, which are chained from the `AddHybridCache` call. By default, it manages `string` and `byte[]` internally and utilizes `System.Text.Json` for other types. It can be configured to use other serializers, such as Protobuf or XML.

Although HybridCache was introduced with .NET 9, its package targets .NET Standard 2.0, so it can be used with older versions of .NET, even .NET Framework 4.6.2 or later.

Now that you've learned the concepts and basic implementation options for caching, let's create a data repository for our web service that caches entities to improve performance and scalability.

Creating data repositories with caching for entities

Defining and implementing a data repository to provide CRUD operations is good practice. We will create a data repository for the `Customers` table in Northwind. There are only 91 customers in this table, so we will cache a copy of the whole table in memory to improve scalability and performance when reading customer records.

Good Practice: In a real web service, you should use a distributed cache like Redis, an open-source data structure store that can be used as a high-performance, high-availability database, cache, or message broker. You can learn about this at the following link: `https://learn.microsoft.com/en-us/aspnet/core/performance/caching/distributed`.

In .NET 9, `HybridCache` was introduced, which automatically switches between in-memory and distributed cache types.

More Information: You can learn more about `HybridCache` at the following link: `https://learn.microsoft.com/en-us/aspnet/core/performance/caching/hybrid?view=aspnetcore-9.0`.

We will follow a modern good practice and make the repository API asynchronous. It will be instantiated by an endpoint using parameter injection, so a new instance is created to handle every HTTP request. It will use a singleton instance of `HybridCache`. Let's go:

1. In the `Northwind.WebApi.csproj` project file, add a package reference for hybrid caching, as shown in the following markup:

```
<PackageReference Include=
  "Microsoft.Extensions.Caching.Hybrid" />
```

2. In `Program.cs`, import the namespace for working with a hybrid cache, as shown in the following code:

```
using Microsoft.Extensions.Caching.Hybrid; // To use
HybridCacheEntryOptions.
```

3. In `Program.cs`, before the call to `Build`, in the section for configuring services, register the hybrid cache service with a default cache entry duration of 60 seconds overall, and 30 seconds for local in-memory caching, as shown in the following code:

```
builder.Services.AddHybridCache(options =>
{
  options.DefaultEntryOptions = new HybridCacheEntryOptions
  {
    Expiration = TimeSpan.FromSeconds(60),
    LocalCacheExpiration = TimeSpan.FromSeconds(30)
  };
});
```

4. In the `Northwind.WebApi` project, create a folder named `Repositories`.
5. Add a new interface file and a class file to the `Repositories` folder, named `ICustomerRepository.cs` and `CustomerRepository.cs`, respectively.
6. In `ICustomerRepository.cs`, define an interface with five CRUD methods, as shown in the following code:

```
using Northwind.EntityModels; // To use Customer.

namespace Northwind.WebApi.Repositories;
```

```
public interface ICustomerRepository
{
  Task<Customer?> CreateAsync(Customer c);
  Task<Customer[]> RetrieveAllAsync();
  Task<Customer?> RetrieveAsync(string id,
    CancellationToken token);
  Task<Customer?> UpdateAsync(Customer c);
  Task<bool?> DeleteAsync(string id);
}
```

7. In `CustomerRepository.cs`, define a class that will implement the interface and uses the hybrid cache (its methods will be implemented over the next few steps, so, for now, ignore the errors you will be shown), as shown in the following code:

```
using Microsoft.EntityFrameworkCore.ChangeTracking; // To use
EntityEntry<T>.
using Northwind.EntityModels; // To use Customer.
using Microsoft.EntityFrameworkCore; // To use ToArrayAsync.
using Microsoft.Extensions.Caching.Hybrid; // To use HybridCache.

namespace Northwind.WebApi.Repositories;

public class CustomerRepository : ICustomerRepository
{
  private readonly HybridCache _cache;

  // Use an instance data context field because it should not be
  // cached due to the data context having internal caching.
  private NorthwindContext _db;

  public CustomerRepository(NorthwindContext db,
    HybridCache hybridCache)
  {
    _db = db;
    _cache = hybridCache;
  }
}
```

8. Implement the method that retrieves all customers to always read the latest customers from the database, as shown in the following code:

```
public Task<Customer[]> RetrieveAllAsync()
{
  return _db.Customers.ToArrayAsync();
}
```

9. Implement the `Retrieve` method to get the customer from the cache if possible, or from the data model, and set it in the cache for next time, as shown in the following code:

```
public async Task<Customer?> RetrieveAsync(string id,
  CancellationToken token = default)
{
  id = id.ToUpper(); // Normalize to uppercase.

  return await _cache.GetOrCreateAsync(
    key: id, // Unique key to the cache entry.
    factory: async cancel => await _db.Customers
      .FirstOrDefaultAsync(c => c.CustomerId == id, token),
    cancellationToken: token);
}
```

10. Implement the `Create` method, as shown in the following code:

```
public async Task<Customer?> CreateAsync(Customer c)
{
  c.CustomerId = c.CustomerId.ToUpper(); // Normalize to uppercase.

  // Add to database using EF Core.
  EntityEntry<Customer> added =
    await _db.Customers.AddAsync(c);
  int affected = await _db.SaveChangesAsync();
  if (affected == 1)
  {
    // If saved to database then store in cache.
    await _cache.SetAsync(c.CustomerId, c);
    return c;
  }
  return null;
}
```

11. Implement the `Update` method to update the database, and if successful, update the cached customer as well, as shown in the following code:

```
public async Task<Customer?> UpdateAsync(Customer c)
{
  c.CustomerId = c.CustomerId.ToUpper();

  _db.Customers.Update(c);
  int affected = await _db.SaveChangesAsync();
  if (affected == 1)
  {
    await _cache.SetAsync(c.CustomerId, c);
    return c;
  }
  return null;
}
```

12. Implement the `Delete` method to delete the customer from the database, and if successful, remove the cached customer as well, as shown in the following code:

```
public async Task<bool?> DeleteAsync(string id)
{
  id = id.ToUpper();

  Customer? c = await _db.Customers.FindAsync(id);
  if (c is null) return null;

  _db.Customers.Remove(c);
  int affected = await _db.SaveChangesAsync();
  if (affected == 1)
  {
    await _cache.RemoveAsync(c.CustomerId);
    return true;
  }
  return null;
}
```

Configuring the customer repository

Now that you've learned the theory, you will put it into practice to configure the repository so that it can be called from within a Minimal API endpoint.

You will register a scoped dependency service implementation for the repository when the web service starts up, and then use constructor parameter injection to get it inside the definition of a new Minimal API endpoint for working with customers.

It will have five action methods to perform CRUD operations on customers—two `GET` methods (for all customers or one customer), `POST` (create), `PUT` (update), and `DELETE`:

1. In `Program.cs`, import the namespace for working with our customer repository, as shown in the following code:

```
using Northwind.WebApi.Repositories; // To use ICustomerRepository.
```

2. In `Program.cs`, add a statement before the call to the `Build` method, which will register the `CustomerRepository` for use at runtime as a scoped dependency, as shown in the following code:

```
builder.Services.AddScoped<ICustomerRepository,
  CustomerRepository>();
```

Good Practice: Our repository uses a database context that is registered as a scoped dependency. You can only use scoped dependencies inside other scoped dependencies, so we cannot register the repository as a singleton. You can read more about this at the following link: `https://learn.microsoft.com/en-us/dotnet/core/extensions/dependency-injection#scoped`.

3. In the `Northwind.WebApi` project, add a new class named `Program.Customers.cs`.
4. In `Program.Customers.cs`, add statements to define two Minimal API endpoint route handlers that respond to HTTP `GET` requests for all customers or customers within a specified country, as shown in the following code:

```
using Microsoft.AspNetCore.Mvc; // To use ProblemDetails.
using Northwind.EntityModels; // To use Customer.
using Northwind.WebApi.Repositories; // To use ICustomerRepository.

static partial class Program
{
  internal static void MapCustomers(this WebApplication app)
  {
    // GET: /customers
    app.MapGet(pattern: "/customers", handler:
      async (ICustomerRepository repo) =>
    {
      return await repo.RetrieveAllAsync();
    });

    // GET: customers/in/[country]
```

```
    app.MapGet(pattern: "/customers/in/{country}", handler:
      async (string country, ICustomerRepository repo) =>
    {
      return (await repo.RetrieveAllAsync())
        .Where(customer => customer.Country == country);
    });
  }
}
```

5. In `Program.Customers.cs`, in the `MapCustomers` method, add statements to map an endpoint route handler that responds to HTTP `GET` requests for an individual customer, as shown in the following code:

```
// GET: customers/[id]
app.MapGet("/customers/{id:regex(^[A-Z]{{5}}$)}",
  async Task<IResult> (string id, ICustomerRepository repo,
    CancellationToken token = default) =>
{
  Customer? c = await repo.RetrieveAsync(id, token);
  if (c is null)
  {
    return TypedResults.NotFound(); // 404 Resource not found.
  }
  return TypedResults.Ok(c); // 200 OK with customer in body.
});
```

6. In `Program.Customers.cs`, add statements to map an endpoint route handler that responds to HTTP `POST` requests to insert a new customer entity, as shown in the following code:

```
// POST: /customers
// BODY: Customer (JSON)
app.MapPost(pattern: "/customers", handler:
  async Task<IResult> (Customer c,
    ICustomerRepository repo) =>
{
  if (c is null)
  {
    return TypedResults.BadRequest(); // 400 Bad request.
  }
  Customer? addedCustomer = await repo.CreateAsync(c);
  if (addedCustomer is null)
  {
    return TypedResults.BadRequest("Repository failed to create 
customer.");
```

```
    }
    else
    {
      return TypedResults.CreatedAtRoute( // 201 Created.
        routeName: "GetCustomer",
        routeValues: new { id = addedCustomer
          .CustomerId.ToLower() },
        value: addedCustomer);
    }
  });
```

7. In `Program.Customers.cs`, add statements to map an endpoint route handler that responds to HTTP `PUT` requests, as shown in the following code:

```
// PUT: /customers/[id]
// BODY: Customer (JSON)
app.MapPut(pattern: "/customers/{id}", handler:
  async Task<IResult> (Customer c,
    string id, ICustomerRepository repo,
    CancellationToken token = default) =>
{
  id = id.ToUpper();
  c.CustomerId = c.CustomerId.ToUpper();
  if (c is null || c.CustomerId != id)
  {
    return TypedResults.BadRequest(); // 400 Bad request.
  }
  Customer? existing = await repo.RetrieveAsync(id, token);
  if (existing is null)
  {
    return TypedResults.NotFound(); // 404 Resource not found.
  }
  await repo.UpdateAsync(c);
  return new TypedResults.NoContent(); // 204 No content.
});
```

8. In `Program.Customers.cs`, add statements to map an endpoint route handler that responds to HTTP `DELETE` requests, as shown in the following code:

```
// DELETE: /customers/[id]
app.MapDelete(pattern: "/customers/{id}", handler:
  async Task<IResult> (string id, ICustomerRepository repo,
    CancellationToken token = default) =>
```

```
{
  Customer? existing = await repo.RetrieveAsync(id, token);
  if (existing is null)
  {
    return TypedResults.NotFound(); // 404 Resource not found.
  }
  bool? deleted = await repo.DeleteAsync(id);
  if (deleted.HasValue && deleted.Value) // Short circuit AND.
  {
    return TypedResults.NoContent(); // 204 No content.
  }
  else
  {
    return TypedResults.BadRequest( // 400 Bad request.
      $"Customer {id} was found but failed to delete.");
  }
});
```

9. In `Program.cs`, before the call to `Run`, call the extension method to map all the customer endpoint route handlers, as shown highlighted in the following code:

```
app.MapCustomers();

app.Run();
```

10. Save all the changes.

When an HTTP request is received by the service, it will create an instance of the `Controller` class, call the appropriate action method, return the response in the format preferred by the client, and release the resources used by the controller, including the repository and its data context.

Specifying problem details

A feature added in ASP.NET Core 2.1 and later is an implementation of a web standard for specifying problem details. If you want to take control, then you can create a `ProblemDetails` instance yourself and include additional information.

Let's simulate a bad request that needs custom data to be returned to the client.

At the top of the implementation of the `Delete` endpoint route handler, add statements to check if the `id` matches the literal string value `"bad"`, and if so, return a custom `ProblemDetails` object, as shown in the following code:

```
// Take control of problem details.
if (id == "bad")
{
```

```
    ProblemDetails problemDetails = new()
    {
      Status = StatusCodes.Status400BadRequest,
      Type = "https://localhost:5151/customers/failed-to-delete",
      Title = $"Customer ID {id} found but failed to delete.",
      Detail = "More details like Company Name, Country and so on."
    };
    return TypedResults.BadRequest(problemDetails); // 400 Bad Request
  }
```

You will try out this functionality later.

Documenting and trying out web services

You can easily try out a web service by making HTTP `GET` requests using a browser. To try out other HTTP methods, we need a more advanced tool.

Trying out GET requests using a browser

You will use Chrome to try out the three implementations of a `GET` request—for all customers, for customers in a specified country, and for a single customer using their unique customer ID:

1. Start the `Northwind.WebApi` web service project using the `https` launch profile.
2. Start Chrome, navigate to `https://localhost:5151/customers`, and note the JSON document returned, containing all 91 customers in the Northwind database (unsorted), as shown in *Figure 15.2*:

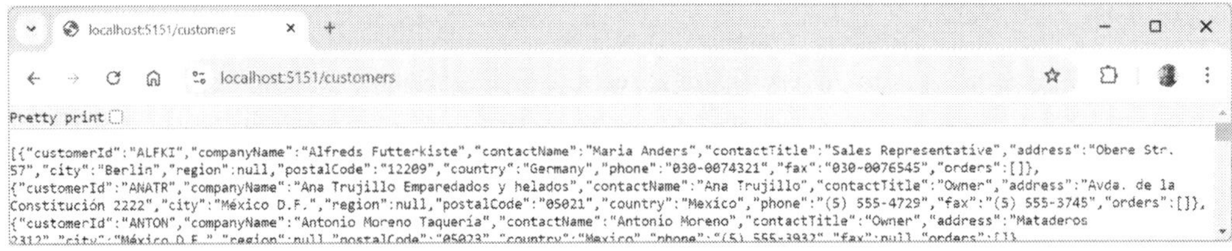

Figure 15.2: Customers from the Northwind database as a JSON document

3. Navigate to `https://localhost:5151/customers/in/Germany` and note the JSON document returned, containing only the customers in Germany.

If you get an empty array [] returned, then make sure you have entered the country name using the correct casing, because the database query is case-sensitive. For example, compare the results of `uk` and `UK`.

4. Navigate to `https://localhost:5151/customers/alfki` and note the JSON document returned containing only the customer named **Alfreds Futterkiste**.

Unlike country names, we do not need to worry about casing for the customer `id` value because, in the customer repository implementation, we normalized the `string` value as uppercase.

But how can we try out the other HTTP methods, such as `POST`, `PUT`, and `DELETE`? And how can we document our web service so it's easy for anyone to understand how to interact with it?

There are many tools for testing web services, for example, **Postman**. Although Postman is popular, I prefer tools like **HTTP Editor** in Visual Studio or **REST Client** in VS Code because they do not hide what is happening. I feel Postman is too GUI-y. But I encourage you to explore different tools and find the ones that fit your style. You can learn more about Postman at the following link: `https://www.postman.com/`.

To solve the first problem, we can use the **HTTP Editor** tool built into Visual Studio and install a VS Code extension named **REST Client**. Rider has its own equivalent. These are tools that allow you to send any type of HTTP request and view the response in your code editor.

To solve the second problem, we can use **OpenAPI**, aka **Swagger**, the world's most popular technology for documenting HTTP APIs. But first, let's see what is possible with the code editor HTTP/REST tools.

Making GET requests using HTTP/REST tools

We will start by creating a file for making `GET` requests:

1. If you have not already installed **REST Client** by Huachao Mao (`humao.rest-client`), then install it for VS Code now.
2. In your preferred code editor, open the `ModernWeb` solution and then start the `Northwind.WebApi` project web service.
3. In **File Explorer**, **Finder**, or your favorite Linux file tool, in the `ModernWeb` folder, create an `HttpRequests` folder.
4. In the `HttpRequests` folder, create a file named `get-customers.http`, and open it in your preferred code editor.
5. In `get-customers.http`, modify its contents to contain an HTTP `GET` request to retrieve all customers, as shown in the following code:

```
### Configure a variable for the web service base address.
@base_address = https://localhost:5151/customers/

### Make a GET request to the base address.
GET {{base_address}}
```

6. Above the HTTP `GET` request, click **Send request**, as shown in *Figure 15.3*.

   Note the response is shown in a new tabbed window.

7. If you are using Visual Studio, then click the **Raw** tab, and note the JSON that was returned, as shown in *Figure 15.3*:

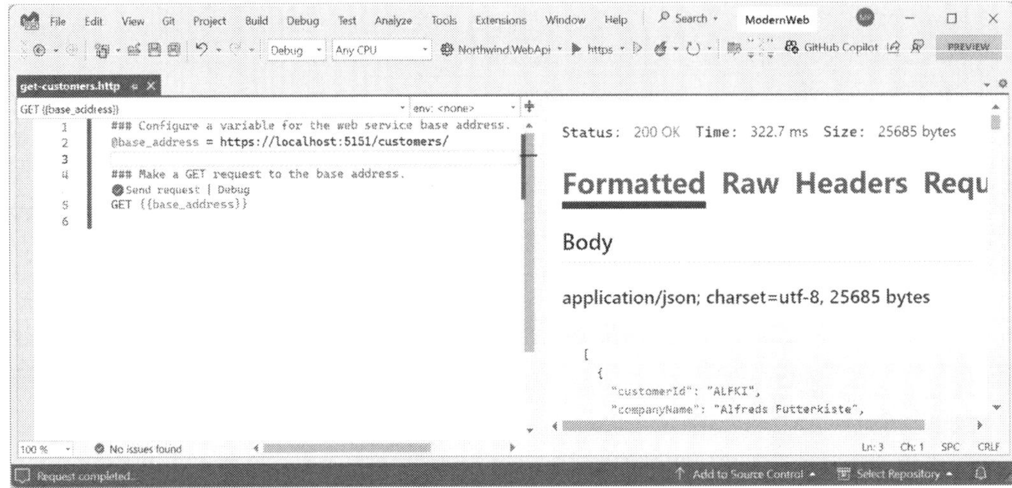

Figure 15.3: Sending an HTTP GET request using Visual Studio

HTTP Editor in Visual Studio version 17.8 and later is a feature designed to add REST client-like capabilities, and its user interface is likely to evolve rapidly as it catches up. You can read its official documentation at the following link: `https://learn.microsoft.com/en-us/aspnet/core/test/http-files`.

8. In `get-customers.http`, add more `GET` requests, each separated by three hash symbols, to test getting customers in various countries and getting a single customer using their ID, as shown in the following code:

```
### Get customers in Germany
GET {{base_address}}in/Germany

### Get customers in USA
GET {{base_address}}in/USA

### Get Alfreds Futterkiste
GET {{base_address}}ALFKI

### Get a non-existent customer
GET {{base_address}}abcxy
```

9. Click the **Send Request** link above each request to send it, and confirm you get the expected response, like a 404 for a non-existent customer.

Making other requests using HTTP/REST tools

Next, we will create a file for making other requests like `POST`:

1. In the `HttpRequests` folder, create a file named `create-customer.http` and modify its contents to define a `POST` request to create a new customer, as shown in the following code:

```
### Configure a variable for the web service base address.
@base_address = https://localhost:5151/customers/

### Make a POST request to the base address.
POST {{base_address}}
Content-Type: application/json

{
  "customerID": "ABCXY",
  "companyName": "ABC Corp",
  "contactName": "John Smith",
  "contactTitle": "Sir",
  "address": "Main Street",
  "city": "New York",
  "region": "NY",
  "postalCode": "90210",
  "country":  "USA",
  "phone": "(123) 555-1234"
}
```

2. Send the request and note the response is `201 Created`. Also note in the **Headers** section that the **Location** (that is, the URL) of the newly created customer is `/customers/ABCXY`, as shown in *Figure 15.4*, and the response included the newly created customer in the response body (not shown in the screenshot but you can see it in the **Formatted** and **Raw** sections):

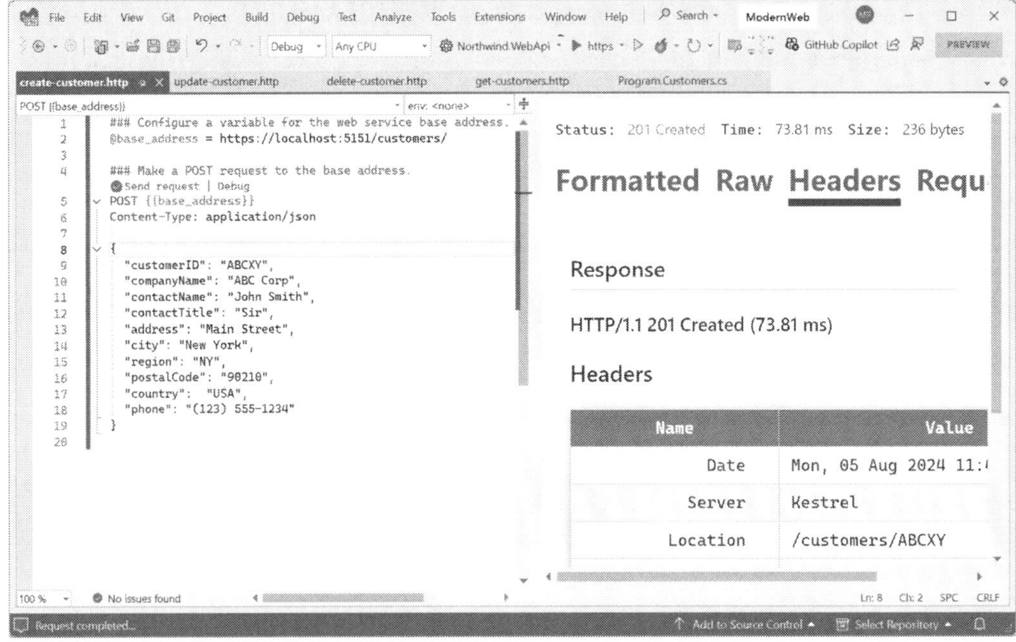

Figure 15.4: Adding a new customer by POSTing to the web service

I will leave you an optional challenge to create `.http` files that try updating a customer (using `PUT`) and deleting a customer (using `DELETE`). Try them on customers that do exist as well as customers that do not. Solutions are in the GitHub repository for this book at the following link:

`https://github.com/markjprice/cs13net9/tree/main/code/ModernWeb/HttpRequests`

Passing environment variables

To get an environment variable in a `.http` script, use `$processenv`, as shown in the following command:

```
{{$processEnv [%]envVarName}}
```

For example, if you have set an environment variable to store a secret value like a password to connect to a SQL Server database that must be kept out of any files committed to a GitHub repository, you can use the following command:

```
{{$processEnv MY_SQL_PWD}}
```

More Information: You can learn more about using environment variables with the REST client at the following link: https://marketplace.visualstudio.com/items?itemName=humao.rest-client#environments. You can learn more about using environment variables and Secret Manager with HTTP Editor at the following link: https://devblogs.microsoft.com/visualstudio/safely-use-secrets-in-http-requests-in-visual-studio-2022/.

Now that we've seen a quick and easy way to test our service, which also happens to be a great way to learn HTTP, what about external developers? We want it to be as easy as possible for them to learn about and then call our service. For that purpose, we will use Swagger.

Understanding the OpenAPI Specification

The **OpenAPI Specification** defines a REST-style contract for your API, detailing all its resources and operations in a human- and machine-readable format for easy development, discovery, and integration.

Developers can use the OpenAPI Specification for a web service to automatically generate strongly typed client-side code in their preferred language or library.

Let's review how OpenAPI is enabled for our web service:

1. If the web service is running, shut down the web server.
2. In `Northwind.WebApi.csproj`, note the package reference for Microsoft's package that implements documentation for OpenAPI that was added by the project template, as shown in the following markup:

```
<PackageReference Include="Microsoft.AspNetCore.OpenApi" />
```

3. In `Program.cs`, in the section for adding services to the container, note the service registered by the project template to use OpenAPI, as shown in the following code:

```
// Learn more about configuring OpenAPI at https://aka.ms/aspnet/openapi
builder.Services.AddOpenApi();
```

4. By default, the document name is `v1`. Set the document name parameter to `v2`, as shown in the following code:

```
builder.Services.AddOpenApi(documentName: "v2");
```

5. In the section that configures the HTTP request pipeline, note the statements for using OpenAPI when in development mode, as shown highlighted in the following code:

```
// Configure the HTTP request pipeline.
if (builder.Environment.IsDevelopment())
{
  app.MapOpenApi();
}
```

6. Start the `Northwind.WebApi` web service project using the `https` launch profile.
7. Start Chrome, navigate to `https://localhost:5151/openapi/v2.json`, and note the JSON document returned, as shown in *Figure 15.5*:

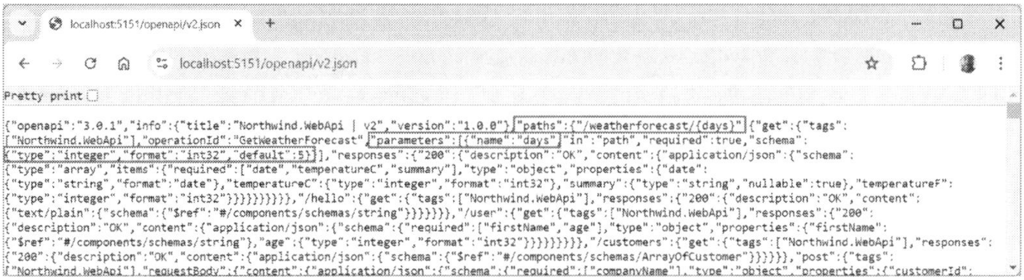

localhost:5151/openapi/v2.json

Pretty print


```
{"openapi":"3.0.1","info":{"title":"Northwind.WebApi | v2","version":"1.0.0"},"paths":{"/weatherforecast/{days}":{"get":{"tags":
["Northwind.WebApi"],"operationId":"GetWeatherForecast","parameters":[{"name":"days","in":"path","required":true,"schema":
{"type":"integer","format":"int32","default":5}}],"responses":{"200":{"description":"OK","content":{"application/json":{"schema":
{"type":"array","items":{"required":["date","temperatureC","summary"],"type":"object","properties":{"date":
{"type":"string","format":"date"},"temperatureC":{"type":"integer","format":"int32"},"summary":{"type":"string","nullable":true},"temperatureF":
{"type":"integer","format":"int32"}}}}}}}}}},"/hello":{"get":{"tags":["Northwind.WebApi"],"responses":{"200":{"description":"OK","content":
{"text/plain":{"schema":{"$ref":"#/components/schemas/string"}}}}}}},"/user":{"get":{"tags":["Northwind.WebApi"],"responses":{"200":
{"description":"OK","content":{"application/json":{"schema":{"required":["firstName","age"],"type":"object","properties":{"firstName":
{"$ref":"#/components/schemas/string"},"age":{"type":"integer","format":"int32"}}}}}}}}},"/customers":{"get":{"tags":["Northwind.WebApi"],"responses":
{"200":{"description":"OK","content":{"application/json":{"schema":{"$ref":"#/components/schemas/ArrayOfCustomer"}}}}}},"post":{"tags":
["Northwind.WebApi"],"requestBody":{"content":{"application/json":{"schema":{"required":["companyName"],"type":"object","properties":{"customerId":
```


Figure 15.5: OpenAPI JSON documentation for the Northwind web service

Note the following about the OpenAPI JSON document:

- It specifies all the relative paths for the service, like `/weatherforecast/{days}`.
- For parameters like `days`, it specifies their type and default value, like `"parameters":[{"name":"days","in":"path","required":true,"schema":{"type":"integer","format":"int32","default":5}}]`.

There are two techniques that you can use to add extra information to the generated documentation, `WithSummary` and `WithDescription`. These are available as either calling extension methods after mapping the endpoint or by decorating the lambda expression with attributes, as shown in the following code:

```
app.MapGet("/helloUsingMethods", () => "Hello world!")
  .WithSummary("This is a summary.")
  .WithDescription("This is a description.");

app.MapGet("/helloUsingAttributes",
  [EndpointSummary("This is a summary.")]
  [EndpointDescription("This is a description.")]
  () => "Hello world!");
```

More Information: You can learn more at the following link: `https://learn.microsoft.com/en-us/aspnet/core/fundamentals/minimal-apis/openapi?view=aspnetcore-9.0#describe-endpoints`.

Enabling HTTP logging

HTTP logging is an optional middleware component that is useful when testing a web service. It logs information about HTTP requests and HTTP responses, including the following:

- Information about the HTTP request

- Headers
- Body
- Information about the HTTP response

This is valuable in web services for auditing and debugging scenarios but beware because it can negatively impact performance. You might also log **Personally Identifiable Information** (**PII**), which can cause compliance issues in some jurisdictions.

Log levels can be set to the following:

- `Error`: Only `Error` level logs
- `Warning`: `Error` and `Warning` level logs
- `Information`: `Error`, `Warning`, and `Information` level logs
- `Verbose`: All level logs

Log levels can be set for the namespace in which the functionality is defined. Nested namespaces allow us to control which functionality has logging enabled:

- `Microsoft`: Include all log types in the `Microsoft` namespace
- `Microsoft.AspNetCore`: Include all log types in the `Microsoft.AspNetCore` namespace
- `Microsoft.AspNetCore.HttpLogging`: Include all log types in the `Microsoft.AspNetCore.HttpLogging` namespace

Let's see HTTP logging in action:

1. In the `Northwind.WebApi` project, `appsettings.Development.json`, add an entry to set the HTTP logging middleware to the `Information` level, as shown highlighted in the following code:

```
{
  "Logging": {
    "LogLevel": {
      "Default": "Information",
      "Microsoft.AspNetCore": "Warning",
      "Microsoft.AspNetCore.HttpLogging.HttpLoggingMiddleware":
"Information"
    }
  }
}
```

Although the `Default` log level might be set to `Information`, more specific configurations take priority. For example, any logging systems in the `Microsoft.AspNetCore` namespace will use the `Warning` level. By making the change we did, any logging systems in the `Microsoft.AspNetCore. HttpLogging.HttpLoggingMiddleware` namespace will now use `Information`.

2. In `Program.cs`, import the namespace for working with HTTP logging, as shown in the following code:

```
using Microsoft.AspNetCore.HttpLogging; // To use HttpLoggingFields.
```

3. In the services configuration section, before the call to `Build`, add a statement to configure HTTP logging, as shown in the following code:

```
builder.Services.AddHttpLogging(options =>
{
  options.LoggingFields = HttpLoggingFields.All;
  options.RequestBodyLogLimit = 4096; // Default is 32k.
  options.ResponseBodyLogLimit = 4096; // Default is 32k.
});
```

4. In the HTTP pipeline configuration section, before the call to `app.UseHttpsRedirection`, add a statement to add HTTP logging, as shown in the following code:

```
app.UseHttpLogging();
```

5. Start the `Northwind.WebApi` web service using the `https` launch profile.
6. Start Chrome and navigate to `https://localhost:5151/customers`.
7. In the command prompt or terminal that shows the output from the web service host, note the request and response have been logged, as shown in the following partial output:

```
info: Microsoft.AspNetCore.HttpLogging.HttpLoggingMiddleware[1]
      Request:
      Protocol: HTTP/2
      Method: GET
      Scheme: https
      PathBase:
      Path: /weatherforecast
      Accept: text/html,application/xhtml+xml,application/
xml;q=0.9,image/avif,image/webp,image/apng,*/*;q=0.8,application/signed-
exchange;v=b3;q=0.7
      Host: localhost:5151
      User-Agent: Mozilla/5.0 (Windows NT 10.0; Win64; x64)
AppleWebKit/537.36 (KHTML, like Gecko) Chrome/115.0.0.0 Safari/537.36
      Accept-Encoding: gzip, deflate, br
      Accept-Language: en-US,en-GB;q=0.9,en;q=0.8,fr-FR;q=0.7,fr;q=0.6
      Upgrade-Insecure-Requests: [Redacted]
...
info: Microsoft.AspNetCore.HttpLogging.HttpLoggingMiddleware[2]
      Response:
      StatusCode: 200
```

```
      Content-Type: application/json; charset=utf-8
info: Microsoft.AspNetCore.HttpLogging.HttpLoggingMiddleware[4]
      ResponseBody: [{"date":"2024-08-06","temperatureC":50,"summary":"
Freezing","temperatureF":121},{"date":"2024-08-07","temperatureC":53,"
summary":"Scorching","temperatureF":127},{"date":"2024-08-08","tempera
tureC":40,"summary":"Mild","temperatureF":103},{"date":"2024-08-09","te
mperatureC":8,"summary":"Bracing","temperatureF":46},{"date":"2024-08-
10","temperatureC":-10,"summary":"Freezing","temperatureF":15}]
info: Microsoft.AspNetCore.HttpLogging.HttpLoggingMiddleware[8]
      Duration: 6.8678ms
```

8. Close Chrome and shut down the web server.

Logging to the Windows-only Event Log

When configuring logging, you might want to enable logging to the Windows Event Log, as shown in the following code:

```
var builder = Host.CreateApplicationBuilder();

// Option 1
builder.Services.AddLogging(logging =>
{
  logging.AddEventLog();
});

// Option 2
builder.Host.ConfigureLogging(logging =>
{
  logging.AddEventLog();
});

// Option 3: .NET 6 or later. Concise and recommended by Microsoft.
builder.Logging.AddEventLog();
```

You will see a code analyzer warning, `CA1416`, because enabling Event Log only works on Windows. If you run this code on any other OS, then a runtime exception would be thrown. To avoid the warning (and runtime error), you should wrap the call to `AddEventLog` with an OS check.

First, import a namespace, as shown in the following code:

```
using System.Runtime.InteropServices; // To use RuntimeInformation.
```

Then, wrap any calls to `AddEventLog`, as shown in the following code:

```
if (RuntimeInformation.IsOSPlatform(OSPlatform.Windows))
{
  // Call the AddEventLog method.
}
```

Support for logging additional request headers in W3CLogger

W3CLogger is a middleware that writes logs in the W3C standard format. You can:

- Record details of HTTP requests and responses.
- Filter which headers and parts of the request and response messages are logged.

Warning! W3CLogger can reduce the performance of an app.

W3CLogger is like HTTP logging, so I will not cover details of how to use it in this book. You can learn more about W3CLogger at the following link: `https://learn.microsoft.com/en-us/aspnet/core/fundamentals/w3c-logger/`.

In ASP.NET Core 7 or later, you can specify that you want to log additional request headers when using W3CLogger. Call the `AdditionalRequestHeaders` method and pass the name of the header you want to log, as shown in the following code:

```
services.AddW3CLogging(options =>
{
  options.AdditionalRequestHeaders.Add("x-forwarded-for");
  options.AdditionalRequestHeaders.Add("x-client-ssl-protocol");
});
```

You are now ready to build applications that consume your web service.

Consuming web services using HTTP clients

Now that we have built and tested our Northwind service, we will learn how to call it from any .NET app using the `HttpClient` class and its factory.

Understanding HttpClient

The easiest way to consume a web service is to use the `HttpClient` class. However, many people use it wrongly because it implements `IDisposable`, and Microsoft's own documentation shows poor usage of it. See the book links in the GitHub repository for articles with more discussion of this.

Usually, when a type implements `IDisposable`, you should create it inside a `using` statement to ensure that it is disposed of as soon as possible. `HttpClient` is different because it is shared, reentrant, and partially thread-safe.

The problem has to do with how the underlying network sockets must be managed. The bottom line is that you should use a single instance of it for each HTTP endpoint that you consume during the life of your application. This will allow each `HttpClient` instance to have defaults set that are appropriate for the endpoint it works with while managing the underlying network sockets efficiently.

Configuring HTTP clients

Microsoft is aware of the issue of .NET developers misusing `HttpClient`, and in ASP.NET Core 2.1, it introduced `HttpClientFactory` to encourage best practices; that is the technique we will use.

In the following example, we will create a Northwind Blazor WebAssembly standalone project as a client for the Northwind Web API service. Let's configure an HTTP client:

1. Use your preferred code editor to open the `ModernWeb` solution and then add a new project, as defined in the following list:
    - Project template: **Blazor WebAssembly Standalone App** / `blazorwasm`
    - Solution file and folder: `ModernWeb`
    - Project file and folder: `Northwind.WebApi.WasmClient`
    - **Authentication type**: None
    - **Configure for HTTPS**: Selected
    - **Progressive Web Application**: Cleared
    - **Include sample pages**: Selected
    - **Do not use top-level statements**: Cleared
2. In the `Northwind.WebApi.WasmClient.csproj` project file, in the package references, remove version attributes.
3. In the `Properties` folder, in `launchSettings.json`, for the `https` profile, for its `applicationUrl`, change the random port number for HTTPS to `5152` and for HTTP to `5153`, as shown highlighted in the following markup:

```
"applicationUrl": "https://localhost:5152;http://localhost:5153",
```

4. Save changes to all modified files.
5. In `Program.cs`, in the call to the `AddScoped` method, add a statement to enable `HttpClientFactory` with a named client to make calls to the Northwind Web API service using HTTPS on port `5151` and request JSON as the default response format, as shown in the following code:

```
builder.Services.AddScoped(sp => new HttpClient {
  BaseAddress = new Uri("https://localhost:5151/") });
```

6. In the `Northwind.WebApi` project, in `Program.cs`, at the top of the file after the namespace imports, declare a string constant for the name of a CORS policy, as shown in the following code:

```
const string corsPolicyName = "allowWasmClient";
```

7. In `Program.cs`, before the call to `Build`, add CORS and configure a policy to allow HTTP calls from clients with different port numbers from the web service itself, as shown in the following code:

```
builder.Services.AddCors(options =>
{
  options.AddPolicy(name: corsPolicyName,
    policy =>
    {
      policy.WithOrigins("https://localhost:5152",
        "http://localhost:5153");
    });
});
```

8. In `Program.cs`, after the call to `UseHttpsRedirection`, enable CORS with the named policy, as shown in the following code:

```
app.UseCors(corsPolicyName);
```

Getting customers as JSON in a Blazor component

We can now create a client page that:

- Makes a `GET` request for customers.
- Deserializes the JSON response using convenient extension methods introduced with .NET 5 in the `System.Net.Http.Json` assembly and namespace.

Let's go:

1. In the `Northwind.WebApi.WasmClient.csproj` project file, add a reference to the entity models project, as shown in the following markup:

```
<ItemGroup>
  <ProjectReference Include=
"..\Northwind.EntityModels.Sqlite\Northwind.EntityModels.Sqlite.csproj"
/>
</ItemGroup>
```

2. In the `Northwind.WebApi.WasmClient` project, in `_Imports.razor`, import the namespace for working with entity models, as shown in the following code:

```
@using Northwind.EntityModels @* To use Customer. *@
```

3. In the `Northwind.WebApi.WasmClient` project, in the `Pages` folder, add a new file named `Customers.razor`.
4. In `Customers.razor`, inject the HTTP client service, and use it to call the Northwind Web API service, fetching all customers, and passing them to a table, as shown in the following markup:

```
@attribute [StreamRendering]
@page "/customers/{country?}"
@inject HttpClient Http
<h3>
  Customers @(string.IsNullOrWhiteSpace(Country)
    ? "Worldwide" : "in " + Country)
</h3>
@if (customers is null)
{
  <p><em>Loading...</em></p>
}
else
{
  <table class="table">
    <thead>
      <tr>
        <th>Id</th>
        <th>Company Name</th>
        <th>Address</th>
        <th>Phone</th>
        <th></th>
      </tr>
    </thead>
    <tbody>
      @foreach (Customer c in customers)
      {
        <tr>
          <td>@c.CustomerId</td>
          <td>@c.CompanyName</td>
          <td>
            @c.Address<br />
            @c.City<br />
            @c.PostalCode<br />
            @c.Country
          </td>
          <td>@c.Phone</td>
        </tr>
```

```
      }
    </tbody>
  </table>
}

@code {
  [Parameter]
  public string? Country { get; set; }

  private IEnumerable<Customer>? customers;

  protected override async Task OnParametersSetAsync()
  {
    if (string.IsNullOrWhiteSpace(Country))
    {
      customers = await Http.GetFromJsonAsync
        <Customer[]>("/customers");
    }
    else
    {
      customers = await Http.GetFromJsonAsync
        <Customer[]>($"/customers/in/{Country}");
    }
  }
}
```

5. In the `Layout` folder, in `NavMenu.razor`, change the **Weather** menu item to show customers instead, as shown in the following markup:

```
<NavLink class="nav-link" href="customers">
  <span class="bi bi-list-nested-nav-menu" aria-hidden="true"></span>
Customers
</NavLink>
```

Starting multiple projects

Up to this point, we have only started one project at a time. Now we have two projects that need to be started, a web service and a Blazor client website. In the step-by-step instructions, I will only tell you to start individual projects one at a time, but you should use whatever technique you prefer to start them.

If you are using Visual Studio

Visual Studio can start multiple projects manually one by one if the debugger is not attached, as described in the following steps:

1. In **Solution Explorer**, right-click on the solution or any project and then select **Configure Startup Projects...**, or select the solution and navigate to **Project | Configure Startup Projects...**.
2. In the **Solution '<name>' Property Pages** dialog box, select **Current selection**.
3. Click **OK**.
4. Select a project in **Solution Explorer** so that its name becomes bold.
5. Navigate to **Debug | Start Without Debugging** or press *Ctrl* + *F5*.
6. Repeat *steps 2* and *3* for as many projects as you need.

If you need to debug the projects, then you must start multiple instances of Visual Studio. Each instance can start a single project with debugging.

You can also configure multiple projects to start up at the same time using the following steps:

1. In **Solution Explorer**, right-click the solution or any project and then select **Configure Startup Projects...**, or select the solution and navigate to **Project | Configure Startup Projects...**.
2. In the **Solution '<name>' Property Pages** dialog box, select **Multiple startup projects**, and for any projects that you want to start, select either **Start** or **Start without debugging**, as shown in *Figure 15.6*:

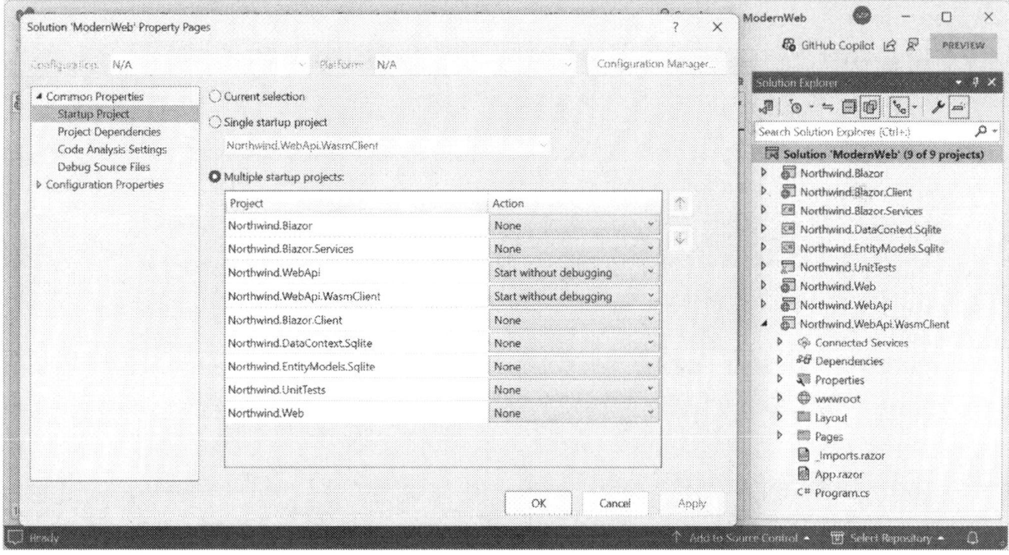

Figure 15.6: Selecting multiple projects to start up in Visual Studio

3. Click **OK**.

4. Navigate to **Debug** | **Start Debugging** or **Debug** | **Start Without Debugging** or click the equivalent buttons in the toolbar to start all the projects that you selected.

You can learn more about multi-project startup using Visual Studio at the following link: `https://learn.microsoft.com/en-us/visualstudio/ide/how-to-set-multiple-startup-projects`.

If you are using VS Code

If you need to start multiple projects at the command line with `dotnet`, then write a script or batch file to execute multiple `dotnet run` commands, or open multiple command prompt or terminal windows.

If you need to debug multiple projects using VS Code, then after you've started the first debug session, you can just launch another session. Once the second session is running, the user interface switches to multi-target mode. For example, in the **CALL STACK**, you will see both named projects with their own threads, and then the debug toolbar shows a drop-down list of sessions with the active one selected. Alternatively, you can define compound launch configurations in the `launch.json`.

You can learn more about multi-target debugging using VS Code at the following link: `https://code.visualstudio.com/Docs/editor/debugging#_multitarget-debugging`.

Starting the web service and Blazor client projects

Now we can try out the web service with the Blazor client calling it:

1. Start the `Northwind.WebApi` project and confirm that the web service is listening on ports `5151` and `5150`, as shown in the following output:

```
info: Microsoft.Hosting.Lifetime[14]
  Now listening on: https://localhost:5151
info: Microsoft.Hosting.Lifetime[14]
  Now listening on: http://localhost:5150
```

2. Start the `Northwind.WebApi.WasmClient` project and confirm that the website is listening on ports `5152` and `5153`, as shown in the following output:

```
info: Microsoft.Hosting.Lifetime[14]
  Now listening on: https://localhost:5152
info: Microsoft.Hosting.Lifetime[14]
  Now listening on: http://localhost:5153
```

3. Start Chrome and navigate to `https://localhost:5152/`.
4. On the home page, in the left navigation menu, click **Customers**, and note the list of customers, as shown in *Figure 15.7*:

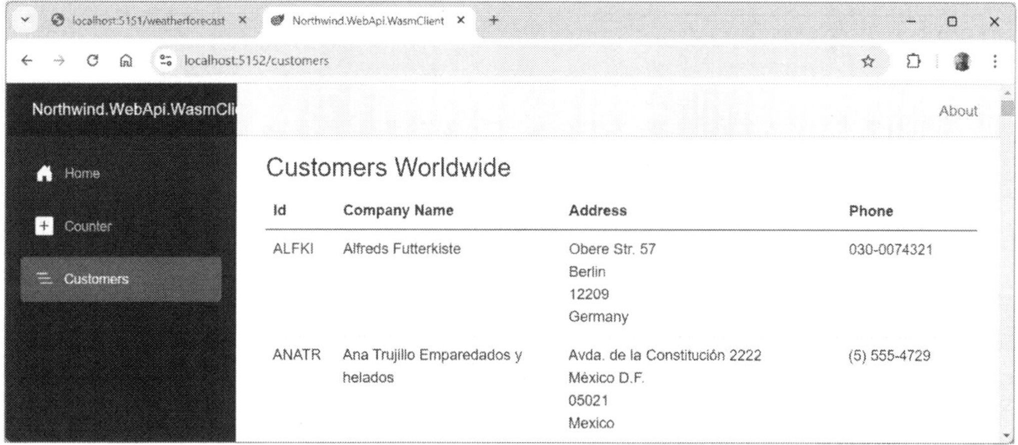

Figure 15.7: Customers worldwide fetched from a web service

5. In the command prompt or terminal for the web service, note that HTTP logging shows that a successful request was made for customers, as shown in the following output:

```
info: Microsoft.EntityFrameworkCore.Database.Command[20101]
      Executed DbCommand (20ms) [Parameters=[], CommandType='Text',
CommandTimeout='30']
      SELECT "c"."CustomerId", "c"."Address", "c"."City",
"c"."CompanyName", "c"."ContactName", "c"."ContactTitle", "c"."Country",
"c"."Fax", "c"."Phone", "c"."PostalCode", "c"."Region"
      FROM "Customers" AS "c"
info: Microsoft.AspNetCore.HttpLogging.HttpLoggingMiddleware[2]
      Response:
      StatusCode: 200
      Content-Type: application/json; charset=utf-8
info: Microsoft.AspNetCore.HttpLogging.HttpLoggingMiddleware[4]
      ResponseBody: [{"customerId":"ALFKI","companyName":"Alfreds
Futterkiste","contactName":"Maria Anders","contactTitle":"Sales
Representative","address":"Obere Str. 57","city":"Berlin","region":null,"
postalCode":"12209","country":"Germany","phone":"030-0074321","fax":"030-
0076545","orders":[]},...
info: Microsoft.AspNetCore.HttpLogging.HttpLoggingMiddleware[8]
      Duration: 1039.4409ms
```

6. In the address bar, change the path to specify a country like `Germany`, `UK`, or `USA`, for example: `customers/UK`. Press *Enter* and note the table updates to only show UK customers.
7. Close Chrome and shut down the two web servers.

Practicing and exploring

Test your knowledge and understanding by answering some questions, getting some hands-on practice, and exploring this chapter's topics with deeper research.

Exercise 15.1 – Online material

Online material could be created by Microsoft or third-parties, or extra content for this book.

Implementing advanced features for web services

If you would like to learn about web service health checks, OpenAPI analyzers, adding security HTTP headers, and enabling HTTP/3 support for `HttpClient`, then you can read the optional online-only section at the following link:

`https://github.com/markjprice/cs13net9/blob/main/docs/ch15-advanced.md`

Minimal APIs parameter binding

You can learn more about how to convert HTTP request data into strongly typed parameters for Minimal APIs endpoints at the following link:

`https://learn.microsoft.com/en-us/aspnet/core/fundamentals/minimal-apis/parameter-binding`

Refit client

Refit is an automatic type-safe REST library for .NET. It was inspired by Square's Retrofit library, and it turns your REST service into a live interface.

`https://github.com/reactiveui/refit`

Web service security using Microsoft Identity

You can learn what's new with Microsoft Identity for authentication and authorization at the following link:

`https://devblogs.microsoft.com/dotnet/whats-new-with-identity-in-dotnet-8/`

Exercise 15.2 – Practice exercises

Practice exercises go deeper into the topics for this chapter.

Creating and deleting customers with HttpClient

Extend the `Northwind.WebApi.ClientWasm` project to have pages where a visitor can fill in a form to create a new customer, or search for a customer and then delete them. The Blazor components should make calls to the Northwind Web API web service to create and delete customers.

Exercise 15.3 – Test your knowledge

Answer the following questions:

1. ASP.NET Core has multiple project templates for building web services. What are they and how do you create them using the CLI?
2. When configuring an HTTP client, how do you specify the format of data that you prefer in the response from the web service?
3. Why did the ASP.NET Core team replace the Swashbuckle package with their own implementation of OpenAPI support?
4. What must you do to specify what responses should be expected when calling a `MapGet` or similar method?
5. List three methods that can be called to return responses with different status codes.
6. List four ways that you can test a web service.
7. Why should you not wrap your use of `HttpClient` in a `using` statement to dispose of it when you are finished even though it implements the `IDisposable` interface, and what should you use instead?
8. What are the benefits of HTTP/2 and HTTP/3 compared to HTTP/1.1?
9. How can you enable clients to detect if your web service is healthy with ASP.NET Core 2.2 and later?
10. What are the main types of object caching and why is the `HybridCache` introduced with .NET 9 the best?

Exercise 15.4 – Explore topics

Use the links in the following GitHub repository to learn more details about the topics covered in this chapter:

`https://github.com/markjprice/cs13net9/blob/main/docs/book-links.md#chapter-15---building-and-consuming-web-services`

Summary

In this chapter, you learned:

- How to build an ASP.NET Core web service using Minimal APIs that can be called by any app on any platform that can make an HTTP request and process an HTTP response
- How to document web service APIs with OpenAPI
- How to test web services using a browser, HTTP editor, or REST client
- How to consume services efficiently

In the *Epilogue*, I will make some suggestions for books to take you deeper into C# and .NET.

Epilogue

I wanted this book to be different from others on the market. I hope that you found it to be a brisk, fun read, packed with practical, hands-on walk-throughs of each subject.

This epilogue contains the following short sections:

- Next steps on your C# and .NET learning journey
- The tenth edition for .NET 10
- Good luck!

Next steps on your C# and .NET learning journey

For subjects that I didn't have space to include in this book but you might want to learn more about, I hope that the notes, good practice tips, and links in the GitHub repository point you in the right direction:

`https://github.com/markjprice/cs13net9/blob/main/docs/book-links.md`

Polishing your skills with design guidelines

Now that you have learned the fundamentals of developing using C# and .NET, you are ready to improve the quality of your code by learning more detailed design guidelines.

Back in the early .NET Framework era, Microsoft published a book that gave good practices in all areas of .NET development. Those recommendations are still very much applicable to modern .NET development.

The following topics are covered:

- Naming Guidelines
- Type Design Guidelines
- Member Design Guidelines
- Designing for Extensibility
- Design Guidelines for Exceptions
- Usage Guidelines
- Common Design Patterns

To make the guidance as easy to follow as possible, the recommendations are simply labeled with the terms **Do**, **Consider**, **Avoid**, and **Do not**.

Microsoft has made excerpts of the book available at the following link:

`https://learn.microsoft.com/en-us/dotnet/standard/design-guidelines/`

I strongly recommend that you review all the guidelines and apply them to your code.

Companion books to continue your learning journey

I have written three more books to continue the learning journey with .NET 8 and .NET 9 that you started with this fundamentals book.

A summary of the .NET 8 and .NET 9 quartet of books and their most important topics is shown in *Figure 16.1*:

C# language, including new C# 13 features, object-oriented programming, debugging, and unit testing.

.NET libraries, including numbers, text, regular expressions, collections, file I/O, and data with EF Core and SQLite.

Modern websites and web services with ASP.NET Core, Blazor, and Minimal APIs.

ASP.NET Core MVC websites, including defining routes, controllers, models, and views.

Caching, web testing, configuration, and containerizing for deployment.

Web API services using controllers and OData, including authentication, authorization, and integration testing.

Umbraco CMS for user-managed content.

More libraries: Internationalization, multitasking, and third-party packages.

More data: SQL Server and Cosmos DB.

More services: Minimal APIs, caching, queuing, GraphQL, gRPC, SignalR, and Azure Functions.

More user interfaces: ASP.NET Core MVC, Blazor, and .NET MAUI.

Tools: IDEs, debugging, memory analysis, and AI assistants.

Tests: Unit, integration, performance, security, and web, including DI and IoC.

Develop: Docker and .NET Aspire.

Design: Patterns, principles, software and solution architecture.

Career: Teamwork and interviews.

Figure 16.1: Companion books for learning C# and .NET

The other books act as companions to this book, and together, they all form a .NET 8 and .NET 9 quartet of books:

1. The first book (the one you're reading now) covers the fundamentals of C#, .NET, and ASP.NET Core for modern web development. The ninth edition for .NET 9 is a small update from the 8th edition because .NET 9 is a Standard Term Support release (only 18 months). It includes the errata and improvements listed since the publishing of the 8th edition, which can be found at the following link: `https://github.com/markjprice/cs12dotnet8/blob/main/docs/errata/README.md`.
2. The second book covers real-world web development with .NET 9. This means it covers technologies that are mature and proven and are based on controller architecture. This includes ASP.NET Core MVC, Web API using controllers, OData, and the most popular .NET CMS, Umbraco.

3. The third book covers important tools and skills you should learn to become a well-rounded professional .NET developer. These include design patterns and solution architecture, debugging, memory analysis, all the important types of testing from unit and integration to performance and web UI testing, and then topics like Docker and .NET Aspire. Finally, it looks at how to prepare for an interview to get the .NET developer career that you want. This book will not be updated for .NET 9.
4. The fourth book covers more specialized topics, like internationalization and popular third-party packages including Serilog and Noda Time. It shows you how to build native AOT-compiled services with ASP.NET Core Minimal APIs and how to improve performance, scalability, and reliability using caching, queues, and background services. It covers implementing more services using GraphQL, gRPC, SignalR, and Azure Functions. Finally, it shows you how to build graphical user interfaces for websites and desktop and mobile apps with Blazor and .NET MAUI. This book will not be updated for .NET 9.

To see a list of all the books that I have published with Packt, you can use the following link:

https://subscription.packtpub.com/search?query=mark+j.+price

Other books to take your learning further

If you are looking for other books from my publisher that cover related subjects, there are many to choose from, as shown in *Figure 16.2*:

Figure 16.2: Packt books to take your C# and .NET learning further

You will also find a list of Packt books in the GitHub repository at the following link:

https://github.com/markjprice/cs13dotnet9/blob/main/docs/book-links.md#learn-from-other-packt-books

The tenth edition for .NET 10

I have already started work identifying areas for improvement for the tenth edition, which we plan to publish alongside updates to the other members of my quartet of .NET books. I expect .NET 10 to make worthwhile improvements to all aspects of .NET and as a **Long Term Support** (**LTS**) release, it will be supported until November 2028.

You can learn how to use .NET 10 with this book, including .NET 10 previews, starting in February 2025, at the following link:

https://github.com/markjprice/cs13net9/blob/main/docs/dotnet10.md

If you have suggestions for topics that you would like to see covered or expanded upon, or you spot mistakes that need fixing in the text or code, then please let me know the details via chat in the Discord channel or the GitHub repository for this book, found at the following link:

```
https://github.com/markjprice/cs13net9
```

Good luck!

I wish you the best of luck with all your C# and .NET projects!

Learn more on Discord

To join the Discord community for this book – where you can share feedback, ask questions to the author, and learn about new releases – follow the QR code below:

```
https://packt.link/csharp13dotnet9
```

Leave a Review!

Thank you for purchasing this book from Packt Publishing—we hope you enjoyed it! Your feedback is invaluable and helps us improve and grow. Please take a moment to leave an Amazon review; it will only take a minute, but it makes a big difference for readers like you.

https://packt.link/r/1835881238

Scan the QR code below to receive a free ebook of your choice.

https://packt.link/NzOWQ

Index

Symbols

C

E

G

H

N

Q

R

T

U

X

Y

Download a free PDF copy of this book

Thanks for purchasing this book!

Do you like to read on the go but are unable to carry your print books everywhere?

Is your eBook purchase not compatible with the device of your choice?

Don't worry, now with every Packt book you get a DRM-free PDF version of that book at no cost.

Read anywhere, any place, on any device. Search, copy, and paste code from your favorite technical books directly into your application.

The perks don't stop there, you can get exclusive access to discounts, newsletters, and great free content in your inbox daily.

Follow these simple steps to get the benefits:

1. Scan the QR code or visit the link below:

https://packt.link/free-ebook/9781835881224

2. Submit your proof of purchase.
3. That's it! We'll send your free PDF and other benefits to your email directly.

Made in the USA
Las Vegas, NV
07 January 2025